HOLOGRAMS – RECORDING MATERIALS AND APPLICATIONS

Edited by **Izabela Naydenova**

Holograms – Recording Materials and Applications
Edited by Izabela Naydenova

Published by InTech

Janeza Trdine 9, 51000 Rijeka, Croatia

Publishing Process Manager Silvia Vlase

Technical Editor Teodora Smiljanic

Cover Designer InTech Design Team

Additional hard copies can be obtained from orders@intechopen.com

Holograms – Recording Materials and Applications, Edited by Izabela Naydenova
p. cm.
978-953-307-981-3

For EU product safety concerns:
IN TECH d.o.o., Prolaz Marije Krucifikse Kozulić 3, 51000 Rijeka, Croatia,
info@intechopen.com or visit our website at intechopen.com.

Meet the editor

Dr. Izabela Naydenova received her MSc in applied optics from the University of Sofia in 1993 and PhD in physics from the Bulgarian Academy of Sciences in 1999. She worked as a Postdoctoral researcher at the Institute for Physical and Theoretical Chemistry, Technical University of Munich, (1999-2002). Dr Naydenova joined the Centre for Industrial and Engineering Optics, Dublin Institute of Technology as an Arnold F. Graves postdoctoral research fellow in October 2002 and took up current position as a Lecturer in the School of Physics, Dublin Institute of Technology in September 2008. Her current research interests are in holographic recording materials, with a particular focus on photopolymer nanocomposites and self-processing photopolymers, and their applications in holographic sensing, holographic data storage, optical micro patterning and micromanipulation, the design and fabrication of diffractive optical devices by dye deposition lithography. Izabela Naydenova is the author and co-author of more than seventy full length articles and two book chapters on holography related topics.

Contents

Preface

The book "Holograms –Recording Materials and Applications" comprises five sections.

The first section has eight chapters on holographic recording materials including ionic liquids in photopolymerisable materials (Chapter 1), Norland optical adhesive 65® as holographic material (Chapter 2), porous glass and polymer nanocomposite (Chapter 3), amorphous chalcogenide films (Chaper 4), azo-dye containing materials (Chapter 5 and 6) and photochromic materials (Chapter 7 and 8). The remaining four sections are dedicated to a variety of holographic applications. Section two has two chapters on further insights to holographic data storage – in depth analysis of collinear holographic storage (Chapter 9) and theoretical analysis of polychromatic reconstruction for volume holographic memory (Chapter 10). Section three is dedicated to holographic devices for application in wavelength division multiplexers in optical fiber systems (Chapter 11), optical circulators based on polarisation – selective substrate-mode volume holograms (Chapter 12) and holograms for beam shaping and generation of diffraction free beams (Chapter 13). Section four has two chapters on security applications – hologram based optical security systems for practical applications in verification of authenticity are described in Chapter 14 and nanophotonic hierarchical holograms are discussed in Chapter 15. The final section is dedicated to reconfigurable phase holograms for microwave signal processing (Chapter 16).

Many of the chapters describe the historical developments leading to the specific topic under discussion and will provide the reader with interesting and useful background information.

The following paragraphs give a brief summary of contents.

Ionic Liquids in Photopolymerizable Holographic Materials identifies ionic liquids suitable for use as additives in photopolymerisable materials. The authors demonstrate the application of these materials in the fabrication of symmetric and asymmetric optical diffusers with directional properties.

Norland Optical Adhesive 65® as Holographic Material presents a photosensitive material comprising Norland Optical Adhesive 65® mixed with crystal violet dye with a high potential for recording holographic elements in real time. The results from recording of Fourier holograms of binary objects are presented.

Light-Sensitive Media - Composites for Recording Volume Holograms Based on Porous Glass and Polymer describes a novel holographic material AgHal-PG. The list of the most important parameters of silver-halide media is supplemented by AgHal-PG-media with new possibilities: obtaining samples with thickness of several millimetres, low shrinkage and limitation of the maximum particle size in the light-sensitive material in pre and post processed form. A high-efficiency hologram is recorded in latent form and, after post-treatment, a distortionless interference structure in a wide dynamic range is obtained.

Digital Holographic Recording in Amorphous Chalcogenide Films discusses the possibilities of hologram recording in As-S-Se chalcogenide films. The authors demonstrate that the As-S-Se chalcogenide films can be successfully used in applied dot-matrix and image-matrix holography as an excellent alternative to organic photoresists for producing high-quality security holograms with high diffraction efficiency.

Azobenzene-Containing Materials for Hologram describes block copolymers with well-defined structures that can eliminate the scattering of visible light by microphase separation and prohibit photoinduced surface deformation when azobenzene blocks form the minority phase. Thick films (> 200 microns) with good optical transparency can be prepared with random copolymers or blended block copolymers, for recording volume holograms.

Holography Based on the Weigert's Effect outlines the advantages of holography based on Weigert's effect and some new approaches in the investigation of the photo processes that occur during holographic recording in photoanisotropic materials.

Holographic Image Storage with a 3-Indoly-Benzylfulgimide/PMMA Film studies the holographic storage applications of 3-indoly-benzylfulgimide/PMMA film including ordinary and polarization holography, which are based respectively on the photochromic and photoinduced anisotropy properties of the material.

Three-Dimensional Vector Holograms in Photoreactive Anisotropic Media describes the principle of vector holography and investigates the optical characteristics of vector holograms recorded in a photoreactive anisotropic medium. Diffraction properties of the holograms recorded in a model medium, are characterised and the results are analyzed with the use of the finite-difference time-domain (FDTD) method. By comparing the experimental and calculated results, the authors elucidate the formation mechanism of the vector holograms.

Diffraction Property of Collinear Holographic Storage System describes the collinear holographic storage system proposed by Optware. The basic theories, including two existing models for collinear systems and Volume Hologram as an Integrator of the Lights Emitted from Elementary Light Sources (VOHIL) model, are introduced. Based on Fresnel transform and the VOHIL model, paraxial solutions to describe the diffraction characteristic of the collinear holographic system are obtained.

Theory of Polychromatic Reconstruction for Volume Holographic Memory develops the theory of holographic reconstruction with polychromatic light. In particular, focusing on its application to holographic memory, the required spectral width, distortion of the reconstructed image, diffraction efficiency, intra- and inter-page crosstalk, and storage capacity are investigated.

Application of Holograms in WDM Components for Optical Fiber Systems discusses the design of a singular device for use both in Coarse and Dense Wavelength Division Multiplexing CWDM/DWDM systems. Applications such as tunable optical filters, demultiplexers and wavelength routers, using holographic SLM technology are reviewed.

Polarization-Selective Substrate-Mode Volume Holograms and Its Application to Optical Circulators introduces polarization-selective substrate-mode volume holograms which are applied in several novel designs of optical circulator. The described optical circulators have a number of advantages such as polarization-independence, compactness, high isolation, low polarization mode dispersion, easy fabrication, and low cost. In addition, the port number of the proposed multi-port device can be expanded easily.

Holographic Synthesis of Diffraction Free Beams and Dark Hollow Beams describes a simple method using holographic techniques in order to generate a variety of diffraction free beams and dark hollow beams. A fundamental part of the study consists in the generation of the boundary condition for the optical field characterized by a transmittance function. This is obtained by interfering two zero order Bessel beams.

Optimization of Hologram for Security Applications studies optical security systems for practical applications in authentication, such as in a card system. The advantages of the optical method in a security are the fast decoding of an encrypted image and the identification of it. Firstly, the authors study a common method of joint transform correlation for optical security systems and the optimization of binary holograms, and prove that the optimization of the hologram can be a powerful tool in the enhancement of system performance. An alternative method employs a phase-coding technique which enables easier realization of practical applications in optical security systems.

Nanophotonic Hierarchical Holograms: Demonstration of Hierarchical Applications Based on Nanophotonics describes the basic concept of the nanophotonic hierarchical hologram with embedded nanophotonic codes and the fabrication of a sample device. One of the most notable characteristics of the proposed approach is embedding a nanophotonic code within the patterns of a hologram composed of one-dimensional grid structures. Because embedding and retrieval of a nanophotonic code require highly advanced technical know-how, this approach can also improve the strength of anti-counterfeiting measures.

Photonic Microwave Signal Processing Based on Opto-VLSI Technology discusses reconfigurable phase holograms for the realization of microwave and RF filters, true-time delays, and beamformers in order to overcome the disadvantages of poor tunability, inflexibility, and low resolution of conventional photonic microwave signal processors. The authors demonstrate that holography is a promising technology for flexible processing of microwave and wideband RF signals with high resolution.

Dr. Izabela Naydenova
Dublin Institute of Technology
Ireland

Part 1

Holographic Recording Materials

1

Ionic Liquids in Photopolymerizable Holographic Materials

Hechun Lin and Peter W. de Oliveira
INM – Leibniz Institute for New Materails, Campus D 2 2, Saarbruecken, Germany

1. Introduction

A variety of materials have been used to record hologram, such as silver halide emulsions, hardened dichromated gelatin, ferroelectric crystals, photochromics, photoresist, photodichroics and photopolymerizable materials [1-3]. Photopolymerizable holographic materials due to their low cost and dry processing have attracted great interest in academics and industry. They have broad applications in holographic memories, recording media, LCD displays, helmet-mounted display, optical interconnects, waveguide couples, holographic diffusers, laser eye protection devices, automotive lighting, and security holograms. The photopolymerizable holographic composite contains mainly a matrix binder, a photopolymerizable monomer, an initiator system, a plasticizer and additives [4-17]. Due to the inter diffusion of the unpolymerized monomers in a holographic film, areas with high and low refractive index are formed during the irradiation with an interference pattern. Many photopolymer systems have been developed including binary photopolymer composites, organic-inorganic nanocomposites, a hybrid organic-inorganic host consisting of porous glass, and a system using monomers capable of cationic ring-opening polymerization.

The addition of a plasticizer or an additive can increase the refractive index modulation and the final diffraction efficiency. Monroe et al. reported that tri(2-ethylhexyl)phosphate, glyceryl tributyrate, polyethylene glycol or functional polyethylene glycol etc. as plasticizers may increase the refractive index modulation [18]. Frank recommended photopolymerizable compositions with triglycerides as additives, which provide a stable holographic material with high refractive index modulation [19]. Tucker et al. used trithiocarbonate as additive to increase the diffraction efficiency, uniformity and reproducibility in the formation of electrically switchable holographic gratings [20]. Finally, one publication reports about an additive to improve the sensitivity of photopolymerizable hologram material [21].

Ionic liquids are organic salts that are liquid at ambient temperatures, preferably at room temperature. They are nonvolatile, thermally and chemically stable, highly polar liquids, high ionic conductivity, large electrochemical window and ease of solubilization of a large organic molecules and transition metal complexes [22-25]. Applications of ionic liquids include their use in synthesis, catalysis, separation, electrochemistry, electrolytes, lubrication, biomass processing, drug delivery and others. The cations of ionic liquids are often large organic cations, like imidazolium, pyridium, piperidium, pyrrolidium, quaternary ammonium, phosphonium, pyrrolidium or pyrazolium etc. The anionic parts can be organic or inorganic anions such as some halides, nitrate, acetate, hexafluorophosphate, tetrafluoroborate, trifluoromethylsulfonate, or bis(trifluoromethanesulfonyl) imide etc (Figure 1). Many combinations of organic cations with different counter anions are already known, and the properties of ionic liquids may be adjusted by the proper selection of the cation and counter anion. The number of possible cation-anion combinations is greater than one million, thus allowing the design of tailor-made ionic liquids for a desired task.

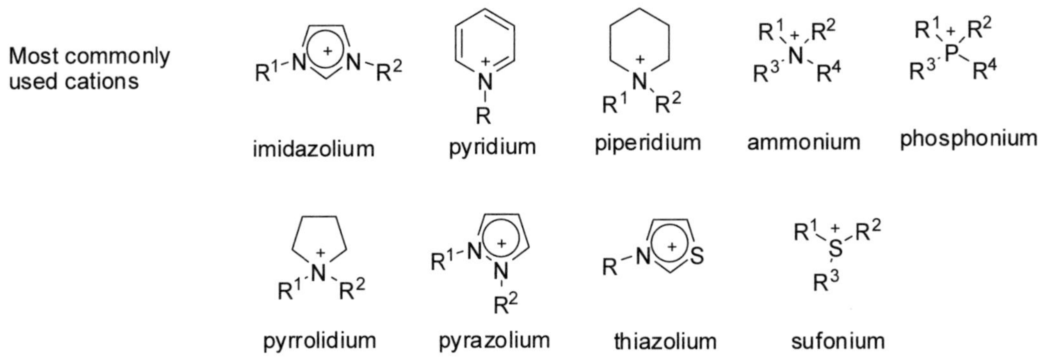

Some possible anions	water-immiscible →		water-miscible
	$[PF_6]^-$	$[BF_4]^-$	$[CH_3CO_2]^-$
	$[NTf_2]$	$[OTf]^-$	$[CF_3CO_2]^-$
	$[BR^1R^2R^3R^4]^-$	$[N(CN)_2]^-$	Br^-, Cl^-, I^-

Fig. 1. Chemical structures of typical ionic liquids.

Ionic liquids have also attracted the attention of polymer chemists [26-30]. Ionic liquids have been used as reaction media in several types of polymerization processes, such as free radical polymerization [31, controlled radical polymerization [32-33}, ring-opening polymerization [34], anionic/cationic polymerization [35], enzymatic polymerization [36], and microwave-assisted polymerization [37] and electrochemical polymerization [38]. The applications of ionic liquids provide several advantages. For instance, in radical polymerization, the k_p/k_t ratio (where k_p is the rate constant of propagation and k_t is the rate constant of termination) is higher than in organic media, and thus better control of the process can be achieved [32-33]. Under mild reaction conditions, the catalytic system can be recycle used [39]. Higher yields [40], high enzyme activity [41], high conductivity polymers [42], etc., have been reported. Ionic liquids have been used as plasticizers of various kinds of polymers [43-44], as templates for porous polymer synthesis [45-46], and as key components in new classes of polymer gels [47-49]. Polymerizable ionic liquids were used to synthesize ionic liquid co-polymers for the applications in ion-conductive polymer film [50], nanostructured liquid crystalline hydrogel [51], or microwave-absorbing polymer composite [52], etc.

Recently, we have explored a new application of ionic liquids in photopolymerizable holographic materials [53-55]. In the chapter, we highlighted our research in detail. The photopolymerizable holographic materials with higher sensitivity, higher resolution and higher diffraction efficiency were synthesized using ionic liquids as additives, which present strong dark diffusion of the monomers during the polymerization process. The materials have been used in fabricating optic diffuser for Liquid-Crystal Displays (LCD). The symmetric and asymmetric diffusers with directional diffusion property were achieved.

2. Experiment

All chemicals were used as received. Ionic liquids were synthesized according to the literature methods [56] or received from IoLiTec GmbH. Poly-(ethylenglycol)-methacrylate (PEGDMA) (average Mn ~ 330) was ordered from Sigma-Aldrich Co., Epoxy L20 and Hardener 3261 from R&G Faserverbundwerkstoffe GmbH. Irgacure 184 was a gift of Ciba Specialty Chemicals (Pty) Ltd. Scanning electron microscopy (SEM) imaging was performed on a JEOL JSM 6400F (JEOL Germany GmbH, Eching, Germany). Optic microscopy imaging was taken with Olympus BH2 equipped with a CCD camera.

2.1 Fabrication of the transmission holographic gratings

The transmission holographic grating was created by means of two-wave interference [57]. The set-up is shown in Figure 2. An argon ion laser was used here as coherent light source. The laser beam with a wavelength of 351 nm of (*power* ~ 32 mW cm^{-2}) was split by a beam splitter into two subsidiary beams of equal intensity and adjusted to obtain an interference pattern on the sample. The beam diameter was about 3 mm. Using He-Ne laser (633 nm) as reference light, the generated first-order light was read with a Lock-in Amplifier M850. The exposure time was controlled by an electronic shutter. The diffraction efficiency (η) was calculated from the ratio of the intensity of the first order laser beam diffracted by the hologram structure (I_1) to the incident intensity I_0 (to minimize the absorption and the scattering effect, I_0 was the incident intensity through the blank sample), $\eta = I_1/I_0$. Several drops of the composite were placed on a glass slide with two pieces of aluminium foil (10 µm) as spacers, on which another glass slide was placed afterwards. By gentle pressing, the drops spread between the two plates to obtain a layer with a thickness of about 10 micrometers. Then the film was exposed to a two-beam laser to create the hologram (the angle between the two beams was 2° and the grating space was approx. 10.0 µm). Other special frequency gratings were fabricating in the similar procedure except to change the angle between the two laser beams.

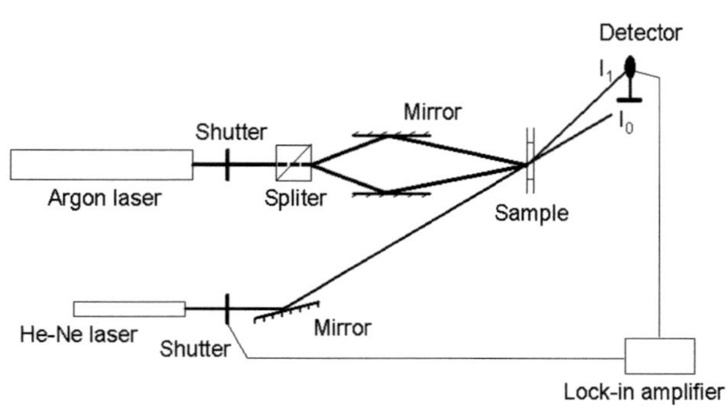

Fig. 2. The optic set-up for the hologram recording.

2.2 The fabrication of optic diffuser

Oriel® Flood Exposure Source (Model 92540-1000, Newport Co.) was used as the UV source. An optic setup was shown in Figure 3. A textured mask with random 2-15 micrometer apertures was used as mask, and polyester (PET) film was used as substrate. 75 μm thickness PET film was used as spacer. The films of photopolymerizable holographic materials were exposed vertically or 60° to collimated UV light through the mask, followed by exposing from another side to complete the polymerization process. The direct transmittance versus the tilt angle was measured with Haze meter (BYK Gardner) and the diffuser was rotated from -60° to +60° around the vertical axis.

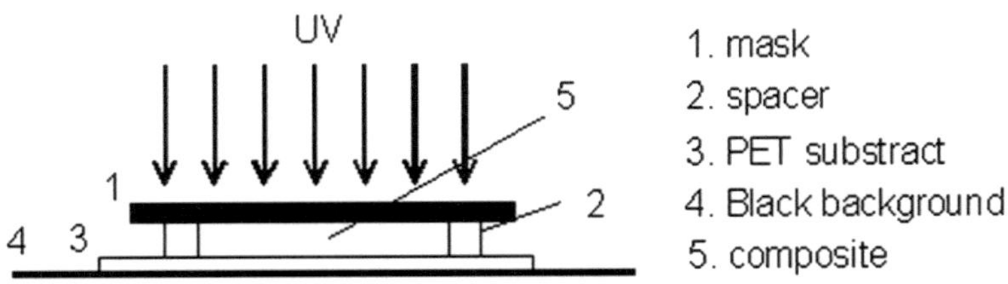

Fig. 3. The optic set-up for fabricating optic diffuser.

3. Results and discussion

3.1 The influence of ionic liquids on photopolymerizable holographic materials

Ionic liquids have significant influence on the kinetics of various polymerization reactions [26-27]. The nature of ionic liquids has strongly influenced the polymerization rate and conversion of oligomer. For instance, Chesnokov et al. reported that the addition of imidazonium salts suppress the polymerization of PEGDMA, however, tetraalkylphosphonium salts improve the photopolymerization [58]. Photopolymerizable holographic materials for practical use need to have high sensitivity, high diffraction efficiency and high resolution. The sensitivity of photopolymerizable holographic material and the photopolymerization rate have different meanings. The sensitivity here is defined as the needed exposure time (or energy) to reach the highest diffraction efficiency. The overall rate of polymerization (R_p) is considered as the rate of disappearance of monomer with respect to time, -d[M]/dt. In our research, we investigated the influence of ionic liquids on the sensitivity, the diffraction efficiency and the resolution of the photopolymerizable holographic materials.

Poly(ethylenylglycol)dimethacrylate (PEGDMA) (average Mn~ 330) was used as monomer, and Irgacure 184 (Irg184) was used as photoinitiator. In some samples, polyvinyl acetate (PVAC) or Epoxy L20/Hardener EPH161 were used as polymer binders. The structures of ionic liquids used are shown in Figure 4. 1,3-dialkylimidazolium, pyridium, and phosphonium with various counter anions were used as additives.

Due to the low solubility of PEGDMA in some ionic liquids, such as BMIMCl, BMIMBr, $BMIMSO_3Me$, $BMIMSO_4Me$, $BMIMSO_3Ph$, and $BMIMHSO_4$, BPMCl, $BPMPF_6$, Bu_4NPF_6, etc, PEGDMA cannot be solved fully in these ionic liquids to form homogenous composites. Table 1 collects the test composites used to record holographic gratings. Figure 5 shows the representative diffraction pattern by probing the hologram with 633 nm He-Ne laser beam. The diffraction efficiencies of the gratings are showed in Table 1 (the angle between the two beams was 2°). Although PEGDMA with IRG184 as initiator gave rise to low diffraction efficiency (Table 1, Sample 1), most of the tested samples gave rise to good diffraction efficiencies (η) except the composites with $BMIMNCN_2$, BMIMSCN or $BMIMFeCl_4$ as additives (Table 1, Sample 2-20), and the formed gratings have better resolution than the sample 1 without ionic liquids (Figure 6a, b). In the presence of PVAC (polyvinyl acetate) (Table 1, Sample 21-32), the diffraction efficiency was further increased except for $C_{32}H_{68}PCl$, $C_{32}H_{68}PPF_6$ and $C_{32}H_{68}PBF_4$, which formed inhomogeneous composites with PEGDMA/PVAC. 34 % of the theoretical maximum diffraction efficiencies for thin hologram [1,2] were obtained. Using Epoxy L20/Hardener EPH161 as polymer binder, satisfying diffraction efficiencies were obtained as well (Sample 35-48). Interestingly, pholymerizable ionic liquids can also be used as additive in photopolymerizable holographic materials. The application of polymerizable ionic liquids may lead to form a more stable hologram. For instance, 1-butyl-3-vinylimidazolium tetrafluoroborate ($BVIMBF_4$) and 1-allyl-3-butylimidazolium tetrafluoroborate ($ABIMBF_4$) were used as additive of photopolymerizable holographic materials. 17% and 19% diffraction efficiency was obtained, respectively (Table 1, Sample 19, 20). Nevertheless, in the presence of PVAC, the theoretical maximum diffraction efficiencies were obtained as well (Table 1, Sample 33, 34). To test the polymerizability of ionic liquids can carry out polymerization under this exposure condition, $BVIMBF_4$ (2.0 g) was mixed with Irg184 (0.05 g) to form a composite. The hologram was formed successfully, but only gave rise to 2% diffraction efficiency.

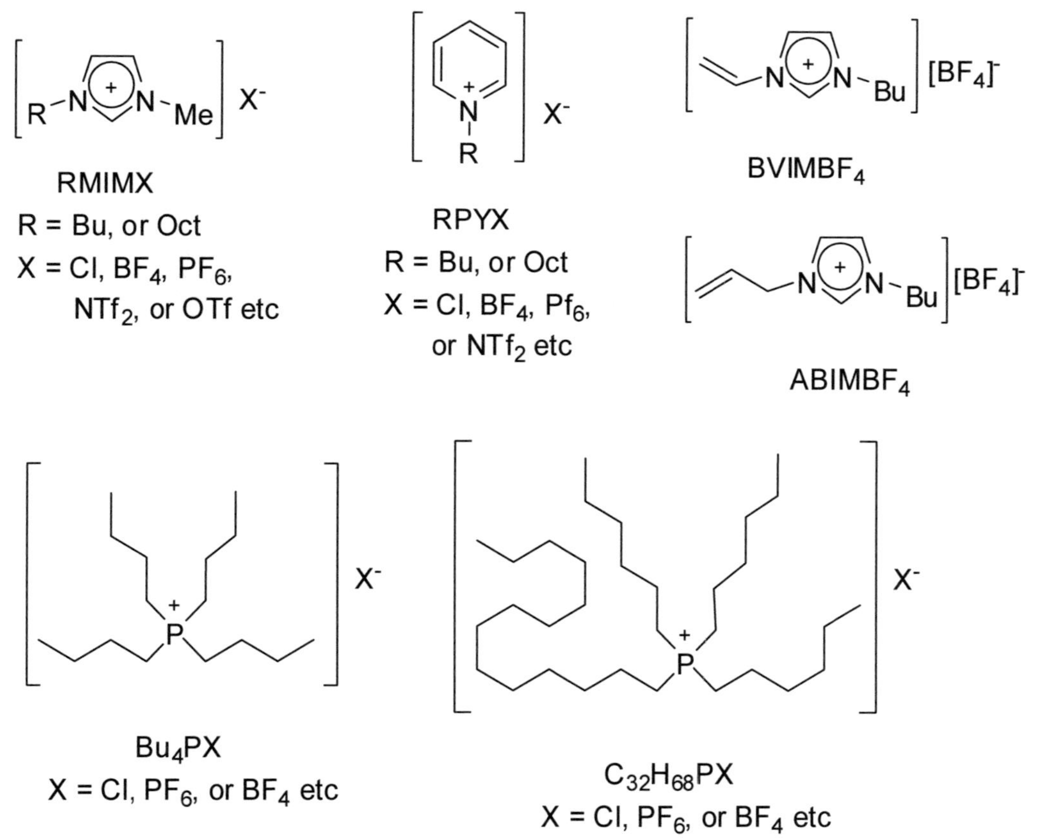

Fig. 4. Molecular structures of ionic liquids.

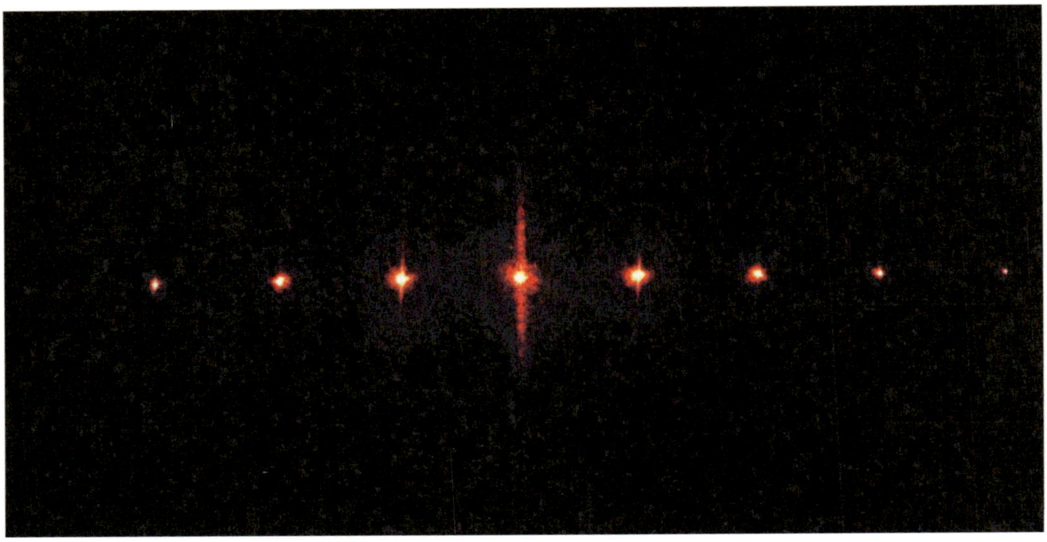

Fig. 5. The diffraction pattern obtained by 633 nm He-Ne laser beam.

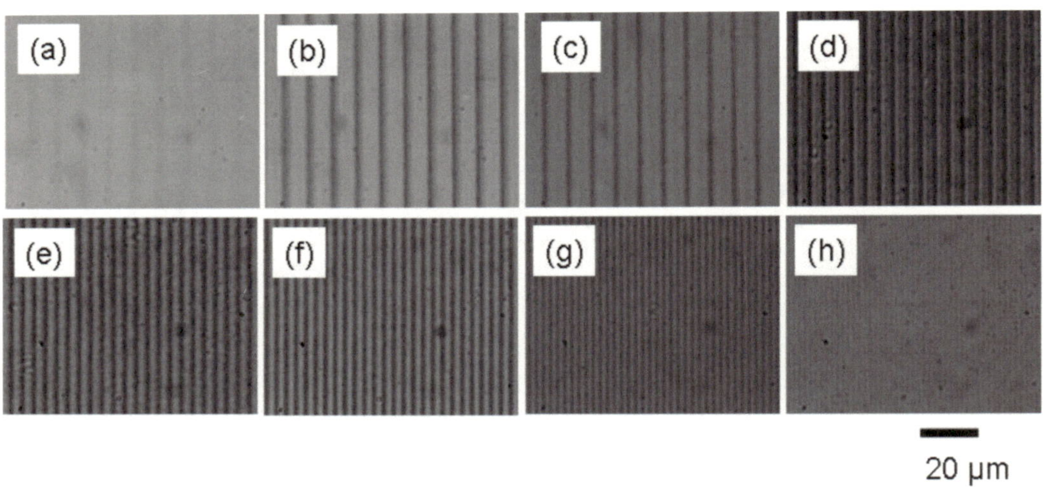

Fig. 6. Comparison of the optic microscopy images of the gratings. (a) Sample 1 (PEGDMA, 4.0g; Irg184, 0.2 g), $\eta = 1\%$, $\Lambda = 10.0$ μm, $\theta = 1.0°$. (b) Sample 2 (PEGDMA, 4.0g; $BMIMBF_4$, 1.0g; Irg184, 0.2 g), $\eta = 16\%$, $\Lambda = 10.0$ μm, $\theta = 1.0°$. (c) Sample 22 (PEGDMA/PVAC, 4.0 g, w/w = 10: 1; $BMIMBF_4$, 1.0 g; Irg184, 0.08 g), $\eta = 34\%$, $\Lambda = 10.0$ μm, $\theta = 1.0°$. (d) Sample 22, $\eta = 14\%$, $\Lambda = 5.6$ μm, $\theta = 1.8°$. (e) Sample 22, $\eta = 11\%$, $\Lambda = 4.7$ μm, $\theta = 2.1°$. (f) Sample 22, $\eta = 8\%$, $\Lambda = 4.0$ μm, $\theta = 2.5°$. (g) Sample 22, $\eta = 3\%$, $\Lambda = 2.9$ μm, $\theta = 3.5°$. (h) Sample 22, $\eta = 1\%$, $\Lambda = 2.2$ μm, $\theta = 4.6°$.

Additive	No matrix Sample[b]	DE (η, %)	With PVAC sample[c]	DE (η, %)	With EPOLH sample[d]	DE (η, %)
-	1	1	21	16	-	-
$BMIMBF_4$	2	16	22	34	35	27
$BMIMPF_6$	3	4	23	34	36	13
$OMIMBF_4$	4	13	24	31	37	24
$OMIMPF_6$	5	9	25	34	38	22
$BMIMNTf_2$	6	16	26	34	39	24
BMIMOTf	7	15	27	34	40	26
$BMIMNCN_2$	8	7	-	-	-	-
BMIMSCN	9	4	-	-	-	-
$BMIMFeCl_4$	10	1	-	-	-	-
$BMIMSO_4Oct$	11	9	28	25	41	21
$BPMBF_4$	12	22	29	34	42	20
$OPMBF_4$	13	15	30	34	43	26
$OPMNTf_2$	14	25	31	34	44	21
Bu_4PBF_4	15	27	32	34	45	29
$C_{32}H_{68}PCl$	16	11	-	-	46	4
$C_{32}H_{68}PPF_6$	17	28	-	-	47	26
$C_{32}H_{68}PBF_4$	18	15	-	-	48	20
$BVIMBF_4$	19	17	33	34	-	-
$ABIMBF_4$	20	19	34	34	-	-

[a] All of the films were exposed for 8 seconds (E ~ 0.26 $J \cdot cm^{-2}$) with 351 nm Ar laser (Power ~ 32 $mW \cdot cm^{-2}$). The diffraction efficiencies were measured with He-Ne laser (633 nm). The thicknesses were about 10 μm, and the grating space is about 10.0 μm. [b] PEGDMA (4.0 g), Irg184 (0.2 g), Ionic liquid (1.0 g); [c] PEGDMA/PVAC (4.0 g, w/w = 10: 1), Irg184 (0.08 g), Ionic liquid (1.0 g); [d] PEGDMA (3.4 g), Epoxy L20/Hardener EPH161 (0.5 g, w/w = 4: 1), Irg184 (0.1 g), Ionic liquid (1.0 g).

Table 1. Properties of the holographic films.[a]

Other different spatial frequency gratings were also fabricated to find that the ionic liquid-photopolymerizable holographic materials perform better at larger grating spacings. For instance, the diffraction efficiency of the grating based on composite 4 is 34% for 10.0 μm, 14% for 5.6 μm, 11% for 4.7 μm, 8% for 4.0 μm, and 3% for 2.9 μm, respectively. Figure 7 shows the spatial frequency response of the material. The spatial frequency increasing leads to the decreasing of the diffraction efficiency. These results are similar to the hologram material based on a photopolymerizable nematic acrylate [20]. Figure 6c-f shows the optic images of the gratings.

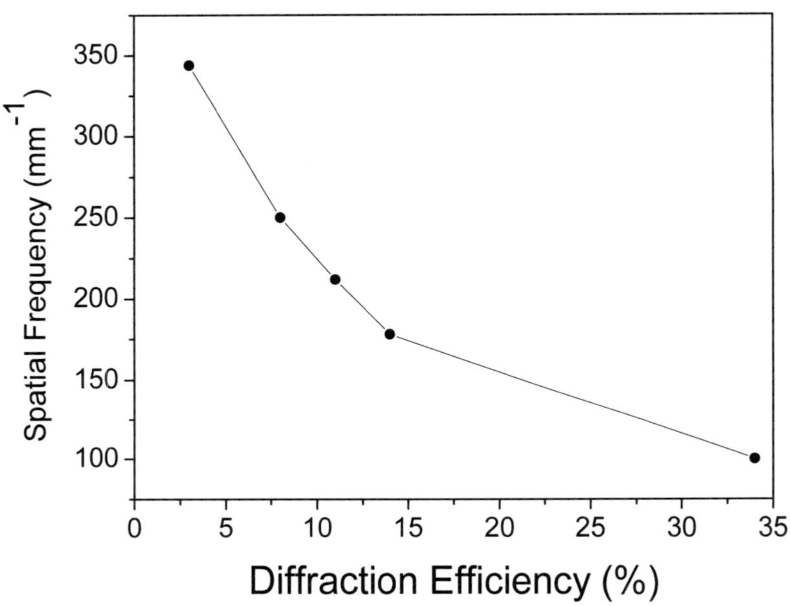

Fig. 7. The spatial frequence (mm^{-1}) vs the diffraction efficiency (%) of composite 4. The spatial frequencies increasing led to the decreasing of the diffraction efficiencies.

Figure 8 shows the representative curves of the diffraction efficiencies with reference to time. In the presence of $BMIMNCN_2$, BMIMSCN or $BMIMFeCl_4$, an unstable hologram was formed (Figure 8f). In contrast, in the presence of $BMIMBF_4$, $OMIMBF_4$, $BMIMPF_6$, $OMIMPF_6$, $BMIMNTf_2$, BMIMOTf, $BMIMSO_4Oct$, $BPMBF_4$, $OPMBF_4$, $OPMNTf_2$, Bu_4PBF_4, $C_{32}H_{68}PCl$, $C_{32}H_{68}PF_6$, $C_{32}H_{68}BF_4$, $BVIMBF_4$ or $ABIMBF_4$, the materials were more sensitive only needing about 5-6 seconds exposure to reach the maximum stable value and had higher diffraction efficiencies (Figure 8 b-e) compared to the composite in the absence of ionic liquid needing about 8 seconds exposure (Figure 8 a). The diffraction efficiency continued to increase to a stable value after stopping the exposure (Figure 8 b-e). According to the diffusion theory for formation of the hologram, the concentration of the monomer (c) is related to the diffraction efficiency (I/I_0): $-dc/dt = kIc$, whereas k depends on the extinction of the quantum yield [59]. For the composites without ionic liquid or with $BMIMNCN_2$, BMIMSCN, or $BMIMFeCl_4$ as additive, the diffraction efficiencies first increase, then drop sharply (Figure 8a, f), which indicates that the diffusion rate is bigger than the polymerization rate at the beginning. At the end triggered by the monomer concentration, the diffusion rate is less than the polymerization rate. This has an impact on the diffraction efficiency. In the presence of some ionic liquids, such as $BMIMBF_4$, $BMIMPF_6$, $OMIMPF_6$, $BMIMNTf_2$, $BPMBF_4$, etc, the diffraction efficiencies increased continually during the hologram formation (Figure 8b-e), which indicates that the diffusion rate is higher than the polymerization rate during the polymerization process. For the formation mechanism of the hologram it can be proposed that, a monomer is polymerized in the exposure region by light activation while writing the information into the material. Since the concentration of

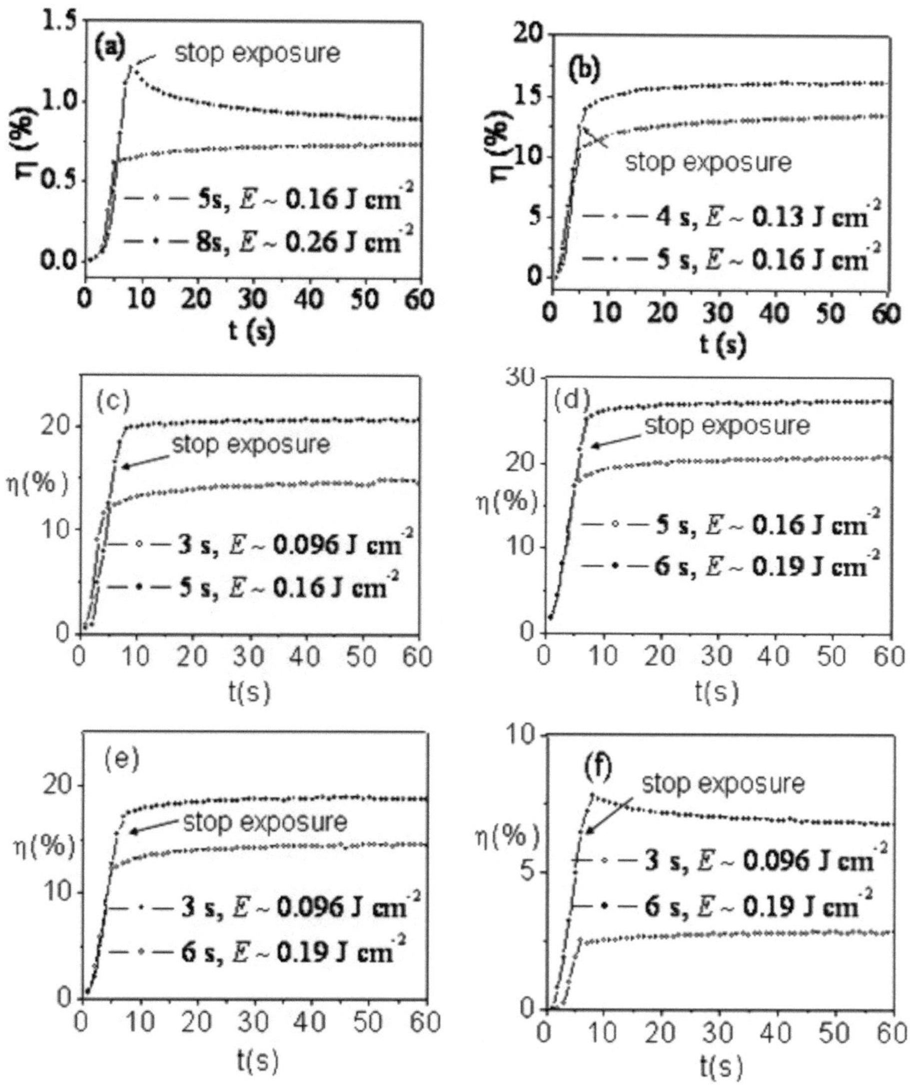

Fig. 8. The diffraction efficiency (%) vs time (second). The power of the laser beam is approximately 32 mW·cm-2. The different exposure time may be seen from the different mark of the skeleton. (a) Sample 1; (b) Sample 2 ($BMIMBF_4$); (c) Sample 12 ($BPMBF_4$); (d) Sample 15 (Bu_4PBF_4); (e) Sample 20 ($ABIMBF_4$); (f) Sample 8 ($BMIMNCN_2$).

the monomer is reduced, monomers in the dark and unexposed regions of the material diffuse to the exposed region. Due to the diffusion-controlled polymerization in the presence of some ionic liquids, the diffraction efficiencies increase continually during the hologram formation. This gives rise to higher diffraction efficiencies and bigger refractive index modulations compared to the other composites, whose diffraction efficiencies first increase, then drop sharply because of the polymerization rate controlled polymerization. On the other hand, the characters of ionic liquid have important effect on the properties of photopolymerizable holographic materials. For example, although both [BMIM][BF_4] and [BMIM][PF_6] can improve the sensitivity of the materials, only the former gave rise to high diffraction efficiency. This may be due to the different solubility property, viscosity or polarity of ionic liquids [27].

Additionally, we have also looked after the morphology of the gratings with scanning electron microscopy. In absence of polymer binder, sample 2 gave rise to a homogeneous grating, where we can not find any obvious phase separation (Figure 9a). But in the presence of epoxy resin, an obvious phase separation occurs which forms droplets of approx. 1.5 μm on the grating (Figure 9b). Possibly these are due to the different solubilization of polymers in ionic liquids [60,61]. Phase separation is often seen in the holographic polymer dispersed liquid crystal (H-PDLC) [62,63].

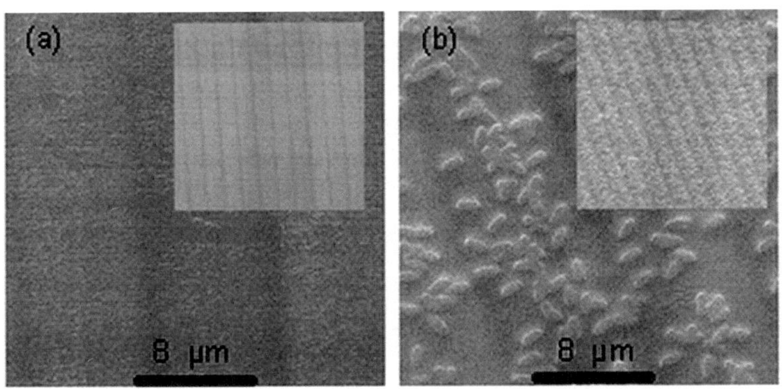

Fig. 9. SEM micrographs of the gratings. (a) sample 2 (PEGDMA, 4.0g; $BMIMBF_4$, 1.0g; Irg184, 0.2 g), (b) sample 35 (PEGDMA, 3.4 g; Epoxy L20/Hardener EPH161, 0.5 g, w/w = 4: 1; Irg184, 0.1 g, $BMIMBF_4$, 1.0g).

3.2 Fabricating optic diffuser using photopolymerizable holographic materials

Optic diffusers are key optic elements in liquid crystal displays (LCDs) which spread the incident light from sources over a wide angle to prevent light sources from being seen directly by viewers and to keep the brightness uniform over the entire display area. Generally, the diffusers can be classified into two types: the particle-diffusing type or the surface-relief type. Particle diffusers rely on the transparent beads inside the plastic films of plates to scatter light [64-66]. The distribution of diffusing beads in the diffuser is non-uniform, which affects the performance of diffusion light. The surface-relief diffusers scatter the light by the microstructures thereon, e. g. microlens diffuser [67,68], random phase diffuser [69], deterministic diffractive diffuser [70] and holographic diffuser [71-77]. Much research has been focused on holographic diffusers, which were produced via exposure of the film of photopolymerizable holographic material to collimated light through a diffuser source or mask. Comparing to other diffusers, the holographic diffusers have unique properties, such as controllable diffusion angle, directional property, volume refractive index variation and high transmittance. Hologram materials such as silver halide sensitized gelatine [72], dichromated gelatine [73], photopolymer [74-76] and azobenzene polymer [77] have been used to fabricate the diffusers. The properties of source diffuser or mask and holographic medium have important effects on the diffuser.

As we discuss in the 3.1 section, ionic liquids can be used as additives to increase the sensitivity, the diffraction efficiency and the resolution of photopolymerizable holographic materials. Interestingly, there is strong dark diffusion of the monomers during the polymerization process. In this section, we described the applications of ionic liquids-photopolymerizable holographic materials in fabricating optic diffusers via lithographic writing process. A textured mask with random 2 - 15 μm apertures was used as the mask. The films of the materials were exposed to collimated UV light through the mask. Figure 10 illustrates the lithographic writing process. The film of ionic liquids-photopolymerizable holographic materials exposes to the UV light. During the exposure to the UV light, the monomers in the bright region were polymerized. Due to the reduction of the monomer concentration in the bright region, the monomers in the dark region diffuse to the bright region to form gradient structure with volume refractive index variation.

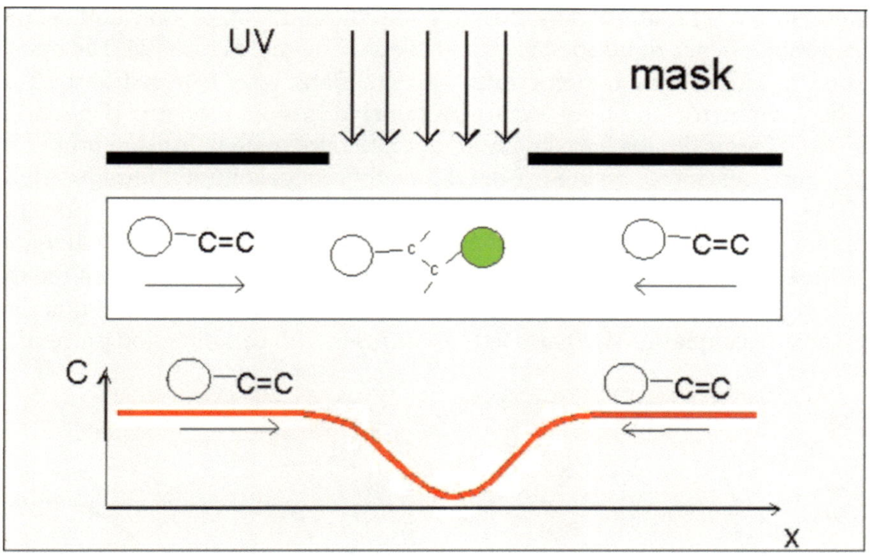

Fig. 10. Illustration of the lithographic writing process.

Symmetric and asymmetric diffusers with directional diffusion properties were both fabricating based on the ionic liquids-photopolymerizable holographic materials. For instance, using $BMIMBF_4$ as additive (sample 2, Table 1), the optic diffusers were obtained successfully with directional diffusion properties. The transmittance values varied from 7-57% within the measured angle (Figure 11a). In comparison, the composites without ionic liquids only afforded a transmittance film. Figure 11b, c show the photos of the diffusion patterns using 633 nm wavelength laser incident to the diffuser (a) a commercial particle-type diffuser and (b) the diffuser fabricated with sample 2. Comparing to the particle-type diffuser, our diffuser can scatter the light more uniformly and effectively.

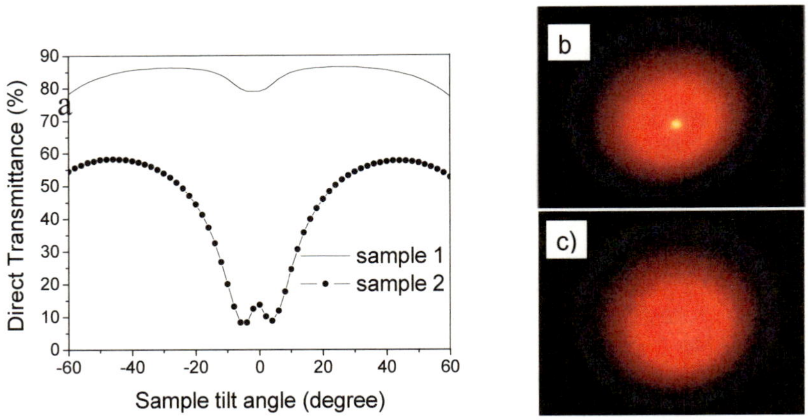

Fig. 11. a) The direct transmittance (%) versus the sample tilt angle (degree) of the symmetric diffuser. The different composites may be seen from the different symbol of the skeleton. Sample 1: PEGDMA, 4.0 g; Irg184, 0.2 g; Sample 2 : PEGDMA, 4.0g; $BMIMBF_4$, 1.0g; Irg184, 0.2 g. b) Photo of the diffusion pattern of the commercial particle-type diffuser. c) Photo of the diffusion pattern of the diffuser fabricated with sample 2.

The transmittance can be adjusted by changing the concentration of ionic liquids. Increasing the concentration of ionic liquids led to more haze as shown in Figure 12a. The materials can also be used to fabricate asymmetric diffusers. The films were exposed to a 60° angle to provide the asymmetric diffusers with directional diffusion property (Figure 12b). The characteristics of ionic liquids have an important influence on the diffusion properties of the diffusers. For instance, using 1-butyl-3-methyl-imidazolium hexafluorophosphate ($BMIMPF_6$) as additive (Table 1, sample 3), which only led to 4% diffraction efficiency in the thin hologram, however representing strong diffusion during the polymerization process. It led to a diffuser with a transmittance value variable from 50% to 79% within the measured angle. Although it had a high diffraction efficiency using polymerizable ionic liquids as additive (Table 1, sample 33, 34), it also led to a diffuser with bad diffusion properties.

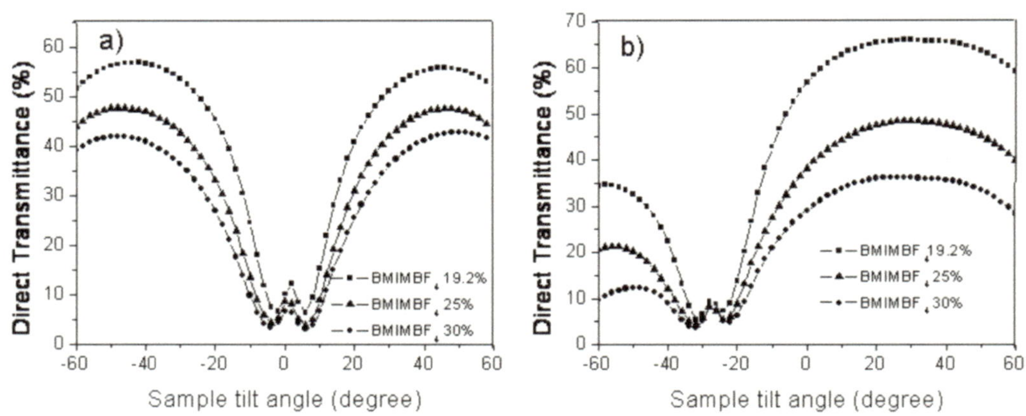

Fig. 12. The direct transmittance (%) versus the sample tilt angle (degree) of the symmetric diffuser. The different composites may be seen from the different symbol of the skeleton.

The cross section of the diffuser was examined with optical microscopy. The modulation of the refractive index was visible as shown in Figure 13 of the fiber structure. The surface of the diffuser was analysed with scanning electron microscopy (SEM). Figure 14 (a, b) shows the surface image of the diffuser based on sample 2. The pattern of the mask has been successfully recorded to form a surface-relief structure. Interestedly, there were many particles in a range tens to hundreds of nanometers on the surface, which possibly arise from phase separation of $BMIMBF_4$ in the bulk during polymerization. After that, we examined the cross section by SEM and found that most nanoparticles appeared in the region near both surfaces and that the bulk was more homogeneous [Figure 14 (c, d)]. The nanoparticles may function as particulate scatterers due to the low refractive index of n = 1.422 of $BMIMBF_4$, compared to n = 1.463 of PEGDMA.

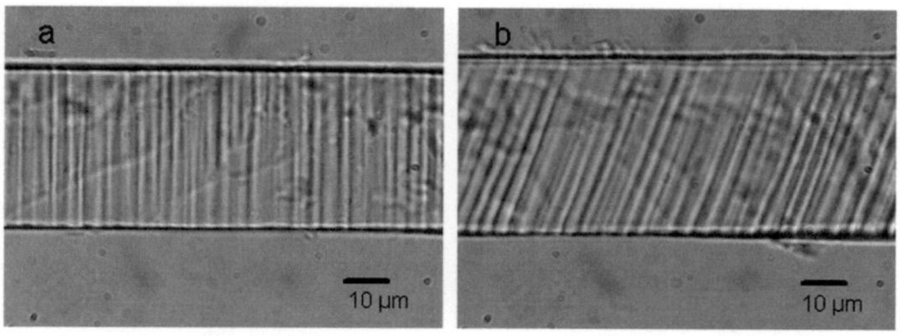

Fig. 13. The cross section optic images of the diffusers based on sample 2. (a) symmetric diffuser, (b) asymmetric diffuser.

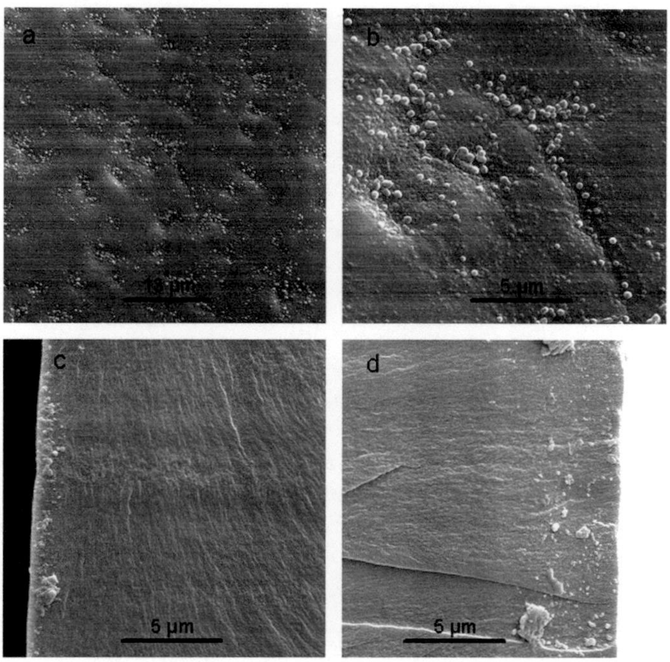

Fig. 14. The SEM images of the diffuser based on sample 2. (a) The surface image in 2000 × magnification. (b) The surface image in 6000 × magnification. (c) The cross section near the mask region. (d) The cross section near the substrate.

For comparison, the diffuser with bad diffusion properties was also observed with optic microscopy and scanning electron microscopy. The fiber structure was successfully formed, but there was no uniform phase separation during the polymerization process as shown the optic images and SEM images in Figure 15.

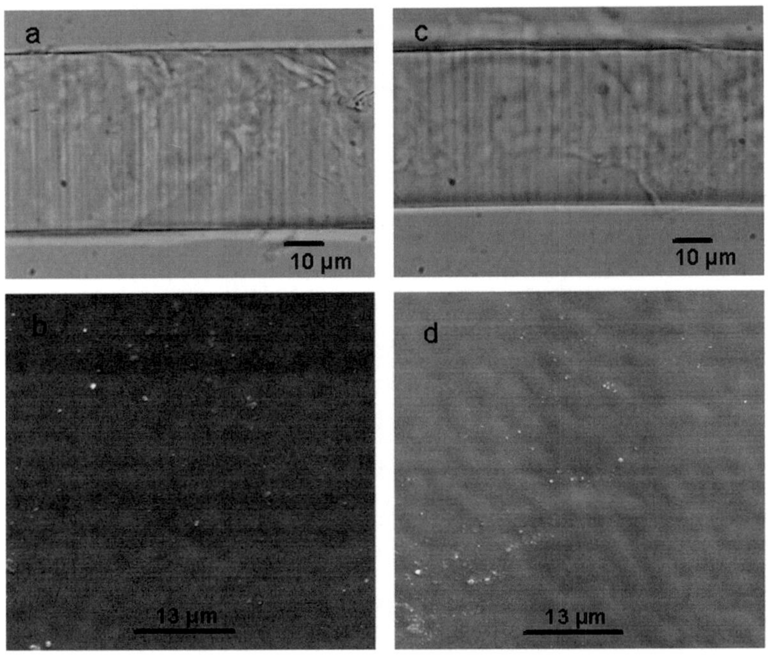

Fig. 15. a) The cross section optic images of the diffusers based on sample 3. b) The cross section optic images of the diffusers based on sample 19. (c) The SEM image in 2000 × magnification of the diffuser based on sample 3. (d) The SEM image in 2000 × magnification of the diffuser based on sample 19.

Most tested composites with ionic liquid as additive formed the fibre structure successfully, which indicates the volume refraction index variation. The diffusion-controlled polymerization in the presence of ionic liquid was beneficial for the formation of the fibre structure. Generally, during the lithographic process, the monomers in the bright region were polymerized. Due to the decreasing of the monomer concentration in the bright region, the monomers in the dark region diffuse to the bright region and polymerize to form the fibre structure. The properties of ionic liquids have an important effect on the diffuser. For instance, $BMIMBF_4$ afforded the better diffusion property than $BMIMPF_6$ and $BVIMBF_4$. One of the reasons is possibly due to the better formation of nanoparticles for the former. Thus, a forming mechanism for the diffuser with good diffusion properties can be proposed, which (during the exposure) leads to photopolymerization of the monomers in the immediate area exposed to ultra-violet light, accompanied and followed by diffusion of monomers from the unexposed regions into the exposed regions. By further polymerization the fibre structure is formed and a phase separation of ionic liquid is observed leading to the formation of nanoparticles. The fibre structure, the surface-relief structure and the formation of nanoparticles altogether are responsible for the directional diffusion property of the diffuser.

4. Conclusion

In summary, we investigated the influence of ionic liquids on photopolymerizable holographic materials. Although not all of the ionic liquids can be used as additives for photopolymerizable holograms, we found that imidazolium, pyridium and phosphonium based ionic liquids with proper counter anions, such as $BMIMBF_4$, $OMIMBF_4$, $BMIMNTf_2$, $BPMBF_4$, $OPMBF_4$, $OPMNTf_2$, Bu_4PBF_4, $C_{32}H_{68}PCl$ etc, can be used as additives to improve the properties of the materials. The sensitivity, resolution and the diffraction efficiency of the materials were increased efficiently. More interestingly, it presented strong dark diffusion of the monomers during polymerization process due to the diffusion controlled polymerization in the presence of some ionic liquids. Polymerizable ionic liquids were also used as additives in the holographic materials. High diffraction efficiencies were obtained as well. The photopolymerizable holographic materials have shown the potential application in fabricating optic diffuser for LCD. The symmetric and asymmetric diffusers with directional properties were successfully produced via lithographic recording method. The diffusion property can be regulated by changing the concentration of ionic liquid. The fiber structure, the surface-relief structure and the formation of nanoparticles lead to the directional diffusion property of the diffuser.
Ionic liquids are often named as so-called green solvents. However, "Greenness" of ionic liquids depends strongly on the structure. It is necessary to mention that ionic liquids exist as a component after the formation of the hologram. Low or no toxicity of ionic liquids is required for the actual application. Ionic liquids are designable. Our results are helpful for designing eco-friendly and successional holographic materials. Further researches on the application of ionic liquids in organic-inorganic nanocomposites and cationic ring-opening polymerization holographic materials are in progress.

5. Acknowledgments

The authors thank the Stiftung Europrofession, the State of Saarland and the Fonds der Chemischen Industrie for financial support. We thank Dr. Peter könig and Dr. Peter Rogin for useful discussions.

6. References

[1] P. Hariharan, *Optical Holography: principles, techniques, and applications.* Cambridge University press, 1996.
[2] P. Hariharan, *Basic of Holography*. Cambridge University press, 2002.
[3] T. J. Trout, J. J. Schmieg, W. J. Gambogi, A. M. Weber, (1998) *"Optical Photopolymers: Design and Applications," Adv. Mater.*, 10 , 1219-1224.
[4] D. J. Lougnot, in: *Radiat. Curing Polym. Sci. Technol.* Vol. 3 (Eds. J. P. Fouassier, J. F. Rabek), Elsevier London, New York, 1993, pp. 65.
[5] M. L. Calvo, P. Cheben, *Fundamentals and advances in holographic materials for optical data storage, in Advances in information optics and photonics* (Eds. A. T. Friberg, R. Dändliker), SPIE Press, Bellingham, 2008, Chapter 15.
[6] J. R. Lawrence, F. T. O'Neill, J. T. Sheridan, (2001) "Photopolymer holographic recording material," *Optic*, 112, 449-463.

[7] N. Suzuki, Y. Tomita, (2004), *"Silica-Nanoparticle-Dispersed Methacrylate Photopolymers with Net Diffraction Efficiency Near 100%."* Appl. Opt., *43, 2125-2129.*

[8] C. Sanchez, M. J. Escuti, C. Van Heesch, C. W. M. Bastiaansen, D. J. Broer, J. Loos, R. Nussbaumer, (2005) "TiO_2 Nanoparticle–Photopolymer Composites for Volume Holographic Recording." *Adv. Funct. Mater.,* 15, 1623-1629.

[9] F. Del Monte, O. Martinez, J. A. Rodrigo, M. L. Calvo, P. Cheben, (2006) "A volume holographic sol-gel material with large enhancement of dynamic range by incorporation of high refractive index species." *Adv. Mater.,* 18 , 2014-2017.

[10] C. Bräuche, H. Anneser, (1990) "Holographic spectroscopy and holographic information recording in polymer matrices." *Lasers in Polymer Sci. Technol. Appl.,* 3, pp.181-210.

[11] N. Suzuki, Y. Tomita, T, Kojima, (2002) "Holograhic recording in TiO_2 nanoparticle-dispersed methacrylate photopolymer films." *Appl. Phys. Lett.,* 81, 4121-4123.

[12] W. S. Kim, Y. Jeong, J. Park, (2006) "Nanoparticle-indused refractive index modulation of organic-inorganic hybrid photopolymer." *Opt. Express,* 14, 8967-8973.

[13] T. J. Trentler, J. E. Boyd, V. L. Colvin, (2002) "Epoxyl resin-photopolymer composites for volume holography." *Chem. Mater.,* 12, 1431-1438.

[14] S. Martin, P. Leclere, Y. Renotte, V. Toal, Y. Lion, (1995) "Characterisation of an acrylamide-based dry photopolymer holographic recording material." *Opt. Eng.,* 33, 3942–3946.

[15] L. M. Goldenberg, O. V. Sakhno, T. N. Smirnova, P. Helliwell, V. Chechik, J. Stumpe, (2008) "Holographic composites with gold nanoparticles: nanoparticles promote polymer segregation." *Chem. Mater.,* 20, 4619-4627.

[16] P. Cheben, M. L. Calvo, (2001) "A photopolymerizable glass with diffraction efficiency near 100% for holographic storage." *Appl. Phys. Lett.,* 78, 1490-1492.

[17] M. Ortuno, E. Fernandez, S. Gallego, A. Belendz, I. Pascual, (2007) "New photopolymer holographic recording material with sustainable design." *Optic Express,* 15, 12425-12435.

[18] B. M. Monroe, W. K. Smotherg, U.S Patent 4942112, Jun. 17, 1990.

[19] K. Frank, CA Patent 2571951, Dec. 29, 2005.

[20] L. J. Tucker, M. B. Sponsler, (2006) "Trithiocarbonate-mediated free-radical photopolymerization: improved uniformity in hologram recording." *Appl. Opt.,* 45, 6973-6976.

[21] Acrylic anhydride was used as an additive to increase the "film speed" in hologram material. A. M. Weber, U.S. Patent 5013632, May 7, 1991.

[22] P. Wasserscheid, T. Welton, *Ionic Liquids in Synthesis,* Wiley VCH, Weinheim, Germany, 2003.

[23] B. Kirchner, *Top Curr Chem: Ionic liquid,* Vol. 290, Spinger-Verlag Berlin Heidelberg, 2009.

[24] P. Wasserscheid, W. Keim, (2000) "Ionic Liquids—New "Solutions" for Transition Metal Catalysis." *Angew. Chem Int. Ed.,* 39, 3772-3789.

[25] M. Freemantle, *An Introduction to ionic liquids,* Royal Society of Chemistry, 2010.

[26] P. Kubisa, (2004) "Application of ionic liquids as solvents for polymerization processes." *Prog. Polym. Sci.*, 29, 3-12.

[27] P. Kubisa, (2005) "Ionic liquids in the synthesis and modification of polymers." *J. Polym. Sci. Part A, Polym. Chem.*, 43, 4675-4683.

[28] P. Kubisa, (2009) "Ionic liquids as solvents for polymerization process-progress and challenges." *Prog. Polym. Sci.*, 34, 1333-1347.

[29] J. Lu, F. Yan, J. Texter, (2009) "Advanced applications of ionic liquids in polymer science." *Prog. Polym. Sci.*, 34, 431-448.

[30] T. Erdmenger, C. Guerrero-Sanchez, J. Vitz, R. Hoogenboom, U. S. Schubert, (2010) "Recent developments in the utilization of green solvents in polymer chemistry." *Chem. Soc. Rev.*, 39, 2217-3333.

[31] A. J. Carmichael, D. M. Haddleton, S. A. F. Bo, K. R. Seddon, (2000) "Copper(I) mediated living radical polymerisation in an ionic liquid." *Chem. Commun.*, 1237-1238.

[32] S. Harrisson, S. R. Mackenzie, D. M. Haddleton, (2002) "Unprecedented solvent-induced acceleration of free-radical propagation of methyl methacrylate in ionic liquids." *Chem. Commun.*, 2850-2851.

[33] S. Harrisson, S. R. Mackenzie, D. M. Haddleton, (2003) "Pulsed laser polymerization in an ionic liquid: strong solvent effects on propagation and termination of methyl methacrylate." *Macromolecules*, 36, 5072-5075.

[34] J. Kadokawa, Y. Iwasaki, H. Tagaya, (2002) "Ring-Opening Polymerization of Ethylene Carbonate Catalyzed with Ionic Liquids: Imidazolium Chloroaluminate and Chlorostannate Melts." *Macromolecular Rapid Commun*, 23, 757-760.

[35] T. Biedron, P. Kubisa, (2004) "Cationic polymerization of styrene in a neutral ionic liquid." *J. Polym. Sci. Part A: Polym. Chem.* 42, 3230-3235.

[36] R. Marcilla, M. Geus, D. Mecerreves, C. J. Duxbury, C. E. Koning, A. Heise, (2006) "Enzymatic polyester synthesis in ionic liquids." *Euro. Polym. J.* 42, 1215-1221.

[37] C. Guerrro-Sanchez, M. Lobert, R. Hoogenboom, U. S. Schubert, (2007) "Microwave-assisted homogeneous polymerizations in water-soluble ionic liquids: an alternative and green approach for polymer synthesis." *Macromolecular Rapid Commun*, 28, 456-464.

[38] E. Naudin, H. A. Ho, S. Branchaud, L. Breau, D. Belanger, (2002) "Electrochemical polymerization and characterization of poly(3-(4-fluorophenyl)thiophene) in pure ionic liquids." *J. Phys. Chem. B.* 106, 10585-1-593.

[39] N. Li, J. Lu, Q. Xu, X. Xia, L. Wang, (2007) "Reverse atom transfer radical polymerization of MMA via immobilized catalysts in imidazolium ionic liquids." *J. Appl. Polym. Sci.*, 103, 3915-3919.

[40] A. J. Carmichael, D. M. Heddleton, S. A. F. Bon, K. R. Seddon, (2000) "Copper(I) mediated living radical polymerization in an ionic liquid." *Chem. Commun.*, 1237-1238.

[41] R. M. Lau, F. Van Rantwijk, K. R. Seddon, (2000) "Lipaze catalyzed reactions in ionic liquids." *Org. Lett.*, 2, 4189-4191.

[42] Y. S. Vygodskii, O. A. Mel'nik, A. S. Shaplov, E. L. Lozinskaya, I. A. Malyshkina, N. D. Gavrilova, (2007) "Synthesis and ionic conductivity of polymer ionic liquids." *Polym. Sci. A.*, 49, 256-261.

[43] M. P. Scott, C. S. Brazel, M. G. Benton, J. W. Mays, J. D. Holbrey, R. D. Rogers, (2002) "Application of ionic liquids as plasticizers for poly(methyl methacrylate)." *Chem. Commum.*, 1370-1371.

[44] M. P. Scott, M. Rahman, C. S. Brazel, (2003) "Application of ionic liquids as low-volatility plasticizers for PMMA." *Eur. Polym. J.* 39, 1947-1953.

[45] [45] P. Snedden, A. Cooper, K. Scott, (2003) "Cross-linked polymer-ionic liquid composite materials." *Macromolecules*, 36, 4549-4556.

[46] L. Zhu, C. Y. Huang, Y. H. Patel, J. Wu, S. V. Malhotra, (2006) "Synthesis of porous polyurea with room-temperature ionic liquids via interfacial polymerization." *Macromol Rapid Commun.*, 27, 1306-1311.

[47] J. Fuller, A. C. Breda, R. T. Carlin, (1997) "Ionic liquid-polymer gel electrolytes." *J. Electrochem Soc.*, 144, L67-69.

[48] P. Wang, S. M. Zakeeruddin, I. Exnar, M. Gratzel, (2002) "High efficiency dye-sensitized nanocrystalline solar cells based on ionic liquid polymer gel electrolyte." *Chem. Commun.*, 2972-2973.

[49] T. Fukushima, K. Asaka, A. Kosaka, T. Aida, (2005) "Fully plastic actuator through layer-by-layer casting with ionic-liquids-based bulky gel." *Angew. Chem. Int. Ed.*, 44, 2410-2413.

[50] H. Ohno, (2007) "Design of ion conductive polymers based on ionic liquids." *Macromol Symp.*, 249, 551-556.

[51] D. Batra, D. N. T. Hay, M. A. Firestone, (2007) "Formation of a biomimetic, liquid-crystalline hydrogel by self-assembly and polymerization of an ionic liquid." *Chem. Mater.*, 19, 1279-1287.

[52] J. Tang, M. Radosz, Y. Shen, (2008) "Poly(ionic liquid)s as optically transparent microwave-absorbing materials." *Macromolecules*, 41, 493-496.

[53] H. Lin, P. W. Oliveira, M. Veith, (2008) "Ionic liquid as additive to increase sensitivity, resolution, and diffraction efficiency of photopolymerizable hologram material." *Appl. Phys. Lett.*, 93, 141101.

[54] H. Lin, P. W. Oliveira, M. Veith, M. Gros, I. Grobelsek, (2009) "Optic diffusers based on photopolymerizable hologram material with an ionic liquid as additive." *Opt. Lett.*, 34, 1150-1153.

[55] H. Lin, P. W. Oliveira, M. Veith, (2011) "Application of ionic liquids in photopolymerizable holographic materials." *Opt. Mater.*, 33, 759-762.

[56] X. Creary, E. D. Wilis, (2005) "Preparation of 1-butyl-3-methylimidazolium tetrafluoroborate." *Organic Syntheses*, 82 , 166-169.

[57] G. W. Stroke, *An introduction to coherent optics and Holography*, Academic Press, New York, 1969, pp. 225.

[58] S. A. Chesnokov, M. Y. Zakharina, A. S. Shaplov, Y. V. Chechet, E. I. Lozinskaya, O. A. Mel'nik, Y. S. Vygodskii, G. A. Abakumov, (2008) "Ionic liquids as catalytic additives for the acceleration of the photopolymerization of poly(ethylene glycol dimethacrylate)s." *Polym. Int.*, 57, 538-545.

[59] H. Sillescu, D. Ehlich, in: *Lasers in Polymer Sci. Technol. Appl.*, Vol 3, (Eds. J. P. Fouassier, J. F. Rabek) CRC-Press Inc., Boca Raton, Florida, 1990, pp. 211-226.

[60] P. Snedden, A. I. Copper, K. Scott, N. Winterton, (2003) "Cross-linked polymer-ionic liquid composite materials." *Macromolecules*, 36, 4549-4556.

[61] N. Winterton, (2006) "Solubilization of polymers by ionic liquids." *J. Mater. Chem.*, 16, 4281-4293.

[62] R. L. Sutherland, L. V. Natarajan, V. P. Tondiglia, (1993) "Bragg gratings in an acrylate polymer consisting of periodic polymer-dispersed liquid-crystal planes." *Chem. Mater.*, 5, 1533-1538.

[63] M. J. Escuti, P. Kossyrev, G. P. Crawford, (2000) "Expanded viewing-angle reflection from diffuse holographic-polymer dispersed liquid crystal films." *Appl. Phys. Lett.*, 77, 4262-4264.

[64] G. H. Kim, (2005) "A PMMA composite as an optic diffuser in a liquid crystal display backlighting unit (BLU)." *Eur. Poly. J.*, 41, 1729-1737.

[65] G. H. Kim, W. J. Kim, S. M. Kim, and J. G. Son, (2005) "Analysis of thermo-physical and optical properties of a diffuser using PET/PC/PBT copolymer in LCD backlight units." *Display*, 26, 37-43.

[66] G. H. Kim, J. H. Park, (2007) "A PMMA optical diffuser fabricated using an electrospray method." *Appl. Phys. A*, 86, 347-351.

[67] S. I. Chang, J. B. Yoon, (2004) "Shape-controlled, high fill-factor microlens arrays fabricated by a 3D diffuser lithography and plastic replication method," *Opt. Express*, 12, 6366-6371.

[68] S. I. Chang, J. B. Yoon, H. Kim, J. J. Kim, B. K. Lee, and D. H. Shin, (2006) "Microlens array diffuser for a light-emitting diode backlight system." *Opt. Lett.*, 31, 3016-3018.

[69] E. E. Garcia-Guerrero, E. R. Mendez, H. M. Escamilla, T. A. Leskova, and A. A. Maradudin, (2007) "Design and fabrication of random phase diffuser for extending the depth of focus," *Opt. Express*, 15, 910-923.

[70] M. Parikka, T. Kaikuranta, P. Laakkonen, J. Lautanen, J. Tervo, M. Honkanen, M. Kuittinen, and J. Turuner, (2001) "Deterministic diffractive diffusers for displays," *Appl. Opt.*, 40, 2239-2246.

[71] C. Gu, J. R. Lien, F. Dai, and J. Hong, (1996) "Diffraction properties of volume holographic diffuser." *J. Opt. Soc. Am. A*, 13, 1704-1711.

[72] S. I. Kim, Y. S. Choi, Y. N. Ham, C. Y. Park, and J. M. Kim, (2003) "Holographic diffuser by use of a silver halide sensitized gelatine process." *Appl. Opt.*, 42, 2482-2491.

[73] S. Wadle, D. Wuest, J. Cantalupo, and R. S. Lakes, (1994) "Holographic diffusers," *Opt. Eng.*, 33, 213-218.

[74] S. Wadle, and R. S. Lakes, (1994) "Holographic diffusers: polarization effects," *Opt. Eng.*, 33, 1084-1088.

[75] C. Joubert, B. Loiseaux, A. Delboulbe, and J. P. Huignard, (1997) "Phase volume holographic optical components for high-brightness single-LCD projectors," *Appl. Opt.*, 36, 4761-4764.

[76] H. Honma, Y. Maekawa, M. Takano, L. M. Murillo-Mora, A. Sato, K. Hirose, and F. Iwata, (2004) "A new type of optical diffuser with the directional property," *Proc. SPIE* 5290, 74-80.

[77] D. Sakai, K. Harada, S. I. Kamemaru, M. A. El-Morsy, M. Itoh, and T. Yatagai, (2005) "Direct fabrication of surface relief holographic diffusers in azobenzene polymer films," *Opt. Rev.*, 12, 383-386.

2

Norland Optical Adhesive 65® as Holographic Material

J.C. Ibarra[1], L. Aparicio-Ixta[2], M. Ortiz-Gutiérrez[2] and C.R. Michel[1]
[1]*CUCEI, Universidad de Guadalajara*
[2]*Universidad Michoacana de San Nicolás de Hidalgo*
México

1. Introduction

Research on photosensitive materials is an active field where the main goal is to find materials with desirable characteristics for optical data storage. Some of these special characteristics are high sensibility, high resolution and wide spectral range, low cost, among others (Smith, 1975). For this purpose many kinds of materials that for this purpose, such as silver halide, photoresist, dichromated gelatin, photopolymers, thermal recording materials, photothermoplastics, photocromics, and photorefractive crystals (Bjelkhangen & Thompson, 1996; Hariharan, 1980; Kang et al., 2004; Koustuk, 1999) have been used. The most widely used at present are photopolymers.
Photopolymers have excellent holographic characteristics, such as high refraction index modulation, real time recording, low cost, etc. The response on these materials depends of parameters such as incident beam intensity, monomers concentration, polymerization velocity, humidity, temperature, thickness of the sample, etc. (Adhami et al., 1991; Gallego et al., 2005; Gleeson, et al., 2005). Recent papers show that photopolymer's thickness is of great importance (Neipp et al., 2003; Ortuño, et al., 2003). The spectral sensibility of these materials can be easily modified if the photopolymers are mixed with dyes such as crystal violet (Luna et al., 1997,1998; Ortiz et al. 2007).
Some photopolymers employed in optical storage are given in (K. & M. Budinski, 1999; Fernandez et al., 2006; Ibarra & Olivares, 2006; Leclere et al., 1995; Naydenova et al., 2006). One of these polymers is an adhesive called Norland Optical Adhesive 65® (NOA 65®). (Pinto & Olivares, 2002) and co-workers report that they have used NOA 65® in its natural form to record computer generated Fourier holograms using microlithography techniques. Recently (Aleksejeva & Teteris, 2010), the photopolymers NOA 60, NOA 61, NOA 63, NOA 65 and NOA 68 were studied as materials for fabrication of volume gratings, they recorded transmission and reflection diffraction gratings and used a He–Cd laser of 325nm line, obtaining diffraction efficiency >80%.
In this work a study became of the holographic material composed by Norland Optical adhesive 65 (NOA 65) mixed with crystal violet dye (CV) was made. In this material we recorded transmission real time phase holographic gratings and Fourier holograms. obtaining diffraction efficiency of 1.85% using a light beam at wavelength 598 nm from a He-Ne laser was obtained. The gratings were recorded changing parameters such as

concentrations between NOA 65 and CV, sample thickness, beams intensity ratio and spatial frequency. The material shows refraction index modulation, which is calculated using the Kogelnik`s theory. The results obtained are show by the behavior of diffraction efficiency versus energy.

2. Materials properties

2.1 Norland Optical Adhesive 65 (NOA 65®)

This polymer is typically used for putting lenses in metal mounts, bounding plastic to glass and cold blocking by cured process. The polymer cure process depends on intensity and wavelength of the UV radiation. Before being exposed to UV radiation, the polymer's adhesive is in liquid state because the monomers and photo initiators will not react with each other. When exposed to UV, the photo initiators undergo a change creating free radicals that react with monomers, producing monomer chains. In the cured state, the monomer chains convert to cross-linked polymer chains.

The absorbance spectra of the NOA 65® obtained with an UV-Vis spectrophotometer is show in Fig. 1, where we can observe that its absorbance displays a plateau in the visible region, showing a maximum absorption in 300 nm. Complementary to this plot, Fig. 2 show the spectral transmission for the UV-Vis-IR regions. The plot was obtained from (Norland Products Incorporate, 1999).

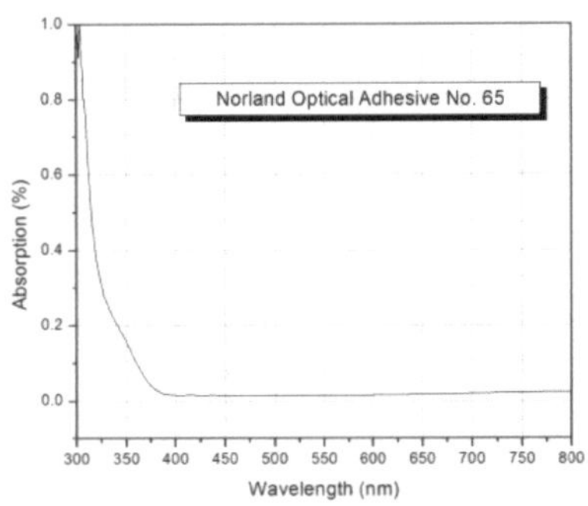

Fig. 1. Absorption spectra of NOA 65 in region UV-Vis.

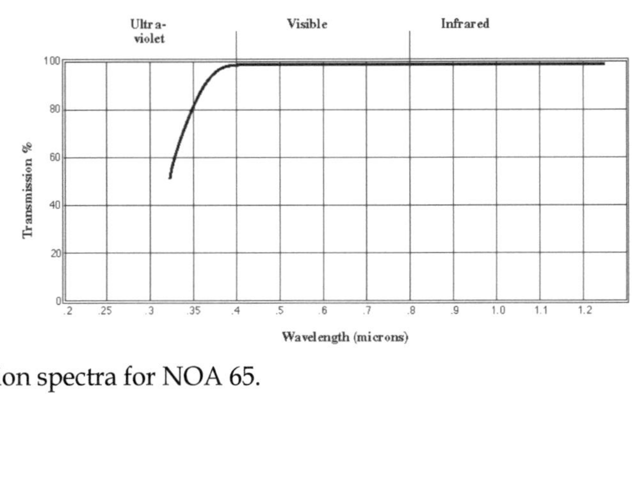

Fig. 2. Transmission spectra for NOA 65.

Table 1 shows some properties of NOA 65 whereas Table 2 shows the typical cure times according to Norland Products instructions.

Solids	100%
Viscosity at 25°C	1,200
Refractive Index of Cured Polymer	1.524
Elongation at Failure	80%
Modulus of Elasticity (psi)	20,000
Tensile Strength (psi)	1,500
Hardness - Shore D	50
Temperature Range	-15 to 60°

Table 1. Typical properties of NOA 65.

LIGHT SOURCE	FILM THICKNESS	PRECURE	FULL CURE
100 Watt Mercury* Spot Lamp at 6 inches	1-10 mil	15 seconds	5 minutes
2-15 Watt Fluorescent* Black Lights at 3 inches	1-10 mil	60 seconds	20 minutes

Table 2. Typical cure times of NOA 65.

Fig. 3 shows the absorbance spectra obtained with a FTIR spectrophotometer showing absorption peaks, indicating the presence of some compounds Table 3 displays brief analysis of the NOA 65 IR spectrum. briefly analysis of the NOA 65® IR spectrum.

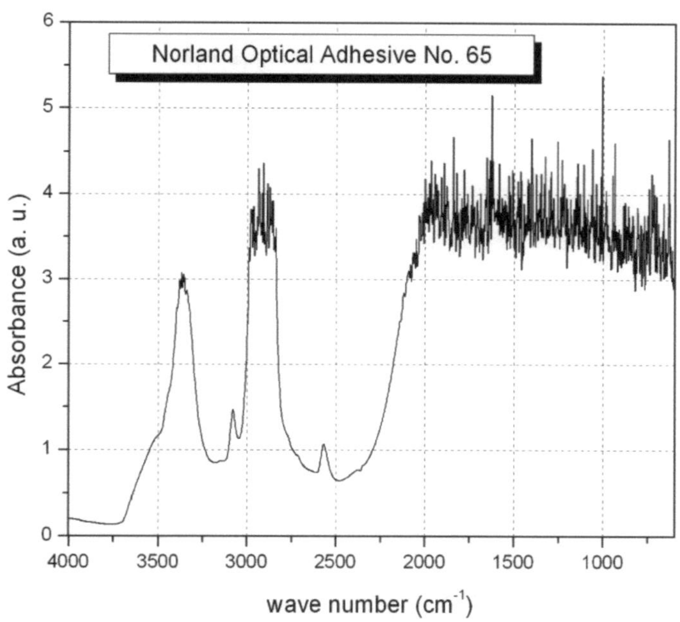

Fig. 3. Absorbance spectra in IR region

Wave number	Group	Strain	Note
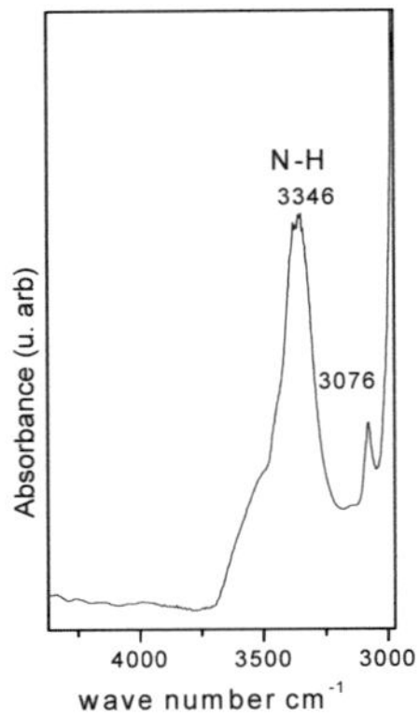	N-H	Asymmetric vibration	For Secondary amide (NH-CO) not associated, they have a sharp and big band to 3460-3300 cm^{-1} Amide in liquid phase, They exhibit a big band to 3270 and a weak band to 3100-3070.
	N-H 3346	Asymmetric vibration	
Possible structure	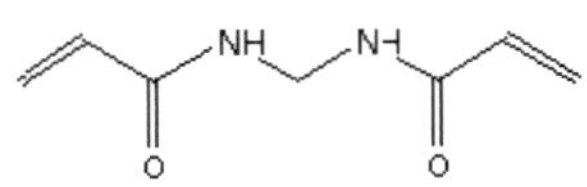		2-Propenamide, n,n'-methylenebis Or metilenebisacrylamide.
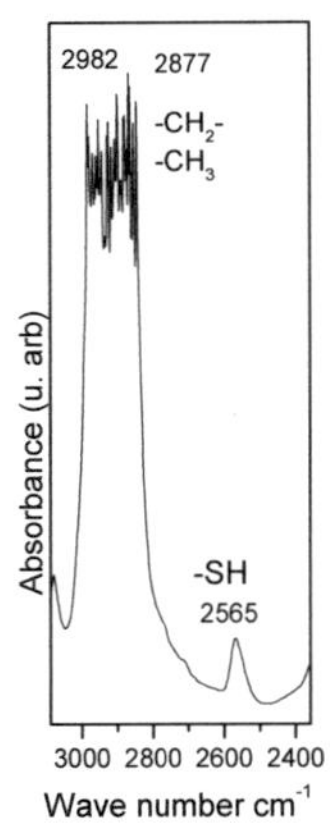	CH_3- 2982	Asymmetric vibration 2980	In 2982-2877 cm^{-1} interval appear aliphatic groups CH3 and CH2.
	CH_2- 2877	Asymmetric vibration 2900±45	The vibration that appears to 255 corresponds to the vibration of the SH, reported as more characteristic band of the thiols. A double SH due to the intensity of this band has been considered.
	Possible structure		

Table 3. IR analysis for NOA 65.

2.2 Crystal violet dye

The crystal violet dye (**CV**) is a dark green powder soluble in water, chloroform, isopropyl alcohol, but not in in ether and ethylic alcohol. The crystal violet dye can be used as antiseptic and a pH indicator for some substances. Its chemical composition is $C_{25}\ H_{30}\ ClN_3$ and molecular weigh 407.98. In Fig 4 we show its absorption spectra showing a peak in the spectral line at 591 nm, making a displacement of the absorption curve towards the yellow and orange color

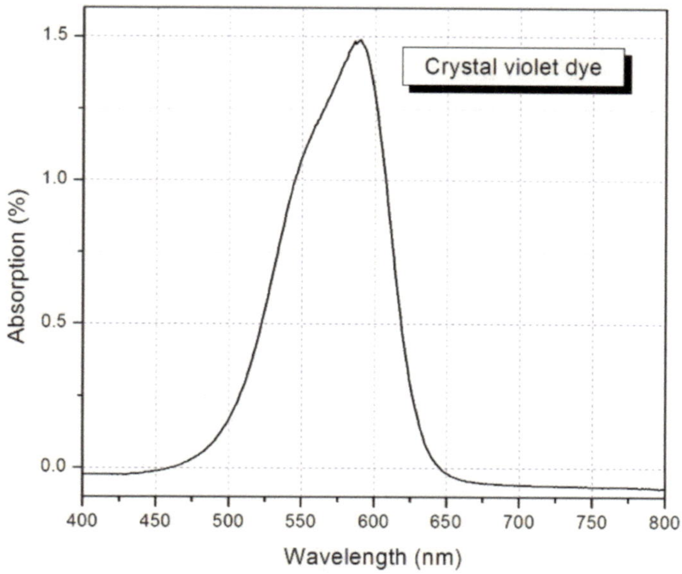

Fig. 4. Absorbance spectra of CV dye showing a peak at 580 nm.

2.3 Photosensitive material

The mix of NOA 65® and CV dye composed the photosensitive material. We prepare three different concentrations varying the quantities of the NOA 65® and CV as mentioned in Table 4 when the compounds are mixed, we deposit them into different glass cells fabricated with two glass substrates separated by mylar (17 μm thickness), cellophane (27 μm thickness) or mica (110μ thickness), All the thicknesses used in this work were measured with an electronic micrometer. The mixture was introduced into the cell by the gravity technique as is shown in Fig. 5

Concentration	NOA 65®	CV
C_1	99.95 %	0.05 %
C_2	99.9 %	0.1 %
C_3	99.85 %	0.15 %

Table 4. Relation of concentration between NOA 65 and CV.

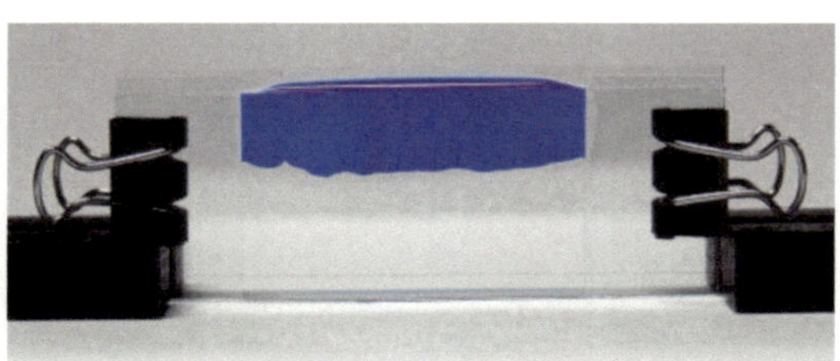

Fig. 5. Photography of the emulsion between two glasses deposited by gravity technique.

Fig. 6 show the absorption spectra in the UV-Vis region of the photosensitive film. The curves correspond to the three concentrations that we prepare and all the concentrations have absorption located in 500 - 630 nm range with a peak in the spectral line at 591 nm.

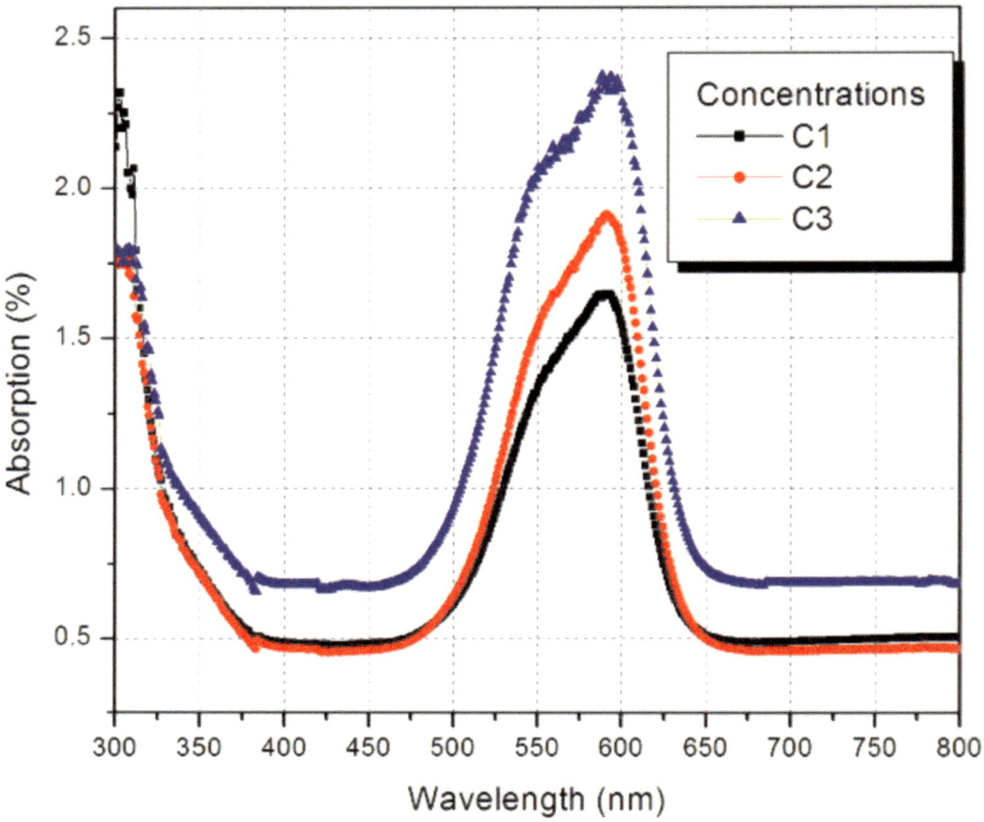

Fig. 6. Absorption spectra of mix NOA 65 and CV for three concentrations of 110 μm thickness cell.

In Table 5 we show the absorption coefficients for the 110μm thickness sample obtained with the Beer´s law considering the spectrum of Fig. 5. In eq. 1 we write the absorption *A*, as function of the molar concentration *c*, thickness *l*, and the absorption coefficient α.

$$A = cl\alpha \tag{1}$$

Concentration	*l (μm)*	α (λ=598 nm)	α (λ=543 nm)
C_1	110	0.004201745	0.001964058
C_2	110	0.005589304	0.003213745
C_3	110	0.00773117	0.006168739

Table 5. Absorption coefficient for the mix NOA 65 and CV.

3. Experimental process

To record the phase holographic gratings in the photosensitive material we use the experimental setup shown in Fig. 7. We use a beam from a He-Ne laser with emission line at λ=598 nm corresponding to yellow color which is separated into two beams by a beam splitter (BS). The beams are reflected by two mirrors (M_1 and M_2) toward the photosensitive material (PM) where an interference pattern is formed and recorded in real time by the period of 3 hours continuous.
In order to realize the measurements of the diffraction efficiency of the holographic grating we use a He-Ne laser as reading beam with emission line at λ=545 nm, which corresponds to green color, it is because this line does not affect the grating´s recording process. The gratings we record in this photosensitive material correspond to phase holographic gratings by refraction index modulation.

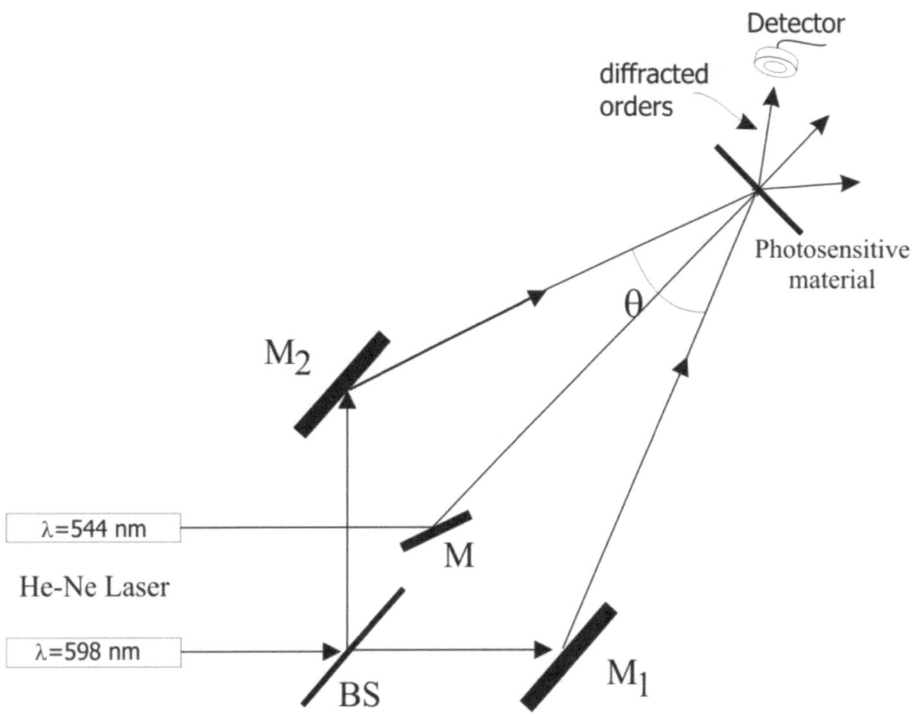

Fig. 7. Setup to make diffraction grating in real time.

The diffraction efficiency is defined as +1 diffracted order intensity and incident beam intensity ratio expressed as percentage as is represented in eq. 2.

$$\eta(\%)=\frac{I_1}{I_i}x100 \quad (2)$$

Where I_1 is the +1 diffracted order intensity and I_i the incident beam intensity. This equation is not considering the Fresnel losses because of the reflection in the photosensitive cell.
The Fig. 8 is a photography of the diffracted pattern produced by the grating recorded on the material with concentration C3, 110 μm thickness sample, an interference angle of 5 degrees

between the beams which according to Bragg Law produces a grating with spatial frequency of 146 *lines/mm* (the spatial frequency of the gratings is defined as the inverse of the period and is measured as lines per millimeter *lines/mm*). The central spot is called zero diffracted order, inside spots are called -1 and +1 diffracted orders (left and right respectively) and outside spots are -2 and +2 diffracted orders (again left and right spots respectively).

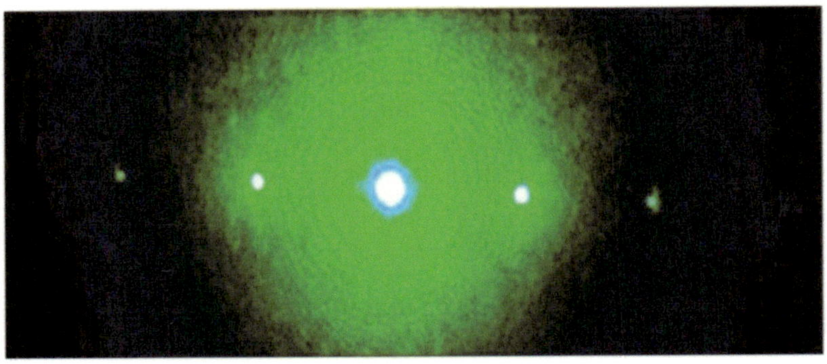

Fig. 8. Real time diffraction pattern showing first and second orders.

3.1 Diffraction gratings in real time

We recorded phase holographic gratings in all material concentrations and thickness and measured the diffraction efficiency. The Fig. 9 shows the diffraction efficiency of the holographic gratings recorded in the 17 μm thickness sample with all concentrations. The diffraction efficiency measurements were taken each 10 minutes after the exposition began. The curves show an increase of the diffraction efficiency concerning exposure energy obtaining η=0.53% as maximum for the concentration C3 (see table 4). These measurements were taken in real time for the +1 diffraction order only.

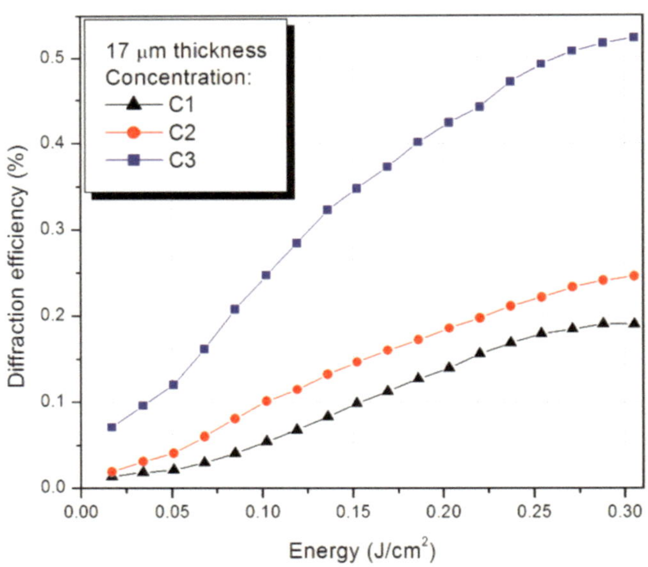

Fig. 9. Diffraction efficiency vs. energy of the three concentrations.

In Fig. 10 we show the diffraction efficiency obtained for all the concentrations again and 27 μm of sample thickness. The maximum diffraction efficiency is for the concentration C3 and its value is 1.1%.

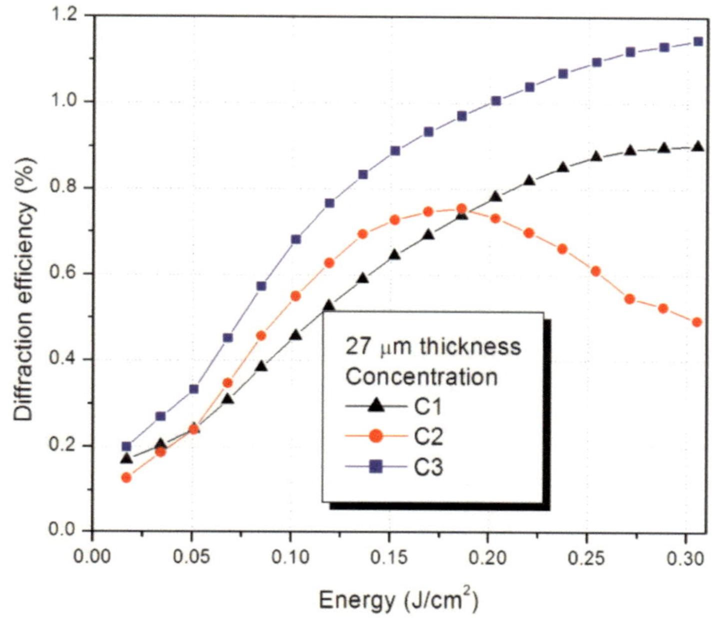

Fig. 10. Diffraction efficiency for the three concentrations and 27 μm sample thickness

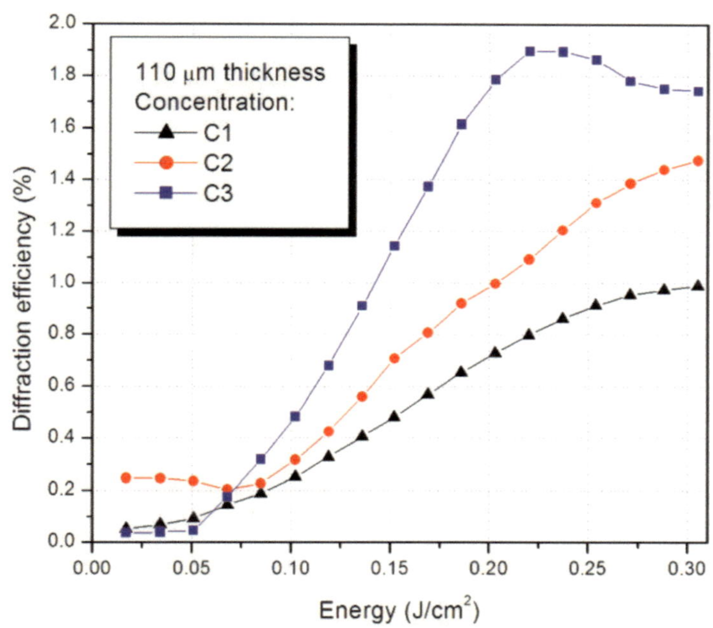

Fig. 11. Diffraction efficiency for the three concentrations, C3 show a major efficiency.

The Fig. 11 shows the results obtained for the 110 μm thickness sample and all the concentrations. Again we record the different gratings in the same conditions. The maximum diffraction efficiency is η=1.85% obtained at 0.225 J/cm² using the concentration C3.

It can be observed in the Figs. 9-11, the thickness of the emulsion plays an important role for the grating modulation. We find that the sample of the thickness of 110 μm is adapted to do diffraction gratings because it presents the highest values of diffraction efficiencies.

Another important parameter for the diffraction gratins recorder is the spatial frequency, which can be modified if we change the interference angle between the beams in the setup shown in Fig. 7 this change allows us to obtain gratings with different period according to the Bragg´s law.

In this sense, we change the interference angle between the two beams and record holographic gratings in the 110 microns thickness sample using the concentration C3. The angles were fixed at 5, 10 and 15 degrees and the results are shown in Fig. 12. The higher diffraction efficiencies are obtained when the interference angle between the beams is 5° producing a grating with frequency of 146 *lines/mm* and has a value η=1.85%. For the 10 and 15 degrees interference angle the diffraction efficiency is very low and produce gratings with 292 *lines/mm* and 436 *lines/mm* respectively. These results suggest us that the material is of low resolution.

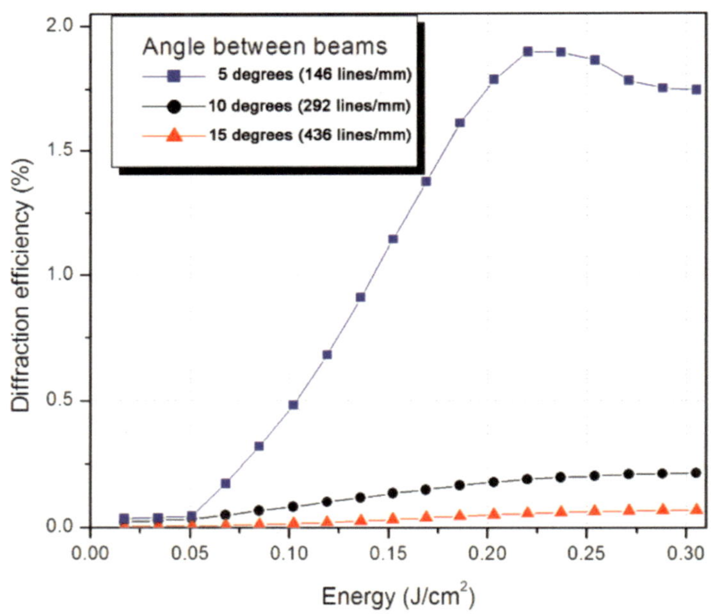

Fig. 12. Diffraction efficiencies for different frequencies.

Finally, we record gratings in the 110 microns thickness sample using the concentration C3 and the interference angle fixed at 5 degrees but we change the beams intensity ratio using the 1:1, 2:1 and 3:1 relations. In Fig. 13 we plot the diffraction efficiency obtained and, as we can see, the diffraction efficiency for the relation 3:1 and 2:1 are very low. The best choices to record holographic gratings in the proposed material are the 110 microns thickness sample prepared with the concentration C3, 5 degrees between the beams and the intensity ratio 1:1.

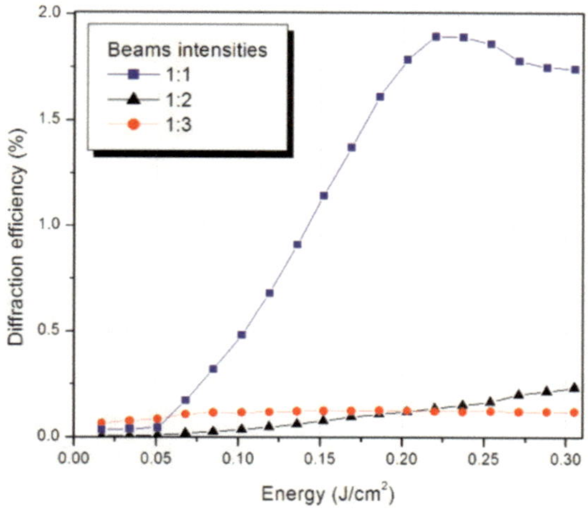

Fig. 13. Diffraction efficiency in function beams intensities ratio. The best diffraction efficiency is obtained at 1:1 relation.

Based on the measured values of the diffraction efficiency η (%) the modulation amplitude of the refraction index Δn can be calculated using the Kogelnik´s theory according to the next equation:

$$\Delta n = \frac{\lambda \cos\theta \arcsin\sqrt{\eta}}{\pi d} \quad (3)$$

Where *d is the* grating´s thickness, λ is the reading beam wavelength (in this case λ = 545 nm) and θ is the incident angle of the reading beam (θ=0) in this case. The Fig. 14 shows the modulation amplitude of the refraction index for three thicknesses of the photosensitive material composed by Norland Optical Adhesive 65® (NOA 65®) mixed with crystal violet dye (CV).

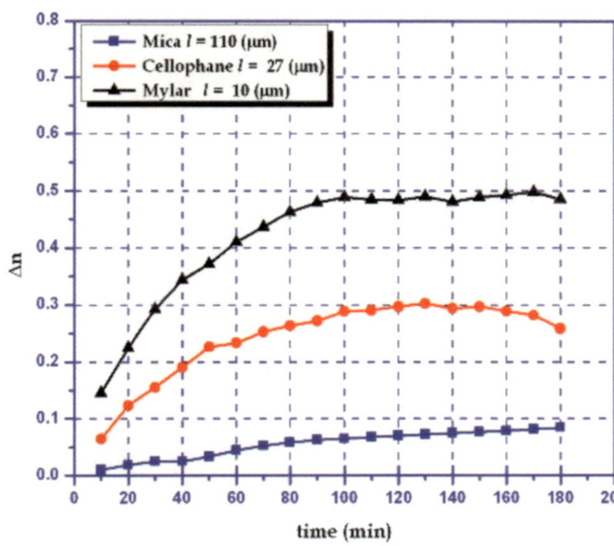

Fig. 14. Modulation amplitude for the refraction index for all thicknesses.

4. The temperature as recording parameter

The diffraction efficiency behavior of the holographic gratings recorded in photopolymers is due to several conditions such as monomer concentration, humidity, temperature, recorder beam intensity, polymerization velocity, mechanical vibrations, etc.
The temperature has an important role during the recording process in our photosensitive material as is shown below. When the material is exposed to interference pattern along 3 hours in small room temperature changes. We measured the temperature each 10 minutes in the photosensitive cell neighborhood and the diffraction efficiency at same time and the results are shown in Fig. 15 we prepare two cells to achieve this proof, so, the red line of the DE in Fig 15(a) corresponds a grating recorded while the room temperature has the behavior showed by red line in Fig. 15(b), using different cell, we record a new grating showing a DE behavior as is indicated by black line in Fig 15(a) while the room temperature has the behavior showed by black line in Fig. 15(b), we repeat this proof for 110 μm, 27 μm and 17 μm thickness cells using the C3 concentration.

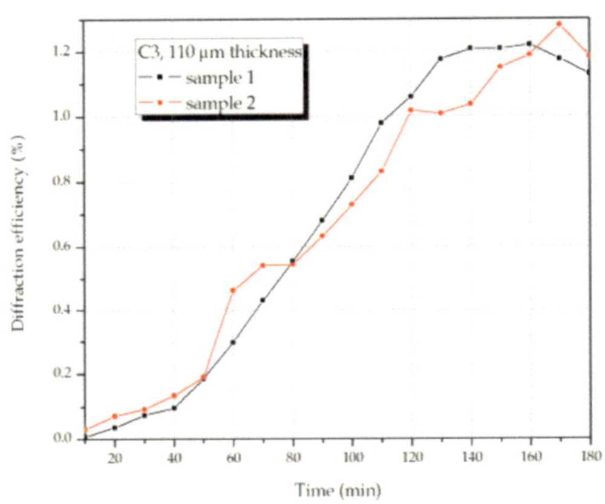

Fig. 15. (a). Diffraction efficiency to certain temperature during the recording process.

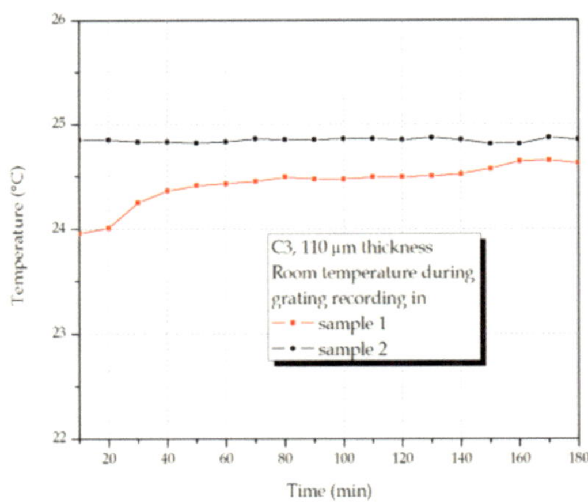

Fig. 15. (b). State of the temperature during the recording process.

Fig. 15 shows the diffraction efficiency and the room temperature during 3 hours. As can be seen, the temperatures have the same behavior and are 25 °C approximately, the DE has a similar values but temperature fluctuations modifies it behavior. When the temperature is almost constant the DE has a softly grow, but when the temperature has some change, the DE show an anomalous behavior. The same behavior for 27 μm and 17 μm was observed as is shown in Fig. 16 and 17.

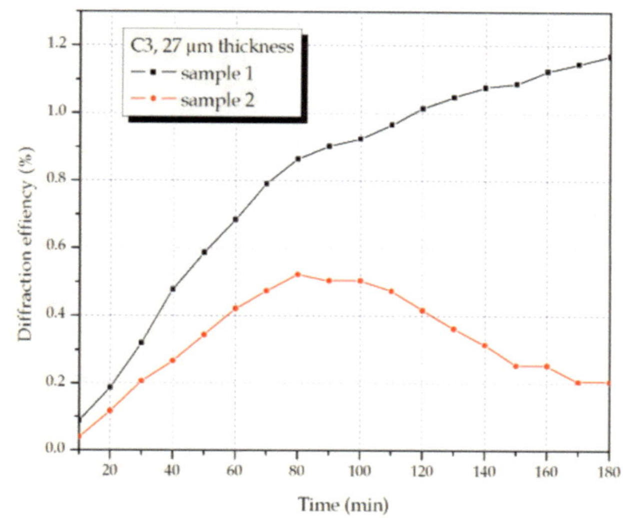

Fig. 16. (a). Plot of the Diffraction efficiency to certain temperature during the recording process.

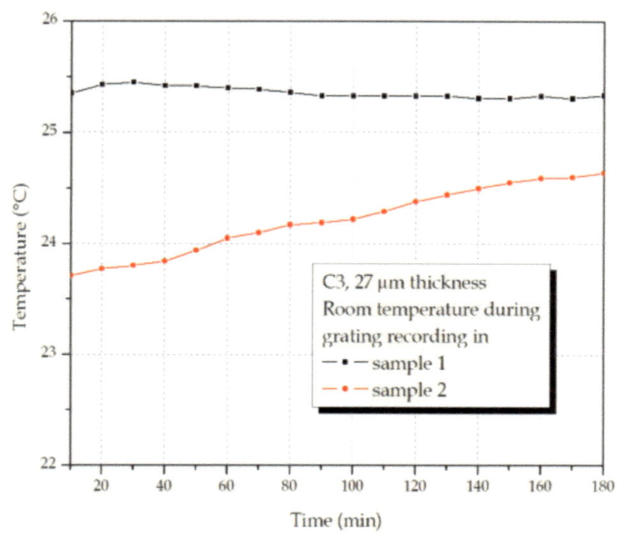

Fig. 16. (b). Temperature behavior during the recording process in the 27 μm thickness cell of two holographic gratings. The red and black line in both plots corresponds to same experiment.

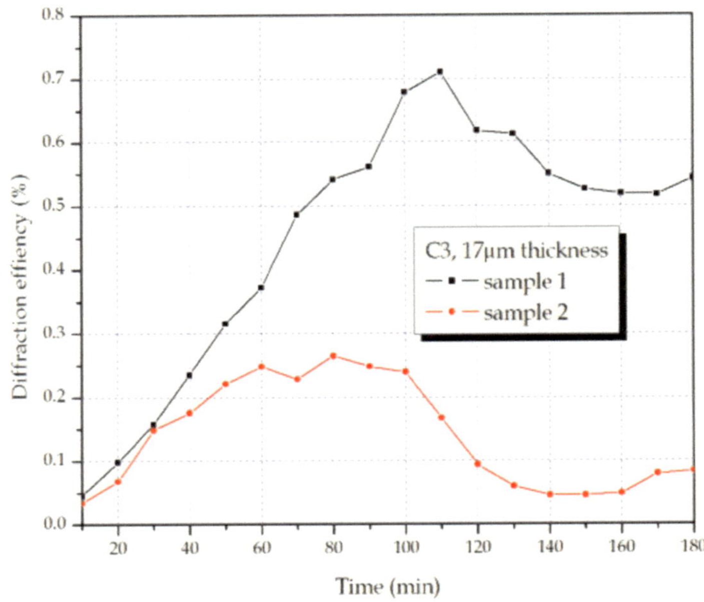

Fig. 17. (a). Plot of the Diffraction efficiency to certain temperature during the recording process.

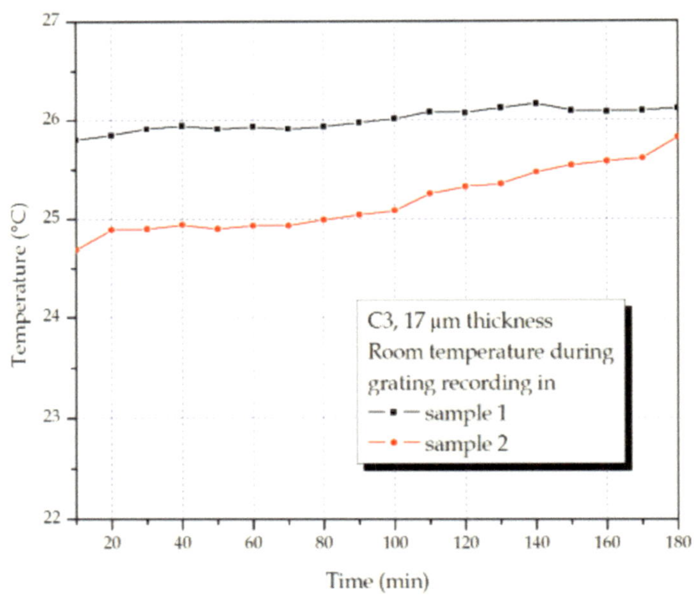

Fig. 17. (b). Temperature behavior during the recording process in the 17 μm thickness cell using the C3 concentration of two holographic gratings.

The temperature changes in the cases showed in Figs 15, 16, and 17 are due that we do not have temperature control in our laboratory. It is important to say; the gratings were recorded along different days. We can conclude that the temperature most be constant during the recording process about 25°C to obtain a softly behavior and the highest DE values.

5. Fourier holograms

Several types of holograms exist; these are of transmission, amplitude, phase, reflection, computer generated holograms, Fourier holograms, etc., (Smith, 1975).
A hologram can be done registering the interference pattern intensity between two beams, called a reference and an object beam, in a photosensitive material or holographic film as is shown in Fig. 18.

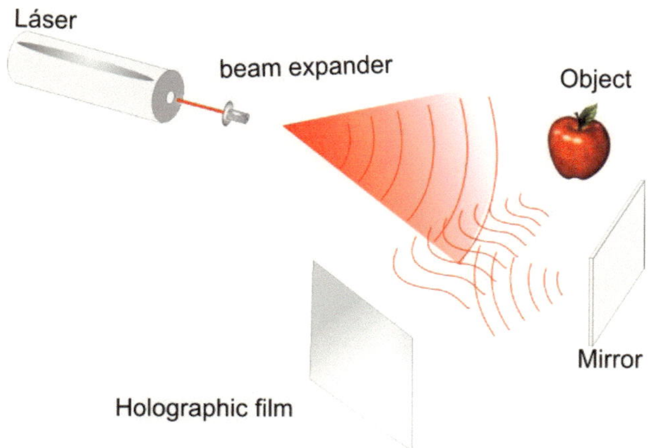

Fig. 18. Basic setup to hologram recording process, the beam reflected by the object is called object beam and the reflected beam by the mirror is the reference beam.

In particular, recording the interference pattern between the reference beam and the Fourier transform of an object produces the Fourier holograms. One of the main characteristics of this type of holograms is that the necessary area to record is small compared with other types of holograms. The Fig. 19 we depicted the scheme to record Fourier holograms.

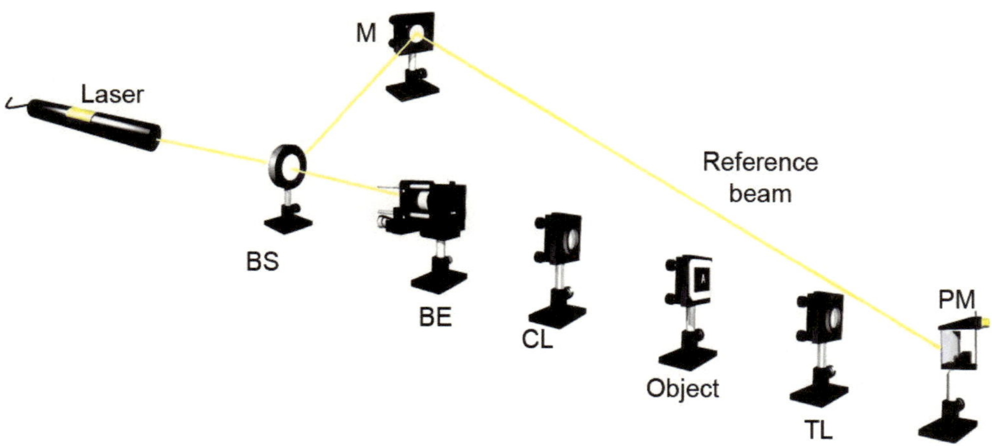

Fig. 19. Schematically representation for Fourier holograms recording. BS: beam splitter, BE beam expander, CL: Collimating lens, TL: transforming lens, PM: photosensitive material, M: mirror.

In the scheme of Fig. 19 a laser beam is expanded by microscope objective and a pinhole (BE) and collimated by the lens (CL) to produce a plane wave that illuminates the object f(x, y); the converging lens (TL) obtains the Fourier transform of the object called object beam (O) and directs it into the photosensitive material (PM) where it interferes with the reference beam (R) coming from a mirror and the pattern is recorded.
Once the hologram was recorded, we use the same reference beam R to reconstruct the images as is depicted in Fig. 20. The reconstruction of the hologram produces two images, namely, real and conjugated images.

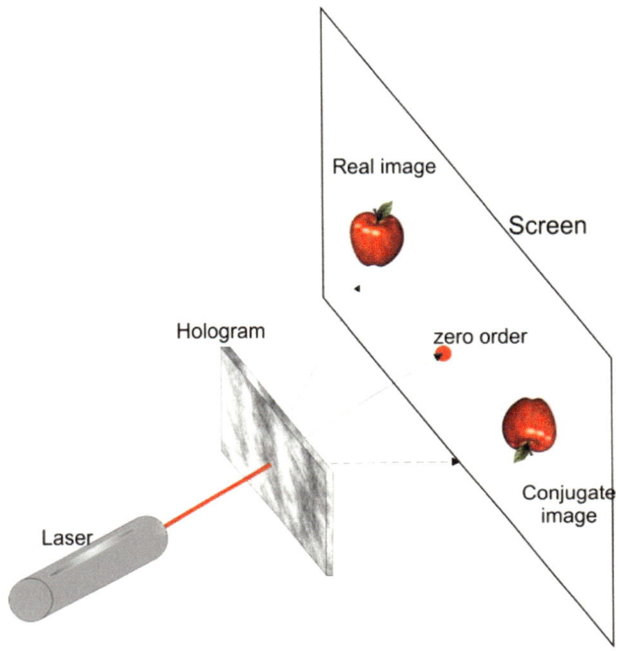

Fig. 20. Reconstruction of the real image of a hologram.

To explain mathematically the Fourier holograms we use the Fig. 21, where a point source is located at coordinates x=-a and y=0. The divergent spherical beams is then transformed at plane wave for the lens and is used as the reference beam R, given by

$$R = R(x,y) = \frac{1}{\lambda f}\int_{-\infty}^{\infty}\int_{-\infty}^{\infty}\delta(x+a)\delta(y)e^{-i2\pi(\nu x+\mu y)}dxdy \tag{4}$$

where λ is the wavelength, f is the focal length of the lens, $\nu = x/\lambda f$ and $\mu = y/\lambda f$ are spatial frequencies. As is indicated in eq.*, the reference beam is a plane wave expressed by the Fourier transform of the point source $\delta(x+a)\ \delta(y)$ located at x=-a. Evaluating the integral in eq. (4), we can express the reference beam as:

$$R = Ae^{i2\pi\nu a} \tag{5}$$

where **A** is the wave amplitude.
The object beam O is the Fourier transform of the object f(x, y) and is given by

$$O = O(x,y) = \frac{1}{\lambda f}\int_{-\infty}^{\infty}\int_{-\infty}^{\infty} f(x,y)e^{-i2\pi(vx+\mu y)}dxdy \quad (6)$$

or

$$O = F\{f(x,y)\} \quad (7)$$

where the Fourier transform is denoted by F.

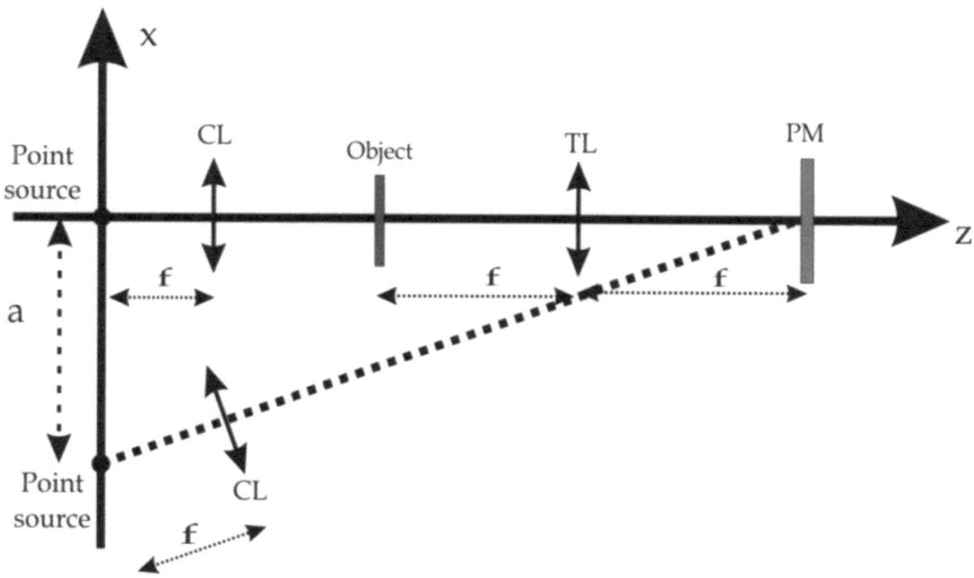

Fig. 21. Schematically representation of the Fourier holograms.

The photosensitive material, PM, registers the interference pattern intensity given by

$$I = |R + O|^2 \quad (8)$$

which can be expressed as

$$I = |R|^2 + |O|^2 + R^*O + RO^* \quad (9)$$

where the symbol * denotes the complex conjugate. Substituting the eqs. (5) and (7) in (9) we can obtain

$$I = A^2 + |F\{f(x,y)\}|^2 + Ae^{-i2\pi va}F\{f(x,y)\} + Ae^{i2\pi va}F^*\{f(x,y)\} \quad (10)$$

$$I = I_1 + I_2 + I_3 + I_4 \quad (11)$$

where

$$I_1 = A^2 \quad (12)$$

$$I_2 = |F\{f(x,y)\}|^2 \tag{13}$$

$$I_3 = Ae^{-i2\pi va}F\{f(x,y)\} \tag{14}$$

$$I_4 = Ae^{i2\pi va}F^*\{f(x,y)\} \tag{15}$$

In order to realize the reconstruction of the hologram, the reference beam R is used to illuminate the exposed film, so we can obtain

$$RI = RI_1 + RI_2 + RI_3 + RI_4 \tag{16}$$

The first two terms in right hand side of equation (15) are constants. The third term is expressed as

$$RI_3 = RAe^{-i2\pi va}F\{f(x,y)\} \tag{17}$$

and substituting eq. (5) in eq. (16) we obtain

$$RI_3 = A\int_{-\infty}^{\infty}\int_{-\infty}^{\infty}\delta(x+a)\delta(y)e^{-i2\pi(vx+\mu y)}e^{-i2\pi a}F\{f(x,y)\}dxdy \tag{18}$$

which can be reduced as

$$RI_3 = A\int_{-\infty}^{\infty}\int_{-\infty}^{\infty}F\{f(x+a,y)\}e^{-i2\pi(v(x+a)+\mu y)}dxdy \tag{19}$$

The result shown in eq. (18) can be interpreted as the double Fourier Transform of the object f(x, y), i. e.

$$RI_3 = F\{F\{f(x+a,y)\}\} = Af(-x-a,-y) \tag{20}$$

where the Fourier transform properties was employed. This term is know as the real image of the object f(x, y) located in the position x=a and correspond to the +1 diffracted order. Similarly, developing the fourth term RI_4 we obtain

$$RI_4 = RAe^{i2\pi va}F^*\{f(x,y)\} \tag{21}$$

and substituting eq. (4) in eq. (21) and using the relation $F^*(f(x, y))=-F(f^*(x, y))$ we obtain

$$RI_4 = -A\int_{-\infty}^{\infty}\int_{-\infty}^{\infty}\delta(x+a)\delta(y)e^{-i2\pi(vx+\mu y)}e^{i2\pi a}F\{f^*(x,y)\}dxdy \tag{22}$$

which can be reduced as

$$RI_4 = -A\int_{-\infty}^{\infty}\int_{-\infty}^{\infty}F\{f^*(x+a,y)\}e^{-i2\pi(v(x-a)+\mu y)}dxdy \tag{23}$$

Equation (22) can be reduced as following

$$RI_4 = -Af^*(-x-a,-y) \tag{24}$$

The term represented by eq. (23) is know as the conjugated image of the object f(x, y) and is the -1 diffracted order.

According to above results, we use the concentration C3, the sample thickness 110 mm, intensity beams ratio 1:1 and 5 degrees between the interference beam to record Fourier holograms. In Fig. 19 we showed the representation of the setup employed, where we use a He-Ne laser at 594 nm wavelength as recording beam and a laser at 544 nm as reading beam. In Fig. 22 we show the exposed area in the cell and, as can be observed, this area is 1mm^2 approximately.

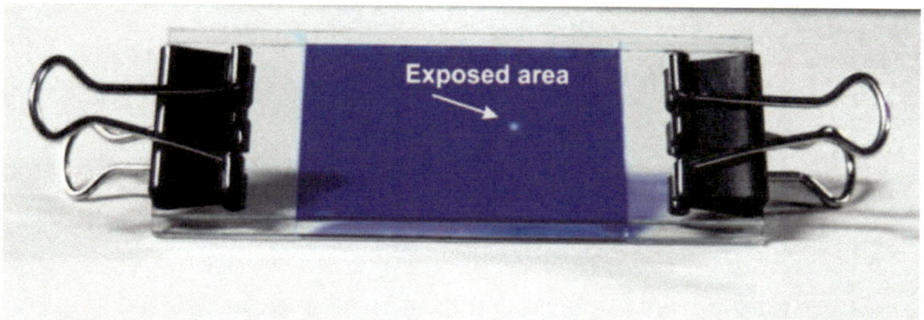

Fig. 22. Photography of the exposed area, the white point is the area where the Fourier hologram was recorded.

In Fig. 23 we show the two binary objects used to make Fourier holograms. The objects are the negative of the Universidad Michoacana de San Nicolás de Hidalgo logo and a text; the objects were photographed with a Nikon camera using a kodalith film and developed in a dark room following the Kodak instructions. After the objects were taken, we put it in the experimental setup and illuminates with a plane wave as is indicated in Fig 19. The reconstructed images are shown in Fig. 24 where only are the real images.

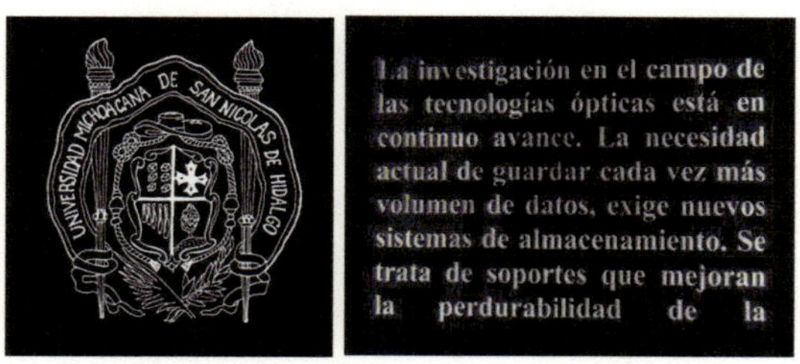

Fig. 23. Binary objects employed to Fourier holograms recording.

Fig. 24. Real images of the Fourier holograms recorded in our photosensitive material.

In order to compare our photosensitive material, we use the SO-253 film from Kodak to make the Fourier hologram of the text and the reconstructed real image is show in Fig 25 It is important to note, that the amount of text in Figs. 24 and 25 are different because we use 1 inch of diameter lens in one optical setup and 2 inch of diameter lens in the optical setup used in order to illuminate the photosensitive material. Another difference is that the SO-253 film has its sensibility at the line λ=633 nm so we record the Fourier hologram with this wavelength and we use the same wavelength to reconstruct the real image.

Fig. 25. Real image of hologram stored in SO 256 film from Kodak.

The advantage of Fourier holograms is that the storage area is small compared with Fresnel holograms and we can apply some multiplexing technique to optimizing the storage area. In table 6 we show the diffraction efficiencies of the holograms showed above measured for the +1-diffracted order.

Hologram	Photosensitive material	Diffraction Efficiency (%)
UMSNH logo	NOA 65 and CV	0.11
Text hologram	NOA 65 and CV	0.16
Text hologram	SO-253 Kodak	1.00

Table 6. Diffraction efficiencies for hologram reconstruction.

6. Conclusions

We present a photosensitive material composed by Norland Optical Adhesive No. 65® mixed with crystal violet dye with a high potential for recording holographic elements in real time, in this work we can emphasize some important characteristics of this material, eg: the phase holographic gratings are refraction index modulated, it is of low resolution; as is expected, a major sample thickness diffraction efficiency is increased, the beam intensity ratio must be 1:1 to obtain a best behavior of gratings. Also we noticed that the room temperature plays an essential role for the registry of holograms, that is to say, temperatures majors to 25 °C and minors to 24 °C its efficiency of diffraction are smaller, finally we have recorder Fourier holograms of binary objects in real time.

7. References

Smith H. (1975). *Principles of holography*, second edition, Johnm Wiley & Sons.

Bjelkhagen H., Thompson B. (1996). *Selected papers on holographic recording materials*, Vol. MS 130, SPIE Optical Engineering Press, Bellingham, USA.

P. Hariharan, (1980). *Holographic recording materials: recent developments*, Optical Engineering, Vol. 19 (5), pp. 636-641.

Kang D., Kim J. & Bae B. (2004). *Simple fabrication of diffraction gratings by two-beam interference method in highly photosensitive hybrid sol gel films*; Optics express, Vol. 12, No. 17, 3947-3953.

Kostuk R. (1999). *Dynamic hologram recording characteristics in DuPont photopolymers*, Applied Optics, Vol. 38, No. 8, 1357-1363.

Adhami R. Lanteigne, D. & Gregory D. (1991). *Photopolymer hologram formation theory*, Microwave and optical technology letters, Vol. 4, No. 3, 106-109.

Gallego S., Ortuño M., Neipp C., Márquez A., Beléndez A., Pascual I., Kelly J. & Sheridan J. (2005). *3 Dimensional analysis of holographic photopolymers based memories*, Optics express, Vol. 13, No. 9, 3543-3557.

Gleeson M., Kelly J., O´Neill F. & Sheridan J (2005). *Recording beam modulation during grating formation*, Applied Optics, Vol. 44, No. 26, 5475-5482.

Neipp C., Gallego S, Ortuño M., Márquez A., Alvarez M., Beléndez A. & Pascual I. (2003). *First-harmonic diffusion-based model applied to a polyvinyl-alcohol-acrylamide-based photopolymer*, J. Opt. Soc. Am. B, Vol. 20, No. 10, 2052-2060.

Ortuño M., Gallego S., Garcia C., Neipp C., Pascual I. (2003). *Holographic characteristics of a 1 mm-thick photopolymer to be used in holographic memories*, Applied Optics, Vol. 42, No. 35, 7008-7012.

Luna D., Olivares A. & Barriel L. (1977). *Photoresist shipley 1350-J with crystal violet tint for holographic optical elements*, Optical Materials, 7, 189-195.

Luna D., Olivares A., Barriel L. & Osorio F. (1998). *Rosin resin with crystal violet tint*, Optical materials 11, 95-100.

Ortiz M., Alemán K., Pérez M. Ibarra J. C. & Olivares A. (2007). *Polyvinyl alcohol and crystal violet as photosensitive film*, SPIE, Vol. 6488.

Fernández E., García C., Pascual I., Ortuño M., Gallego S. & Beléndez A. (2006). *Optimization of a thick polyvinil alcohol-acrylamide photopolymer for data storage using a combination of angular and peristrophic holographic multiplexing*, Applied optics, Vol. 45, No. 29, 7661-7666.

Martin S., Leclere P., Renotte Y., Toal V. & Lion Y. (1994). *Characterization of an acrylamide-based dry photopolymer holographic recording material*, Optical Engineering; Vol. 33, No. 12, 3942-3946.

Naydenova I., Sherif H. , Mintova S, Martin S. & Toal V. (2006). *Holographic recording in nanoparticle-doped photopolymer*; Holography 2005, International Conference on Holography, Optical Recording and processing of Information, Proc. of SPIE Vol. 6252.

Ibarra J. C. & Olivares A. (2002). *New holographic recording material: bromothymol blue dye with rosin*, Optical materials, Vol. 20, 73-80.

Budinski K., M. K. Budinski (1999). *Engineering Materials Properties and Selection*, sixth ed., Prentice Hall, NJ.

Pinto B., Olivares A. & Fuentes I. Holographic material film composed by NOA 65® adhesive; Optical Materials; 20 (2002); 225-232.

Aleksejeva J. &, Teteris J. (2010). *Volumen Grating Recording in Acrylate-Based Photopolymers*, Latvian Journal of Physics and Technical Sciences: Vol. 43 Number 3, pp 13-2. DOI: 10.2478/v10047-010-0010-5.

Norland Products Incorporated, Norland Optical Adhesive 65 (1996, 1999). 695 Joyce Kilmer Ave. New Brunswick, NJ (980) 545-7828 Catalog 0906-S01, 0906-S02.

3

Light-Sensitive Media-Composites for Recording Volume Holograms Based on Porous Glass and Polymer

O.V. Andreeva and O.V. Bandyuk
The National Research University of Information Technologies, Mechanics and Optics
Russia

1. Introduction

Volume holography, or holography in three-dimensional media, dates back to Yu. N. Denisyuk's works (Denisyuk, 1962), who implemented the idea of hologram recording in a three-dimensional medium by means of recording a hologram in counterpropagating beams, using traditional silver-halide light-sensitive materials. Adaptation of traditional photomaterials for purposes of image (pictorial) holography consisted in reconstruction of Lippman photographic layers with the size of light-sensitive grains less than 25 nm and use of photochemical processing techniques that allow obtaining amplitude-phase high-efficiency holograms in the visible region (Denisyuk & Protas, 1963).

Three-dimensional holograms with Klein parameter (Q) that describes the degree of three-dimensionality of a hologram grating on order of 10 (Kogelnik, 1969), obtained on traditional photomaterials of thickness on order of 10 µm, are referred to as 3D-thin holograms. Hologram gratings with $Q > 1000$ are commonly considered 3D-volume holograms. To meet the condition, the thickness of the recording medium should be by 2-3 orders of magnitude greater than in the case of 3D-thin holograms and amount to a value on order of millimeters.

Recording 3D-volume holograms made use at different times of different recording media: crystals, photochromic glasses, etc., their main features in high demand in 3D-volume holography being large thickness and negligible shrinkage. Yet it became clear that the available media fail to meet the set of requirements placed on media for hologram recording, and foremost, for recording static holograms, intended to be used as hologram optical elements and holograms for long-term information storage. The need to develop the theoretical and experimental research in the field of volume holography called for creation of recording media of great thickness and with corresponding properties. Creating such media required new approaches to their development and corresponding measurement techniques for parameters.

The present section introduces media for 3D-volume holography, which were created on the basis of principles developed in 80s-90s of the XX century (Sukhanov, 1994a). The media demonstrated the potentiality of the implementation of the proposed theoretical principles of creation of volume recording media in practice and also proved conductive to refining parameter measurement techniques for volume holograms, to studying and understanding the relationship between physical-chemical processes of manufacturing technology for light-sensitive samples of volume recording medium and parameters of obtained holograms.

2. Light-sensitive media for volume hologram recording: Basic requirements and principles of design

2.1 Basic requirements to recording media to construct static volume holograms

The interest to static volume holograms with thickness of several millimeters is due, mainly, to their use in research and in creation of holographic optical elements (HOE) of various purposes with unique parameters (Sukhanov et al., 1984; Ludman at al., 1997; Popov, et al., 2000; Luo, at al., 2008). Requirements to materials for recording of such holograms are rather specific and cannot be met when using traditional light-sensitive materials.

Recording media for construction of static 3D-volume holograms should have the following properties:

- high spatial resolution (greater than 3000 mm^{-1});
- large physical thickness (100-10000 μm);
- high physical and mechanical performance;
- sensitivity in the wavelength region of present-day lasers;
- transparence at the operating wavelength ($\tau > 0.5$);
- non-destructive reconstruction of recorded information.

Of greatest practical interest are media with photo-induced alteration of refractive index, which are suitable to record high-efficiency phase holograms, a special place among them belonging to media with latent image. Such media are most preferable in creation of HOE, as the recording process generates holograms of low efficiency, so-called "latent image" that leaves the recorded interference pattern undistorted. High efficiency of recorded holograms is in the case achieved through multiple magnification of modulation amplitude of a latent image hologram with the help of post-exposure treatment.

2.2 The principle of composite structure

The most natural way of realizing light-sensitive media with great physical thickness and high physical and mechanical performance is creation of medium-composite. The requisite set of parameters is in the case ensured by virtue of the medium components fulfilling different functions. Rigid framework has the necessary physical and mechanical properties and ensures the negligible shrinkage of samples; light-sensitive composite situated inside the framework ensures the properties needed for hologram construction (Sukhanov, 1991, 1994b).

The principle of creating the composite structure of light-sensitive medium by using a rigid framework was implemented in two ways. Firstly: the framework is created and formed independently; the light-sensitive composite is introduced into finished mold of the sample. The way was followed in realization of recording medium (RM) on the base of porous glasses with different light-sensitive composites (Sukhanov et al., 1988), one of them being the AgHal-based composite. Under such manner of RM realization, the framework effect on photochemical properties of samples is minimal and results from interaction of light-sensitive composite with the internal surface of the framework. Secondly: the framework is created and formed simultaneously with the light-sensitive composite - formation of sample-composite and rigid framework takes place at the same time. The manner was implemented by using organic dye phenanthrenequinone (PQ) as a light-sensitizing agent and forming the rigid framework by means of radical polymerization of composite on the base of methylmethacrylate monomer (MMA). The properties of the framework itself in such a RM are not limited only to maintenance of sample rigidity; the frame plays a certain part in recording, enhancement and fixation of holograms.

2.3 The dispersion refraction principle

The approach to creation of volume light-sensitive media, proposed in work (Lashkov & Sukhanov, 1978) and termed the dispersion refraction principle, consists in the following. It is necessary that the impact of light caused a sharp change of the absorption band of the material in the spectral region very far away from the operating wavelength range of recording radiation. The change of the absorption band is accompanied by change of dispersion, which leads to photo-induced refractive index appearing in the operating spectral range. Sensitivity of created RM in given wavelength interval can be ensured by using sensitized phototransformations. The change of refractive index due to the above causes was proposed to term as sensitized dispersion refraction. Purposeful use of given principle allows selecting substances with corresponding spectral parameters and calculating the possible modulation amplitude of constructed hologram. The principle became the base for realization of the first volume polymeric medium "Reoksan", whose light sensitivity is due to sensitized photo-oxidation of compounds of anthracene structure (Sukhanov, 1986, 1994a). It was used to create a medium with diffusion enhancement and is applicable to description and analysis of photophysical processes in most media intended for recording and reading the information in the optical spectral range.

2.4 The diffusion enhancement principle

The principle of diffusion enhancement supplements previous ones and opens up new opportunities to combine properties of the framework and light-sensitizing agent, thus setting direction of the search for new substances and photochemical and photophysical processes in creation of volume light-sensitive materials. According to the principle, particles of light-sensitizing agent of medium-composite should have no binding with the rigid framework: when a latent image hologram is recorded, two antiphase gratings are formed in such medium, one of them being made up of photochemically transformed molecules (photoproduct), the other one (supplementary to the first) being made up of light-sensitive particles remaining unchanged (unexposed). The formation of two antiphase gratings is a necessary, but not a sufficient condition: the main special feature of the principle is that after the medium exposure the photoproduct particles should be rigidly bound to the framework (here, unexposed particles remain unbound). The hologram recorded, the particles, unbound to the framework (free), diffuse with the passage of time, spreading uniformly across the sample bulk and causing the supplementary grating to degrade and the photoproduct grating to "develop" (be enhanced), since particles are not subject to diffusion.

This principle was successfully implemented in practice by using polymeric medium on the base of polymethylmethacrylate with PQ (Cherkasov et al., 1991; Veniaminov et al., 1991, 1996; Steckman et al., 1998; Lin et al., 2000, 2006; Luo at al., 2008; Liu at al., 2010; Yu at al.,

2010). Attempts to implement the principle using other compositions was less successful. However, experts consider the principle today among the most promising in creation of volume recording materials for holographic memory (Ashley et al., 2000; Shelby, 2002; Liu at al., 2010).

3. Study of volume light-sensitive media using recorded holograms

Volume recording media of about a millimeter thickness for holography exhibit, as a rule, very small values of modulation amplitude of optical parameters ($\Delta n \approx 10^{-3} \div 10^{-5}$) and small spatial size (fractions of a micron) of behavior of physico-chemical processes, taking place under impact of radiation and post-exposure treatment. These particular properties make it difficult to use standard instruments and procedures in study of such media and their changes under impact of radiation and other factors. The basic technique for research into volume RM parameters is the holographic one - investigation of media through construction of holograms and study of their parameters. The technique comprises task-oriented alteration of controlled variables in recording and post-exposure treatment of holograms; measurement of hologram parameters at each stage of their construction; theoretical analysis, enabling to correlate the changes of hologram parameters with experimental conditions and RM parameters.

As a test object, we used hologram-gratings, which were theoretically analyzed with the help of coupled wave theory (Kogelnik, 1969). This theory is an indispensable tool in study of RM using recorded holograms, as it allows relating the measured hologram parameters with the amplitude of modulation of optical characteristics of a medium.

The work dealt with phase and amplitude-phase transmission hologram-gratings with different thicknesses and constants, constructed, as a rule, in a symmetrical (or close to that) configuration of interfering beams relative to the sample surface. Main measured holographic parameters were the diffraction efficiency (DE) and selectivity contour of holograms, constructed under different recording and post-exposure treatment conditions.

Diffraction efficiency (η) of a hologram was defined as the ratio of diffracted beam intensity (I_d) to the sum of beam intensities behind the hologram: $\eta = I_d / (I_d + I_0)$, where I_0 is the intensity of the zero-order diffraction beam. Maximum values of DE for a hologram are attained at reconstruction under Bragg conditions. According to the coupled wave theory, the DE of a volume phase hologram-grating at reconstruction under Bragg conditions can be represented by the formulas: $\eta = \sin^2 \varphi_1$ for transmission holograms and $\eta = \tanh^2 \varphi_1$ reflection holograms. Phase modulation amplitude, φ_1, is defined by the expression:

$$\varphi_1 = \pi n_1 T / \lambda \cos \theta \tag{1}$$

where n_1 is the amplitude of variation of the first harmonic of refractive index of the medium; T is the hologram thickness; λ is the radiation wavelength; 2θ is the angle of diffracted beam (I_d) to zero beam (I_0). This equations relate the measured parameter of transmission hologram (η) to RM parameters (n_1, T) and experimental conditions (λ, $\cos \theta$). Fig. 1a shows dependences of DE on the magnitude of phase modulation for phase transmission (curve 1) and reflection (curve 2) holograms in the variation range of phase modulation of media under study ($0 < \varphi_1 < 2\pi$).

In the presence of the amplitude component, the DE of an amplitude-phase transmission hologram is defined taking into account that the modulation amplitude of the first harmonic of hologram absorption index (α_1) and the average hologram absorption index (α_0). Fig. 1a presents the dependence (curve 3) for the case $\alpha_1 = \alpha_0 = 0.05$. The oscillatory nature of dependence $\eta(\varphi_1)$ is seen to persist. The data on Fig. 1a characterize variation of DE of holograms at their reconstruction under Bragg conditions. To describe the deviations from Bragg conditions, mismatch parameter (ξ) is used. Dependence of hologram DE, or diffracted beam intensity, on mismatch parameter ξ is the selectivity contour of a hologram. One recognizes the spectral selectivity contour of a hologram - dependence $I_d(\lambda)$ at $\theta = \theta_{Br}$ and the angular selectivity contour of a hologram - $I_d(\theta)$ at $\lambda = \lambda_{Br}$. The halfwidths of spectral and angular selectivity contours, $\Delta\lambda$ and $\Delta\theta$, are a measure of hologram selectivity. Comparison of selectivity of hologram-gratings with different DE by measured values of $\Delta\lambda$ and $\Delta\theta$ is appropriately done in the variation range of phase modulation of a medium $0.1\pi < \varphi_1 < 0.5\pi$, where the halfwidth values for a selectivity contour are practically independent of the DE of a hologram and are defined by its constant and thickness.

As have been noted, the dependence of DE of transmission volume holograms on phase modulation value (φ_1) is of oscillatory nature. Here, for phase holograms without absorption, $\varphi_1 = k\pi \pm \sin^{-1}\sqrt{\eta}$, where $k = 0, 1, 2, 3$; therefore, different sections of DE variation range (as indicated on Fig. 1a) are to correspond to different formulas for calculation of φ_1 by measured values of η. Fig. 1b shows selectivity contours of transmission phase holograms with DE = 50% (at reconstruction under Bragg conditions), which were obtained at different values of phase modulation amplitude. It is clearly seen that φ_1 for high-efficiency transmission holograms can be unambiguously found by DE values only with account of selectivity contour shape as opposed to thin-layer holograms, where, as a rule, $\varphi_1 < 0.5\pi$. The study of amplitude-phase holograms should involve not only the hologram selectivity contour (the dependence of diffracted beam intensity on parameter ξ), but also the dependence of intensity of zero beam that passed the hologram without changing the direction, on parameter ξ (Fig. 1c).

Dependences $I_d(\xi)$ and $I_0(\xi)$, given on Fig. 1 (b,c) for hologram-gratings with different values of phase modulation, display unsymmetrical nature of dependence $I_0(\xi)$ of amplitude-phase holograms (Fig. 1c) as opposed to phase holograms (Fig. 1b). Thus, symmetry of dependence $I_0(\xi)$ with respect to Bragg conditions is evidence of absence of the amplitude component in the hologram under study, which is necessary for correctness of calculations and estimation.

When the coupled wave theory is used to analyze experimental results, account is to be taken of the measurements of hologram parameters taking place, as a rule, in the air, and the formula-defined relationship of the studied parameters being established inside the medium. Comparison of experimentally measured hologram parameters to those calculated theoretically for given experimental conditions allows finding the amplitude of modulation of optical parameters of the medium in a hologram and their variation in processes under study as well as estimating some other characteristics, e. g., the uniformity of the light-sensitive agent distribution in the sample bulk, the effective hologram thickness against the geometric dimensions of the sample and so on.

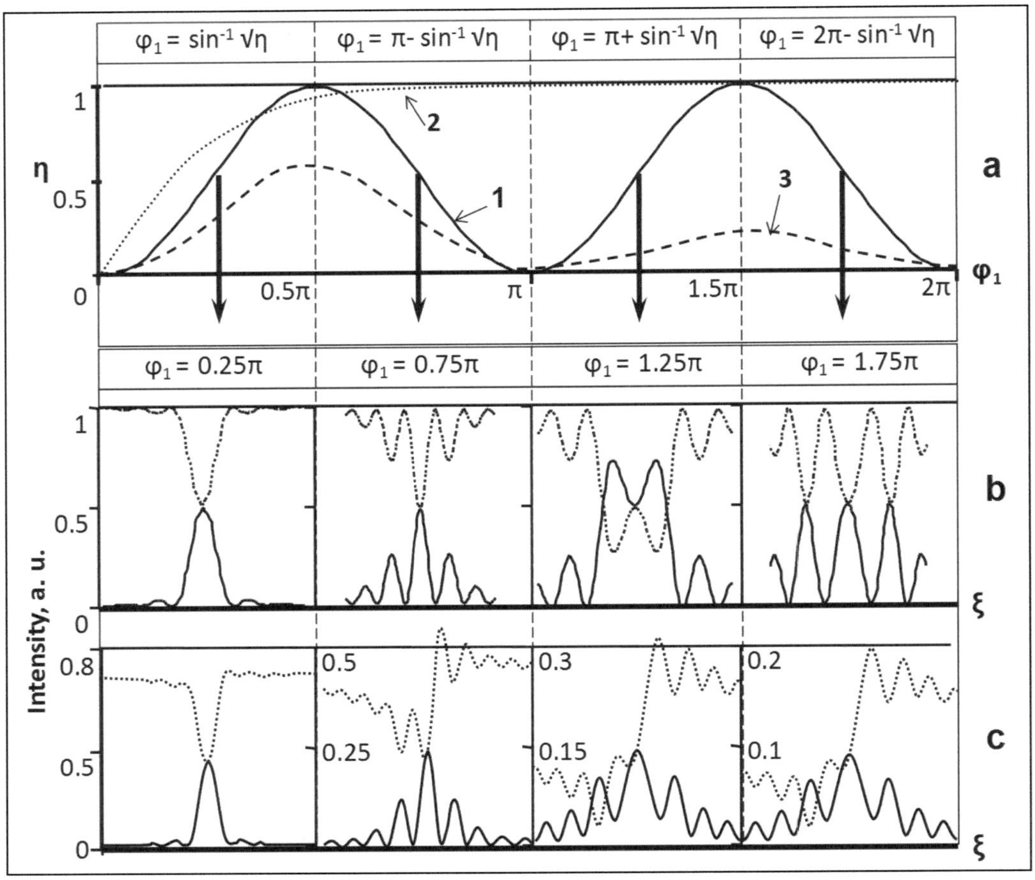

Fig. 1. a - dependence of diffraction efficiency (η) on phase modulation amplitude (φ_1) for volume phase transmission (curve 1) and reflection (curve 2) holograms; amplitude-phase transmission hologram with absorption index $\gamma_0 = \gamma_1 = 0.1$ (curve 3). b,c - intensity distribution in diffracted (solid lines) and zero (dotted lines) beams at deviation from Bragg conditions (ξ) at reconstruction of transmission phase hologram (b) and transmission amplitude-phase hologram (c) at $\gamma_1 = \gamma_0 = 0.1$ with phase modulation: 1 – $\varphi_1 = 0.25\pi$, 2 – $\varphi_1 = 0.75\pi$, 3 – $\varphi_1 = 1.25\pi$, 4 – $\varphi_1 = 1.75\pi$.

4. Light-sensitive AgHal-porous glass medium-composite

4.1 AgHal-porous glass samples: Manufacture and main features

The rigid framework of porous volume media provides porous glass, obtained from two-phase glass by treating samples in acid and alkali solution. The need to secure the requirements to rigid matrix during development of recording media led to creation of samples NPG-17 with stable and reproducible parameters, which are obtained from initial two-phase glass DV-1 by the developed technology. Samples NPG-17, used to create RM, are polished plates or disks 1-3 mm thick. The average pore diameter is 17 nm, the pore-occupied free volume of a sample is (52 ± 4) %. The light absorption in the short-wave region by air-dry samples is quite considerable because of scattering by porous structure and absorption by framework components (see Fig.2, curves 1). Under impregnation of pores with an immersion liquid, a 1 mm thick sample exhibits high transparence practically in all visible region (see Fig.2, curves 2).

The light sensitivity of porous silver-containing recording medium is provided by the silver-halide component with gelatin as a protective colloid. Light-sensitive component is formed as a solid-phase shell that is rigidly bound with framework walls (see Fig.2c,d), while central regions of internal framework cavities remain unfilled, thus forming a network of through capillaries, which provides access for liquid and gaseous chemical agents. The synthesis of light-sensitive composite is performed directly inside a sample with the use of KBr, KJ, $AgNO_3$ and gelatin solution. The synthesis process is typical for manufacturing high-resolution RM.

The solid-phase shell occupies less than 10 % of the total volume of an air-dry sample. Under impregnation of water solutions, the shell swells and fills the entire internal pore volume without changing its localization relative to framework walls, to which it is rigidly bound. In synthesis of AgHal, the size of formed particles and the variation of their localization in the course of post-exposure treatment cannot exceed the maximum size of porous ducts, which amounts when using matrices NPG-17 to 20 nm. The feature makes AgHal-porous glass (AgHal-PG) media fundamentally distinct from AgHal film plates, where it is practically impossible to create an ensemble of particles with limitation of maximum sizes and to ensure their rigid localization during the hologram construction.

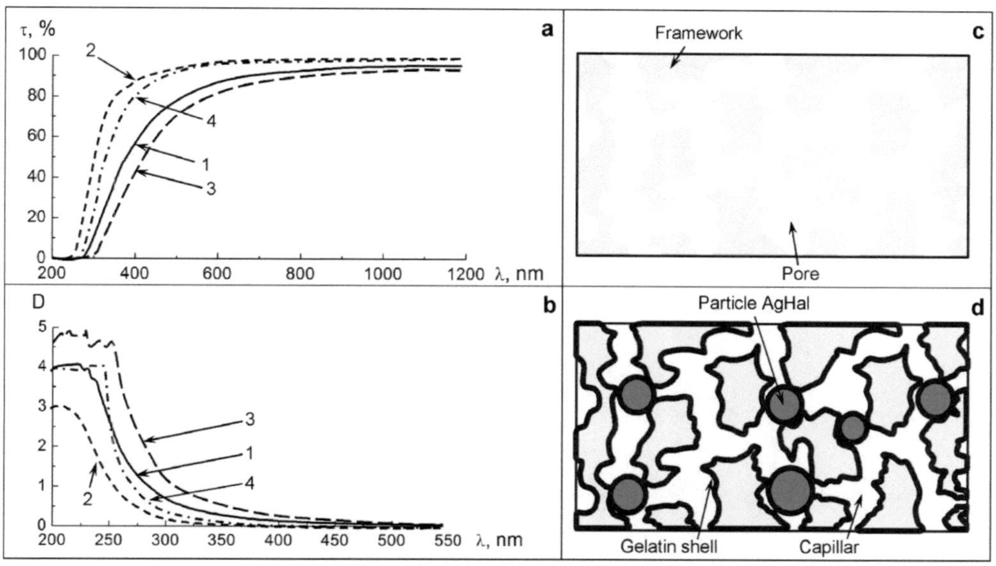

Fig. 2. Spectral dependence of transmission (a) and optical density (b) of porous samples in air (curves 1, 3) and in water immersion (curves 2, 4); 1, 2 – a sample NPG-17 of thickness 1 mm; 3, 4 – a sample with gelatin shell of internal capillaries: measurements in the air are relative to the air; those in water are relative the water. Schematic drawing of the cross-section of porous glass (c) and light-sensitive medium AgHal-PG (d).

Unsensitized samples of medium-composite are sensitive in the intrinsic light sensitivity of AgHal in spectral region $\lambda < 510$ nm. The developed synthesis process allows performing optical sensitization of AgHal composite with the use of dyes for the visible and near IR regions. During experiments, hologram recording with the use of Ar-ion laser (488 nm) was assured by intrinsic sensitivity of AgHal and that with the use of He-Ne laser (633 nm) was due to optical sensitization. Hologram parameters were measured in the red (633-655 nm) and near IR regions (1.5 μm).

4.2 Basic stages of hologram construction process on samples of AgHal-porous glass

The stages of hologram construction on samples of AgHal-PG medium are typical for holograms on traditional AgHal media: pre-exposure treatment of samples (e. g., impregnation of an immersion into a sample to reduce scattering); exposure; development; additional post-exposure treatment - stop bath, fixation, bleaching and so on; rinsing; drying. The network of through capillaries allows using water solutions for pre- and post-exposure treatment of porous samples, but essentially retards and modifies the physico-chemical processes, developed to construct holograms on traditional AgHal materials.

Recording. During hologram recording in a medium under effect of radiation, formation of the so-called "latent image", typical for AgHal media, takes place. Latent image centers (LIC) cause practically no changes in optical properties of the sample and form a hologram that has, immediately after recording, low diffraction efficiency (< 1%). LICs are as a rule also the development centers that determine the formation of developed particles. Fig. 3a (curve 1) shows angular selectivity contours of "latent image" hologram. The width of measured contours is in agreement with the theoretical values, found for a hologram with constant-depth amplitude modulation, but the sidelobes exceed the calculated values. The fact is evidence that the LIC concentration near the sample surface is somewhat higher than on average in the sample.

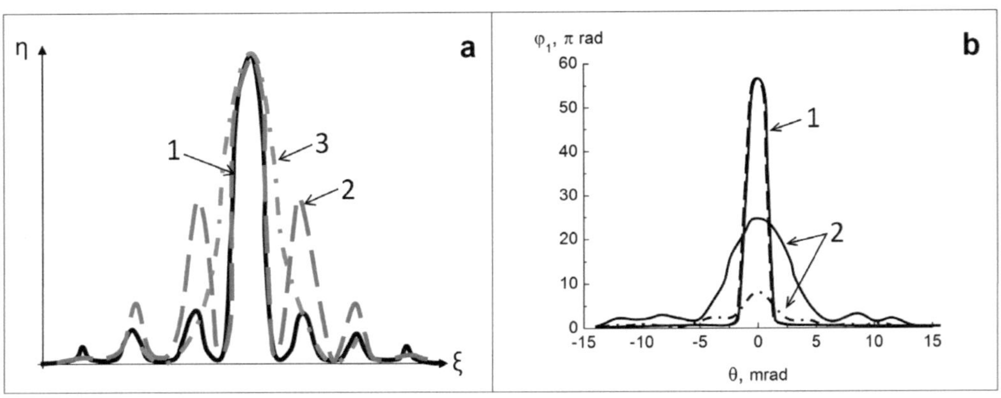

Fig. 3. Angular selectivity contours of holograms at different stages of their construction: a - a hologram at the latent image stage (curve 1), after development (curve 2), after thickness reduction owing to grinding away of surface layers of the sample (curve 3); b - developed hologram (curve 1) after the bleaching stage (curve 2) in air-dry condition (solid curves) and in water immersion (dashed curves).

Development. The basis of the elaborated process of development was taken to be that of construction of holograms on film plates with the use of developer PRG-1. The elaboration of the process of hologram construction took account of special features of the elaborated process of synthesis of light-sensitive component in nano-porous matrix and minimized the effect of the chemical activity of quartz-like framework on silver recovery process. A modification of developer PRG-1 was proposed, which was used in the performed experiments: anhydrous sodium sulfite - 0.2 g, hydroquinone - 0.2 g, KBr - 0.15 g, water - 100 ml. Development time is 8-20 hrs at temperature 20 °C. Stop bath is dipping in 0.2 % solution of citric acid for 30 min. Rinsing is in distilled water for 20 hrs. Drying is at room temperature.

Developed particles of AgHal-porous glass medium-composite, like those of AgHal film plates, represent particles of recovered metallic silver of colloid structure. Fig. 4a presents attenuation spectra of developed film plates PFG-03 (curve 1) and AgHal-PG (curve 2) after development to formulation PRG-1 so as to obtain developed silver particles of colloid structure. Curve 3 represents attenuation spectrum of diluted water preparation of Ag-PG sample of type 2, which is dispersed (powdered) and spread in water solution.

As seen from the given results, all attenuation spectra are of highly marked selective nature with a maximum in the short-wave part of the visible region (0.39÷0.43 μm), which is evidence of colloid structure of studied particles, close in shape to a sphere. When comparing spectra, a shift of the maximum of attenuation spectra of Ag-PG samples to the short-wave region is clearly seen, which is evidence of reduced average developed particle size in AgHal-PG as compared to film plates PFG-03 .

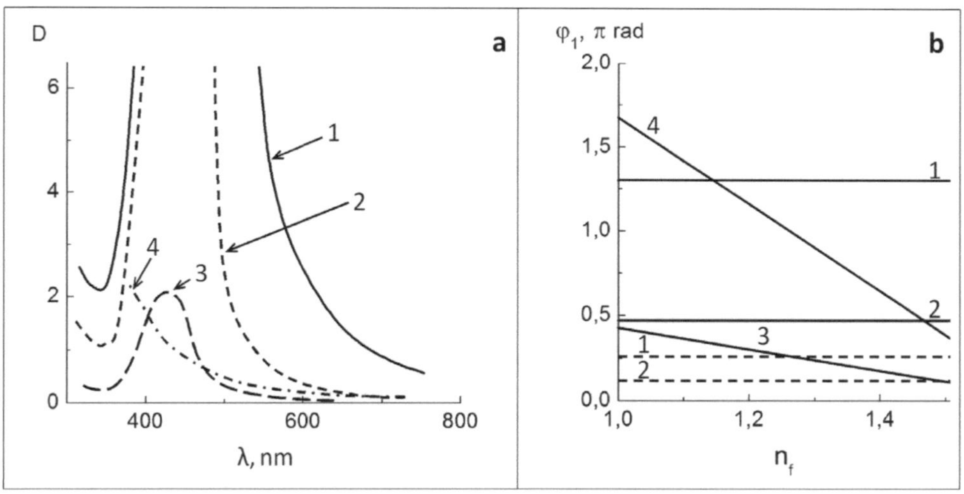

Fig. 4. a - attenuation spectra of AgHal media after the development stage with creation of particles of colloid structure (curves 1-3) and after the bleaching stage (curve 4): 1 – film plate PFG-03); 2 – AgHal-PG samples with thickness 0.3 mm; 3 – preparation of dispersed sample 3, diluted with water; 4 – AgHal-PG sample after the bleaching stage. b – phase modulation amplitude φ_1 of PG-holograms as function of the filler of the free volume of pore n_f for $\lambda = 0.63$ μm (solid curves) and $\lambda = 1.5$ μm (dashed curves): AgHal-PG after development (1,2) and after the bleaching (3); porous glass with dichromated gelatin (4).

The development process results are affected by a lot of factors, related to the conditions of performing the synthesis of light-sensitive composite and hologram recording, i. e. "life history" of samples prior to their development. The angular selectivity contours holograms on Fig. 3b (curve 1) demonstrate the influence - of immense interest is the possibility to obtain a developed hologram contour, having no sidelobes. Angular selectivity contours of developed holograms (see Fig. 3a, curve 2) have as a rule higher sidelobes, which is evidence of increasing non-uniformity of hologram modulation amplitude across the sample depth in the course of development. Grinding away of surface layers of the developed sample allows constructing a hologram less thick, but more uniform in distribution of modulation amplitude (Fig. 3a, curve 3).

Bleaching. The bleaching process, when colloid particles of metallic silver are transformed to those of silver halide, transparent in the visible region (see Fig. 4a, curve 4), changes also the hologram structure due to changes of optical parameters of the components and their distribution across the sample depth, which is manifested in alteration of the selectivity contour of the bleached hologram as compared to the initial developed hologram (Fig. 3b, curves 2). Impregnation of an immersion into free volume of pores of bleached holograms causes their DE to lower, which is to be kept in mind as the use of bleached holograms without impregnation of immersion is often impossible because of noticeable scattering as compared to unbleached ones.

4.3 Investigation of developed samples and Ag-porous glass holograms

Developed Ag-PG holograms are amplitude-phase ones: they exhibit absorption in the visible region, especially high in its short-wave part, which strongly affects the potential of usage and application of such holograms. The presence of amplitude modulation is evidenced by non-symmetric dependence of $I_0(\delta\theta)$ on conditions of hologram reconstruction, namely, $I_0(-\delta\theta) \neq I_0(+\delta\theta)$ at $\lambda = 0.63$ µm. The amplitude modulation decreases noticeably with growing wavelength of reconstructing radiation (see Fig. 4b) and at $\lambda = 1.5$ µm the Ag-PG hologram can be considered a purely phase one, which is evidenced by experimentally measured $I_0(\delta\theta)$ and $I_d(\delta\theta)$: dependence $I_0(\delta\theta)$ is symmetric and typical for a phase hologram, $I_0(-\delta\theta) = I_0(+\delta\theta)$.

The relation between amplitude and phase components in such holograms is determined by spectral dependence of permittivity (ε) of developed silver particles, which is in the case of finely dispersed particles different from the corresponding dependences for bulk silver. This was the reason for performing theoretical estimates and calculations of spectral dependences of effective optical parameters of porous model medium, containing Raleigh silver particles, and parameters of constructed Ag-PG holograms (Sukhanov et al., 1996).

According to calculations based on using the concept of limitation of free path length of electron in a small particle (Kreibig, 1970, 1978, 1981), composite medium has the minimum width of attenuation band at the size 15-20 nm of its constituent silver particles that have no defects of crystalline structure. Free path length (l) of electron in a particle of such ideal medium is $l = 8$ nm. The value of free path of electron in silver particles, composing a real medium with attenuation spectrum shown on Fig. 4 (curve 3), is found to be $l = 2$ nm.

This free path value is not an estimate of the average particle size, as in addition to size there are other unmanageable factors: contamination, crystalline structure defects etc. But it is quantity l that determines the effective optical constants of a medium, which are used in further calculations to estimate the parameters of AgPG holograms. Comparison of effective optical parameters of “ideal” (with particles $l = 8$ nm) and real (with particles $l = 2$ nm) media with identical structure parameters, whose calculated values are given on Fig. 5a, shows the attenuation spectrum of real medium to be essentially wider than that of the “ideal” one, whereas the change of refractive index due to the presence of silver particles (Δn_{Ag}) is practically the same and is determined by their concentration, C_{Ag}, (the quoted calculations used the volume silver concentration in the medium equal to 10^{-4}).

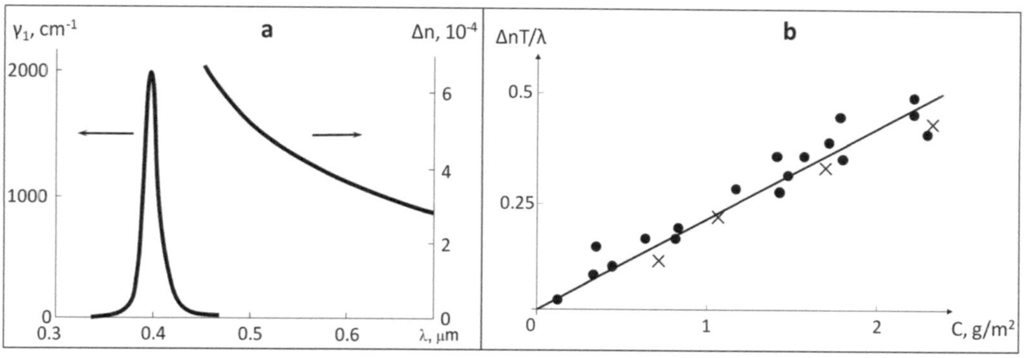

Fig. 5. a - theoretically found spectral dependences of optical constants of porous model medium, due to the presence of "ideal" silver particles (l = 8 nm) and "real" particles (l ≈ 2 nm); γ and Δn are, respectively, absorption constant and refractive index. b - dependence of phase incursion (ΔnT/λ), due to the presence of silver particles in the medium, on silver coverage, C, of the studied sample: experimental measurements with the use of developed holography film plates PFG-02 (dots); calculations by measured attenuation spectra with the use of dispersion Kramers-Kronig relations (crosses).

The dependence of diffraction efficiency of transmission amplitude-phase hologram gratings, formed by colloid silver particles, on silver concentration, given on Fig. 6, is of oscillatory nature because of presence of strong phase modulation and is dependent on the operating wavelength and optical properties of its constituent silver particles. The given data show the maximum values of DE of transmission AgPG holograms to be attained at a certain value of silver concentration: as the wavelength grows, so do both the maximum values of DE and the silver concentration needed to attain them. The DE values for holograms in a porous medium with ideal particles can be thought of as a theoretical limit for parameters of Ag-PG holograms.

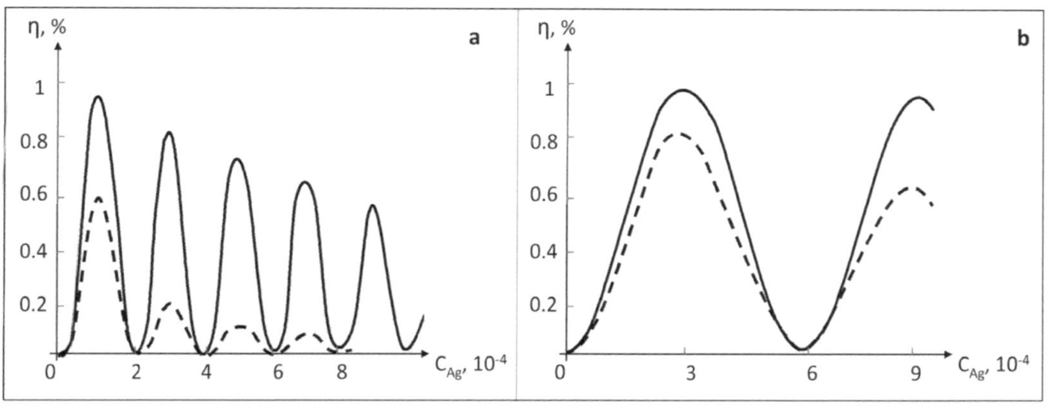

Fig. 6. Theoretically calculated dependences of DE of transmission Ag-PG holograms on developed silver concentration C_{Ag} for porous medium 1 mm thick with ideal (l = 8 nm, solid curves) and real (l ≈ 2 nm, dashed curves) particles: λ = 0.63 μm (a) and λ = 1.5 μm (b).

The presence of direct proportionality of refractive index of developed RM, containing colloid silver particles, on developed silver concentration, is confirmed by measurements of refractive index of developed high-resolution AgHal materials (film plates PFG-02) in the red region $\lambda \sim 650$ nm. Experimental measurements, their results given on Fig. 5b, were performed under different conditions of formation of developed particles and at different developed silver concentrations; the resulting dependence can be represented by the empirical formula:

$$\Delta n_{Ag} T/\lambda = 0.28\ C\ (g/m^2) \quad (2)$$

where Δn_{Ag} is the change of the medium refractive index due to the presence of silver particles; T is the medium thickness; C is the surface silver concentration (silver coverage in g/m^2), measured by rhodanine method. Eq. (2) enables estimation of developed silver concentration by measured values of phase incursions in the red region $\lambda \sim 650$ nm due to the presence of colloid silver particles, without refinement of their optical parameters.

Experimental measurements of a hologram at $\lambda = 1.5$ µm are performed in the absence of the amplitude component and allow finding the phase modulation of a hologram at given wavelength, which allows estimating the spectral dependence of amplitude and phase modulation of the studied Ag-PG hologram, $\varphi_1(\lambda)$ and $\gamma_1(\lambda)$, by using known wavelength dependence of optical constants of real medium. Calculations of angular selectivity contours of amplitude-phase holograms at 0.63 µm, carried out by experimentally found phase modulation at $\lambda = 1.5$ µm, have shown satisfactory agreement with the results of experimental measurements.

This situation enables using known value φ_1 (0.63-0.65 µm) to estimate the concentration of silver, forming given hologram, with the help of empirical dependence Eq. (2). The performed calculation have shown the volume concentration of silver, forming the Ag-PG hologram, to be $C_1 = (1 \div 3)\ 10^{-4}$, which corresponds to surface concentration $1 \div 5\ g/m^2$, comparable to a similar value in developed holographic film plates PFG-02 and PFG-03. It should be noted that average hologram absorption γ_0 (hence average silver concentration in the sample, C_0) exceed the estimates found by values γ_1 and C_1 due to the presence of veil, formed by colloid silver particles, which take no part in hologram construction, yet lead to an increase of sample absorption in the spectral region of the absorption band of colloid particles.

4.4 Distinctive features of AgHal-porous glass medium-composite in hologram construction

The proposed version of light-sensitive porous medium-composite on the AgHal base significantly enhances the potential of using traditional AgHal media in holography, which is evidenced by its main features.

1. Porous glasses are close in physical and mechanical properties to silicate glass, have close thermal expansion coefficient, and are shrink-proof. Upon placing air-dry samples into liquid medium and using water solutions in post-exposure treatment, geometrical dimension of the framework remain unchanged, while local deformations of solid-phase shell inside a pore are substantially less than light wavelength, lack any regular pattern and cause no distortion of hologram structure. The experiments were carried out with samples in the shape of plane-parallel plates or disks 1-3 mm thick.
2. Size of light-sensitive particles and those forming a finished hologram cannot exceed the maximum size of porous framework ducts. When using matrices NPG-17, the size amounts to 20 nm.

3. Light sensitivity of unsensitized samples across the spectrum is limited by intrinsic sensitivity of AgHal (λ<510 nm) and amounts to 10^{-3} J/cm^2 when recording holograms at 488 nm with DE = 10 %. The possibility to perform optical sensitization of AgHal-PG was demonstrated and samples, sensitive in the red region, were obtained: holograms with DE = 50 % were constructed when recording by He-Ne laser radiation (λ = 633 nm) at exposure 10^{-2} J/cm^2.
4. Formation_of low-efficiency holograms (DE <1 %) during recording leaves the structure of recorded interference pattern undistorted, and substantial hologram enhancement is possible in the course of post-exposure treatment.
5. Post-exposure treatment is performed with the use of traditional photochemical water solutions owing to the presence of through capillary network. Formulation can be modified, conditions of running the processes can be optimized, and new stages can be introduced.
6. Impregnation of an immersion into free volume of pores to reduce light scattering by porous samples is possible both at stages of recording and operation of constructed holograms. When using holograms formed by colloid silver particles, impregnation into free pore volume of an immersion with refractive index close to that of the framework (water, acetone, alcohol, CCl_4) produces practically no change in their phase modulation (the change lies within the experimental error), as distinct from holograms, obtained with the use of bleaching or dichromated gelatin.
7. Concentration of developed silver in the samples is close on order of magnitude to the corresponding parameter in AgHal film plates (1÷5 g/m^2). The distance between particles that form a finished hologram is several times their diameter.

5. Polymeric material Difphen

5.1 The principle of action

There are at present several modifications of polymeric light-sensitive media on the base of phenanthrenequinone (PQ), which implement the diffusion enhancement principle. The authors devised the technology to obtain a material, whose samples have certain holographic and physical-mechanical parameters, conditioned by modes of sample synthesis and hologram construction. The name Difphen (from words DIFfusion and PHENanthrenequinone) allows singling given material out of variety of other modifications of PQ-based polymeric medium with diffusion enhancement.

Samples of material Difphen (like some other materials of given group) represent a solid solution of organic dye PQ, uniformly distributed in polymethylmethacrylate (PMMA). Light sensitivity of the material results from capacity of PQ to bond to polymer under irradiation, transforming into 9,10-disubstituted derivative of phenanthrene (HPQR) according to schematic diagram (Chercasov et al., 1991):

$$PQ \xrightarrow{h\nu} {}^3PQ \xrightarrow{RH} HPQ\bullet \xrightarrow{R\bullet} HPQR \qquad (3)$$

where ^{3}PQ is the triplet-excited PQ molecule, HPQ• is the semiquinone radical, RH и R• are, respectively, the polymer molecule and radical.

Samples of recording medium 1÷5 mm thick are obtained by means of bulk polymerization of PQ solutions in methylmethacrylate (MMA) between molding glass plates. Fig. 7a (curve 1) shows PQ absorption spectrum, measured in a sample with concentration $8{,}5 \cdot 10^{-3}$M, with clearly defined long-wave maximum of absorption in the visible region ($\lambda = 405 \div 410$ nm). Under the impact of radiation, PQ changes its chemical structure, while the formed photoproduct (semiquinone radical) bonds to a PMMA molecule and loses its mobility. Absorption spectrum of photoproduct differs from that of PQ: the long-wave maximum vanishes (Fig.7a, curve 2). The dissimilarity of absorption spectra of PQ and its photoproduct preconditions the difference in their refraction indices and determines the efficiency of recorded latent image hologram (without effect of post-exposure treatment) at given wavelength.

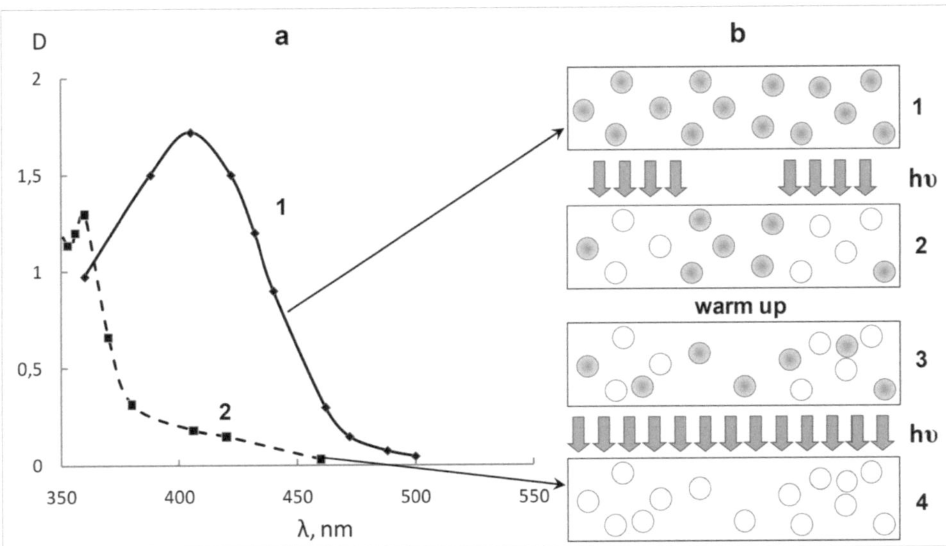

Fig. 7. a – absorption spectra of PQ (curve 1) and photoproduct (curve 2) in polymeric matrix. b – diagram to explain hologram construction process in polymeric medium with diffusion enhancement: distribution of molecules of PQ (○) and photoproduct (●) across the sample bulk in the initial state (1), after recording of hologram-grating (2), after warm-up (3), after fixation (4).

The process of post-exposure warm-up leads to redistribution of concentration of unexposed PQ molecules uniformly in the sample bulk, which ensures enhancement of recorded interference pattern, formed by photoproduct, that is, hologram "development". The sample after fixation becomes non-light-sensitive, the unreacted PQ, distributed uniformly across the sample volume, transforms into non-light-sensitive photoproduct with the same distribution, therefore, the fixation process leaves n_1 unchanged. The presence of photoproduct molecules in the medium makes the maximum values of n_1 of constructed hologram dependent on recording radiation intensity, spatial frequency of interference pattern being recorded and initial PQ concentration.

Thus, hologram construction on given material includes the following basic stages:

- hologram recording by radiation with wavelength $400 < \lambda < 530$ nm;
- warm-up of samples for 50 hrs at temperature 50°C;
- hologram fixation by incoherent radiation with wavelength in the PQ absorption region $\lambda = 430$-490 nm.

The work used samples obtained by means of casting polymerization in the shape of disks with diameter 20-40 mm and thickness 1-4 mm. Fixation of samples under study was affected by radiation of mercury lamp at 436 nm or that of "blue" LED with maximum of radiation band at 470 nm. Schematic diagram of information recording in such medium is shown on Fig. 7b by the example of constructing a hologram-grating.

5.2 Relation of holographic characteristics to conditions of sample synthesis

To obtain a medium with required holographic characteristics, it is necessary to know the effect of synthesis conditions on constructed hologram parameters and be in a position to purposefully change and control them.

The performed experiments have shown the quality of constructed holograms to be strongly affected by physical and mechanical properties of samples, which are determined by their hardness, and the most important holographic characteristic, determined by synthesis conditions for medium samples, to be the dependence of diffraction efficiency, of recorded holograms on post-exposure warm-up time. The data given on Fig. 8a, curve 1, show a typical dependence, which is specific for optimal synthesis conditions and can be approximated with two straight lines, as shown on the figure (dashed line).

The first dashed line describes the DE growth from the beginning of post-exposure warm-up (the DE value for latent image hologram) to attainment of the maximum values of DE, its slope being defined by PQ diffusion rate at given warm-up temperature (that is, degradation rate of the grating, formed by PQ unreacted with light). The second dashed line describes the hologram behavior after the maximum values of DE are attained and is, for the optimal synthesis samples, parallel to the abscissa axis. The intersection point of dashed approximation straight lines defines the characteristic sample warm-up time, t_{ch}, necessary to attain the maximum efficiency of recorded holograms.

With synthesis conditions, leading to "soft" sample structure, a latent image hologram has high DE values, quickly attains maximum efficiency at warm-up, is unstable and partially degrades at elevated temperature (Fig. 8a, curve 2). With synthesis conditions, leading to "hard" sample structure, PQ molecule diffusion proceeds too slowly, and high values of phase modulation are unattainable under a reasonable temperature-time regime of post-exposure treatment (curve 3). The clear-cut dependence of characteristic sample warm-up time on sample synthesis conditions allowed dividing all studied samples into three main groups: optimal synthesis samples, "soft" synthesis samples, and "hard" synthesis samples. Because of necessity to estimate the hardness of samples, a procedure was devised to allow relating physical and mechanical properties of samples to synthesis conditions and holographic process parameters.

Hardness was found by resistance to indentation of diamond pyramid into a material (Vickers test). The said procedure is notable for its relative simplicity, reproducibility and is supported with standard industrial instruments. Samples were tested with the help of PMT-3 hardness tester. Measured for each sample was the diagonal of the diamond pyramid indent in the material, its length depending on the sample hardness. It is the diagonal length, expressed in relative units (tied up with the use of a specific instrument), that is denoted as hardness index, K_h.

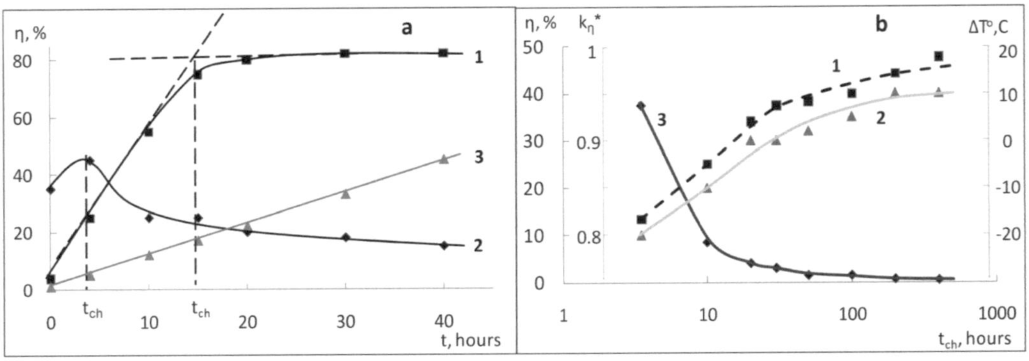

Fig. 8. a - dependence of diffraction efficiency η of holograms, recorded on samples of different synthesis, on post-exposure warm-up time (t_{ch} is characteristic warm-up time); curve 1 - optimal synthesis, curve 2 - soft synthesis, curve 3 - hard synthesis, spatial frequency of hologram 640 mm^{-1}. b - dependences of relative hardness index K_h* of samples (curve 1), deviations of synthesis temperature from the given one ΔT (curve 2), latent image hologram DE (curve 3) on characteristic warm-up time t_{ch} of recorded holograms.

Measurements of K_h for samples with different warm-up conditions and shelf lives (up to 10 years) have shown that there is limiting value $(K_h)_{lim}$, which defines the maximum possible hardness of samples, manufactured by given process. This allowed estimating the degree of sample hardness with the help of relative hardness index $K_h^* = (K_h)_{lim}/K_h$ (curve 1 on Fig. 8b). Sample synthesis conditions were varied by changing some parameters: polymerization initiator concentration, PQ concentration, temperature-time regime of polymerization etc. Fig. 8b (curve 2) shows the dependence of impact of deviation of synthesis temperature from the optimal one ($\Delta T = T - T_{opt}$) on characteristic sample warm-up time to be in very good correlation with similar dependence $K_h^*(t_{ch})$.

Dependence $\eta(t_{ch})$ of latent image holograms (Fig. 8b, curve 3) supplements the notion of existence of a rather narrow range of parameters of sample synthesis, wherein one can obtain holograms with required characteristics: stable parameters in the course of operation, low efficiency of latent image holograms, and high efficiency after warm-up under given temperature-time regime. Samples, obtained under optimal synthesis conditions, are seen from the data of Fig. 8b to exhibit a certain degree of hardness, which allows carrying out preliminary sample quality control without staging a labor-consuming holographic experiment. (Diamond pyramid diagonal is no longer than tenths of a millimeter, which allows making hardness measurements without deterioration of the working sample.)

The fact of existence of limiting values of hardness for studied material samples along with measurements of sample hardness at each stage of hologram construction process have shown that post-exposure sample warm-up can be divided into two basic stages: warm-up-1 till specified hologram parameters are achieved (at temperature 50 °C) and warm-up-2 till limiting values of sample hardness are achieved (additional warm-up at temperature 60-65 °C, which is higher than at warm-up-1, but below the vitrifying point). Warm-up-2 can take place both before and after sample fixation under conditions that leave hologram parameters unchanged.

5.3 Main characteristics

5.3.1 Modulation transfer function

Resolving power of recording material under study is from traditional viewpoint determined by dimensions PQ molecules, whose diameter is less than 2 nm. Hologram-gratings on given material were obtained at maximum spatial frequency about 3000 mm^{-1} with efficiency over 80% (Sukhanov, 1994a), yet it can be said with certainty that the potential of the material in recording of interference structure of high spatial frequency is far from exhausted.

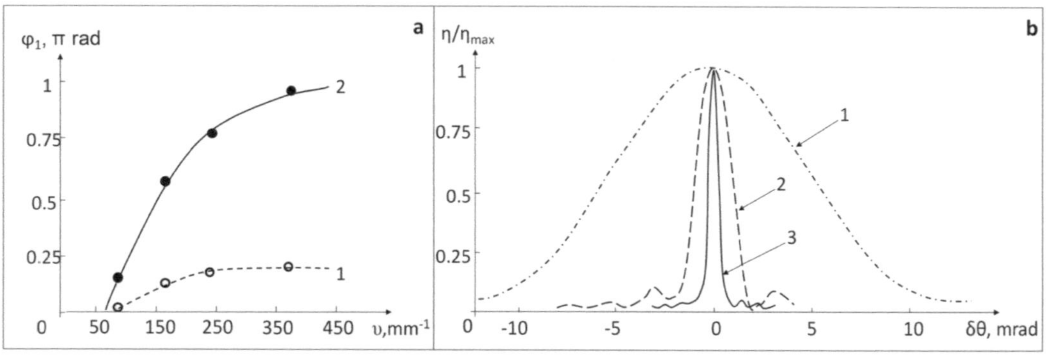

Fig. 9. a – dependence of phase modulation of transmission hologram-gratings φ_1 on spatial frequency of recorded interference pattern γ: 1 – latent image hologram (after exposure); 2 – hologram after complete cycle of post-exposure treatment. b – angular selectivity contours of transmission holograms with different spatial frequency, recorded with $\varphi_1 < 0.5\pi$: 1 – $\gamma = 70$ mm^{-1}, 2 – $\gamma = 320$ mm^{-1}, 3 – $\gamma = 1100$ mm^{-1}.

Maximum attainable phase modulation amplitude of hologram-gratings, recorded in a medium with diffusion enhancement, depends on grating constant, post-exposure warm-up conditions and is defined by diffusion rate of PQ molecules, which leads to a decrease of phase modulation in the region of low spatial frequencies.

Measurement results for parameters of holograms, recorded in spatial frequency region 70÷350 mm^{-1}, are given on Fig.9a. Phase modulation of latent image holograms (curve 1) is less than 0.2π and practically independent of spatial frequency at $v > 200$ mm^{-1}. For holograms with completed post-exposure treatment cycle (curve 2), phase modulation increases with growing spatial frequency. Thus, it can be concluded that material Difphen with sample thickness on order of 1 mm allows obtaining transmission hologram-gratings with DE over 70 % ($\varphi_1 > 0.3\pi$) in spatial frequency region $v > 100$ mm^{-1}. When dealing with higher spatial frequency region, the maximum attainable values of hologram phase modulation are strongly affected by factors, determined by the manner of treatment and conditions of hologram recording, such as stability of recording scheme, design features of hologram attachment and so on. Contours of angular selectivity of hologram-gratings with different spatial frequency are given on Fig. 9b and demonstrate agreement with theoretically calculated contours for hologram-gratings of given thickness.

5.3.2 Dynamic range

When using optimal-synthesis samples of material Difphen, the exposure range, wherein the linear mode of recording of latent image holograms can be ensured, is limited by values

$H < 1\ J/cm^2$, as seen from the data of Fig. 10a that shows exposure dependences of hologram parameters after completion of recording (curve 1) and after completion of post-exposure treatment cycle (curve 2). At $H > 1\ J/cm^2$, the recording process is of non-linear character, as high-efficiency latent image holograms are formed.

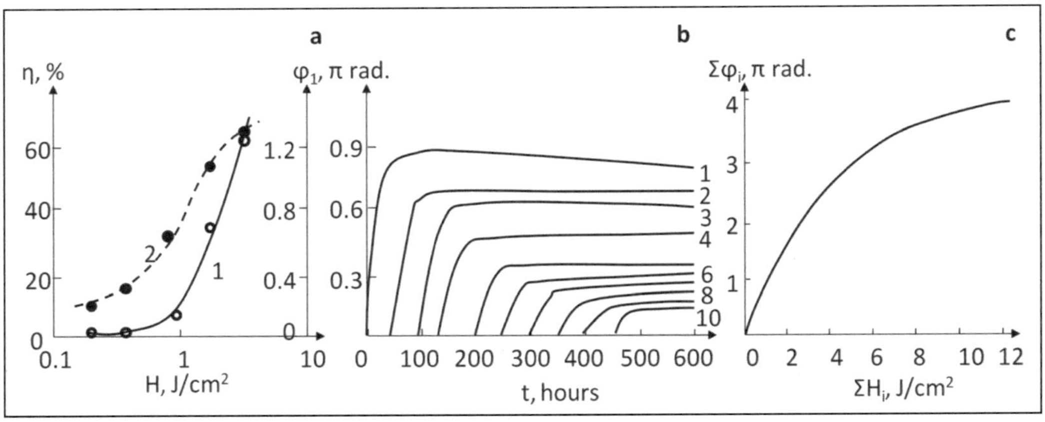

Fig. 10. a - exposure dependences of hologram parameters: diffraction efficiency η after completion of recording (curve 1) and phase modulation φ_1 after completion of post-exposure treatment (curve 2); spatial frequency 640 mm^{-1}. b - dependence of phase modulation amplitude φ_1 of superimposed holograms, recorded on the same sample section, on sample warm-up time; for each hologram $H_i = 0.8\ J/cm^2$ (holograms are numbered according to order of recording, i = 1, 2, 3,...10). c - dependence of total phase modulation of multiple hologram $\Sigma\varphi_1$ on total exposure of recorded holograms ΣH_i.

The performed experiments on recording of a single hologram on samples 1 mm thick have shown attainability of high enough values of modulation amplitude of the medium, in excess of π radians. Limitations in the case are due to impossibility to obtain latent image holograms with low DE, rather than dynamic range.

Dynamic range of volume RM can be estimated and used in superimposed hologram recording. The potential was demonstrated by superimposed recording of more than 1000 low-efficiency holograms on the same sample section with the view of using it as archive memory (Steckman et al., 1998). Specifics of superimposed recording of high-efficiency holograms is linked with deterioration of holograms, constructed in a medium with non-uniform distribution of light-sensitive particles due to recording of previous holograms. The features of given material allowed overcoming the drawback by way of construction of superimposed holograms with individual post-exposure warm-up: recording 1 - warm-up; recording 2 - warm-up and so on. Fig. 10b shows dependence of phase modulation of superimposed holograms, constructed by angular multiplexing technique, on total time of sample warm-up. Each hologram, regardless of recording number, is recorded in a medium with uniform distribution of PQ in the sample bulk: PQ concentration falls with hologram number growing, which leads, as seen on Fig. 10b, to smaller attainable values of phase modulation of single hologram.

Of much interest in studying the volume RM is quantity $\Sigma\varphi_1$, which is a sum of values of phase modulation of all superimposed holograms that constitute given multiple hologram, and describes the dynamic range of used light-sensitive medium. Fig. 10c shows dependence of total phase modulation of superimposed holograms on their total exposure, obtained in experiment on construction of 10 superimposed holograms with individual warm-up. The dependence describes the dynamic range of used RM, being a kind of characteristic curve of light-sensitive material. Maximum values of total phase modulation, attained in the experiment, are as high as $\Sigma\varphi_1 \cong 12$ rad, which was obtained on a sample 2.3 mm thick. With the use of polymeric media on the base of PMMA with PQ, other authors in work (Steckman et al., 1998) attained values $\Sigma\varphi_1 \cong 4.8$ rad for a sample 3 mm thick, and in work (Lin et al, 2000) – $\Sigma\varphi_1 \cong 14$ rad for a sample 8 mm thick. The cited works point out that $\Sigma\varphi_1$ grows linearly with growing sample thickness. Comparison of experimental data in the cited works and those by the authors, testifies that using the individual post-exposure warm-up of superimposed holograms can increase dynamic range of the material and obtain higher values of $\Sigma\varphi_1$ at given sample thickness.

5.3.3 Other characteristics

Modulation amplitude of refractive index in holograms. Experimental data on the magnitude of the amplitude of refractive index variation in a hologram and its dependence on wavelength of reconstruction radiation were obtained in study of hologram-gratings, recorded on optimal-synthesis samples of material Difphen with different thickness. Angular selectivity contour of hologram-grating under study was measured using lasers with different wavelengths; at each wavelength, phase modulation amplitude, φ_1, was determined and modulation amplitude of the first harmonic of refractive index, n_1, of a hologram was calculated by formulas of Kogelnik's theory. Fig. 11a shows calculation results for $n_1(\lambda)$ of studied samples. Dashed lines illustrate the process character and allow making some estimates. Modulation amplitude of refractive index of studied holograms in the visible region is within limits $n_1 < 10^{-3}$; dependence $n_1(\lambda)$ in the studied spectral range (473÷808 nm) displays normal behavior of dispersion, conditioned by absorption band of the material in the UV region.

The effect of ambient humidity variation on hologram performance. Polymeric RM on the PQ base are commonly considered to demonstrate negligible shrinkage, yet it is known that PMMA samples in atmosphere of increased humidity are capable of absorbing up to 1.5 % of water, which causes the sample thickness and average refractive index to change. Stability of hologram parameters at variations of ambient humidity was estimated by the change in the position of the maximum and in the shape of angular selectivity contour under conditions of different ambient humidity. Use was made of measurement design for angular selectivity contour of volume holograms with divergent radiation beam and data recording on CMOS-matrix. The collected data, processed according to devised procedure, are given on Fig. 11b.

In the study of holograms with width of angular selectivity contour 1.3 mrad, recorded on a sample ≈ 1 mm thick, the shift of contour maximum position at an abrupt change of ambient humidity from 85 % to 50 % at temperature 21 °C corresponds to a change of radiation diffraction angle by 0.1÷0.2 mrad.

Stability of hologram parameters over a long period of time of storage (6÷9 years) and operation. During storage and observation time, the holograms stayed in a room with temperature (15-40) °C and were used for occasional experiments as optic circuitry elements, for students' laboratory works etc. Used as measured and controlled hologram parameters were DE and angular selectivity contour. The data show that over the time of storage and operation of studied holograms there is no noticeable and systematic degradation of their parameters in a wide range of used exposures and spatial frequencies.

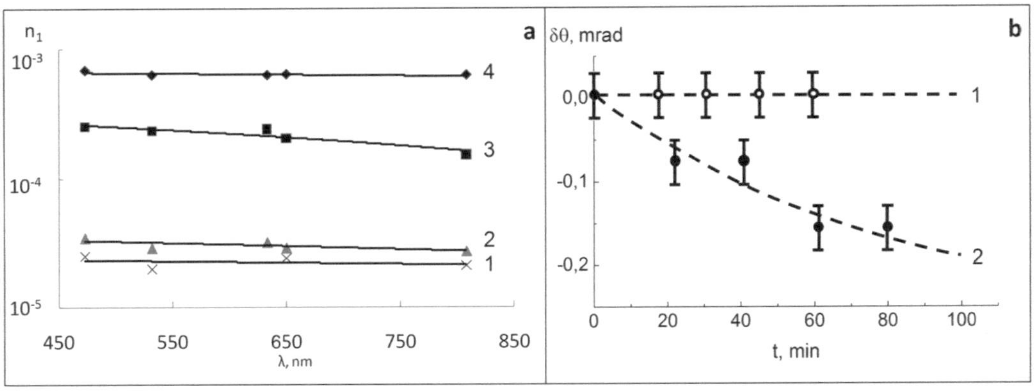

Fig. 11. a - wavelength dependence of modulation amplitude of the first harmonic of refractive index n_1 for holograms constructed on different samples of material Difphen: 1-3 – bulk-synthesis samples 2.3 mm, 2.0 mm, 1.3 mm thick, 4 – film sample 0.1 mm thick. b – position of the maximum of contour of studied hologram-grating under stable conditions (1) and at abrupt change in ambient humidity from 85 % to 50 % (2). Dashed lines illustrate the character of processes.

5.4 Film-type samples of material Difphen

Of great interest is to obtain samples of given material, which have thickness in the range 50÷500 μm, since the available product line of RM for holography has such samples represented by isolated laboratory-made specimens (e.g., Mahilny et al, 2006). The devised technological mode is unsuitable to obtain samples about 100 μm thick by means of bulk polymerization.

Film samples were obtained by the technique of pouring from solution, which is in some cases used to accomplish similar tasks under laboratory conditions. The basic components were PQ, PPMA and organic solvent. This resulted in homogeneous, uniformly colored films of size greater than 10×10 cm and thickness from 80 to 350 μm, whose elasticity allowed samples of various shape to be cut out of them. Here, volume PQ concentration in film samples was several times that in bulk ones. Performed tests have shown that samples, manufactured by given process, retain light-sensitive properties after storing at room temperature for a year.

To carry out holographic tests of film samples, custom designed cartridges were used: hologram recording, post-exposure treatment and measurement of parameters proceeded in the stable-state mode of the film sample, fixed between glass plates to exclude local deformations. Measurement results for DE of hologram-gratings, recorded at $\lambda = 488$ nm at spatial frequency 360 mm^{-1} with different exposures, have shown that values of DE on order of 50 % were achieved on test samples of film 180 μm thick (measurements at 633 nm). In a number of cases, a decrease of hologram DE under fixation was observed, which never occurred in working with bulk samples.

Obtained film samples permitted to run hologram recording in the "latent image" mode (with low values of DE), enhance holograms in the course of post-exposure warm-up and fix the constructed holograms. Comparative characteristics of bulk and film samples (with the same optical density at the recording wavelength), derived by using a unified procedure of holographic testing, are given in Table 1. Quantitative estimation of enhancement of hologram-gratings, Q, used the value of modulation amplitude of refractive index, n_1, obtained after warm-up and after fixation relative to the value of n_1 of latent image hologram.

Sample type	Thickness, mm	Transmission (λ = 488 nm), %	DE (λ = 633 nm)	Enhancement, Q (λ = 633 nm)	
				After warm-up	After fixation
Bulk	1.45	~ 20	0.94	5.1	5.0
Film	0.18	~ 20	0.43	4.0	4.5

Table 1. Comparison of parameters of holograms, constructed on bulk and film samples of polymeric material with PQ (measurements at λ = 633 nm).

Investigation of stability characteristics of parameters of holograms, recorded on film-type samples with thickness on order of 100 μm, in the course of long-term storage and operation has shown that some studied holograms, recorded at exposures over 0.3 J/cm^2, exhibited a DE increase beyond the limits of predicted experimental errors (see Table 1). The effect of DE growth during storage of holograms, recorded on film samples, may be due to a host of causes, including, apparently, plastic nature of RM samples used.
Performed experiments have shown the fruitfulness of the devised approach to obtaining polymeric holographic film-type materials and the necessity to carry on the research and to improve parameters of film samples and procedure for their use with the view of creating an assortment of materials for hologram recording with wide spectrum of parameters.

5.5 Main special features of samples of volume polymer medium

Material Difphen belongs to the group of polymeric recording materials, implementing the diffusion enhancement principle, which is at present fruitfully used by various authors in a number of science and engineering projects. Material samples have certain holographic, and physical and mechanical parameters, determined by the devised modes of sample synthesis and of hologram construction. The following main special features can be noted.

1. Resolving power of material exceeds 3000 mm^{-1} and is limited by PQ molecule size and photoproduct molecule mobility.
2. The modulation transfer function is untypical for traditional light-sensitive materials. In the low spatial frequency region (less than 50 mm^{-1}), hologram construction on given material is impossible; phase modulation amplitude of recorded holograms grows with increasing spatial frequency of a hologram. The use of such media can be very helpful in information recording that requires canceling the low-frequency component of spatial spectrum of radiation.
3. Feasibility of construction of latent image holograms and their subsequent enhancement under conditions of unchanged interference structure, formed at the recording stage, which ensures linearity of information recording in a wide dynamic range.
4. Wide dynamic range and feasibility of construction of high-efficiency superimposed holograms owing to the use of individual post-exposure warm-up.

5. Feasibility of producing samples of thickness from tens of microns to several millimeters.

6. Specifics of parameter measurements of high-selectivity volume holograms

6.1 Diffraction efficiency

Principal difficulties in measurement of parameters of high-selectivity volume holograms stem from the rigid requirements on the illuminating beam in its divergence (spatial frequency spectrum of radiation) and monochromaticity (radiation wavelength spectrum). To correctly find the maximum values of diffraction efficiency of a hologram under Bragg conditions, meeting the following conditions is necessary.

- Divergence of reading radiation beam ($\Delta\psi$) should be much narrower than angular selectivity of a hologram ($\Delta\theta$): $\Delta\psi < \Delta\theta$. The contour measured in the case is a convolution of the true angular selectivity contour of a hologram and the contour, defined by spatial frequency spectrum of the reading radiation beam (see Fig. 12a).
- Spectral distribution of reading radiation beam ($\Delta\Lambda$) should be much narrower than spectral selectivity of a hologram ($\Delta\lambda$): $\Delta\Lambda < \Delta\lambda$.
- At given experiment geometry and transmission hologram thickness (T) the region of overlap of zero and diffracted beams on the exit surface of the sample (A_{out}) with respect to that on the entry surface (A_{in}) should satisfy relationship: $A_{out}/A_{in} > 0.8$.

6.2 Angular selectivity contour

A very common measurement procedure for angular selectivity contour of volume holograms provides for angular scanning of parameters, i. e. rotation of a hologram relative to the reading beam when using collimated monochromatic laser radiation (Fig. 12b). Advantages of the technique are obvious:

- comparative simplicity of optical scheme for measurements; ease of data processing operation;
- possibility to carry out measurements in a strictly localized area of a hologram.

Drawbacks of the technique are less obvious, but render the measurements of high-selectivity holograms rather involved:

- rigid requirements on divergence and geometry of the reading radiation beam (see 6.1);
- long duration of measurements of a contour, which increases with growing accuracy of construction;
- complexity of forming a collimated radiation beam with divergence on order of fractions of a milliradian.

For a single measurement act to yield information, incorporating complete data of angular selectivity contour, use can be made of a divergent beam of monochromatic radiation (Fig. 12c) rather than collimated one. Hologram-reading radiation beam is to have spatial frequency spectrum ($\Delta\psi$) in an interval, greater than the width ($\Delta\theta$) of spatial spectrum of diffracted radiation. A radiation detector, placed into diffracted or zero beam, as Fig. 12c shows, records spatial distribution of beam intensity in given spatial domain. Used in practice to this end are CMOS-matrices (including matrices of digital cameras). Fig. 13 (bottom) shows distribution of intensity in diffracted (a) and in zero (b) radiation beams, recorded on camera matrix.

Advantages of the technique are:

- recording of selectivity contour takes place in a single measurement act;
- acquisition of information, using both diffracted and zero beams, and formation of optical scheme of an experiment to achieve necessary measurement accuracy;
- possibility to trace the dynamics of contour variation for the hologram under study both in rapid processes and in sufficiently slowly varying conditions.

Drawbacks of the technique are also obvious:

- labor intensity of data processing;
- necessity to stabilize and fine adjust the scheme elements for experiments.

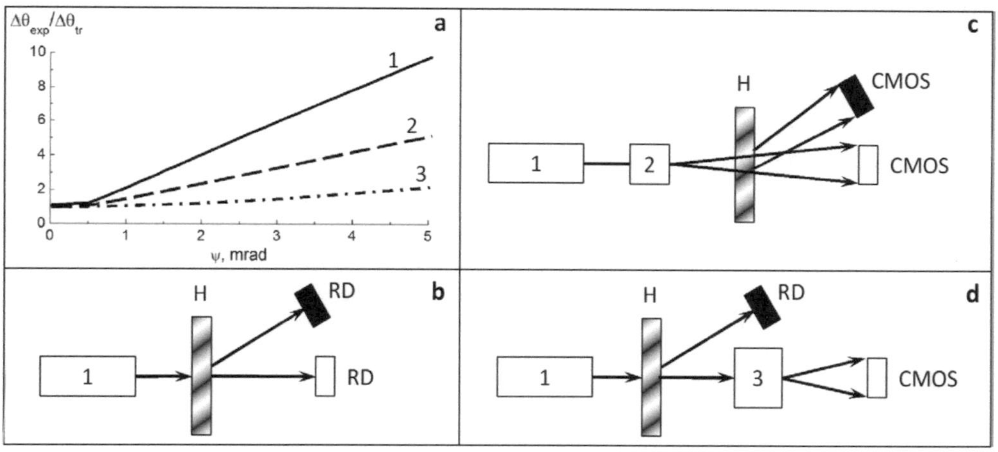

Fig. 12. a - effect of halfwidth of spatial spectrum of reading radiation ($\Delta\psi$) on ratio $\Delta\theta_{exp}/\Delta\theta_{tr}$ at different values of $\Delta\theta_{tr}$: 1 - $\Delta\theta_{tr}$ = 0.5 mrad; 2 - $\Delta\theta_{tr}$ = 1 mrad; $\Delta\theta_{tr}$ = 3 mrad. $\Delta\theta_{exp}$ is the halfwidth of experimentally measured contour; $\Delta\theta_{tr}$ is the halfwidth of true contour of the hologram under study. b,c,d - schematic diagram of measurement of angular selectivity contour by collimated (b) and divergent (c) beams of monochromatic radiation; measurement of spectral selectivity contour by collimated radiation beam with a wide wavelength spectrum (d): 1 – radiation source, 2 – optical system for formation of divergent radiation beam; 3 – spectral instrument for expansion of wavelength spectrum in spatial frequencies; H – hologram; RD – radiation detector.

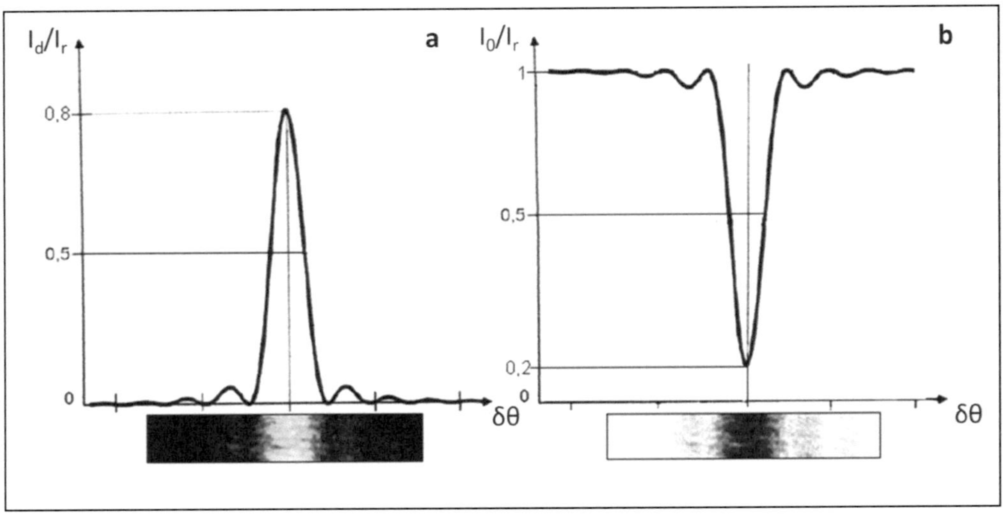

Fig. 13. Distribution of radiation intensity in diffracted (a) and zero (b) radiation beams, recorded on a camera matrix (bottom); processing result for experimental data (top).

6.3 Spectral selectivity contour

When measuring the spectral selectivity contour of a volume hologram according to schematic diagram, shown on Fig. 12b, the hologram is to be scanned with a collimated radiation beam of a wavelength, varying within the spectral range of the contour. At different times, this was carried out in different ways. In work (Denisyuk et al., 1970), the measurements involved a monochromator, with a collimated beam of radiation with a wide spectral range being formed in front of its entrance slit. In work (Sukhanov et al., 1984), the reading radiation wavelength was changed with the help of frequency-tuned dye laser with excimer pump: in this case, the laser radiation divergence was less than 0.5 mrad and the spectral width of scanning radiations was 0.01 nm. The halfwidth of spectral selectivity contour, measured in the present work for a reflection hologram, was $\Delta\lambda = 0.16$ nm (at DE = 80%). Such measurements, naturally, require special equipment and cannot be accomplished using standard techniques for measuring spectral characteristics of optical elements.

Applying the collimated laser radiation with wide spectral interval of wavelengths to measurement of the spectral selectivity contour of volume holograms was proposed by the authors and was implemented with the use radiation of femtosecond laser and semiconductor laser. The schematic diagram of measurements is given on Fig. 12d. A collimated beam of laser radiation with a wide spectrum illuminates hologram, which can be placed in positions "outside Bragg conditions" ($I_d = 0$) and "under Bragg conditions" (I_d is at the maximum for wavelength λ_{Br} within the reading radiation wavelength range).

A radiation beam, having passed a hologram with the direction unchanged (I_0), arrives at the entrance slit of spectral instrument (3 on Fig. 12d), the spatial pattern of wavelength spectrum expansion being recorded behind it on a CMOS-matrix as a spectrogram. With the hologram placed "outside Bragg conditions", reading radiation spectrum is recorded (curves 1 on Fig. 14); with the hologram "under Bragg conditions", spectrum of zero diffraction beam (curves 2 on Fig. 14) is recorded, in similarity to the distribution, given on Fig. 13c. Spectral selectivity contour of a hologram represents in this case a differential contour (curves 3 on Fig. 14), resulting from comparison of two spectrograms.

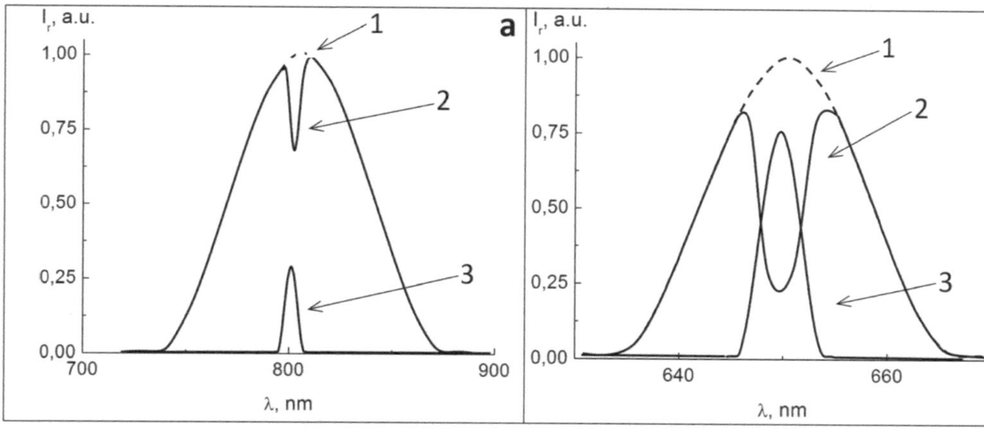

Fig. 14. Spectral distribution of intensity of laser radiation, passed the hologram outside Bragg conditions (curves 1) and under Bragg conditions of hologram reading (curves 2); spectral selectivity contour of a hologram (curves 3): a – femtosecond laser (λ_{max} = 808 nm); b – semiconductor radiation source (λ_{max} = 654 nm).

7. Conclusion

One of the cardinal problems of 3D holography is provision of research in the field with recording materials (Denisyuk, 1980). Volume recording media for holography are at present manufactured in laboratory conditions in the form of isolated specimens or small batches. Obtaining samples with stable and reproducible performance is, as the authors' experience shows, still possible even in such conditions.

The current studies reveal those properties of devised materials, which open up new application opportunities far beyond narrow professional use of recording media for holography. A number of special features of recording media, considered in the paper, can be quite in demand to accomplish unconventional tasks in various fields of science and engineering.

AgHal-PG-media exhibit a set of parameters, pertaining to commonly used traditional AgHal-media: possibility to achieve high sensitivity, the width of spectral sensitization, the variety of techniques of post-exposure treatment etc. The list of the most important parameters of silver-halide media is supplemented by AgHal-PG-media with new opportunities: obtaining samples with thickness of several millimeters; shrinkproof; limitation of the maximum particle size in the light-sensitive agent and post-treatment products.

Polymeric medium with diffusion enhancement has a modulation transfer function, which is untypical for traditional light-sensitive materials and allows excluding the region of low spatial frequencies during the information recording. A no less important and rather unique property is the possibility to obtain the structure of high-efficiency hologram as a latent image at the recording stage and thus achieve a distortionless recorded interference structure in a wide dynamic range after post-treatment. It should be also noted that enhancement and fixation of holograms recorded on such a medium require no treatment in water solutions.

Advancement of volume holography and provision of this line of research with experimental base for comprehensive studies makes it necessary to investigate the processes taking place in the bulk of recording media during hologram construction, which, in its turn, calls for improvement of research techniques and methods to control the parameters of target processes.

Volume recording media and methodology of their investigation, presented in the paper, contribute in the authors' unpresuming opinion to the solution of the problems in question.

8. References

Ashley, J. et al. (2000). Holographic data storage, *IBM J. Res. Dev.* Vol. 44(3), pp. 341-368

Cherkasov, A.S. et al. (1991). *Opt. Spectrosc*, Vol. 71(2), pp. 344-350

Denisyuk, Yu.N. (1962). *Doclady AN SSSR*, Vol. 144(6), pp. 522

Denisyuk, Yu.N. & Protas, I.R. (1963). *Opt. and Spectrosc.*, Vol.14, pp.721, USSR

Denisyuk, Yu.N. et al. (1970). *Opt. Mech. Prom.*, Vol. 2, pp. 29-31

Denisyuk, Yu.N. (1980). Holography and its prospects, *J. Appl. Spectrosc*, Vol. 33, pp. 397-414

Kogelnik, H. (1969). *The Bell System Technical Journal*, Vol. 48, No. 9, pp. 2909-2947.

Kreibig, U. (1970). *Z.Phys. B,Condens. Matter*, Vol. 234, pp. 307-318

Kreibig, U. (1978). *Z.Phys. B, Condens. Matter*, Vol. 31(1), pp. 39-47

Kreibig, U. et al. (1981). *Surf. Sci.*, Vol. 106, pp. 308-312

Lashkov, G.I. & Sukhanov, V.I. (1978). *Opt. Spectrosc.*, Vol. 44, pp. 590-594

Lin, S.N. et al. (2000). *Opt. Lett.*, Vol. 25(7), pp. 451-453

Lin, S. H. et al. (2006). Doped photopolymers for volume holographic applications, *Proceedings of SPIE*, Vol. 6314

Liu, D. Yu. et al. (2010). Diffusional enhancement of volume gratings as an optimized strategy for holographic memory in PQ-PMMA photopolymer, *Optics Express*, Vol. 18, Issue 7, pp. 6447-6454. doi:10.1364/OE.18.006447

Ludman, J.E. et al. (1997). Very thick holographic nonspatial filtering of laser beams, *Opt. Engng.*, Vol. 36(6), pp. 1700-1705

Luo, Yu. et al. (2008). Optimization of multiplexed holographic gratings in PQ-PMMA for spectral-spatial imaging filters, *Optics Letters*, Vol. 33(6), pp. 566-568. doi:10.1364/OL.33.000566

Mahilny, U.V. et al. (2006). *Applied Physics B: Lasers and Optics*, Vol. 82(2), pp. 299 - 302

Popov, A., et al. (2000). Spectrally selective holographic optical elements based on a thick polymer medium with diffusional amplification, *J. Opt. A: Pure Appl. Opt.*, Vol. 2(5), pp. 494-499

Shelby, R.M. (2002). Materials for holographic digital data storage, *Proceedings of SPIE*, Vol. 4659, pp. 344-360

Steckman, G.J. et al. (1998). Characterization of phenanthrenequinone-doped poly(methyl methacrylate) for holographic memory, *Opt. Lett.*, Vol. 23(16), pp. 1310-1312

Sukhanov, V.I. et al. (1984). Three dimensional hologram on reoxan as narrow band spectral selector, *Sov. Tech. Phis. Lett.*, Vol. 10, pp. 387-389

Sukhanov, V.I. (1986). *Physics-Uspekhi (Advances in Physical Sciences)*, Vol. 148(3), pp. 541-543.

Sukhanov, V.I. et al. (1988). *Opt. and Spectrosc.*, Vol. 65(2), pp. 282-284, (USA)

Sukhanov, V.I. (1991). *Proceedings of SPIE*, Vol. 1238, pp. 226-230

Sukhanov, V.I. (1994a). Porous-Glass as a Storage Medium, *Opt. Appl.*, Vol. 24(1-2), pp. 13-26

Sukhanov, V.I. (1994b). 3-Dimensional Deep Holograms and Materials for Recording Them, *J. Opt. Technol.*, Vol. 61(1), pp. 49-56

Sukhanov, V.I. et al. (1996). *Opt. and Spectrosc.*, Vol. 81(5), pp. 779-783, (USA)

Veniaminov, A.V., et al. (1986). *Optics and Spectroscopy*, Vol. 60(1), pp. 142-147

Veniaminov, A.V. et al. (1991). *Opt. Spectrosc.* Vol. 70, pp. 505-508

Veniaminov, A.V. et al. (1996). *Opt. Spectrosc.* Vol. 81, pp. 617-621

Yu, D., et al . (2010). Holographic storage stability in PQ-PMMA bulk photopolymer, *Optics Communications*, 283, pp. 4219–4223

4

Digital Holographic Recording in Amorphous Chalcogenide Films

Andrejs Bulanovs
Innovative Microscopy Center, Daugavpils University
Latvia

1. Introduction

After the invention of holography and development of hologram micro-embossing, mass replication of hologram became possible. Use of a hologram as a security option for valuable documents and products was soon realized and won a large-scale application, which did not hinder development of other security features though. A great variety of other diffractive microstructures commonly known as diffractive optically variable image devices (DOVID) has appeared. Alongside the development of computer technology, digital holographic recording in applied holography has acquired a wide range of application. Nowadays, various types of digital holograms are used for protection and identification of industrial products and documents against counterfeiting, as well as for packing and decorating.

Now two optical technologies for digital recording of protection holograms and DOVID, having similar principle of image formation, are widely used: dot-matrix and image-matrix holographic recordings (Pizzanelly, 2004). A dot-matrix hologram consists of an array of fine diffractive dots holding an image together (**Fig. 1a**). Each dot in such a hologram consists of a uniform diffraction pattern in which the grating pitch and grating orientation

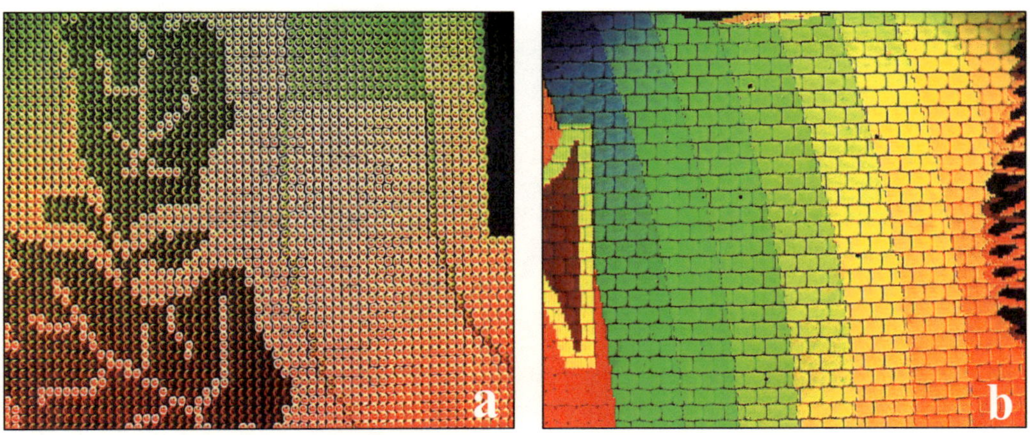

Fig. 1. Photos of microstructure of dot-matrix (**a**) and image-matrix (**b**) holograms recorded on $As_{40}S_{20}Se_{40}$ photoresist.

may vary from dot to dot. In contrast, an image-oriented hologram is composed of an array of microscopic images (**Fig. 1b**). Positive organic photoresists are conventionally used in both technologies for recording relief-phase holograms.

In its turn, formation of surface relief in the process of selective dissolution of amorphous chalcodenide films is well known. The phenomenon is based on difference in dissolution rate of the exposed and unexposed areas of chalcogenide film surface in alkali developers (Teteris, 2002). This phenomenon of photo-induced changes in the dissolution rate of a large group of amorphous chalcogenide semiconductor films was the basis for an extensive development of a new class of inorganic resists.

Holographic recording in amorphous chalcogenide films is the subject of investigation for many scientists. In 1970, the first articles about the recording of holographic diffraction grating in the amorphous films As_2S_3 were published. Due to essential photoinduced changes of optical properties, thin films of amorphous chalcogenide semiconductors are very promising media for phase hologram recording. It is worth mentioning that diffraction efficiency of phase hologram in chalcogenide media reached up to 80%, and optical resolution exceeds 5000 mm^{-1}.

Organic photoresists are mainly used for recording relief-phase holograms; they are sensitive enough only in the ultraviolet part of spectra $\lambda < 480$ nm. For lasers with the wavelength λ=400-650 nm, use of photoresists based on chalcogenide films is very promising. Amorphous chalcogenide films have been recently used as a material for holography in visible spectrum ($\lambda \leq 650$ nm) with high resolution (> 5000 lines/mm) and light sensitivity in the range 1-10 J/cm^2.

Successful results of recording holograms on As-S-Se films have long been known. Photoresists based on amorphous chalcogenide films are effectively employed by some holographic companies in the production of rainbow holograms and diffraction gratings. But use of these films for recording dot-matrix and image-matrix holograms has not yet been studied. There seem to be two main reasons for this. First, As-S-Se films have a low sensitivity (~1-20 J/cm^2) in comparison with organic photoresists (~1-100 mJ/cm^2). Second, image-matrix technologies of hologram recording result in great losses of laser radiation energy (up to 99 %) at optical and Fourier filtrations.

To overcome the above-specified problems in recording dot-matrix and image-matrix holograms, a device specially adapted for the use of As-S-Se films has been developed. Optimum modes of optical recording are investigated, and relief-phase holograms with high diffraction efficiency (DE) were recorded by means of the created device. Another positive result was that the time for recording holograms on the As-S-Se based photoresist was comparable to the time for recording them on organic materials.

The present article contains data of an experimental recording of relief-phase digital holograms and DOVID on chalcogenide thin films. The obtained holograms can be duplicated by standard modern technologies.

2. Dot-matrix holographic recording

Intense research into optical characteristics of photoresists based on chalcogenide amorphous glasses has brought about the need for a device that would allow for such materials to be studied and adapted for the use in manufacturing technologies.

An experimental digital dot-matrix (DDM) device for holographic recording was developed and assembled at the laboratory in Daugavpils University. As distinct from the known

DDM devices in which a laser with an intensity modulator and an optical scheme are separate parts, our device has a laser and an optical scheme integrated into one unit. This approach simplifies the overall system construction and adjustment, reduces number of optical components in design and increases reliability. This allows for recording a two-dimensional array of diffracting pixels with preset parameters on the surface of a photosensitive media thus making it possible to produce a holographic image using pixels with parameters that are computer-controlled during the recording.

The fundamental concept of dot-matrix holograms lies in writing the image pixel-by-pixel varying the grating orientation, grating pitch and, if possible, pixel size. Many other approaches to the creation of a similar dot matrix hologram are known. One of them consists in applying laser beam or e-beam lithography to create each grating pixel. Dot-matrix holograms generated by the mentioned type of lithography are commonly referred to as e-beam holograms. They generally achieve an even better resolution and brightness than the mentioned dot-matrix holograms. However, the use of an e-beam machine is extremely expensive. Since the cost of dot matrix holography is much lower than that of e-beam or laser beam holography, it is expected that the former will gain a broad range of application.

A dot matrix hologram comprises a two-dimensional array of micron-sized diffractive elements or pixels. Each element contains a diffraction grating formed by two-plane coherent wave interference. The period of interference fringes is determined by the angle between the interfering laser beams and their wavelength.

The recording is made by converging two focused laser beams to a point of a preset size on the photoresist surface. Thousands of such closely spaced pixels form a dot matrix hologram.

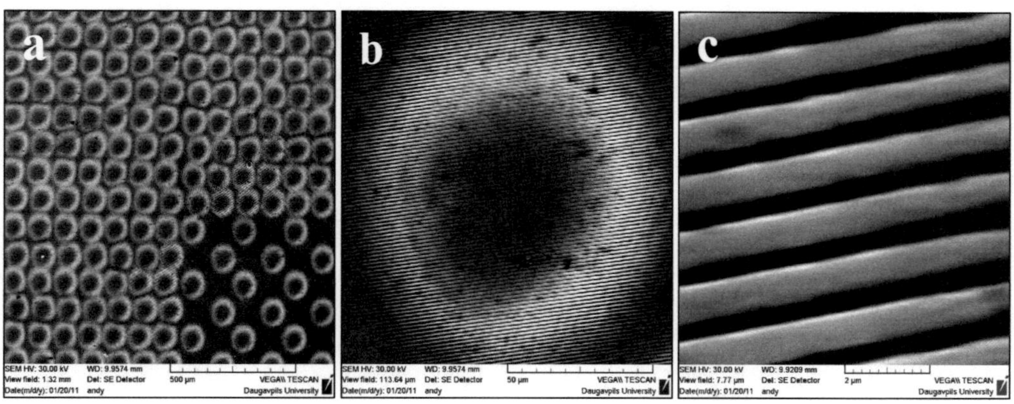

Fig. 2. SEM images of dot-matrix hologram recorded on $As_{40}S_{20}Se_{40}$ photoresist at different magnification. **a**) Structure of a dot-matrix hologram. **b**) Image of one pixel. **c**) Diffraction grating in the pixel.

On illumination, each pixel of the hologram diffracts light at a specific angle chosen in the process of making. The angle and the direction in which the diffraction grating of a pixel reflects light are determined by the following two factors:

- Grating orientation in a pixel. This is achieved by rotating the convergence plane of two laser beams before pixel recording.
- The spatial frequency of the grating in pixel. This is achieved by changing the angle between two laser beams.

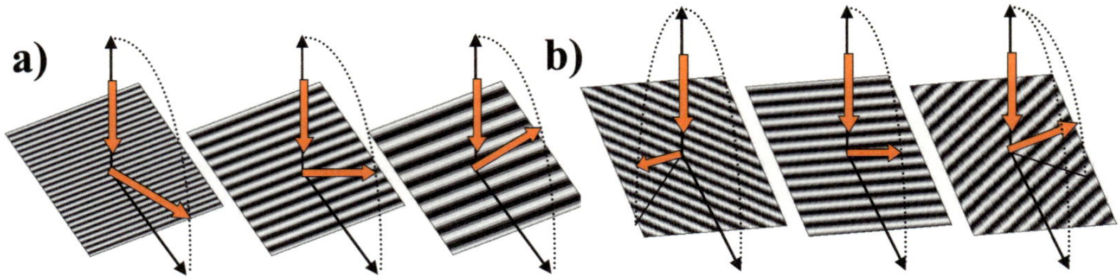

Fig. 3. The changes of the diffraction depend on the period and orientation of the diffraction grating. **a**) Varies the period of the grating. **b**) Varies the orientation of the grating.

By changing the orientation of a grating, the diffracted light can be made skew horizontally left or right with regard to the vertical axis of the initial beam (**Fig.3b**). Only pixels with the same grating orientation are seen in a certain light every time the hologram is rotated. This creates an active 'kinetic' effect, which is difficult to obtain in a conventional holography. The spatial frequency of grating in diffractive pixel determines the vertical diffraction angle (**Fig.3a**). By selecting definite spatial frequencies, each pixel can be assigned a different colour of the visible spectrum. The grating orientation and spatial frequency together determine the wavelength of diffracted light and the direction in which the observer can see the pixel. Each pixel in the computer image usually corresponds to a diffractive pixel in the hologram. The program uses the computer image to calculate the orientation and grating frequency of every element in the hologram according to a developed algorithm. The value of the pixel exposure time is optimized in order to obtain the maximum diffraction efficiency at a given intensity of laser beam. Prior to each exposure, it is possible to set a pause of 10-300 ms.

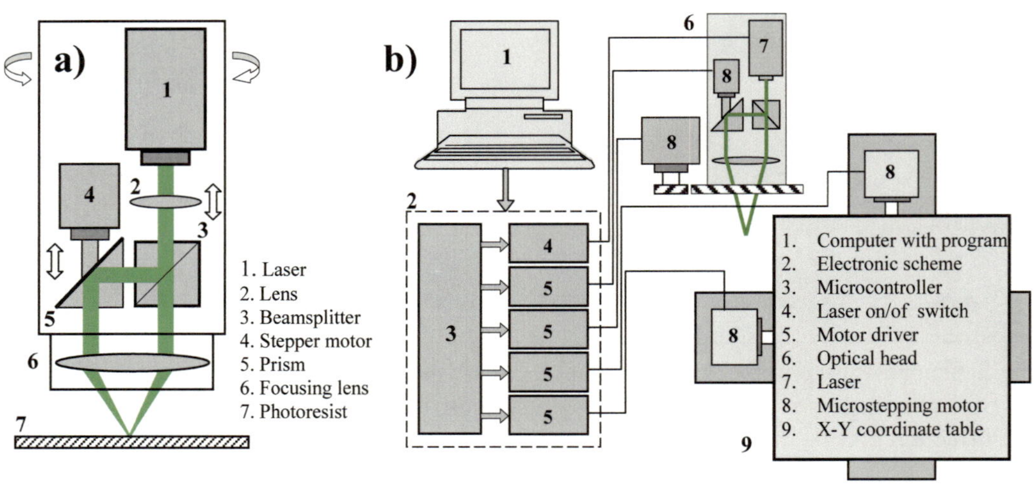

Fig. 4. a) Scheme of the optical head. b) Scheme of the dot-matrix recording device.

This allows for decreasing the vibration that inevitably arises with the use of stepping motors for mechanical moving. The distance between the pixels determines the pixel density and hologram size. The optimal distance is considered equal to the pixel diameter. At smaller distances, pixels overlap, whereas at bigger distances, the pixel density decreases. A significant loss of the diffraction efficiency is observed in both cases.

The overall structure of a dot-matrix holographic recording device and its photo are shown in **Fig. 4b** and in **Fig. 5a**, respectively. A glass plate coated with photoresist is positioned with the help of a accurate coordinate table **9** (see **Fig.4b**).

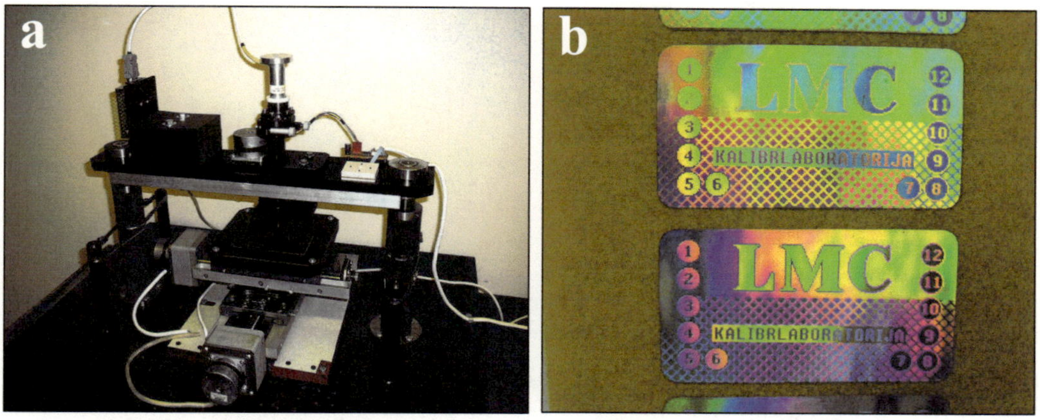

Fig. 5. **a)** Photo of a dot-matrix recording device. **b)** Photo of a protective dot-matrix hologram recorded on $As_{40}S_{20}Se_{40}$ photoresist with the help of the created device.

The size of X-Y movable area was 75/75 mm, and the positional accuracy was about 3 μm. Stepping motors **8** were used for rotating the optical head, moving the reflecting prism in optical scheme for changing grating pitch size. Also stepping motors **8** were used for changing position of the coordinate X-Y table. The permissible minimum level of the motor vibration was achieved by the use of the microstepping motor driver with operation in eighth-step mode (we use A3977 microchip). The electronic module **2** shown in the figure was assembled on the basis of a microcontroller ATMega16, the parts of which were connected with driver microschemes of stepping motors **8** and electronic key of the switching laser **4** (on-off). The electronic module is intended to receive commands and data from the computer **1** via an RS-232 interface and to produce electronic signals on the basis, of the data and commands to control the drivers of subordinate devices. Such coordination of the program with peripherals allows for speeding up and optimizing the calculation of hologram in the main program, whereas the microcontroller program provides the command over all mechanical motion.

An interference pixel with the size of several tens of microns is formed in the rotating optical scheme **Fig.4a.** As a source of coherent radiation, a compact diode pumped Nd: YAG SHG laser (model LSR532-ML50, wave length 532 nm, max. power 50 mW) was used, which can operate in radiation modulation regime by the external TTL signal (frequency of modulation up to 25 kHz). In this case, an exterior electro-optical shutter, which is commonly used with gas lasers, is not needed. For the input of a laser beam in the rotated optical scheme, it is usually necessary to centre the beam along the rotation axis. Also for preservation of constant polarization during the hologram recording it is necessary to use a $\lambda/4$ phase plate before and after the input of a laser beam in the optical head **(Fig. 6a)**. The disadvantage of similar designs is complexity of adjustment, a great number of additional optical elements, and necessity of using a symmetric optical scheme. **Fig. 6** shows the fundamental difference between design of created dot-matrix recording device and similar devices

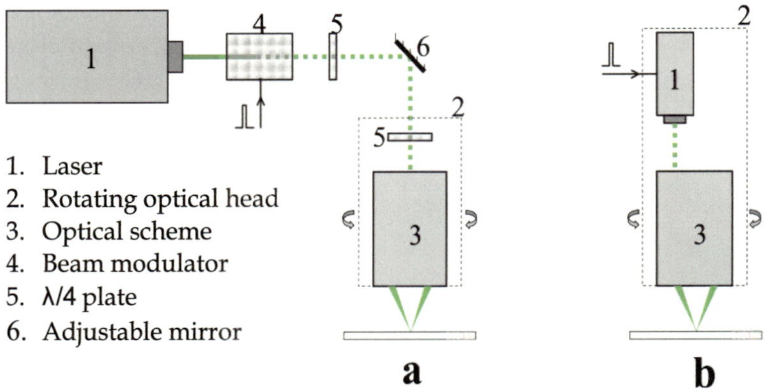

Fig. 6. Entering of the laser beam into a rotating optical part of the dot-matrix recording device. **a)** The laser is located outside of optical scheme. **b)** Laser is integrated into the optical scheme.

Compactness of the laser allowed us to include it in the design of the optical head **(Fig.6b)**. Such a design minimised the number of optical elements and enabled utilization of an asymmetric optical scheme, which is very stable and easily adjusted.

Figure **Fig.4a** shows the optical scheme, which was used for the formation of diffraction pixels with preset pitch size and orientation angle. The beam splitter divides the laser beam into two parts, equal in intensity and parallel to the rotation axis, which are reduced with the help of a focusing lens **6** (f=20 mm) and converge in the point of a certain size (30-150 μm). The size of the point is adjusted manually by shifting a far focusing lens **2** (f=300 mm). The stepper motor **4** moves a reflecting prism **5** in the vertical direction and allows for changing distance between the divided beams and, accordingly, frequency of diffraction grating in the pixel within the range of 750-1000 mm^{-1}. The orientation angle of grating in the pixel is changed by turning the optical head.

The disadvantage of this optical design is a considerable path-length difference between the interfering wave fronts (10-15 mm), which imposes stricter requirements on the laser coherence. The coherence length of the used laser radiation was therefore measured with a Michelson interferometer and exceeded 3 cm, which was enough for interference conditions. Rather a large difference of the optical path between beams did not allow us to record pixels less than 30 μm in size because the focusing points of beams are slightly different.

The developed device for dot-matrix holographic recording was successfully used for research purposes and for producing protective holographic labels **(Fig. 5b)**.

For recording a dot-matrix hologram, $As_{40}S_{15}Se_{45}$ chalcogenide thin films were used. Films were obtained by thermal evaporation onto glass substrates in the vacuum of ~10^5 Torr. The method of film evaporation in vacuum allows for obtaining homogeneous films on the substrates of different size and form. The effect of light on As-S-Se films gives the possibility to carry out selective etching of differently exposed parts with a great resolution of up to 5000 mm^{-1}; that allows for the use of these films in applied holography. These films possess a differential chemical etch rate between exposed and unexposed areas; the exposed portions have a slower etch rate in non-aqueous amine-base solution due to structural changes. Structural changes arise due to the absorbtion of light with bandgap energy, this results to the creation of the electron-hole pairs, whose subsequent recombinations leads to the processes of atoms bond reforming. The wide spectral sensitivity range of the material (from ultraviolet up to 650 nm) makes the task of choosing a laser easier. Optical recording on a photoresist and the following chemical etching result in the creation of a relief-phase hologram (**Fig. 7b**).

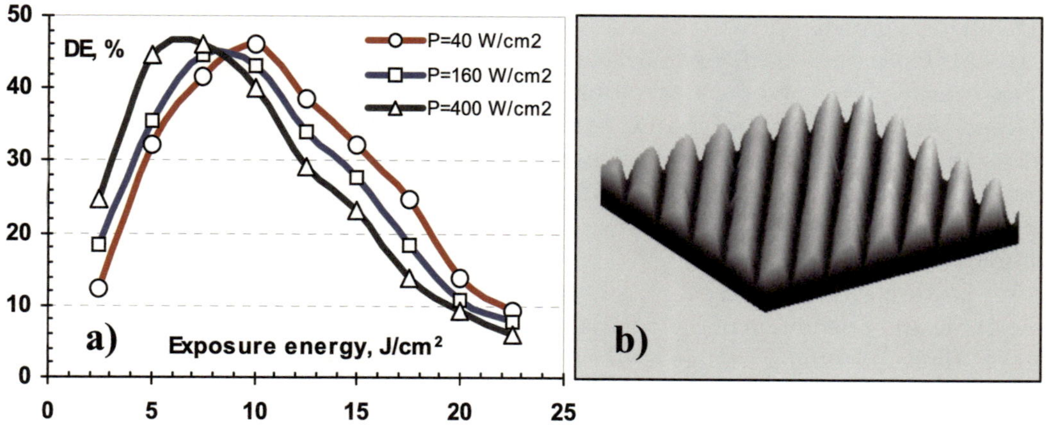

Fig. 7. **a)** Diffraction efficiency of dot-matrix holograms recorded on $As_{40}S_{20}Se_{40}$ photoresist as a function of exposure dose. **b)** 3D model of the gratings profile based on an SEM image.

The relief microstructure of the photoresist surface is copied onto a hard metallic material, usually nickel, by the galvanic method. Once a hologram pattern has been created, the traditional hologram manufacturing process, including electroforming, mechanical recombining, and embossing, can be used in the mass production of holograms.

The diffraction efficiency (DE) of diffraction grating covered by a ~20 nm layer of Al by thermal deposition in vacuum is shown on **Fig. 7a**. Determination of diffraction efficiency was performed at a normal incidence of a laser beam onto the surface of a hologram. The intensites of diffraction light was measured in both first maxima and diffraction efficiency was calculated with the help of the formula $DE=(I_{+1}+I_{-1})/I_0$, where I_0 is an intensity of the incident beam, and I_{+1} and I_{-1} are intensities of the first diffraction maxima.

Relief-phase dot-matrix holographic gratings were obtained in the composition of $As_{40}S_{20}S_{40}$ with the diffraction efficiency about 50%. In course of determining dependence of the DE on exposure, it was observed that sensitivity of the material increased when the exposure power density increases. So at the power density 40 W/cm^2, the optimum exposure was close to 10 J/cm^2; when the power density increases up to 400 W/cm^2, the optimum exposure decreases approximately twofold, reaching ~ 5 J/cm^2 (**Fig. 7a**). The Gaussian distribution of intensity in a laser beam gives an uneven distribution of the exposure in a pixel when recording it on a photoresist (**Fig. 2b**). As a result, the effective diameter of pixel changes varying the exposure time. This effect makes it possible to handle the pixel density by a programme when holograms are recorded in spite of certain losses in the diffraction efficiency due to deviation from optimal exposure dose.

3. Image-matrix holographic recording

With the development of spatial light modulators and computer technology, applied digital holographic recording has acquired a wide range of use. After the appearance of translucent

Liquid Crystal (LC) and reflecting Liquid Crystal on Silicon (LCoS) spatial light modulators (SLM) of a new generation, an interest regarding their use in optical schemes for recording computer-generated holograms (CGH) and diffractive optically variable image devices (DOVID) has grown.

An SLM chip has a twisted or parallel-aligned nematic liquid crystal layer to modulate light. It changes the phase and polarization state of the incident light. Parameters of modulation depend on the alignment of the LC and polarization state of the incident light. The LC alignment is controlled, pixel-by-pixel, using DVI (Digital Visual Interface) signals from the computer. Electro-optical effects in liquid crystal displays make them suitable for amplitude and phase modulation of coherent wave fronts so that they can be used as programmable diffractive elements.

After passing through the spatial light modulator, a wave undergoes a local change of the phase and orientation of the polarization plane. The size of the minimum area of the wave front undergoing change is determined by the size of the modulator pixel. The degree to which the wave parameters change depends on the angle between the optical axis of the modulator and polarization direction of a laser beam, as well as on the level of electrical signals on the modulator pixels. Two extreme cases are possible. In the first the modulator changes only the phase of the transmitted wave leaving polarization and amplitude unchanged. This is a phase modulation regime. In the second case, the modulator changes the direction of polarisation plane, with small changes of phase in all sections of the wave front. The wave becomes modulated in amplitude after going through the polarizer.

For recording CGH and DOVID, an image-matrix optical set-up shown in **Fig. 8a** was used.

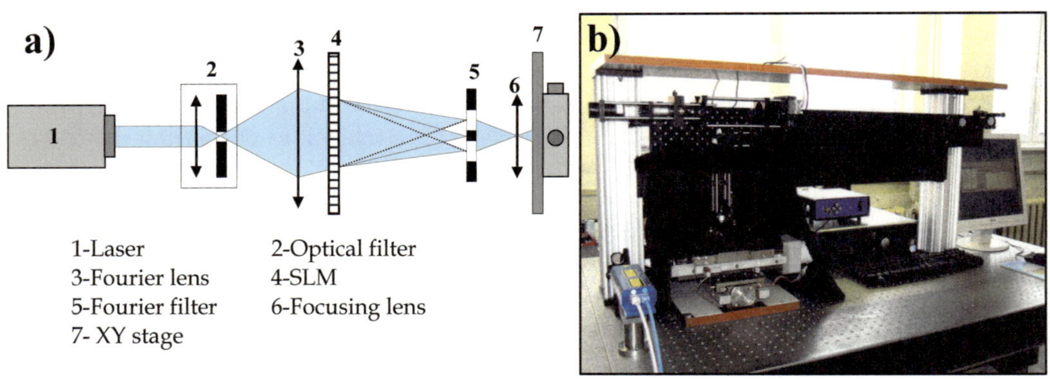

Fig. 8. a) Optical scheme of the image-matrix holographic recording device. b) Photo of the device.

It utilizes a diode laser **1** (100mW, 405nm), whose spatially filtered and expanded beam illuminates a transparent SVGA spatial light modulator **4** located behind the Fourier lens **3**. The lens **3** in the plane conjugated to the pinhole **2** creates the Fourier spatial spectrum of the laser wavefront diffracted on the amplitude-phase image displayed on SLM. Intensity peaks of the spectrum, that are the farthest from the centre, correspond to the higher orders of diffraction. A spatial filter **5** is positioned in the Fourier plane behind the SLM in order to block the zero and higher than first orders. After the inverse Fourier transformation, a hologram with a doubled frequency of diffraction grate is created with the help of an objective **6** by interference of the I_{+1} and I_{-1} orders of the wave diffracted at the modulator. This means that the image on the photoresist surface for periodic structures is formed only due to the interference of the second harmonics of spatial spectra. For example, if we will send on the modulator the image corresponding to N black-and-white stripes, on the photoresist surface we will get a sinusoidal intensity distribution consisting of 2N peaks. An image is displayed on the SLM then reduced to microscopic size (140x105 μm) by a lens system **6** and recorded on the photoresist plate placed on the XY-stage **7**. An important advantage of such set-ups is that they do not require vibration damping.

A general principle of the formation of a large size hologram is shown in **Fig. 9.** The recording is performed sequentially by rectangular micron-sized areas as shown in **Fig. 9a**. The sections tightly fill the entire area of the hologram, and a human eye does not perceive a discrete structure of the hologram consisting of micron-sized sections. Optical recording of the interference pattern is performed on each individual area of the hologram.

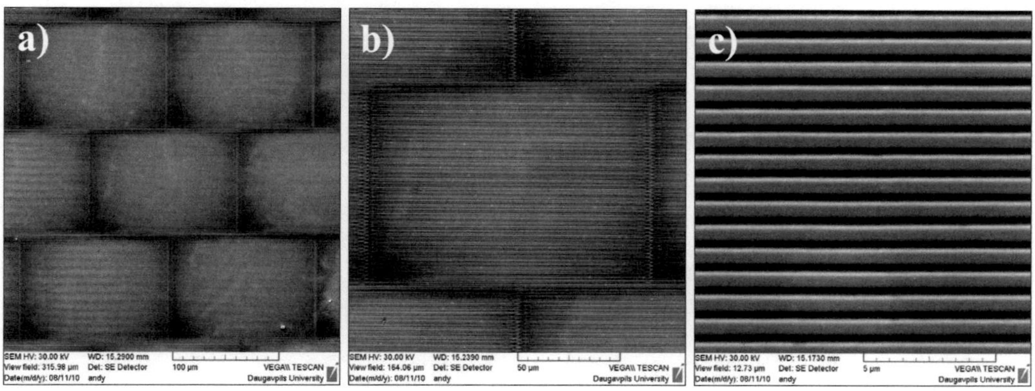

Fig. 9. SEM images of image-matrix hologram at different magnification.
a) Image-matrix structure of hologram. **b**) Image of one frame. **c**) Diffraction grating in the frame.

The pattern is set by the amplitude-phase image on the modulator and corresponds to the graphic figure on the computer monitor. In order to lessen the perceived discreteness of the hologram structure, the neighbouring rows of its micro-images are shifted the length of half-frame relative to each other. Each separate image from the matrix structure contains a computer-generated hologram (CGH); in a simple case, it may be a set of gratings with a grating pitch and orientation adjusted to the requirements (**Fig. 9c**).

The phase modulation regime with Fourier filtration is promising for use in applied holography for it offers a way of obtaining diffraction gratings with the frequency 300-2000 mm^{-1}. The maximal frequency is estimated according to the following formula:

$F_{max} = 2\ n.a./\lambda$, where n.a. is a numerical aperture of objective, and λ is a laser wave length.

The minimum frequency is determined by the size of the central diaphragm (L ~ 1mm) in the mask **5**, which blocks the zero order of diffraction and is calculated by the formula $F_{min}=L/(\lambda\ f)$, where λ is a laser wave length, f focal length of the objective **6**, and L size of the central diaphragm.

Lessening the laser energy losses in case of an optimal optical scheme and reducing the frame area two times in comparison with the conventional (200x150 µm) one allowed for a considerable increase of illumination intensity comparing to analogous devices employing organic photoresists. This, in its turn, made it possible to record holograms on chalcogenide films of low sensitivity without any considerable loss of rate.

In many practical cases the principle of formation of the holographic image with the help of image-matrix technology is close to the dot-matrix technology considered above. The recording of the image-matrix hologram is made by frames of a certain size which depends on the adjustment of optical system. Every frame contains an array of diffraction pixels, which fill all of the frame area and usually have a square form. A diffraction grating with the given period and orientation or functional microstructure (for example, shown in Fig. 15) is recorded in every such pixel.

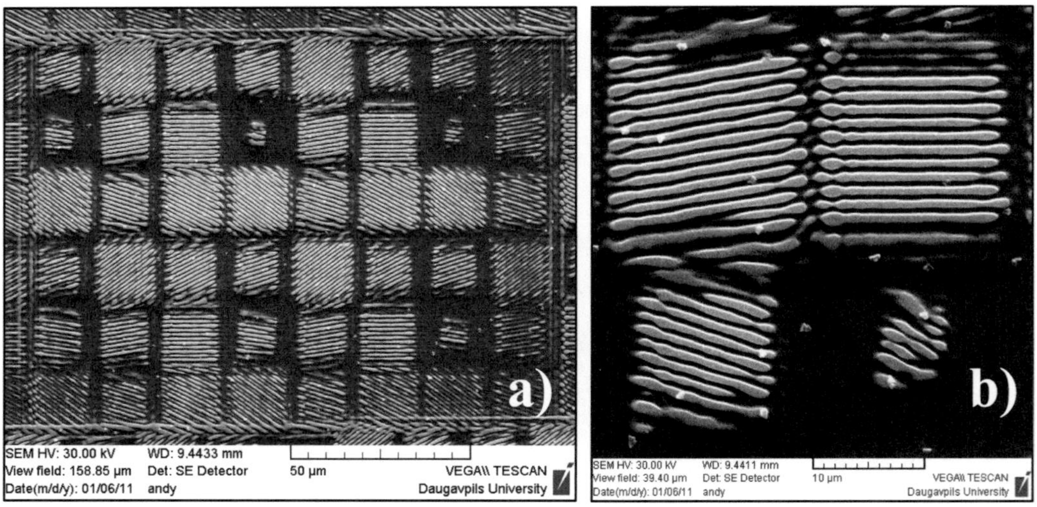

Fig. 10. **a)** SEM image of one frame with the matrix of 8x6 different pixels. **b)** SEM image of pixels with different grating orientation and modulated by size.

When the pixel size changes, its relative visual brightness varies. This perception is useful for recording of grey-scale images or photos. The programmed realization of kinetic visual effects is possible when the turn of the hologram results in the change of visible images. Another example is the possibility of imitating three-dimensional images when an individual image is formed for each eye according to the changed viewing angle of an object (perspective).

Fig. 10a shows a SEM image of one frame of the stereogram containing 9 combined images for creating an illusion of a 3D object. In this case, each frame consists of an array of 8x6 pixels of size 15 µm. Every pixel in the array corresponds to a graphic point in one of the combined images. Each of the combined images is visible from a certain angle which is determined by the orientation of diffraction grating of pixels, appropriate to this image. The size of a pixel is proportional to a level of grey color of the appropriate graphic point (**Fig. 10b**). Realization of a holographic image in real colors is also possible. In this case, each dot of the graphic image corresponds to three diffraction elements and an R-G-B model of the basic color mix is realized in the direction of the observer.

The designed and assembled device for image-matrix holographic recording was successfully used for research purposes as well as for producing holograms for the protection and identification of industrial products and documents. A sample of image-matrix hologram recorded on inorganic As-S-Se photoresist is shown in **Fig.11b**. Optimization of the relief formation processes makes it possible to produce gratings with high values of diffraction efficiency. A 3D model of grating profile corresponding to the DE~50% is shown in **Fig.11a.**

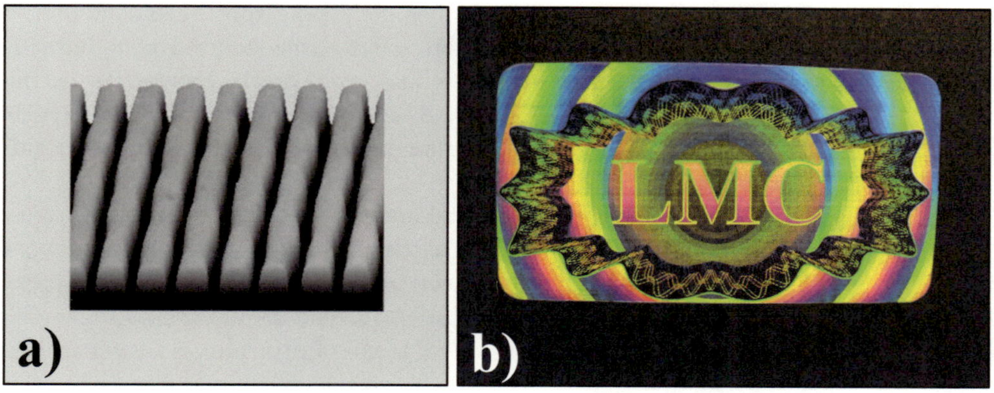

Fig. 11. **a)** 3D model of the gratings profile based on a SEM image. **b)** Photo of a protective image-matrix hologram recorded on $As_{40}S_{20}Se_{40}$ photoresist with the help of the created device.

For recording an image-matrix hologram, we used a $As_{40}S_{15}Se_{45}$ chalcogenide thin film and organic photoresist of the AZ1800 series. Micro relief of the hologram surface in these cases is shown in **Fig. 12**.

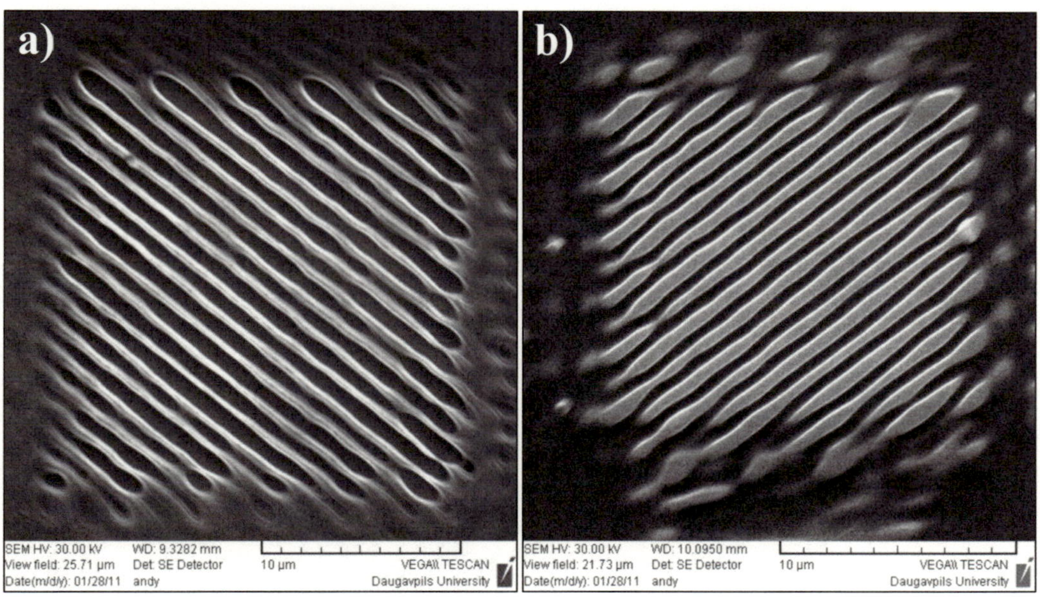

Fig. 12. SEM image of pixel with diffraction grating recorded by image-matrix technology **a)** positive organic photoresist of the AZ1800 series **b)** inorganic negative $As_{40}S_{20}Se_{40}$ photoresist.

Evaporated As–S-Se films undergo photostructural changes in the regions exposed to the light of the energy higher than the bandgap energy. When the thermal deposition method is used, the vapours are condensed on a room-temperature substrate, the chemical reactions among the fragments of the original compounds are not quick enough, and the high-temperature state is then partly or fully fixed and preserved. A micro-heterogeneous film, which is far from thermodynamic equilibrium, is then obtained. Exposure of such films can induce photochemical reactions, e.g. polymerization or depolymerization among individual fragments of the film, changes of local structure, and, as a result, change the chemical reactivity, i.e. dissolution rate in various alkaline and base organic solvents. After exposure, the samples were chemically treated in non-water amine-based organic solutions (negative etching) to form a relief pattern.

When placed in an appropriate solvent, the exposed areas of the film are protected from the etchant in proportion to the irradiation dose. The dependence of etching speed on the exposure dose of such negative wet etching is shown in **Fig. 13**, where the exposed parts of the samples are more resistant to the amine-based solvents than the unexposed regions. The highest selectivity of etching can be achieved after 2 J/cm^2 of exposure. A longer exposure slightly increases the etching ratio.

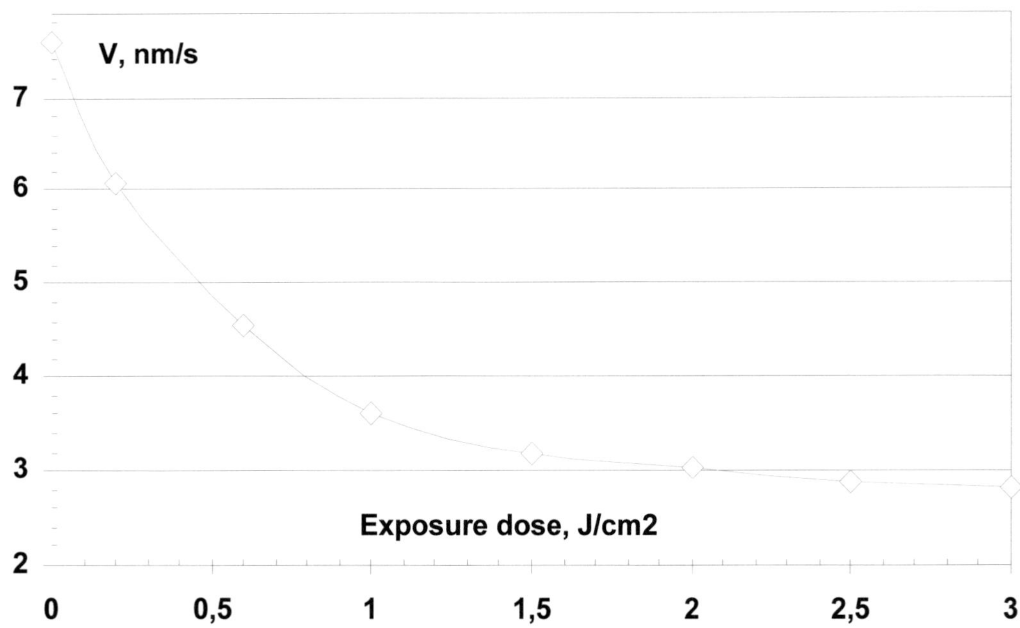

Fig. 13. Dependence of etch rate of $As_{38}S_{20}Se_{42}$ films in non-aqueous amine-base solution on exposure (λ=405 nm).

The effect, whereby the speed with which amorphous As-S-Se films undergo chemical etching, directly depends on the exposure dose, allows for obtaining relief-phase holograms suitable for mass production by conventional technologies and widely used in applied holography.

For relief depth of ~ 150 nm (λ/4 for wavelength 550nm) photo induced structural changes in the film surface layer i.e. the layer that actually undergoes chemical etching, which is ~

250 nm thick, are necessary. In such a case, it appears optimal to use a laser with the wavelength corresponding to the section of the As-S-Se film spectrum with a low optic penetration (λ= 400-550 nm for $As_{38}S_{20}Se_{42}$ films) in order to minimize the exposure time; a higher penetration results in a lower photo induced changes at the film surface.

For obtaining reflective relief-phase holograms on As-S-Se films, lasers with wavelengths 400-650 nm can be used. But for a higher speed of recording image-matrix holograms, it is preferable to employ short wavelength lasers. The smallest depth of structural changes corresponding to λ=*405nm* is sufficient for obtaining relief-phase holograms with the required depth of micro-relief by means of selective etching.

The experimental curves showing dependence of diffraction efficiency on exposure dose for different grating frequences can be seen in **Fig. 14**. The maximum values of diffraction efficiency are approximately equal for all grating frequences and reach 55-65% in exposures in the range 0.8-1.1 J/cm^2.

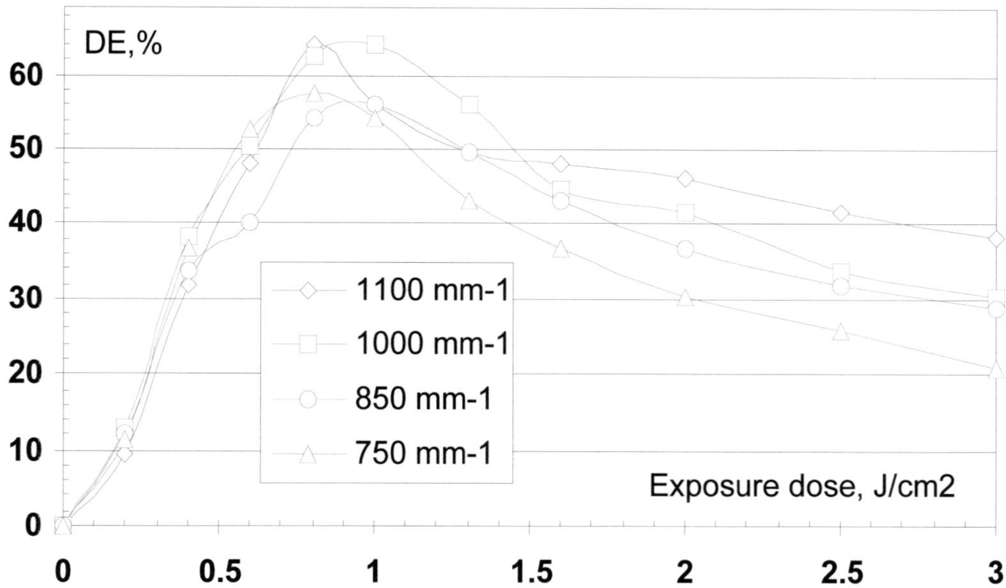

Fig. 14. Dependence of diffraction efficiency of image-matrix hologram on exposure dose at different grating frequencies.

One of the major advantages of image-matrix technology over other ones is that it allows for recording a huge amount of microstructures that differ from simple diffraction gratings. They can be used as additional security elements embedded in holograms. **Fig. 15** shows some relief microstructures that were recorded on As-S-Se photoresist with the help of image-matrix technology. The structure of resist relief was investigated by using a scanning electron microscope.

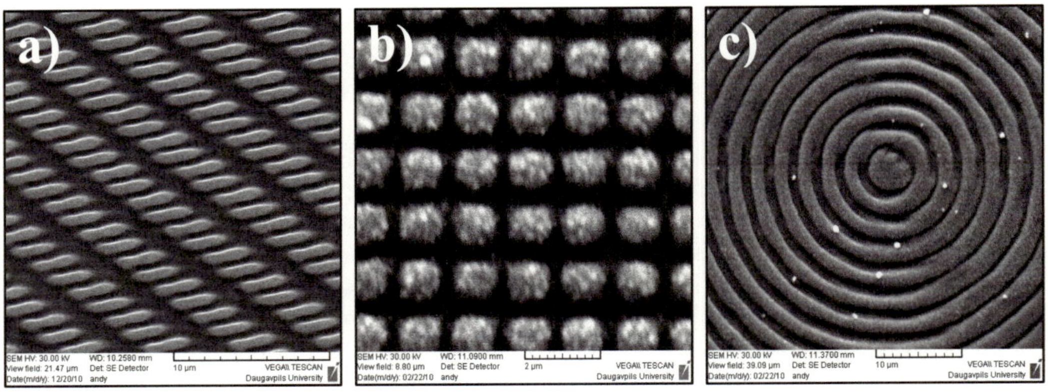

Fig. 15. Samples of microstructure that were recorded on As-S-Se photoresist with the help of image-matrix technology: **a**) complex grating, **b)** 2D diffraction grating, **c)** axicon.

4. Image-matrix recording device based on LCoS spatial light modulator

At the current state of technological development, LCoS (Liquid Crystal on Silicon) optical modulators have characteristics excelling those of the modulators created by the LCD (Liquid Crystal Display) and DLP (Digital Light Processing) technologies. From the point of view of their practical application, LCoS devices have the best of such important parameters as small pixels (down to 8 μm), a great number of pixels (HDMI resolution), high fill factor (till 95%), high frame rate (up to 180Hz), high reflectivity (better than 70%), weakly coupled amplitude or phase modulation, high contrast, and depth of phase modulation above 2π in the visible range. These improved parameters of recent LCoS displays make them appropriate for high quality digital holography and many other related optical applications.

Fig.16 shows a picture and a overall scheme of the experimental setup for holographic recording using the LC-R 2500 LCoS modulator. The basic principle of the operation is the same as described in third part of the chapter. With the help of the aligned mirror **2**, radiation of the laser **1** is delivered onto the optical filter **4**. The λ/2 plate **3**, placed in front of the optical filter, makes it possible to vary the angle of the laser beam polarization relative to the modulator **6** optical axis.

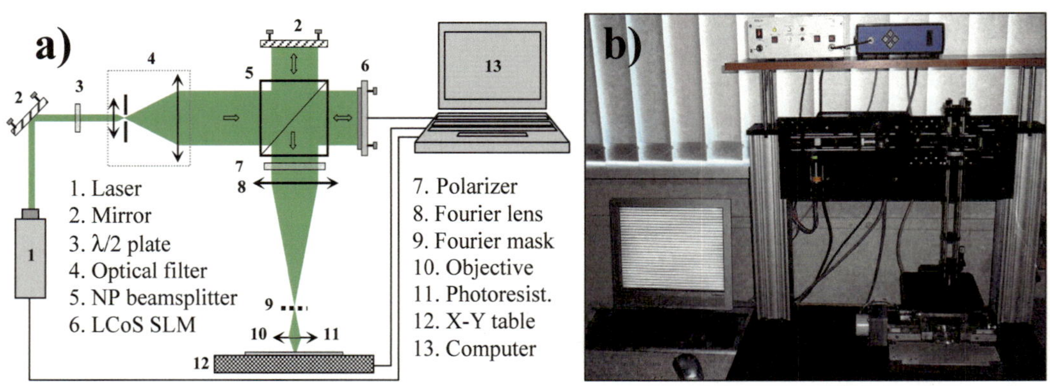

Fig. 16. The scheme (a) and photo (b) of the experimental setup.

The widened and spatially filtered laser beam with a flat wave front falls on the non-polarizing beam splitter **5** (with the coefficient of amplitude division 50/50). The first part of the split beam is incident on the optical modulator LC-R 2500, and after reflecting from its surface, undergoes a local change of phase and polarization plane in accordance with the image on the computer monitor. After reflecting from the mirror **2**, the second part of the beam is used as a reference beam in the interference mode for transition from phase to amplitude modulation of the wave front in optical recording.
The modulation regime (amplitude and phase) is determined by the configuration of the optical scheme (**Fig. 17a, b**), as well as by the position of the polarizer **7** and by the angle between laser radiation polarization plane and modulator optical axis. The long focal length lens **8** (f = 450 mm) forms in its focal plane the Fourier spatial spectrum of the laser radiation diffraction on the SLM image. The mask **9** is situated in the focal plane of the lens **8** and is meant for modifying the Fourier spectra. The inverse Fourier transformation is performed by a micro objective lens (50x, n.a. = 0.5, f = 3.6mm) **10** onto the surface of a plate with photoresist **11**. A compact diode pumped SHG ns laser (model STA-01SH, λ=532 nm, P=50 mW) modulated electronically through TTL signals was used in the optical scheme. In order to form sequential light impulses with a given energy, a regime of the laser **1** operation with generation of optical radiation by external synchronizing pulses is used. Laser pulse with duration ~1 ns and energy ~1.5 μJ corresponds to one external TTL synchronizing pulse. Optical exposure is obtained by a series of electric synchronizing pulses with repetition rate 30 kHz. In the amplitude modulation regime (**Fig. 17a**), polarization of the incident laser radiation coincides with the optical axis of modulator, and the orientation angle of the polarizer relative to SLM optical axis is $\beta=90^0$. After reflection from the SLM surface, at the local area of the wave front the polarization plane is rotated proportional to the signal on the of the modulator pixel. While going through the polarizer **7**, the wave front undergoes amplitude modulation.

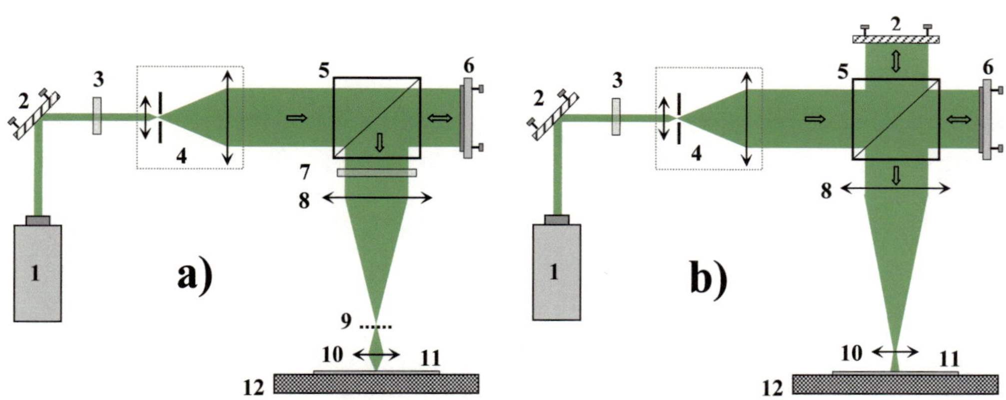

Fig. 17. Optical schemes corresponding to different modes of operation.a) Regime of wave front amplitude and phase modulation. **b)** Amplitude modulation by the phase contrast method.

After the direct and inverse Fourier transformation, which is performed by the lens **8** and objective **10**, the wave front is projected on the photoresist surface with the about 120 times minimization for forming a frame. To form an image in the frame, the full Fourier spectrum is used and in this case the mask is absent. The amplitude modulation regime allows for recording holograms with a given profile of diffraction grating (rectangular, triangular, sinusoidal, etc.) and frequency ~ 0-800 mm^{-1}.

In the phase modulation regime (also **Fig. 17a**), the polarization plane of the incident wave is at angle $\alpha=25^0$ with the optical axis of modulator and the polarizer is orientated at the angle $\beta=145^0$. In this regime, the polarization of wave front reflected from the modulator undergoes slight changes of direction. After the polarizer, the intensity changes by not more than by 20% of the maximum. The changes of phase in the range from 0 to 2π are proportional to the level of the incoming signal. The mask **9** blocks the zero diffraction order and the orders higher than the first one.

After the inverse Fourier transformation, as it was described in section **3**, a hologram with a doubled frequency of diffraction grating is created with the help of the objective **10** by interference of I_{+1} and I_{-1} orders of the wave diffracted on the modulator. A polarizer **7** is not needed in this case. The phase modulation regime with Fourier filtration is promising for use in applied holography for it gives an opportunity to obtain diffraction gratings with frequency 300-2000 mm^{-1}.

Interference techniques for phase visualisation are shown in **Fig. 17b**, where the signal and reference beams travel along the same optical axis and interfere in the output of the optical system. In this case, the LCoS modulator works in the phase modulation regime. A reference wave front allows for the visualization of phase information in the original wave front.

The common path interference method is also known as the phase contrast method. From the point of view of Fourier optics, a reference beam in the focal plane of the lens **8** changes amplitude and phase of the zero order component in the Fourier spectra of the diffracted object wave, and the SLM image after the reverse Fourier transformation on the photoresist surface by the objective **10** has an amplitude component changing in proportion to the initial phase. Another way of realizing the phase contrast technique allows for visualization of phase perturbations by the use of the Fourier plane phase shifting filter **9** as shown in **Fig. 17a**. In this case, an efficient filter might have near to 100% transmission and a $\pi/2$ phase shift in the central region. But making such a filter in laboratory conditions is a technologically difficult task.

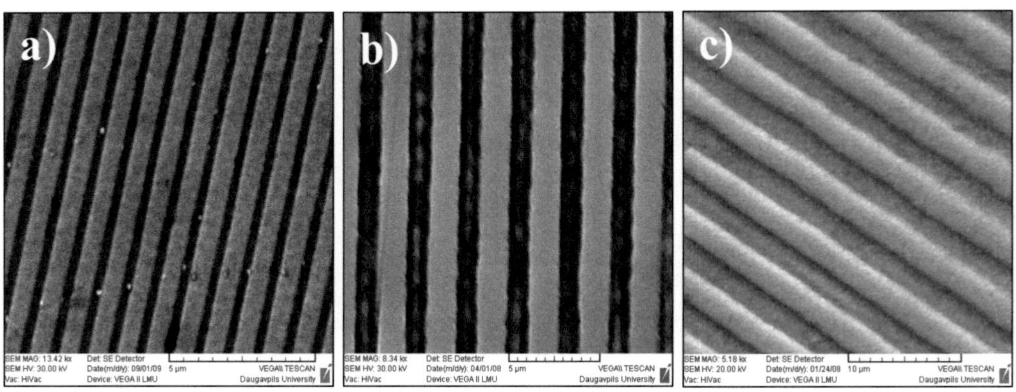

Fig. 18. SEM images of diffraction gratings recorded in different modes of operation on $A_{40}S_{20}Se_{40}$ thin films: **a)** phase modulation regime with Fourier filtration (period 1 μm), **b)** amplitude modulation regime (period 2.5 μm), and c**)** phase contrast method (period 5 μm).

5. Dot-matrix holographic recording on As_2S_3-Al films

For the recording of relief-phase holograms, organic photoresists are mainly used; they are sensitive enough only in the ultraviolet part of spectra $\lambda < 480$ nm. For lasers with the wavelength λ=480-630 nm, it is possible to use chalcogenide film photoresists. Due to the high values of the photo-induced changes of optical properties, the thin films of amorphous chalcogenide semiconductors make a very promising medium for holographic recording. But the use of such films for recording dot-matrix holograms is complicated because of the low sensitivity E=5-30 J/cm^2. For example, the time for producing dot-matrix holograms consisting of 10^6 pixels is measured by tens of hours. That is why a possibility of increasing sensitivity of films As_2S_3 in the system As_2S_3-Al was studied. Usually, in order to get a relief hologram, a film of photoresist is put onto the substrate with an absorbing layer. In this case the interference pattern in the photoresist layer is made only by incident waves. The interference pattern in the As_2S_3-Al system during the holographic recording can be considered as a sum of incident waves and those reflected by the aluminum sub layer. In case of their overlap, the maximum diffraction efficiency (DE) and sensitivity of the medium are obtained. One of the main conditions for coincidence of interference patterns of incident and reflected waves is equality of incidence angles upon the film surface. If this condition is met, any one of the incident beams after reflecting by the aluminum layer goes in the direction of the incidence of another one (**Fig. 19**). Optical schemes having equal incidence angles of beams are mainly used for the dot-matrix and image-matrix holographic recordings. Taking into consideration the possibility of selective etching of As_2S_3 films, the As_2S_3-Al system is promising from the point of view of obtaining relief-phase holograms.

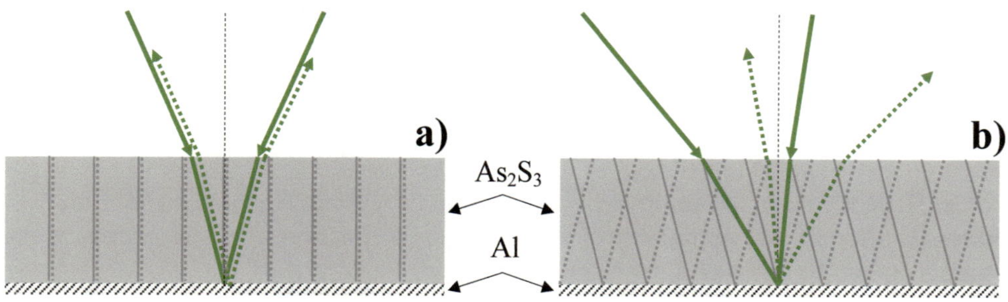

Fig. 19. Scheme of interference field in As_2S_3-Al films.
a) The incidence angles are equal. **b)** The incidence angles are diffrent.

Interference pattern of reflected waves is less stable in time, since before the interference the waves pass the As_2S_3 layer, which has some changes of optical properties during exposure. The aim of the experiments was to determine the conditions ensuring the maximum overlap of interference patterns of incident and reflected waves. A relatively easy way of obtaining the minimum reflection coefficient in a definite part of spectra is viewed as an important feature of the system As_2S_3-Al. Monitoring interference minimum of reflection during the evaporation of film makes it possible to get the system As_2S_3-Al with a minimum value of initial reflection, i.e. 10-15%. Thus, maximum energy absorption of incident waves is obtained and, as a result, the sensitivity of composition during the recording of dot-matrix holograms increases up to 40%.

As_2S_3 films were obtained by thermal evaporation in vacuum onto aluminum (~100 nm) coated glass substrates. The thickness of the As_2S_3 film was closely controlled during the process of evaporation. Monitoring reflectivity with the interference technique at the wavelength of 532 nm during evaporation of the As_2S_3 layer allowed for obtaining a As_2S_3-Al pattern with the initial reflection of 10-15%. The process of evaporation was stopped immediately as soon as the interference minimum of reflection was reached.

Laser irradiation of As_2S_3 films causes modulation of optical refractive index, which brings about modification of interference conditions, that is modification of the reflectivity coefficient of the As_2S_3-Al system. It was experimentally established that the minimum value of reflection coefficient in the As_2S_3-Al system depends on the thickness of As_2S_3 film and reaches R_{min}=10% at d=2.4 μm (**Fig. 20a**). Dependence of the pixel hologram DE on film thickness is presented in **Fig. 20b**. The maximum diffraction efficiency DE ~40 % was obtained when the film thickness was d=2.4 μm (**Fig. 21**). With an increase of the film thickness, absorption in the As_2S_3 layer at the wavelength λ=532 nm becomes high and lessens the influence of Al-reflected waves on the formation of the interference pattern. In the case of thick films (d > 5 μm), holographic recording takes place only due to interference of incident waves.

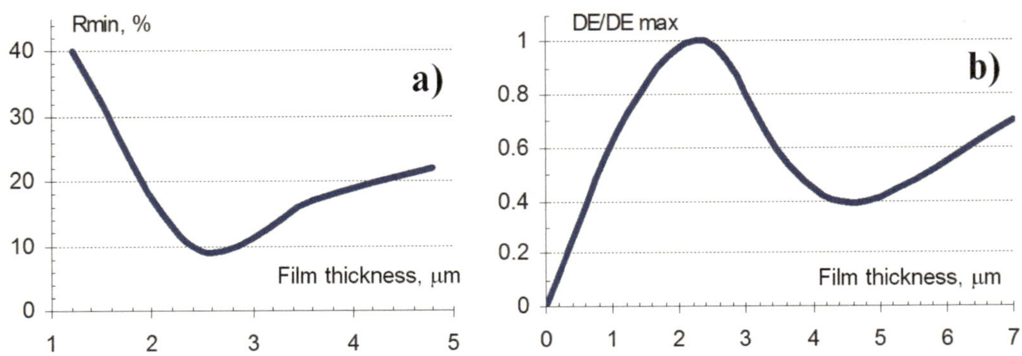

Fig. 20. **a)** Dependence of minimum reflection R_{min} of the system As_2S_3-Al on the film thickness (λ=532 nm). **b)** Dependence of the maximum diffraction efficiency of dot-matrix holograms on the thickness of films As_2S_3-Al

Fig. 21. Photos of the test dot-matrix holograms recorded on As_2S_3-Al films at different magnifications.

During holographic recording, local changes of optical properties of the recording medium, corresponding to the distribution of interference maxima, take place. Changes in local

absorption make it possible to observe the result of holographic recording in thin films. The interference pictures in As_2S_3-Al films of different thickness after the recording of dot-matrix holograms are shown in **Fig. 22a-e**. As seen in **Fig. 22c,d,** two systems of interference bands have equal periods and are mutually shifted. One system of bands corresponds to the interference of incident waves; the other one corresponds to the waves which are reflected from the aluminum surface. Interference maxima of the incident and reflected waves in the films of thickness 2.4 μm are situated very near to each other (see **Fig. 22c**). This case corresponds to the maximum diffraction efficiency of a holographic pixel recording. The optimal exposure for the recording of holograms in the As_2S_3-Al system is E=12-15 J/cm^2. This half that needed for the As_2S_3 films of the same thickness on substrates with absorbing layer of iron oxide. There are two factors significantly influencing the sensitivity of the As_2S_3-Al system during the recording of dot-matrix holograms.
The first is reduction of exposure dose due to superposition with matching of interference patterns for the incident and Al layer reflected waves. This happens when the film thickness is ~2.4 μm (λ=532 nm). The second factor is the possibility of production of As_2S_3-Al films with the minimum reflection coefficient in the part of spectra where the optical recording is made. The main problem in practical implementation of the As_2S_3-Al system in dot-matrix holography is that of depositing an As_2S_3 layer of high uniformity on large substrates.

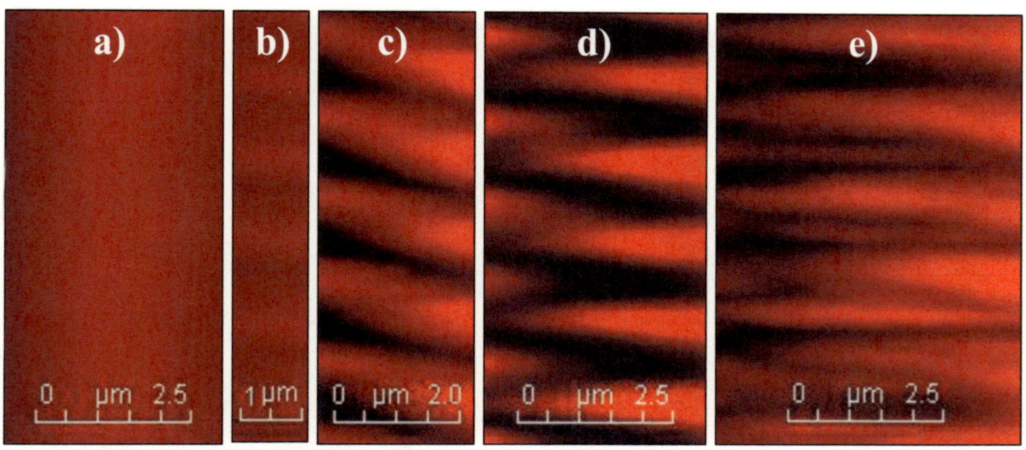

Fig. 22. Image of optical absorption in volume of As_2S_3-Al films after dot-matrix hologram recording (λ=532 nm, N=800 mm^{-1}, E=10 J/cm^2). The pictures were obtained while performing Z-scanning (λ=632 nm) by the confocal microscope LEICA TCS SP5. The left part of each picture corresponds to the layer Al, the right one to the film-air surface; **a)** without recording, **b)** film 1.2 μm, **c)** film 2.4 μm, **d)** film 3.6 μm, **e)** film 4.8 μm.

6. Calculation principles for dot-matrix holograms

The principle of dot-matrix holography is based on decomposition of a hologram image into a two-dimensional array of elementary pixels containing diffraction gratings with parameters that need mathematical calculation. Their size usually lies in the range of 10-100 μm and depends on the technology of optical recording that is used. In each elementary pixel of the hologram there is a diffraction grating with certain period *d*and orientation

angle φ relative to horizontal axis in the plane of hologram (**Fig 23a.b**). Size and shape of pixels may also change. The shape influences the fill factor of the whole hologram. For example, the fill factor of rectangular pixels can reach 100% and for round ones it is only 80%. In the former case, consequently, the diffraction efficiency of the hologram will be about 20% higher than in the latter. It is possible to modulate diffracted light intensity by changing the size of pixels or exposure time during optical recording.
The calculation task is to define the condition that a particular pixel of a hologram would direct a given spectral part of diffracted light towards the observer. The observer can see diffracted light from each element of a hologram only at a certain angle; and the total perception of diffraction of all elements makes a visual effect that corresponds to the initial graphic design. It is obvious that the main factors for reconstruction of the whole hologram are orientation and period of diffraction grating in each pixel, as well as position of an observer and the light source during the reconstruction. In these conditions, the holographic image will correspond to the initial graphic design. If the positions of an observer and the light source are defined, period d and orientation angle φ (Fig. 23a) of the grating completely determine the conditions when an incident white light can be diffracted with the given spectral region towards the observer. In this case, the color of the hologram element corresponds to the color of a dot in the initial graphic image.

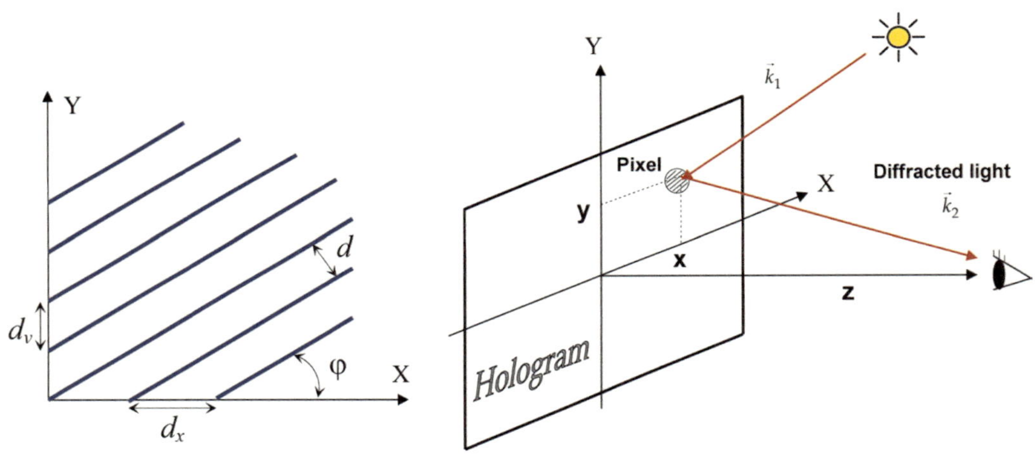

Fig. 23. **a)** Parameters of diffraction grating. **b)** An image element formation by diffraction pixel.

The coordinate system XYZ is chosen in such a way that the hologram is lying in the plane XY, and the Z axis directed to the observer from the center of the hologram (Fig. 23b). Assume that the observer's eye is on the Z axis at the distance of about 30 cm. The light source illuminating the hologram has a continuous spectrum in the visible region, and the direction of the incident radiation is given by the unit vector $\vec{k}_1$. We must define the parameters (d,φ) of the diffractive pixel with the coordinates (x,y), which the observer can see in the color with the wavelength λ_0. If the light with the wavelength λ falls on a flat diffraction grating along the unit vector $\vec{k}_1 = (\cos\alpha_1, \cos\beta_1, \cos\gamma_1)$ the direction of diffraction, which is defined by the unit vector $\vec{k}_2 = (\cos\alpha_2, \cos\beta_2, \cos\gamma_2)$, can be determined by the formulae (Yaotang et al., 1998):

$$\cos\alpha_2 = \cos\alpha_1 \pm \frac{\lambda}{d_x};\ \cos\beta_2 = \cos\beta_1 \pm \frac{\lambda}{d_y};\ \cos\gamma_2 = \sqrt{1 - \cos^2\alpha_2 - \cos^2\beta_2} \tag{1}$$

We will make calculations in order to form image in the +1 order of diffraction. We take the position of a point source with continuous spectrum of light in the plane YZ. If the angle between the Z axis and direction from the light source to the hologram center is θ, then $\vec{k}_1 = (0, -\sin\theta, \cos\theta)$, where the change of the Z-component of incedent light after reflection from the hologram surface is taken into account. The direction of the +1 diffraction order from the pixel with the coordinates (x, y) to the eye of the observer is defined by the unit vector $\vec{k}_2 = (-\frac{x}{L}, -\frac{y}{L}, \frac{z}{L})$, where $L = \sqrt{x^2 + y^2 + z^2}$ is the distance between the pixel and the eye of the observer. Taking into account the above mentioned conditions, the formulas in (1) will look like the following:

$$\cos\alpha_2 = -\frac{\lambda \sin\varphi}{d};\ \cos\beta_2 = -\sin\theta + \frac{\lambda\cos\varphi}{d};\ \cos\gamma_2 = \sqrt{1 - \cos^2\alpha_2 - \cos^2\beta_2}\ . \tag{2}$$

For the central pixel of the hologram with coordinates $(0, 0)$ and the color λ_0 based on the formulas in (2), we obtain the following parameters of the diffraction grating: $(d, \varphi) = (d_0, 0) = (\frac{\lambda_0}{\sin\theta}, 0)$, where $d_0 = \frac{\lambda_0}{\sin\theta}$.

Usually, in practical cases $\theta = 30^0 - 45^0$, for example, if $\theta = 30^0$, the desired interval of spatial frequencies of diffraction gratings for all colors in spectral range $\lambda_0 = 440 - 650\,nm$ will be $f = d_0^{-1} = 1135 - 650\,mm^{-1}$. At the next step, we can define the parameters (d, φ) of a pixel with the coordinates (x, y) having the same visual characteristics (color, brightness) for the observer as in the case of the central pixel.

This requires solving a system of equations for the variables d and φ:

$$\cos\alpha_2 = -\frac{\lambda_0 \sin\varphi}{d};\cos\beta_2 = -\sin\theta + \frac{\lambda_0\cos\varphi}{d};\cos\alpha_2 = -\frac{x}{L};\cos\beta_2 = -\frac{y}{L}. \tag{3}$$

From the equations in (3) we can find the formulas for calculating the parameters of diffraction gratings of the pixel:

$$tg\,\varphi = -\frac{\cos\alpha_2}{\cos\beta_2 + \sin\theta} \quad \Rightarrow \quad \varphi(x, y) = arctg\left\{\frac{x}{L\sin\theta - y}\right\} \tag{4}$$

$$d = -\frac{\lambda_0 \sin\varphi}{\cos\alpha_2} \quad \Rightarrow \quad d(x, y) = \frac{d_0\, L\sin\theta}{\sqrt{(L\sin\theta - y)^2 + x^2}} \tag{5}$$

The size of security holograms is usually not bigger than 3x3 cm and corresponds to the conditions $z >> x$ and $z >> y$; in this case, the following simplified formulas can be used for calculations:

$$d(x,y) \approx \frac{d_0}{1-\dfrac{y}{z\sin\theta}}, \text{ or the equivalent } f(x,y) = \frac{1}{d(x,y)} \approx f_0\left(1-\frac{y}{z\sin\theta}\right), \text{ where } f_0 = d_0^{-1} \quad (6)$$

and

$$\varphi(x,y) \approx \frac{x}{z\sin\theta - y}, \quad (7)$$

with the distance from hologram to observer, as mentioned above, $z \approx 30\,cm$.
In the extreme case, when $z \to \infty$ (or $x^2 + y^2 \to 0$) equations (6 and 7) become quite simple:

$$d(x,y) \approx d_0 \text{ and } \varphi(x,y) \approx 0 \quad (8)$$

Equations (8) can be used for holograms of up to 1.5x1.5 cm in size, with holograms of larger sizes demonstrating a noticeable color distortion and irregularity of the brightness of pixels in the hologram area.
In the photos presented in Fig.24, one can see two holograms of the same size recorded on the basis of the same original image, a uniform background of a certain color. In case of Fig.24a, calculation was made according to the formulas in (8), and in case of Fig.24b, taking into account the formula in (7). It is clearly seen that in the first case, the brightness of pixels in the direction of observation strongly depends on the position of pixels in the hologram.

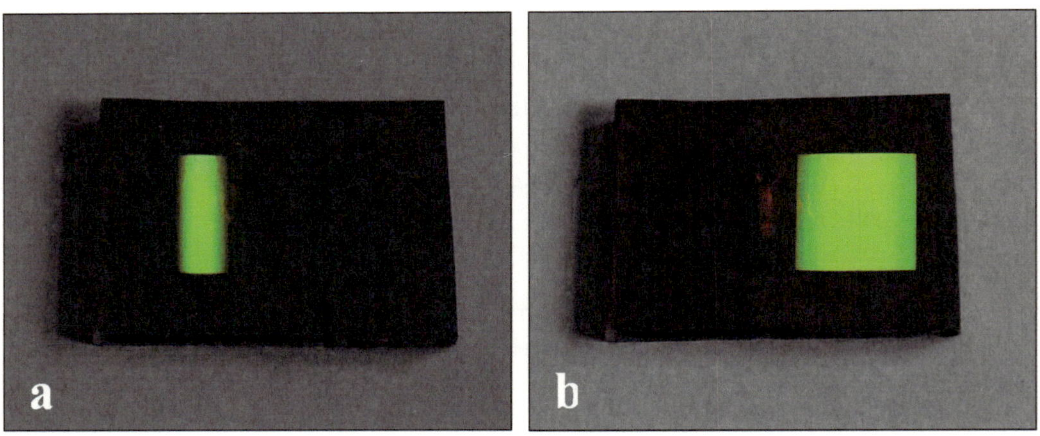

Fig. 24. Photos of a recovered image from a dot-matrix hologram recorded on $As_{40}S_{20}Se_{40}$ photoresist. a) Orientation angle of diffraction grating of all pixels $\varphi = 0$
b) Orientation angle of diffraction grating depending on the position of pixels $\varphi = K x$

Calculation of pixel parameters for stereograms or kinematic effects is similar to the considered case. It is only necessary to take into account a new position of observation of the hologram that can be achieved by replacing the old variables (x,y) in formulas (4)-(7) by the new ones $(x-\Delta x, y-\Delta y)$, where $(\Delta x, \Delta y)$ indicate displacement of the observer in the observation plane.

7. Conclusion

The developed and assembled optical devices for dot-matrix holographic and image-matrix recording have been successfully used for scientific purposes as well as for producing holograms for the protection and identification of industrial products and documents. Its compact dimensions, reliability and low cost price may be interesting for the needs of small and medium size business enterprises.

The possibilities of hologram recording on As-S-Se chalcogenide films have been studied. The obtained results show that the above mentioned chalcogenides may be successfully used in applied dot-matrix and image-matrix holography as an excellent alternative to organic photoresists for producing high-quality security holograms with high diffraction efficiency up to 65%. The increase in film sensitivity with increase of the exposure power density has been discovered. It makes the application of pulse recording attractive.

8. Acknowledgments

This research was partly supported by the ESF project "Starpdisciplinārās zinātniskās grupas izveidošana jaunu fluorescentu materiālu un metožu izstrādei un ieviešanai" Nr. 2009/0205/1DP/1.1.1.2.0/09/APIA/VIAA/152

9. References

Pizzanelly D. (2004). The development of direct-write digital holography. Technical review, Holographer.org.

Teteris J. (2002). Holographic recording in amorphous chalcogenide semiconductor thin films, JOAM Vol.4, No. 3, 687.

Teteris J. (2003). Holographic recording in amorphous chalcogenide thin films. Current Opinion in Solid State and Materials Science 7, 127.

Chih-Kung L., Wen-Jong W., Sheng-Lie Y., (2000) Optical configuration and color representation range of a variable-pitch dot matrix holographic printer. Applied Optics Vol.39, No.1, 40.

Yaotang.L, Tianji W., Shining Y., Shining Y., Shichao Z. (1998). Theoretical and experimental study of dot matrix hologram, Proc. SPIE Vol. 3569, 121.

Frumar M., Cernosek Z., Jedelsky J., Frumarova B., Wagner T. (2001). Photoinduced changes of structure and properties of amorphous binary and ternary chalcogenides, JOAM Vol.3, No. 2, 177.

Ozols A.,Reinfelde M., Nordman O. (2001). Photoinduced Anisotrophy and holographic recording in amorphous chalcogenides, Proc. of SPIE Vol.4415

Reinfelde M., Teteris J., Kuzmina I. (2003). Amorphous As-S-Se films for holographic recording. Proc. of SPIE Vol.5123, 128.

Kostyukevych S. Moskalenko N. (2001). Using non-organic resist based on As-S-Se chalcogenide glasses for combined optical/digital security devices, Semiconductor Physics, Quantum Electronics & Optoelectronics, Vol.4, N 1, 70.

Kohler C., Schwab X., (2006). Optimally tured sptial light modulators for digital holography, Applied Optics, Vol.45, No.5, p. 960-967

Gerbreders V., Teteris J., Sledevskis E. (2007). Photoinduced changes of optical reflectivity in As_2S_3-Al system, JOAM, Vol.9, No.10, 3153.

5

Azobenzene-Containing Materials for Hologram

Haifeng Yu and Takaomi Kobayashi
Top Runner Incubation Center for Academia-Industry Fusion and Department of Materials Science and Technology, Nagaoka University of Technology, Japan

1. Introduction

With the development of information technology, quick rate of data transfer and high capacity of data storage are expected for advanced recording media. Although data storage media and technique have been developed from soft discs, hard disk, CD-ROM and DVD-ROM to multi-layered blue-ray discs, undoubtedly, holography is one of the most fascinating and attractive techniques. As shown in Figure 1, which illustrates comparison of media in data storage and holographic ones. The optical holography is based on a three-dimensional storage method (Bieringer, 2000; Kogelnik, 1969). This provides unique opportunities for the next-generation storage technique by a simple recording and reading process.

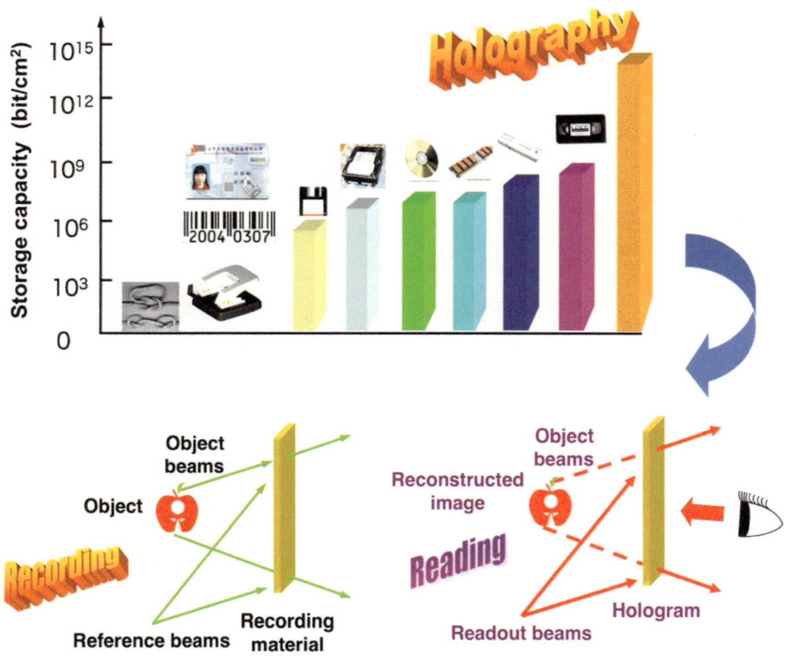

Fig. 1. Comparison of several data storage media (above) and scheme of holographic storage (below).

In 1948, Gabor first reported the holography method (Gabor, 1948), but this was not widely studied until the discovery of laser technique in 1960s (Kogelnik, 1969). The laser technology can easily supply two light beams with the same frequency and a stable phase difference for interference, which is necessary for recording information of one object (wavefronts, in Figure 2). Therefore, holography becomes to be a unique technique that enables simultaneous recording of both phases and amplitudes of light waves.

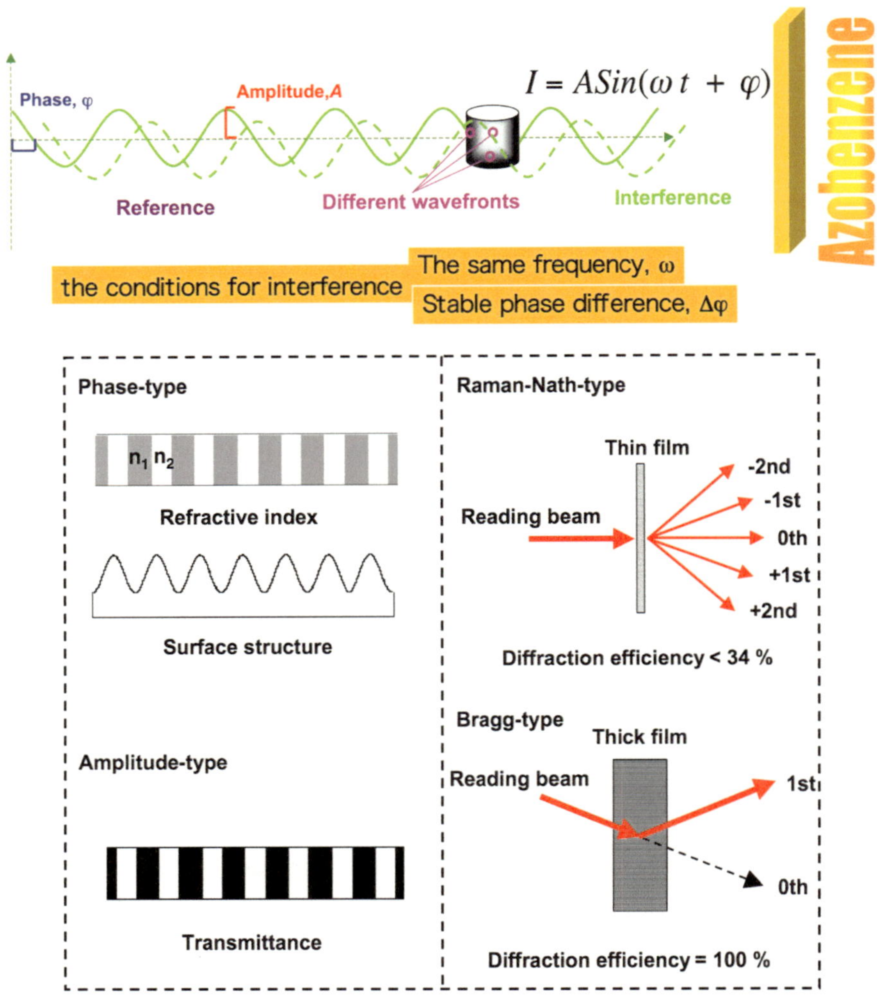

Fig. 2. Recording of one object (wavefronts) with interference of two coherent laser beams (above) and classification of holograms.

One of the most fascinating features of holography is that it is capable of recording and displaying a complete three-dimensional image of an object (a lot of wavefronts in Figure 2). In holography, recording the phase and amplitude of light waves are performed by periodic alternation of physical properties of materials. According to the manner of recording of interference patterns, holograms are mainly classified into two types (Collier, 1971; Smith, 1977). As shown in Figure 2, one is an amplitude-type hologram, in which the interference pattern is recorded as a density variation in recording media. The other is a phase-type hologram, in which fringe patterns are recorded as a change in surface structure or refractive index. Theoretically, the diffraction efficiency of phase-type holograms is always higher than that of amplitude-type ones. Accordingly, the phase-type holograms for data recording are superior to the latter and most studies on holography are related to the phase-type ones. On the other hand, as shown in the bottom illustration of Figure 2, according to the thickness of recording films, the holograms can be categorized into Raman-Nath type (thin films) and Bragg type (thick films). The Raman-Nath hologram recorded in a thin film causes multiple diffraction of an incident beam, leading to low diffraction efficiency of the first diffraction beam with the maximum of about 34%. The Bragg hologram shows a single diffraction, which enables 100% diffraction efficiency.

Generally, storage media materials are among the most important for holographic applications (Smith, 1977). In the past two decades, much attention has been focused on the inorganic crystals as erasable holographic media. While organic materials offer incomparable advantages, such as simple processing, low cost and versatility. Desirable materials for holograms should exhibit high diffraction efficiency, fast response, high resolution, stable and reversible storage, low-energy consuming in the recording and reading processes as well as easy mass production. Therefore, introduction of organic materials is needed. Until now, none of them meets entirely the above-mentioned requirement, and absence of novel materials with high performance has become bottleneck of the holographic technique.

As one of the best known chromophores, azobenzene (AZ) with two benzene rings connected with an azo (-N=N-) bridge in its chemical structures (Figure 3) has attracted much attention of materials chemists for their potential applications in holography. AZ-containing materials with photoresponsive functions can be easily modulated into hierarchical patterns by adjusting the input light properties - wavelength, intensity, polarization, phase, interference pattern, etc. As a result, holograms can be recorded in AZ-containing materials by photoinducing an orientation change of molecules in a periodic pattern obtained from interference of two coherent light beams, the object and the reference beams. Since photo-modulation of refractive index can be obtained in AZ-containing materials, the recorded hologram is phase-type. In this chapter, recent progress in AZ-containing liquid crystalline (LC) or amorphous polymers, polymer-dispersed low-molecular weight compounds or glassy oligomers will be discussed.

2. AZ-containing materials

An AZ moiety is well-known for its reversible photoisomerization, which may act as both a mesogen and a photoresponsive moiety when it is attached with soft substituents. Both photoisomerization and photoinduced LC-to-isotropic phase transition are involved in AZ-containing LC materials. A myriad of AZ derivatives can be tailored by modifying the substituents in the two benzene rings, resulting in their maximum absorption from the ultraviolet to visible regions. According to the electron-absorption properties, three kinds of AZ chromophores have been summarized as shown in Figure 3 (Natansohn & Rochon, 2002; Kumar & Neckers, 1989). The left side "AZ" has relatively poor π-π^* and n-π^* absorbance overlap and the lifetime of the cis-isomer is relatively long. The middle one is "amino-AZ", and there is significant overlap of the two bands and the cis-isomer lifetime is shorter. The right side AZ is "pseudostilbene", where the AZ is usually substituted with electron-donor and electron-acceptor substituents in both ends of the framework.

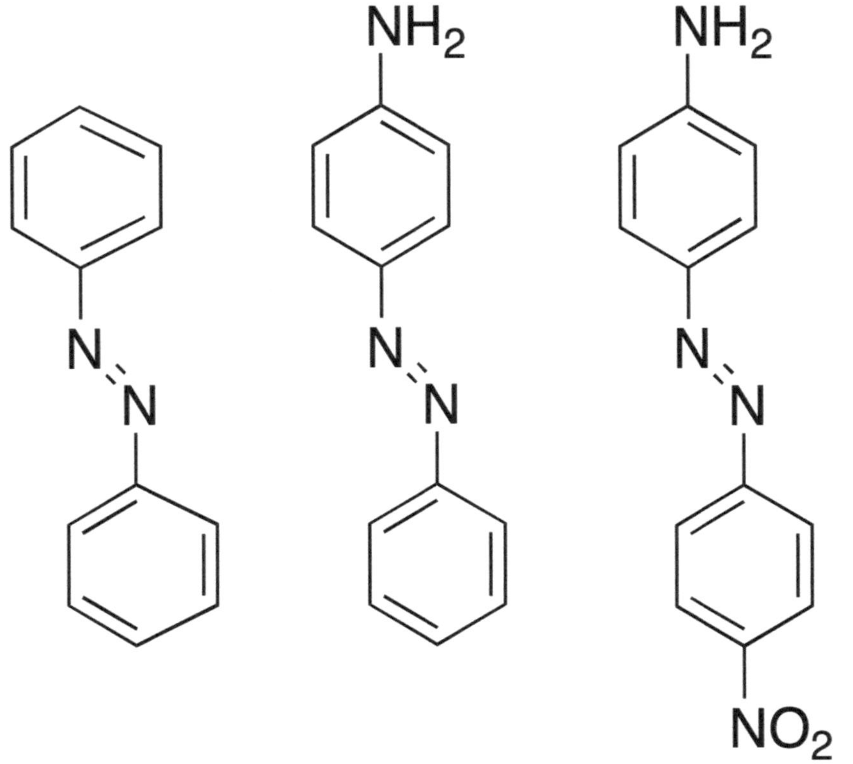

AZ amino-AZ pseudustilbene

Fig. 3. Examples of three kinds of AZs classified by Natansohn and Kumar.

Generally, AZ-containing materials show several photoresponsive features like trans-to-cis photoisomerization, polarization-selective photoalignment, photochemical phase transition and photoinduced cooperative motions, as shown in Figure 4 (Yu & Kobayashi, 2010; Chen et al., 2010a). Upon UV irradiation, trans-to-cis photoisomerization often occurs for AZ-containing materials in solutions or solid states, leading to a large change in molecular shape and polarizability. The cis-isomers can return to their trans-isomers by thermal treatment or visible light irradiation. When a linearly polarized light is used, AZs selectively absorb light with the polarization direction parallel to their transition moments. The probability of the absorption is proportional to the $\cos^2 \theta$ (Figure 4), where θ is the angle between the transition moment of an AZ and the light polarization direction. Combining the polarization-selective trans-to-cis isomerization and un-selective back cis-to-trans isomerization, the number of AZ moieties with their transition moments normal to the light polarization direction gradually increases, resulting in the light-selective alignment, with transition moments of AZs almost perpendicular to the polarization direction of the actinic light. This is well known as the Weigert effect (Weigert, 1919). Such a photoalignment process is reversible, which means that AZs can be re-orientated to any controlled directions by choosing appropriate polarization direction of light (Yu & Ikeda, 2011).

When AZ-containing materials possess an LC phase, photoinduced phase transition can be caused in the LC phase since the trans-AZ can be a mesogen because its molecular shape is rod-like, whereas the cis-AZ never shows any LC phase due to its bent shape. When photoinert mesogens (like cyanobiphenyl groups) exist with dopant AZs, photoinduced molecular cooperative motion can be observed (Ikeda, 2003). If a small proportion of AZ molecules change their alignment in response to an external light stimulus, the other LC molecules also alter their alignment, coinciding with the ordered AZs pre-aligned. Actually, only a small amount of energy as to induce an alignment change of about 1 mole% of LC molecules is enough to bring about the alignment change of the whole system of AZ and LC mixtures. In other words, a huge amplification of light signals is possible in LC photonic systems owing to the molecular cooperative motion. As shown in Figure 4, both photoalignment and photochemical phase transition can be isothermally caused although the photoinert mesogens do not absorb the actinic light. These properties have been widely utilized in display, optical devices and holograms.

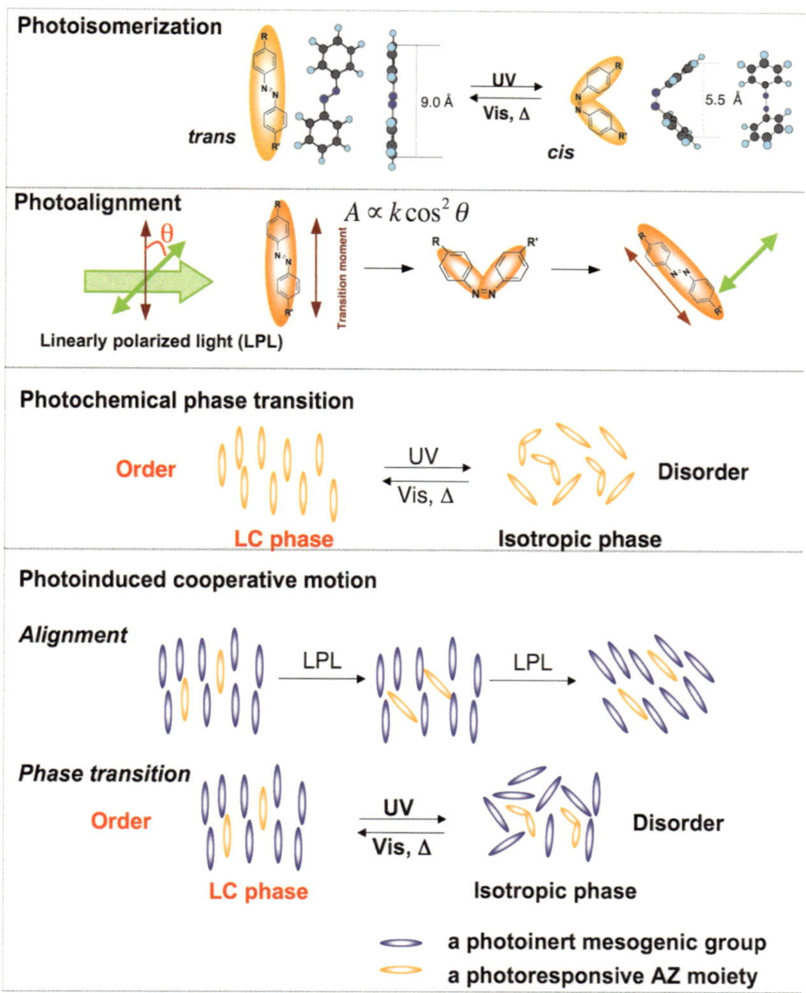

Fig. 4. Summarized properties of AZ-containing materials.

3. Low-molecular-weight AZs

Because of easy preparation and high purity of low-molecular-weight compounds, the application of AZ materials for recording holograms began with them. Formation of holographic gratings in films of poly(vinyl alcohol) doped with low-molecular-weight AZ compounds was first reported by utilizing photoisomerization of AZs (Todorov et al., 1983, 1984, 1985). The holographic gratings recorded in the polymer-based samples were self-developing and auto-erasing or re-writable since the thermally induced back isomerization of AZs could be easily obtained (Pham et al., 1997). To increase the formed birefringence $|n_2-n_1|$ shown in Figure 2 and enhance the performance of recorded holograms, polymer-dispersed AZ-containing low-molecular-weight LC compounds were also employed as recording media (Tondiglia et al., 1995). Due to the existence of low-molecular-weight LCs, the diffraction efficiency of the gratings could be electrically controlled (Sutherland et al., 1994). Moreover, dynamic holography was achieved in AZ-doped nematic LCs (Chen & Brady, 1993).

Although many research focused on the low-molecular-weight AZ–doped polymers, the stability of the induced anisotropy was low. To improve stability of the recorded gratings, molecular glasses fabricated with low-molecular-weight AZ compounds were developed (Shirota et al., 1998). This kind of materials can form stable amorphous and homogenous films without scattering of visible light (Nakano et al., 2002). Surface-relief gratings were successfully inscribed although the AZ derivatives were confined in the glassy sates. More interestingly, low-molecular-weight AZ crystals and co-crystals also showed capability of formation of surface-relief gratings (Nakano et al., 2005). It was believed that photoreaction only occurred in the surface.

For holograms recorded in low-molecular-weight AZ materials, gain effect often occurs. It was reported that diffraction efficiency of gratings recorded in calamitic LC materials with low-molecular-weight compounds was increased sharply after annealing (Stracke et al., 1999). In recording holograms using pentaalkinylbenzene derivatives, a strong gain effect was obtained by the selective growth of crystals in the non-irradiated areas (Frese et al., 2003). For smectic trisazomelamine, the gain effect was acquired accompanied by the growth of surface modulation. At elevated temperatures, the free volume available to mesogens increased and the thermal movement led to a collective orientation along the direction given by those groups pre-aligned by the recording beams. This caused an increase of the order degree in the pre-orientated domain, which contributed to the increase of diffraction efficiency of holograms.

4. AZ-containing polymers

4.1 AZ-containing amorphous polymers

Low-molecular-weight AZ materials are superior in terms of mobility. However, it is often difficult to prepare holographic gratings with narrow fringe spacing (i.e., high resolution) and high stability because of the high mobility of AZ molecules with low molecular weight. From the application point of view, AZ-containing polymer materials might be one of the best choices for holograms because of their simple processability and high stability. On the other hand, the ordered structures or photoinduced alignment of chromophores can be easily fixed in polymer matrix.

In 1995, two groups independently reported surface-relief gratings inscribed in AZ-containing polymer materials almost at the same time (Rochon et al., 1995 ; Kim et al., 1995).

In films of poly[4'-(2-acryloxy)ethylamino-4-nitroazobenzene] (pDR1A, Figure 5a), surface-relief gratings with a sinusoidal shape was recorded with an interference pattern of two light beams at 514 nm. The obtained grating structures were stable but could be erased by heating the polymer above its glass transition temperature. In addition, no permanent damage of the film was observed. Multiple gratings can be simultaneously written and gratings can be overwritten (Rochon et al., 1995). Using an epoxy-based amorphous polymer containing AZ side groups (Figure 5b), surface-relief gratings with relatively large amplitude was successfully inscribed with two laser beams at 488 nm. Furthermore, recording perpendicular gratings on the same film was also achieved. Such surface-relief gratings in amorphous polymer films showed uniform and controllable morphologies, like the depth of relief, the grating periodicity, and so son. Moreover, more complicated topological surfaces were tailored by superimposing several surface-relief gratings. Recently, the recorded surface-relief structures were studied for potential applications as LC alignment (Li et al., 1999), polarization discriminator, waveguide couplers (Viswanathan et al., 1999) and antireflective coatings (Natansohn & Rochon, 2002).

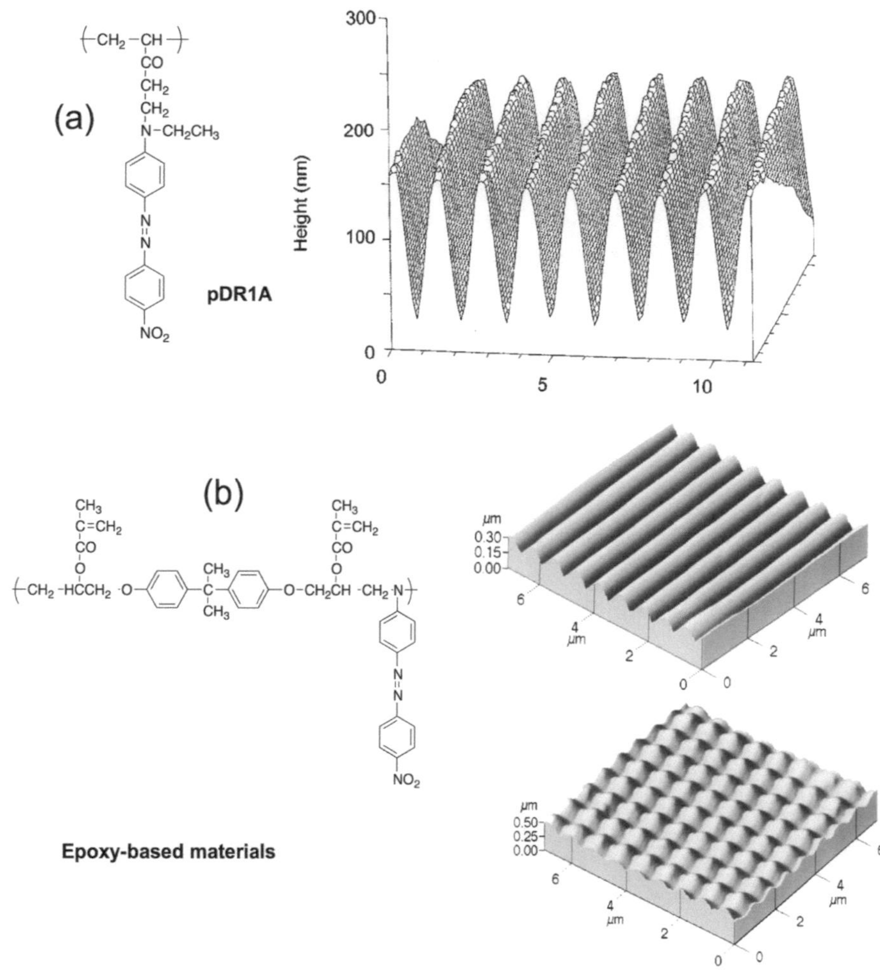

Fig. 5. Surface-relief gratings recorded in AZ-containing amorphous polymer materials.

4.2 AZ-containing LC polymers

Comparing with amorphous materials, LC materials possess unique features like self-organization, fluidity with long-range order, molecular cooperative motion, formed large birefringence and anisotropy in various physical properties (optical, mechanical, electrical and magnetic), and alignment change induced by external fields at surfaces and interfaces (Ikeda, 2003; Yu & Kobayashi, 2010). Therefore, LC materials can bring about a larger change in refractive index relative to amorphous ones. An alignment change of LC materials in a periodic fashion can be easily induced upon irradiation with interference patterns by overlapping two coherent beams, resulting in a large refractive-index modulation. This can contribute to a high diffraction efficiency of recorded gratings. In films of AZ-containing LC polymers, both surface-relief and refractive-index gratings can be recorded correspondingly.

Wendorff et al. showed that holographic gratings could be inscribed in LC polymers composed of AZ moieties and photoinert mesogenic groups (Eich et al., 1987; Eich & Wendorff, 1987; Anderle & Wendorff, 1994). Hvilsted et al. achieved holographic gratings with diffraction efficiencies of about 40% in AZ-containing LC polyesters (Hvilsted et al., 1995). Recently, formation of holographic gratings in side-chain LC polymers containing AZ moieties by means of photochemical phase transition was intensively studied. Hasegawa et al. achieved dynamic holographic gratings by means of photochemical nematic-to-isotropic phase transition in LC copolymers containing an AZ moiety with strong donor-acceptor substituents in side chain (pseudostilbene-type AZs in Figure 3). As shown in Figure 6, the inscription and removal of holographic gratings with a narrow fringe spacing of 1.4 μm was obtained within 150 and 190 ms, respectively (Hasegawa et al., 1999a , 1999b). Moreover, the optical switching behaviors of the holographic diffraction were observed repeatedly by turning the writing beams on and off. Using a siloxane group as a spacer in preparation of LC polymers decreased the glass transition temperature of the designed materials. This was because of the flexibility of the siloxane unit, which resulted in an effectively photoinduced nematic-to-isotropic phase transition at room temperature and formation of real-time holographic gratings (Hasegawa et al., 1999c).

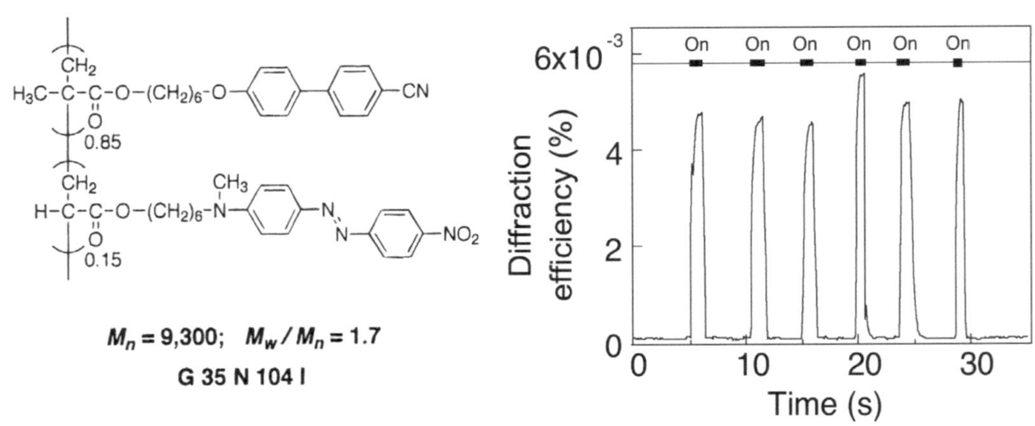

Fig. 6. Dynamic holographic gratings by means of photochemical phase transition in AZ-containing LC polymers.

In a nematic LC polymer with AZs as side mesogens, both surface relief and refractive index modulations were recorded, as shown in Figure 7. In nematic LC phase, the obtained gratings exhibited a higher diffraction efficiency than that recorded in the glassy state even though the amplitude of surface-relief structures of the former was lower than that of the latter one (Yamamoto et al., 1999). These indicated that the contribution of refractive-index gratings to the diffraction efficiency was larger than that of surface-relief gratings. Based on the same materials, holographic image storage using a photomask as an object was successfully obtained (Yamamoto et al., 2000).

To enhance the refractive-index modulation, both cyanobiphenyl and tolane groups were used as photoinert moieties to prepare LC copolymers with AZs as photoresponsive mesogens (Yoneyama et al., 2001, 2002). However, the photoresponse was not high enough due to a low content of AZ moieties in the copolymers. To improve the photosensitivity and induce a larger birefringence for holographic applications, a concept of molecular architecture of AZ-containing LC materials was proposed in Figure 8 (Okano et al., 2005). A tolane group was directly attached onto the 4- or 4'- position of AZ molecules to prepare an azotolane mesogen, which shows far longer molecular conjugation length than one single tolane group or an AZ moiety (Okano et al., 2005). Figure 8 lists the molecular structures and LC phase behaviors of a series of azotolane nematic LC polymers synthesized based on this principle, in which three, four and five benzene rings are included in the mesogens, respectively. All the prepared azotolane-containing LCPs showed a wide nematic LC range and exhibited a phase transition temperature higher than 200 °C (Okano et al., 2006). Obviously, the photoresponse of the designed LC polymer with high birefringence was greatly enhanced because of high density of AZ mesogens, resulting in formation of holographic grating in thin films within short time under low-intensity irradiation of recording beams.

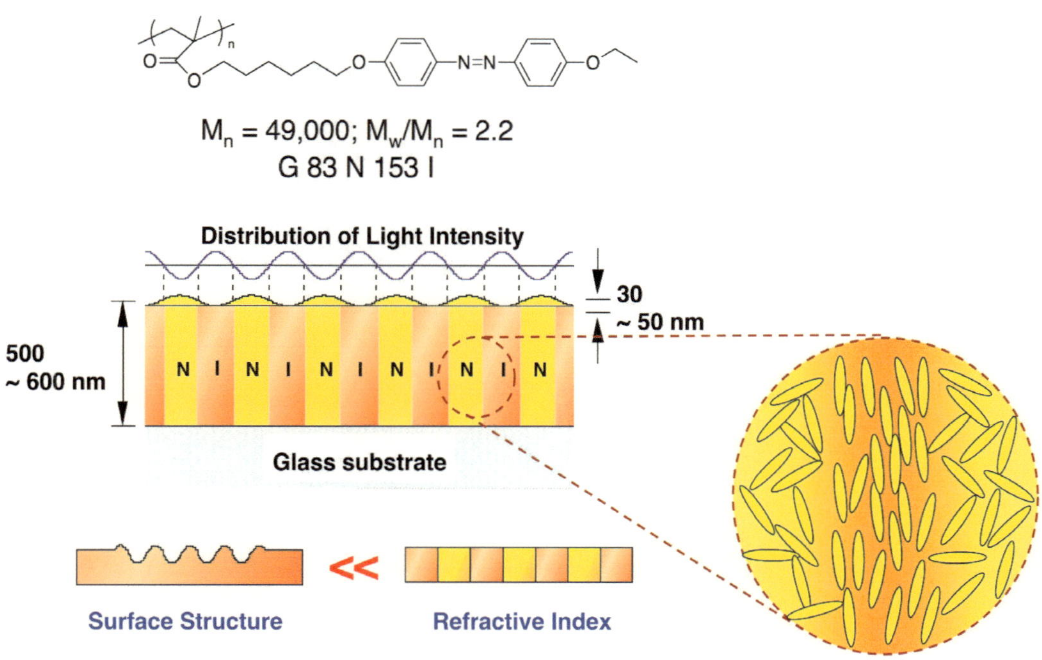

Fig. 7. Holographic gratings recorded in nematic LC polymer with AZs as side mesogens.

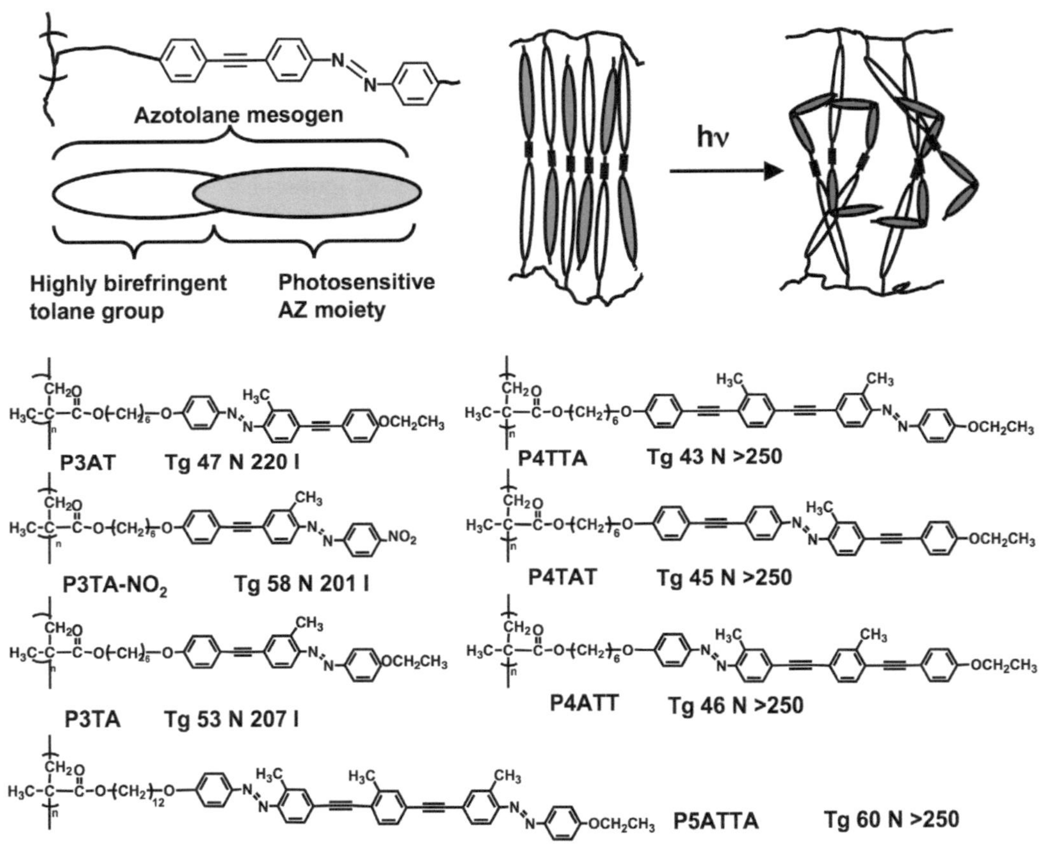

Fig. 8. Azotolane LC polymers with high birefringence for holograms.

Using two phase-type gratings recorded in LC cells (Figure 9), grating waveguide couplers with a flat surface were fabricated as shown in Figure 9 (Bang et al., 2007). When a probe beam at 633 nm was incident to one grating, the beam propagated in the waveguide and an output beam came out from the other grating with the throughput coupling efficiency of about 5%. Upon irradiation of the film between the two gratings with UV light to cause trans–cis photoisomerization and order-disorder transition of the AZ moiety, the intensity of the output beam was repeatedly switched upon alternating irradiation of visible light. It was found that the alternating irradiation at 366 and 436 nm induced reversible changes in the intensity of the output beam.

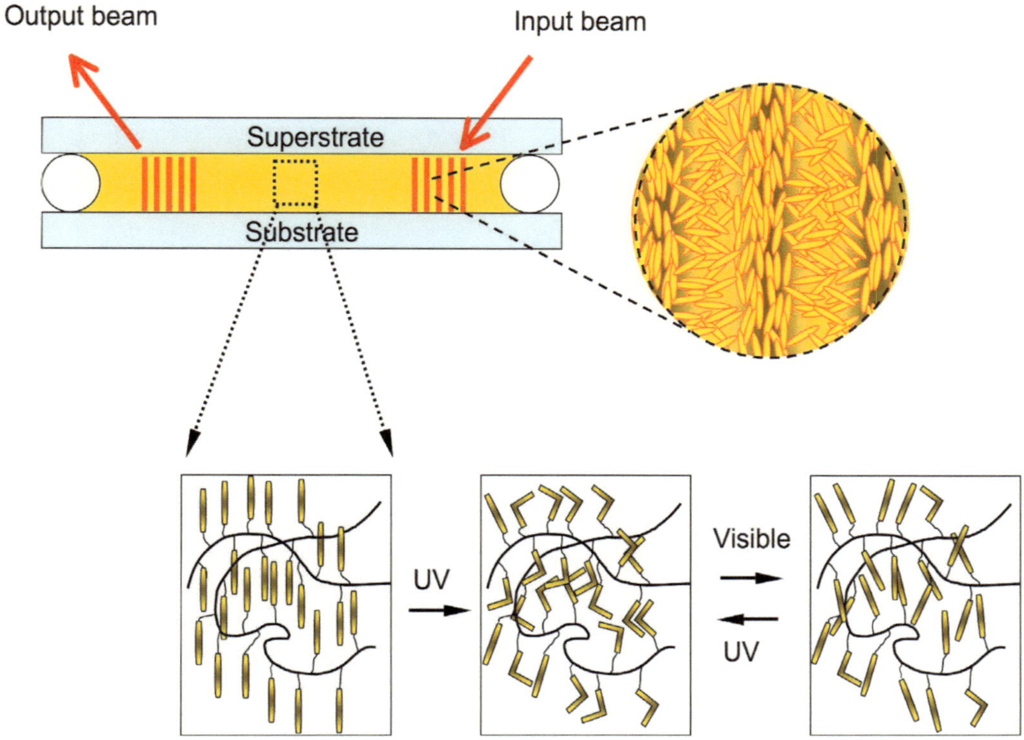

Fig. 9. Periodically flat-structured grating waveguide couplers and photo-switching behaviors.

Due to molecular cooperative motion of mesogens in LC polymers, it is difficult to record holographic gratings with a narrow periodicity in a subwavelength scale (Yu et al., 2008a). Pre-treatment with UV irradiation to induce cis-azobenzene-rich isotropic phase can illiminate such molecular cooperative effect of mesogens and enhance the photoresponse in an a lowly viscous state. These enabled us to obtain the subwavelength modulation of surface relief and refractive index with interference patterns in Figure 10 (Yu et al., 2008a). The surface relief of less than 10 nm and the refractive-index modulation were detected by atomic force microscopy in tapping and phase modes, respectively. A large phase retardation and formed birefringence were observed in the recorded subwavelength gratings.

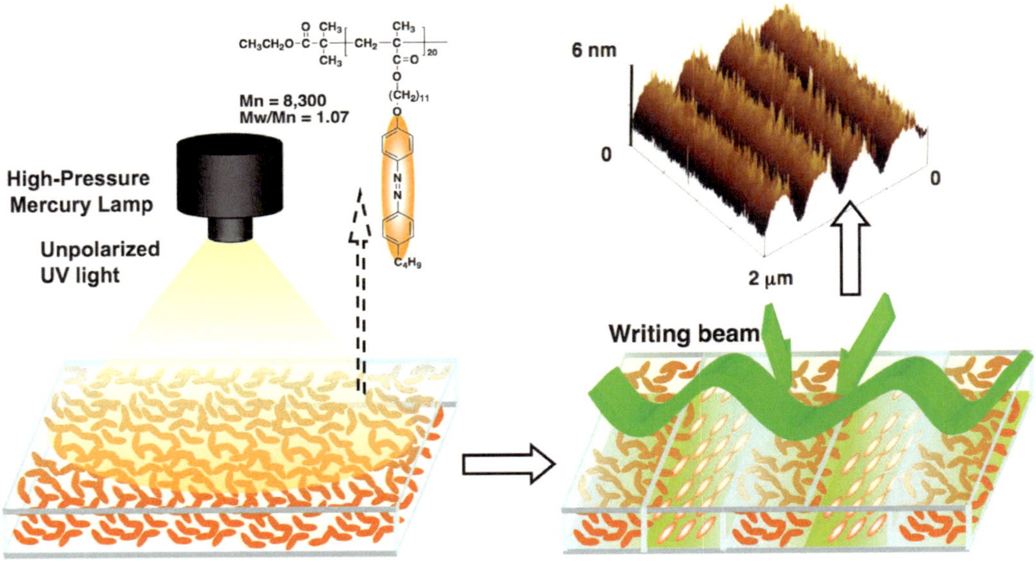

Fig. 10. Subwavelength modulation of surface relief and refractive index in LC polymer films.

4.3 AZ-containing LC block polymers with well-defined structures

In LC block copolymers with well-defined structures, microphase separation often occurs because of the inherent immiscibility of the different blocks, resulted in diverse of nanostructures like nanospheres and nanocylinders (Yu et al., 2006a, 2006b, 2007d, 2007e, 2009b, 2011; Chen et al., 2010b). Generally, the phase-segregated nano domains are far smaller than the wavelength of the visible light (Yu et al., 2007b, 2007c), making it possible to get rid of the scattering effect. Such kinds of AZ-containing LC block polymers with nanoscale phase separation have been regarded as one of ideal materials candidates for holograms.

Generally, the diffraction efficiency is one of the most important parameters for holographic gratings. In amorphous polymer materials, a surface-relief grating contributes mainly to diffraction efficiency. Recently, microphase separation of AZ-containing LC block copolymers was used to control diffraction efficiency by enhancement of surface relief upon thermal annealing (Yu et al., 2005a, 2009a; Naka et al., 2009). As shown in Figure 11, both surface-relief and refractive-index gratings were recorded upon irradiation of an interference pattern, in which selective photoisomerization and the isotropic-to-LC phase transition were induced in the bright areas of pre-treated films. The diffraction efficiency of the gratings depended strongly on the polarization of the reading beam because of the photoalignment of mesogens in the bright area of the writing patterns. After grating formation, the surface-relief structures with a sinusoidal shape were clearly observed in the AFM images (Figure 11). The fringe spacing of the surface relief was 2.0 μm, which was identical to that of the refractive-index gratings. Then nanoscaled microphase separation was proceeded by annealing the grating sample. As a result, the surface relief was increased to about 110 nm (18.3 % of the film thickness), almost one order of magnitude larger than that before annealing. The peak-to-valley contrast became more explicit after annealing, due to the enhancement of the surface modulation. Furthermore, the sinusoidal shape of the surface profile became a little irregular, indicating that the LC alignment was disturbed upon microphase separation. Together with the enhancement of surface relief, the diffraction efficiency increased to about 9.0 %, almost two orders of magnitude larger than the diffraction efficiency before annealing. This increased diffraction efficiency might be ascribed mainly to the enhancement of surface modulation. Furthermore, multi-processes of the refractive-index gratings were successfully achieved as presented in Figure 11. The obtained grating structures were clearly observed in the polarizing optical microscopic pictures, which were also verified by their diffraction patterns (Yu et al., 2007a).

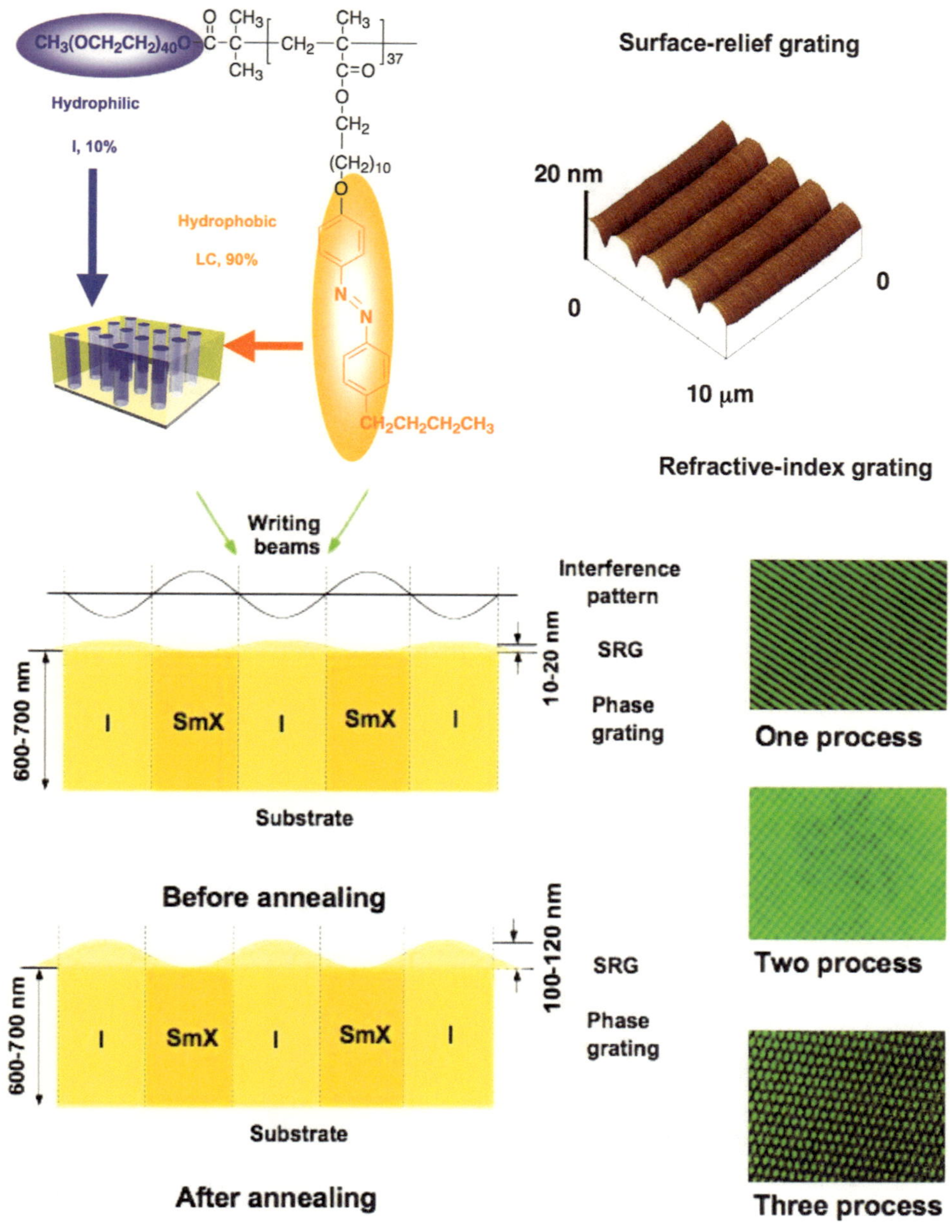

Fig. 11. Holographic gratings recorded in AZ-containing LC block copolymers with well-defined structures and enhancement of gratings upon microphase separation.

Comparing to other methods to control diffraction efficiency, such as gain effects, mechanical stretch, electrical switch, self-assembly, mixture with LC and cross-linking, the microphase-separation method had advantages of being simple and convenient (Yu et al., 2005a, 2009a). To precisely control diffraction efficiency of recorded gratings in the amphiphilic block copolymers, the effect of recording time on grating formation and enhancement was systematically studied. The best enhancement effect was obtained at 10 s recording upon microphase separation. By adjusting the recording time, the diffraction efficiency was finely controlled from 0.13% to about 10% (Yu et al., 2009a).

As described in section **2**, a small external stimulus can induce a large change in refractive index of the materials by the cooperative effect between photoresponsive AZ moieties and photoinert groups. These propertied have been widely used in holographic recording, which is especially useful in AZ-containing block copolymers with mesogens in the minority phase dispersed in glassy substrates (Yu et al., 2007b). It was reported that the photoinduced mass transfer was greatly prohibited due to the microphase separation in grating recording, lack of surface-relief structures was observed (Breiner et al., 2007). Thus, refractive-index modulation plays an important role in the grating formation in such well-defined block copolymers.

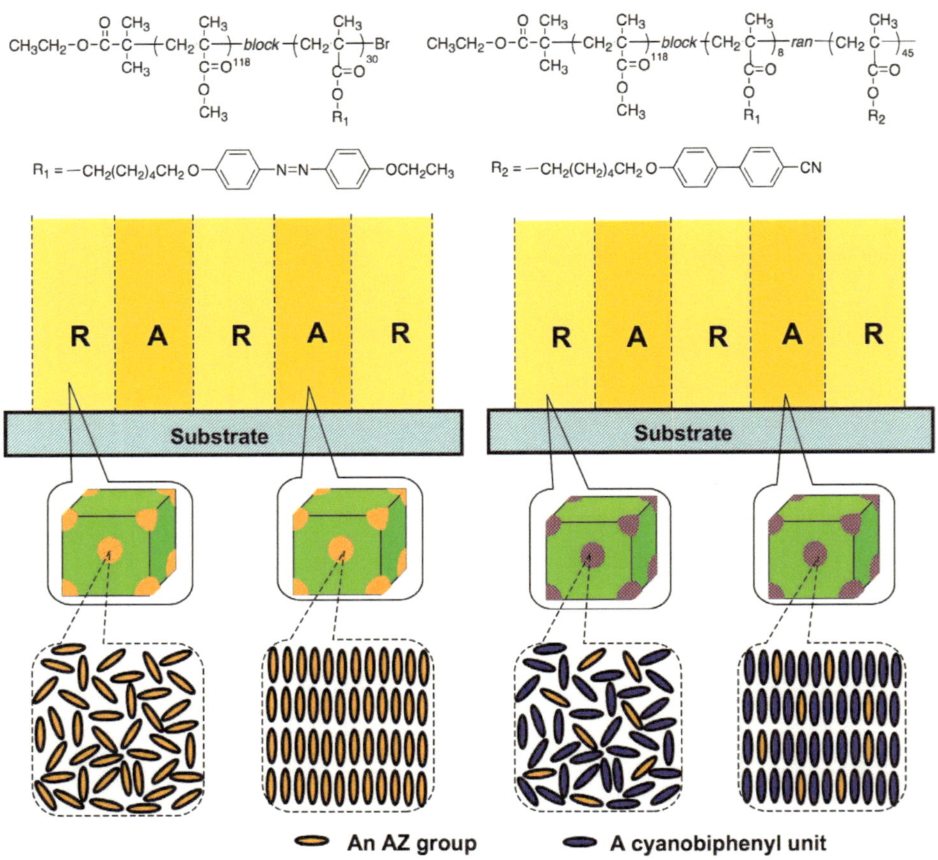

Fig. 12. Enhancement of surface-index modulation and prohibition of surface-relief gratings in block copolymers with photoresponsive AZ groups in the minority phase. (A, aligned, R, random).

As shown in Figure 12, holographic gratings were recorded in films of two polymethyl methacrylate (PMMA)-based block copolymers containing AZ moieties (Yu et al, 2008b). One was a well-defined diblock copolymer, and the other sample was a diblock random copolymer. Here, the diblock random copolymer consisted of two blocks, in which one segment was PMMA and the other mesogenic block was statistically random distributed. After grating formation, both films showed no formation of surface-relief gratings, and only refractive-index gratings were obtained. Upon irradiation of two coherent laser beams, refractive-index gratings in the diblock copolymer containing AZs were recorded by photoalignment of AZs dispersed in phase-separated domains. In contrast, the photoalignment of the AZ moieties was amplified by the photoinert cyanobiphenyl moieties as a result of the cooperative effect in the diblock random copolymer. This led to a similar refractive-index modulation, although the AZ content was lower in the diblock random copolymer. The cooperative motion was confined within the nanoscale phase domains, unlike the case of random copolymers with statistically molecular structures.

Being one of commercially available products, an ABA-type triblock copolymer, polystyrene (PS)-*b*-polybutadiene-*b*-PS, is famous for its thermoplastics. The hard block of PS with a content of 20-30 wt% forms the minority phase upon microphase separation, which acts as physical crosslinks for the majority phase of the soft block of rubbery polybutadiene (PB). Mechanical stretching can induce a large elastic deformation with recoverable properties. By applying this concept, block copolymers with thermoplastics were prepared (Bai & Zhao, 2001, 2002; Zhao et al., 2002). Upon stretching-induced elastic deformation of grating samples recorded in the thermoplastic block copolymers, fringe spacing or grating periodicity was successfully adjusted, as shown in Figure 13.

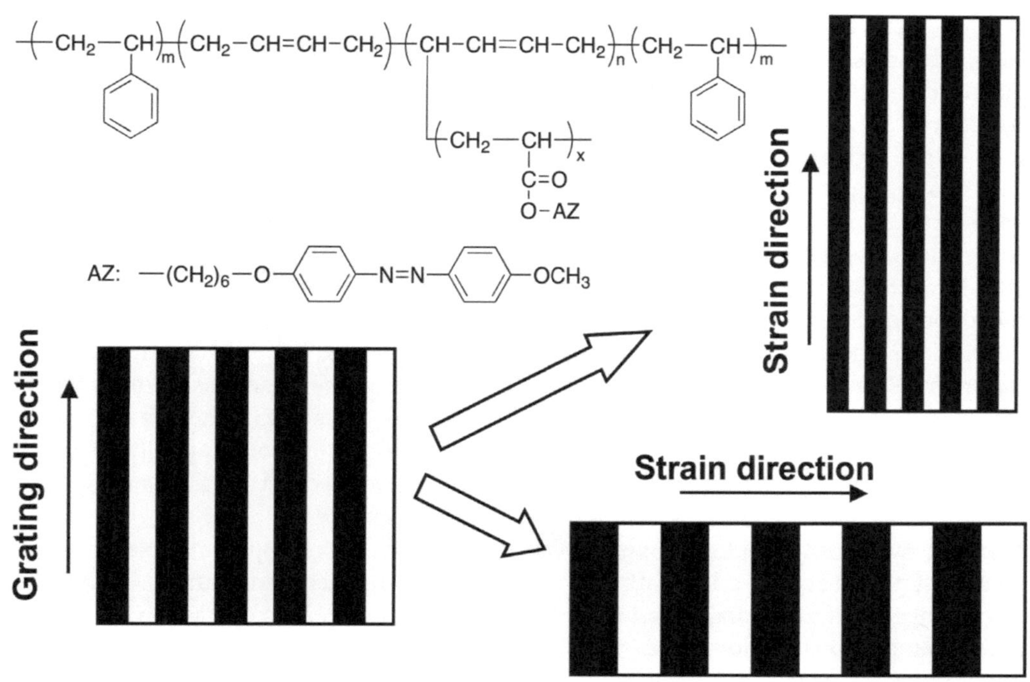

Fig. 13. Mechanically tunable fringe spacing of gratings recorded in an AZ-containing block copolymers with thermoplastics.

Generally, the fringe spacing could be decided by the pattern of the used photomask, when the grating was recorded with one writing beam. On the other hand, holographic gratings were also inscribed in block copolymer films by two coherent laser beams with an equal intensity. The recorded fringe spacing can be evaluated by $\Lambda = \lambda w/(2\sin\theta)$, where λw and θ are the wavelength and the incident angle of the writing laser beams, respectively. Once the writing beams are obtained, the fringe spacing is fixed. Tunable features of the fringe spacing were achieved in films of the thermoplastic block copolymers (Figure 13). When the strain direction was parallel to the grating direction, the fringe spacing was decreased. On the contrary, the fringe spacing was increased when the strain direction was perpendicular to the grating direction. By the stretching, diffraction efficiency of the gratings was adjusted accordingly (Zhao et al., 2002). Recently, mechanically tunable fringe spacing was also obtained in gratings recorded with ABA–type triblock copolymers showing properties of conventional thermoplastic elastomers, in which rubbery poly(n–butyl acrylate) was designed as middle soft block and photoresponsive polymers acted as hard block (Cui et al., 2004).

5. Novel AZ-containing materials

Recent progress in chemistry and material science enables one to freely design functional materials with suitable processes to satisfy the need of advanced materials for a variety of applications. Integration of AZ materials with other functionalized materials and processing provides the designed materials for holograms with special features. For instance, graphene nanosheet grafted with AZ-containing polymer brushes was prepared via a "grafting-from" approach, which was used as dopant for an AZ-containing molecular glass in recording gratings (Wang et al., 2011). The diffraction efficiency of the inscribed gratins was enhanced due to the mass transfer of the graphene nanosheets showing a high refractive index. Chemically crosslinking of AZ-containing polymer films with surface-relief gratins could fix the obtained surface modulation, producing permanent shape change (Zettsu et al, 2008).

As shown in Figure 2, the Raman-Nath hologram recorded in a thin film exhibited a theoretically maximum diffraction efficiency of about 34%. This was far lower than the Bragg-type hologram in a thick film with a maximum diffraction efficiency of 100%. Furthermore, angular multiplicity could be easily obtained in the Bragg hologram, enabling thick films to be suitable for volume storage. To prepare thick and transparent films, amorphous and highly transparent PS blended with their block copolymers containing an AZ block as the minority phase to eliminate the scattering by utilizing the nanoscale microphase separation (Häckel et al., 2005 , 2007). Since all of the mesogens were confined in nanospheres dispersed in PS matrix, thick and optically transparent films were obtained by injection-molded method. These films showed good angle-multiplexing capability, in which 200 holograms were superimposed and reconstructed independently at the same spot and more than 1000 write/erase cycles were successfully obtained.

However, the density of photoresponsive mesogens in the block copolymer blend systems was very low, leading to a low diffraction efficiency of the recorded gratings. To induce a large change in refractive index and record Bragg gratings with high diffraction efficiency, a series of amorphous random copolymers with both AZ moieties and photoinert mesogens was prepared, as shown in Figure 14 (Saishoji et al., 2007 ; Ishiguro et al., 2007). The highly transparent PMMA was utilized as substrate, when $x > 50$ mol% in the materials preparation. Cyanobiphenyl and tolane groups were selected as the photoinert part to enhance the refractive-index modulation. AZ mesogens with a low content to the Z = about 5 mol%, which acted as photoresponsive moieties. Transparent and thick films (> 100 μm) were fabricated using a melting and pressing process.

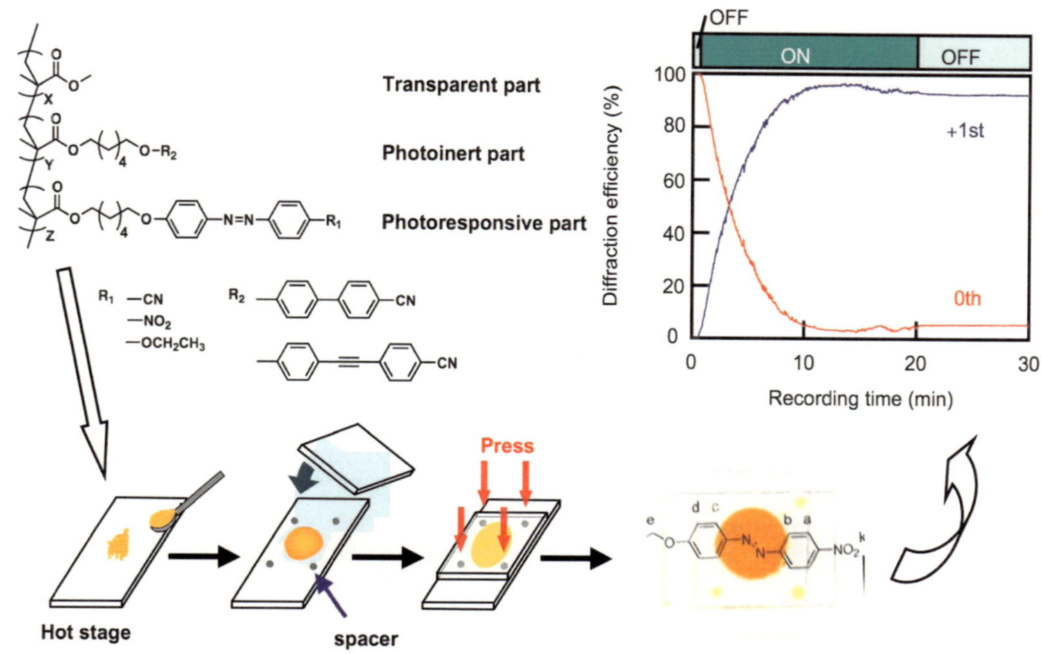

Fig. 14. Transparent and thick films of AZ-containing copolymers for Bragg gratings.

Upon irradiation with the writing beams, the first-order diffracted beam (+1st) appeared immediately and the intensity of the zeroth-order beam (0th) decreased (Figure 14). The maximum diffraction efficiency reached above 98 % in all the prepared films. Furthermore, the polarization-selective multiple holographic data storage could be obtained using the photoinduced anisotropy as well as rewritable holographic recording with about 100% diffraction efficiency. These materials prepared by a simple but balanced formulation would provide a new guideline for the construction of high-performance holographic devices.

6. Conclusion

Recent advance in AZ-containing materials has enabled one to record both Raman-Nath and Bragg holograms. Among these materials, low-molecular-weight compounds and amorphous polymers showed good capability of surface-relief grating formation. In LC polymers, refractive-index gratings are often recorded, accompanying with surface-relief modulation. Block copolymers with well-defined structures can eliminate the scattering of visible light by microphase separation and prohibit surface deformation when AZ blocks forms the minority phases. Furthermore, thick films (> 200 microns) with good optical transparency can be prepared with random copolymers or blended block copolymers, providing substrates for recording volume holograms.

7. References

Anderle, K.; Wendorff, J. H. (1994). Holographic Recording, Using Liquid Crystalline Side Chain Polymers.*Mol. Cryst. Liq. Cryst.*, 1994, 243, 51-75.

Bai, S.; Zhao, Y. (2001). Azobenzene-Containing Thermoplastic Elastomers: Coupling Mechanical and Optical Effects. *Macromolecules,* 2001, 34, 9032–9038.

Bai, S.; Zhao, Y. (2002). Azobenzene Elastomers for Mechanically Tunable Diffraction Gratings. *Macromolecules,* 2002, 35, 9657–9664.

Bang,C.-U. ; Shishido, A. ; Ikeda, T. (2007). Azobenzene Liquid-Crystalline Polymer for Optical Switching of Grating Waveguide Couplers with a Flat Surface. *Macromol. Rapid Commun.*, 2007, 28, 1040-1044.

Bieringer,T. (2000). in *Holographic Data Storage* (eds. Coufal, H. J., Psaltis, D., Sincerbox, G. T.) Springer Series in Optical Sciences, Springer 209 (New York, 2000).

Breiner, T.; Kreger, K.; Hagen, R.; Hackel, M.; Kador, L.; Muller, A.; Kramer, E.; Schmidt, H. (2007). Blends of poly(methacrylate) Block Copolymers with Photoaddressable Segments. *Macromolecules,* 2007, 40, 2100–2108.

Chen, A. ; Brady, D. J. (1993). Two-wavelength Reversible Holograms in Azo-dye Doped Nematic Liquid Crystals. *Appl. Phys. Lett.*, 1993, 62, 2920/1-3.

Chen, D.; Liu, H.; Kobayasi, T.; Yu, H. F. (2010a). Multiresponsive Reversible Gels Based on a Carboxylic Azo Polymer.*J. Mater. Chem.*, 2010, 20, 3610-3614.

Chen, D.; Liu, H.; Kobayasi, T.; Yu, H. F. (2010b). Fabrication of Regularly-Patterned Microporous Films by Self-Organization of an Amphiphilic Liquid-Crystalline Diblock Copolymer under a Dry Environment. *Macromol. Mater. Eng*. 2010, 295, 26-31.

Collier, R. J. ; Burckhardt, C. B. ; Lin, L. H. (1971). *Optical Holography,* (Ed: R. Colier), Academic Press: New York, 1971.

Cui, L.; Tong, X.; Yan, X.; Liu, G.; Zhao, Y. (2004). Photoactive Thermoplastic Elastomers of Azobenzene-Containing Triblock Copolymers Prepared Through Atom Transfer Radical Polymerization. *Macromolecules,* 2004, 37, 7097–7104.

Eich, M.; Wendorff, J. H.; Reck, B.; Ringsdorf, H. (1987). Reversible Digital and Holographic Optical Storage in Polymeric Liquid Crystals. *Makromol. Chem. Rapid Commun.* 1987, 8, 59-63.

Eich, M.; Wendorff, J. H. (1987). Erasable Holograms in Polymeric Liquid Crystals. *Makromol. Chem. Rapid Commun.* 1987, 8, 467-471.

Frese, T. ; Wendorff, J. H. ; Janietz, D. ; Cozan, V. (2003). Optical Storage in Pentaalkinylbenzene Derivatives. Strong Gain Effect by Recrystallization Processes. *Chem. Mater.* 2003, 15, 2146-2152.

Gabor, D. (1948). A New Microscopic Principles. *Nature,* 1948, 161, 777-778.

Häckel, M ; Kador, L. ; Kropp, D. ; Frenz, C. ; Schmidt, H. (2005). Holographic Gratings in Diblock Copolymers with Azobenzene and Mesogenic Side Groups in the Photoaddressable Dispersed Phase. *Adv. Funct. Mater.*, 2005, 15, 1722-1727.

Häckel, M. ; Kador, L. ; Kropp, D .; Schmidt, H. (2007). Polymer Blends with Azobenzene-Containing Block Copolymers as Stable Rewritable Volume Holographic Media. *Adv Mater.*, 2007, 19, 227-231.

Hasegawa, M. ; Yamamoto, T. ; Kanazawa, A. ; Shiono, T. ; Ikeda, T. (1999a). Dynamic Grating Using Photochemical Phase Transition of Polymer Liquid Crystals Containing Azobenzene Derivatives. *Adv. Mater.*, 1999, 11, 675-677.

Hasegawa, M. ; Yamamoto, T. ; Kanazawa, A. ; Shiono, T. ; Ikeda, T. (1999b). Photochemically -Induced Dynamic Grating by Means of Side-Chain Polymer Liquid Crystals. *Chem. Mater.*, 1999, 11, 2764-2769.

Hasegawa, M. ; Yamamoto, T. ; Kanazawa, A. ; Shiono, T. ; Ikeda, T. ; Nagase, Y. ; Akiyama, E. ; Takamura, Y. (1999c). Real-Time Holographic Grating by Means of Photoresponsive Polymer Liquid Crystals with Flexible Siloxane Spacer in the Side Chain. *J. Mater. Chem.*, 1999, 9, 2765-2772.

Hvilsted, S. ; Andruzzi, F. ; Kulinna, C. ; Siesler, H. W. ; Ramanujam, P. S. (1995). Novel Side-Chain Liquid Crystalline Polyester Architecture for Reversible Optical Storage. *Macromolecules*, 1995, *28*, 2172-2183.

Ikeda, T. (2003).Photomodulation of Liquid Crystal Orientations for Photonic Applications. *J. Mater. Chem.*, 2003, 13, 2037-2057.

Ishiguro, M. ; Sato, D. ; Shishido, A.; Ikeda, T. (2007). Bragg-type Polarization Gratings Formed in Thick Polymer Films Containing Azobenzene and Tolane Moieties. *Langmuir*, 2007, 23, 332–338.

Kim, D. Y. ; Tripathy, S. K. ; Li, L. ; Kumar, J. (1995). Laser-Induced Holographic Surface Relief Gratings on Nonlinear Optical Polymer Flms. *Appl. Phys. Lett.* 1995, 66, 1166-1168.

Kogelnik, H. (1969). Coupled Wave Theory for Thick Hologram Gratings. *Bell Syst. Technol. J.* 1969, 48, 2909–2946.

Kumar, G. S.; Neckers, D. C. (1989). Photochemistry of Azobenzene-Containing Polymers. *Chem. Rev.*, 1989, 89, 1915-1925.

Li, X.; Natansohn, A. ; Rochon, P. (1999). Photoinduced Liquid Crystal Alignment Based on a Surface Relief Grating in an Assembled Cell. *Appl. Phys. Lett.*, 1999, 74, 3791-3793.

Liu, H.; Kobayasi, T.; Yu, H. F. (2011). Easy Fabrication and Morphology Control of Supramolecular Liquid-Crystalline Polymer Microparticles. *Macromol. Rapid Commun.*, 2011, 32, 378-383.

Naka, Y.; Yu, H. F.; Shishido, A.; Ikeda, T. (2009) Photoalignment and Holographic Properties of Azobenzene–Containing Block Copolymers with Oxyethylene and Alkyl Spacers. *Mol. Cryst. Liq. Cryst.*, 2009, 513, 131–141.

Nakano, H. ; Takahashi, T. ; Kadota, T. ; Shirota, Y. (2002). Formation of Surface Relief Grating Using a Novel Azobenzene-based Photochromic Amorphous Molecular Material. *Adv. Mater.*, 2002, 14, 1157-1160.

Nakano, H. ; Tanino, T. ; Shirota, Y. (2005). Surface Relief Grating Formation on a Single Crystal of 4-Dimethylaminoazobenzene. *Appl. Phys. Lett.*, 2005, 87, 061910/1-3.

Natansohn, A. ; Rochon, P. (2002). Photoinduced Motions in Azo-Containing Polymers. *Chem. Rev.*, 2002, 102, 4139-4175.

Okano, K. ; Shishido, A. ; Tsutsumi, O. ; Shiono, T. ; Ikeda, T. (2005). Highly Birefringent Liquid-Crystalline Polymers for Photonic Applications: Synthesis of Liquid-Crystalline Polymers with Side-Chain Azo-Tolane Mesogens and Their Holographic Properties. *J. Mater. Chem.*, 2005, 15, 3395-3401.

Okano, K. ; Shishido, A. ; Ikeda, T. (2006). Photochemical Phase Transition Behavior of Highly Birefringent Azo-Tolane Liquid-Crystalline Polymer Films: Effects of the Position of the Tolane Group and the Donor-Acceptor Substituent in the Mesogen. *Macromolecules*, 2006, 39, 145-152.

Pham, V.P. ; Galstyan, T. ; Granger, A. ;Lessard, R.A. (1997). Novel Azo Dye-Doped Poly(methyl methacrylate) Films as Optical Data Storage Media. *Jpn. J. Appl. Phys.*, 1997, 36, 429-438.

Rochon, P. ; Batalla, E. ; Natanhson, A. (1995). Optically Induced Surface Gratings on Azoaromatic Polymer Flms. *Appl. Phys. Lett.*, 1995, 66, 136-138.

Saishoji, A., Sato, D., Shishido, A. ; Ikeda, T. (2007). Formation of Bragg Gratings with Large Angular Multiplicity by Means of the Photoinduced Reorientation of Azobenzene Copolymers. *Langmuir*, 2007, 23, 320–326.

Shirota, Y.; Moriwaki, K. ; Yoshikawa, S. ; Ujike, T. ; Nakano, H. (1998). 4-[Di(biphenyl-4-yl)amino]azobenzene and 4,4'-Bis[bis(4'-tert-butylbiphenyl-4-yl) amino] azobenzene as a Novel Family of Photochromic Amorphous Molecular Materials. *J. Mater. Chem.*, 1998, 8, 2579-2581..

Smith, H. M. (1977). *Holographic Recording Materials*, (Ed: H. M. Smith), Springer-Verlag: Berlin, 1977.

Stracke, A. ; Wendorff, J. H. ; Janietz, D. ; Mahlstedt, S. (1999). Functionalized Liquid-Crystalline Donor–Acceptor Triple Compounds Containing Azobenzene for Optical Storage. *Adv. Mater.* 1999, 11, 667-670.

Sutherland, R. L. ; Tondiglian, V. P. ; Natarajan, L. V.; Bunning, T. J.; Adams, W. W. (1994). Electrically Switchable Volume Gratings in Polymer-Dispersed Liquid Crystals. *Appl. Phys. Lett.*, 1994, 64, 1074-1076.

Todorov, T.; Tomova, N.; Nikolova, L. (1983). High-Sensitivity Material with Reversible Photo-induced Anisotropy. *Opt. Commun.* 1983, 47, 123-126.

Todorov, T.; Nikolova, L.; Tomova, N. (1984). Polarization Holography. 1: A New High-Efficiency Organic Material with Reversible Photoinduced Birefringence. *Appl. Opt.* 1984, 23, 4309-4321.

Todorov, T.; Nikolova, L.; Tomova, N. (1985). Polarization Holography. 3: Some Applications of Polarization Holographic Recording. *Appl. Opt.* 1985, 24, 785-788.

Tondiglia, V.P.; Natarajan, L.V.; Sutherland, R.L.; Bunning, T.J.; Adams, W.W. (1995). Volume Holographic Image Storage and Electro-Optical Readout in a Polymer-Dispersed Liquid-Crystal Film. *Opt. Lett.*, 1996, 20, 1325-1327.

Viswanathan, N.; Kim, D.; Bian, S.; Williams, J.; Liu, W.; Li, L.; Samuelson, L.; Kumar, J.; Tripathy, S. (1999). Surface Relief Structures on Azopolymer Flms. *J. Mater. Chem.*, 1999, 9, 1941-1955.

Wang, D.: Ye, G.; Wang, X.; Wang, X. (2011). Graphene Functionalized with Azo Polymer Brushes: Surface-Initiated Polymerization and Photoresponsive Properties. *Adv. Mater.*, 2011 (DOI: 10.1002/adma.201003653).

Weigert, F. (1919). Dichroism Induced in a Fine–Grain Silverchloride Emulsion by a Beam of Linearly Polarized Light. *Verh. Dtsch. Phys. Ges.*, 1919, 21, 479–483.

Yamamoto, T. ; Hasegawa, M. ; Kanazawa, A. ; Shiono, T. ; Ikeda, T. (1999). Phase-Type Gratings Formed by Photochemical Phase Transition of Polymer Azobenzene Liquid Crystals: Enhancement of Diffraction Efficiency by Spatial Modulation of Molecular Alignment. *J. Phys. Chem. B*, 1999, 103, 9873-9878.

Yamamoto, T. ; Hasegawa, M. ; Kanazawa, A. ; Shiono, T. ; Ikeda, T. (2000). Holographic Gratings and Holographic Image Storage by Photochemical Phase Transition of Polymer Azobenzene Liquid-Crystal Films. *J. Mater. Chem.*, 10, 337-342 (2000).

Yoneyama, S. ; Yamamoto, T. ; Hasegawa, M. ; Tsutsumi, O. ; Kanazawa, A. ; Shiono, T. ; Ikeda, T. (2001). Formation of Intensity Grating in a Polymer Liquid Crystals with a Side-Chain Azobenzene Moiety by Photoinduced Alignment Change of Mesogens. *J. Mater. Chem.*, 2001, 11, 3008-3013.

Yoneyama, S. ; Yamamoto, T. ; Tsutsumi, O. ; Kanazawa, A. ; Shiono, T. ; Ikeda, T. (2002). High-Performance Material for Holographic Gratings by Means of a Photoresponsive Polymer Liquid Crystal Containing a Tolane Moiety with High Birefringence. *Macromolecules*, 2002, 35, 8751-8758.

Yu, H. F.; Okano, K.; Shishido, A.; Ikeda, T.; Kamata, K.; Komura, M.; Iyoda, T. (2005a). Enhancement of Surface–Relief Gratings Recorded in Amphiphilic Liquid–Crystalline Diblock Copolymer by Nanoscale Phase Separation. *Adv. Mater.* 2005, 17, 2184–2188.

Yu, H. F. ; Shishido, A. ; Ikeda, T. ; Iyoda, T. (2005b). Photoinduced Alignment and Multi-Processes of Refractive-Index Gratings in Pre-Irradiated Films of an Azobenzene-Containing Liquid-Crystalline Polymer. *Mol. Cryst. Liq. Cryst.*, 2007, 470, 71-81.

Yu, H.F.; Li, J. ; Ikeda, T. ; Iyoda T. (2006a). Macroscopic Parallel Nanocylinder Array Fabrication Using a Simple Rubbing Technique. *Adv. Mater.*, 2006, 18, 2213-2215.

Yu H. F. ; Iyoda, T. ; Ikeda, T. (2006b). Photoinduced Alignment of Nanocylinders by Supramolecular Cooperative Motions. *J. Am. Chem. Soc.*, 2006, 128, 11010-11011.

Yu, H. F. ; Shishido, A. ; Iyoda, T. ; Ikeda, T. (2007a). Photoinduced Alignment and Multiprocesses of Refractive-Index Gratings in Pre-Irradiated Films of an Azobenzene-Containing Liquid-Crystalline Polymer. *Mol. Cryst. Liq. Cryst.*, 2007, 470, 71-81.

Yu, H. F. ; Asaoka, A. ; Shishido, A. ; Iyoda, T. ; Ikeda, T. (2007b). Photoinduced Nanoscale Cooperative Motion in a Novel Well-Defined Triblock Copolymer. *Small*, 2007, 3, 768-771.

Yu H. F. ; Shishido, A. ; Iyoda, T. ; Ikeda, T. (2007c). Novel Wormlike Nanostructure Self-Assembled in a Well-Defined Liquid-Crystalline Diblock Copolymer with Azobenzene Moieties. *Macromol Rapid Commun.*, 2007, 28, 927-931.

Yu, H. F. ; Li, J. ; Shishido, A. ; Iyoda, T. ; Ikeda, T. (2007d). Control of Regular Nanostructures Self-Assembled in an Amphiphilic Diblock Liquid-Crystalline Copolymer. *Mol. Cryst. Liq. Cryst.*, 2007, 478, 271-281.

Yu, H. F.; Shishido, A.; Li, J.; Iyoda, T.; Ikeda, T. (2007e). Stable Macroscopic Nanocylinder Arrays in An Amphiphilic Diblock Liquid-Crystalline Copolymer with Successive Hydrogen Bonds. *J. Mater. Chem.*, 2007, 17, 3485-3488.

Yu, H. F. ; Shishido, A. ; Ikeda, T. (2008a). Subwavelength Modulation of Surface Relief and Refractive Index in Pre-Irradiated Liquid-Crystalline Polymer Films. *Appl. Phys. Lett.*, 2008, 92, 103117/1-103117/3.

Yu, H. F. ; Naka, Y. ; Shishido, A. ; Ikeda, T. (2008b). Well-Defined Liquid-Crystalline Diblock Copolymers with an Azobenzene Moiety: Synthesis, Photoinduced Alignment and their Holographic Properties. *Macromolecules*, 2008, 41, 7959-7966.

Yu, H. F.; Shishido, A.; Iyoda, T.; Ikeda, T. (2009a). Effect of Recording Time on Grating Formation and Enhancement in an Amphiphilic Diblock Liquid-Crystalline Copolymer. *Mol. Cryst. Liq. Cryst.*, 2009, 498, 29–39.

Yu, H. F.; Kobayasi, T. (2009b). Fabrication of Stable Nanocylinder Arrays in Highly Birefringent Films of an Amphiphilic Liquid-Crystalline Diblock Copolymer. *ACS Appl. Mater. Interfaces*, 2009, 1, 2755-2762.

Yu, H. F.; Kobayashi, T.; Ge, Z. (2009c). Precise Control of Photoinduced Birefringence in Azobenzene-Containing Liquid-Crystalline Polymers by Post Functionalization" *Macromol. Rapid Commun.*, 2009, 30, 1725-1730.

Yu, H. F.; Kobayashi, T. (2010). Photoresponsive Block Copolymers Containing Azobenzene and Other Chromophores. *Molecules*, 2010, 15, 570-603.

Yu, H. F.; Ikeda, T. (2011). Photocontrollable Liquid-Crystalline Actuators. *Adv. Mater.*, 2011, DOI: 10.1002/adma.201100131

Yu, H. F.; Kobayashi, T.; Hu, G. (2011). Macroscopic Control of Microphase Separation in an Amphiphilic Nematic Liquid-Crystalline Diblock Copolymer. *Polymer*, 2011, (DOI 10.1016/j.Polymer, 2011, 52, 1154-1161).

Zhao, Y.; Bai, S.; Dumont, D.; Galstian, T. (2002). Mechanically Tunable Diffraction Gratings Recorded on an Azobenzene Elastomer. *Adv. Mater.*, 2002, 14, 512–514.

Zettsu, N. ; Ogasawara, T. ; Mizoshita, N. ; Nagano, S. ; Seki, T. (2008). Photo-Triggered Surface Relief Grating Formation in Supramolecular Liquid Crystalline Polymer Systems with Detachable Azobenzene Units. *Adv. Mater.*, 2008, 20, 516–521.

6

Holography Based on the Weigert's Effect

Zurab V. Wardosanidze
Institute of Cybernetics of the Georgian Technical University
Georgia

1. Introduction

Despite the numerous scientific and technological achievements in the field of holography, its main task still remains the registration and the reconstruction of the complete information of the wave scattered by the object.

The wave of the light, scattered by an object, is characterized by four main parameters: Intensity, Phase, Wavelength and Polarization. The primary task of holography is solved, if all four parameters are recorded and reproduced adequately.

The problem of holographic recording and reconstruction of the first three parameters was practically solved by using ordinary light-sensitive materials, which respond only to the Intensity of light (Gabor, 1948; Leith &Upatnieks, 1962; Denisiuk, 1974). As to the fourth parameter - the polarization the problem of its adequate recording and reconstruction was not solved. It is natural that, in the case of the usual photosensitive material, all the characteristics of the wave, except polarization, are displayed as the distribution of isotropic optical properties. However, in the general case of holographic recording the light wave, scattered on the object, changes the polarization. Therefore the total interference pattern will have the modulation also with the polarization, the information about of which is lost during the holographic recording in the ordinary photographic materials.

It should be noted that from the four basic parameters of the light the polarization can be regarded as the most complex. Particularly, to assess the polarization of the light, the measurement of four values is needed (so called Stokes parameters, S_0; S_1; S_2; S_3)(Born&Wolf, 1974). However, it turned out that the question of the polarization is more difficult for the unpolarized light. For example, the light remains unpolarized during the passing of anisotropic and gyrotropic objects, but its statistical structure is altered, which radically changes the result of interference. The interference of unpolarized waves gives a usual pattern of light and dark bands, which are also unpolarized. But, if on the path of one of the interfering waves we put an anisotropic half-wave or gyrotropic half-wave (rotator at 90^0) phase plate, the interference pattern disappears and the field becomes evenly lighted. However, the polarization analysis shows, the interference pattern is polarized in this case (Langsdorf&Du Bridge, 1931; Vavilov, 1932). Therefore, the unpolarized light has a certain polarization parameter which can be detected and, accordingly, can be recorded in appropriate conditions.

Thus, the problem of the holographic registration and the reconstruction of the polarization state of the wave can be considered as the most difficult problem in the holography. For registration of the polarization characteristics of light a photosensitive material, which responds to the polarization is necessary. Such materials are known for a long time, and the phenomenon of creation of the anisotropy in the photosensitive material under the influence of the linearly polarized light is called as Weigert's effect (Weigert, 1919). Later similar effect was observed in a variety of photosensitive materials, including the azo-dyes (Kondo, 1932). The Weigert's effect is a particular case of the sensitivity to the polarization, because in addition to the linear polarization the elliptical and circular polarizations exist as well. About the gyrotropic response of the Halide-silver materials under the influence of circularly polarized light was reported in 1928 (Zocher&Coper, 1928). However, since that time, the gyrotropic response of the photosensitive materials was not confirmed unambiguously.

Holographic recording and reconstruction, on the basis of Weigert's effect, for the first time was implemented in the particular cases of the orthogonal linear and circular polarizations of the reference and object waves (Kakichashvili, 1972;). Weigert's effect has played an important role in the development of holography. Having no analogies holographic diffractive optical elements of the anisotropic structure with complex properties were created. Holographic recording and reconstruction were implemented in the particular cases of linear, circular, partially polarized and unpolarized light. The Weigert's effect has stimulated the search of new photosensitive materials and on the other hand, expanded the range of the studies of holographic processes.

2. Weigert's effect in azo-dye materials

From the point of view of Weigert's effect the azo-dyes can be considered as the most interesting materials because of their photosensitivity, reversibility and high value of photo anisotropy (Kondo, 1932). The Azo-dyes are the organic compounds containing one or more group of double chemical bonds of nitrogen , the so called — N = N — chromophore.

In general, the azobenzene molecule is the basic kernel of the azo-dyes. Besides, it is known that azo-dyes undergo E/Z photo isomerization around — N = N —, the classical example of which is Azobenzene (Eltsov, 1982)

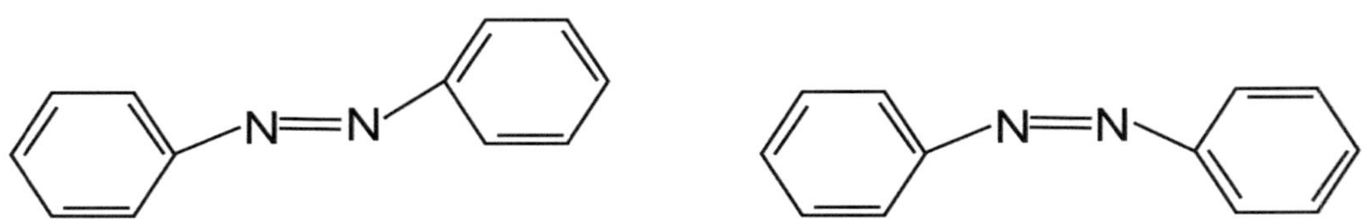

E - isomer Z - isomer

Analogous transformation will have more complex nature for the azo-dyes containing more than one **— N = N —** chromophore and different radicals. Currently it is accepted, that the basic factor of Weigert's effect is photo E/Z isomerization in azo-dye materials. But, it is proved, that E/Z isomerization is not necessary and sufficient requirement process for the initiation of Weigert's effect in the azo-dye materials and its mechanisms may be different.

Some authors attempted to explain Weigert's effect in the azo-dye materials, but they were not confirmed uniquely (Shatalin, 1989; Ebralidze, 2002).

The Azo-dyes are the most promising photosensitive materials with the Weigert's effect.

The known photosensitive materials have the following principal disadvantages:

1. The effect of two or more different physical factors is necessary for the realization of the multiple recording and the deletion processes;
2. The spontaneous relaxation takes place;
3. Fatigue of the material during the multiple recording and deleting processes.

From this point of view azo-dye materials with Weigert's effect have obvious advantages:

1. Possibility of multiple recording and deletion with the same actinic light with different polarization;
2. The fatigue of the azo-dye materials is not observed;
3. The dark relaxation practically does not take place after the discontinuation of the impact of actinic light and the generated anisotropy maintains stability in the wide range of temperatures for a long time.

2.1 One more possible mechanism of the Weigert's effect in azo-dye colored materials

Three azo-dyes with symmetric structures were chosen for the investigation: Azobenzene, Fernandes, and Mordant Pure Yellow

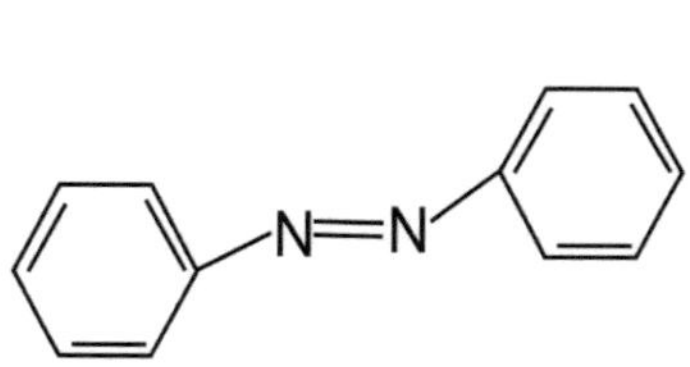

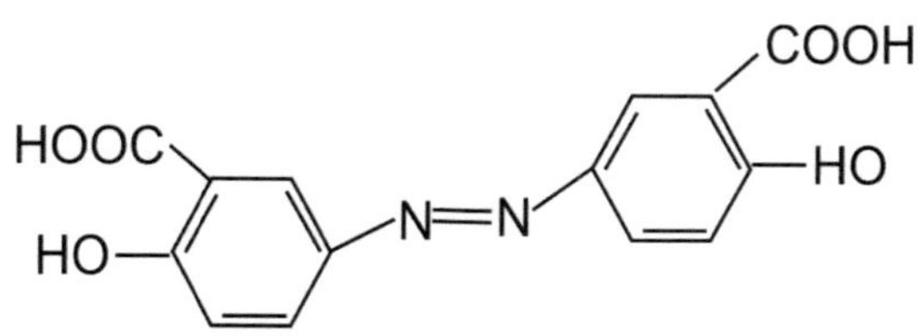

Azobenzene Frnandes

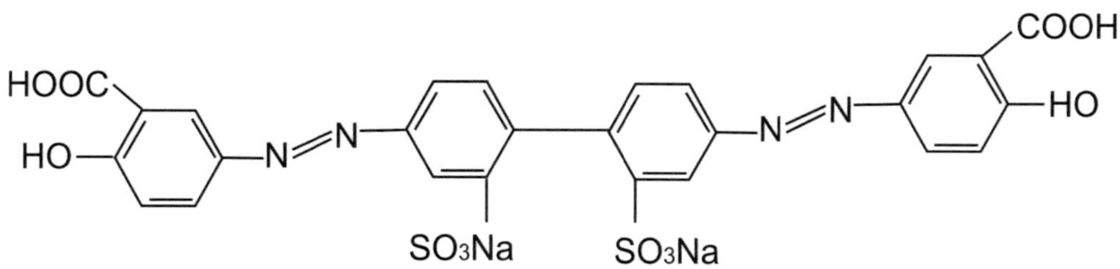

Mordant Pure Yellow

Linear polarized radiation of He-Cd laser (λ = 441,6 nm) was used as an actinic light. The birefringence and the dichroisme induced by the linear polarized actinic light can be calculated by means of formulae (Kakichashvili&Shaverdova, 1982):

$$\Delta n\tau \equiv (n\tau)_x - (n\tau)_y = -\frac{1}{2\kappa d}\ln\frac{I_0}{I_{90}}$$
$$\Delta n \equiv (n_x - n_y) = \frac{1}{\kappa d}\arccos\frac{4I_{45} - (I_0 + I_{90})}{2\sqrt{I_0 I_{90}}} \quad (1)$$

where $\Delta n\tau$ is the anisotropy of absorption, Δn - birefringence, $k=2\pi/\lambda$, I_0, I_{45}, I_{90} are the optical transmittance of the exposed samples for the given wavelength when the probe beam is polarized at 0^0, 45^0, 90^0 relative to the electric vector of the linearly polarized actinic light.

Fig. 1 shows spectral transmittance characteristics of azobenzene doped in nonpolar matrix of polystirol and in polimethilmetacrilat, before and after the exposure (curves I, II).

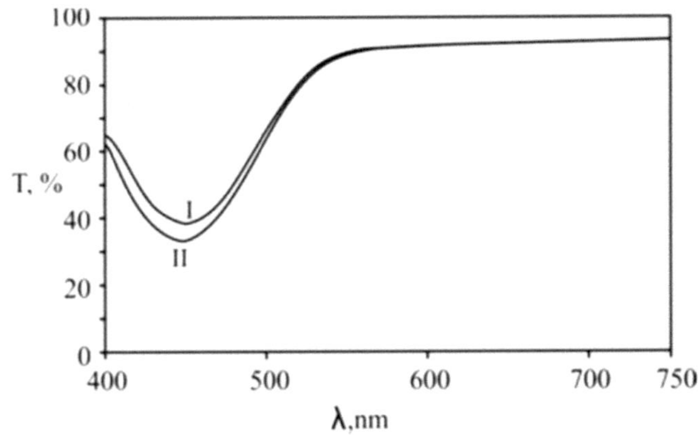

Fig. 1. Spectral transmittance characteristics of azobenzene

The increase of the absorption in the wave range of 400 - 560 nm as a result of the influence of the actinic light confirms E/Z isomerization, the sole possible photo-process in this instance. Weigert's effect does not take place in the given case. Consequently, the E/Z photoisomerisation is not sufficient condition for it's emergence. The water-soluble Azo-dyes Fernandes and Mordant Pure Yellow were introduced into gelatin. Fig. 2 shows spectral transmittance characteristics of azo-dye Mordant Pure Yellow before (curve I) and after (curves II,III,IV) exposure to linearly polarized actinic light (λ = 441,6 nm).

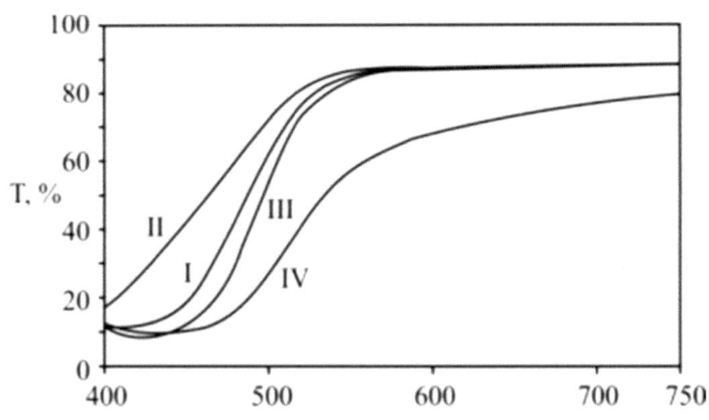

Fig. 2. Polarization spectral transmittance characteristics of azo-dye Mordant Pure Yellow

Curve (II) corresponds to the spectrum of transmittance of the irradiated sample, when the polarization of sensing light coincides with the polarization the actinic light. Curve (III) corresponds to the transmittance spectrum of the same sample, for orthogonal polarization of sensing light and curve (IV) corresponds to transmittance spectrum, when the angle is 45^0, between the polarizations of actinic and sensing light. As can be seen from Fig. 3, in contrast to the azobenzene case, Weigert's effect takes place. Analogous results were obtained for monoazo-dye Fernandes. Thus, one of the characteristics of observed Weigert's effect is that the absorption of these materials decreases in the direction of electric vector of actinic light and, on the contrary, increases in orthogonal direction. This fact indicates that the number of absorbing oscillators parallel to polarization of actinic light decreases and the number of absorbing oscillators with the orthogonal directions, increases. However, decrease in the absorption of the material along the direction of actinic light polarization of is greater than its increase in orthogonal direction in the wave range of 400-560 nm. It means that the decrease in the number of absorbing oscillators oriented along the electric vector of actinic light must be greater, than the increase in the number of their orthogonal absorbing analogous oscillators, which is inadmissible. One of the explanations of this phenomenon may be that the initial oscillator and the oscillators, which were obtained after transformation, must be different. Some anomalies of azo-dyes with hydrogen — H and hydroxyl — OH groups in the isomerization process confirms such approach (Eltsov, 1982). Under the influence of actinic light — N = N — chromophore is breaking down and one of its nitrogen atom creates — N - H — auxochrome oscillator with perpendicular direction. In this case the strength of the — N - H — oscillators is much lower than — N = N — oscillator (Eltsov, 1982). As gelatin contains many endings H, OH, NH_2 and some amount of water (HOH), there always can exist analogous transformation of molecules of the azo-dyes.

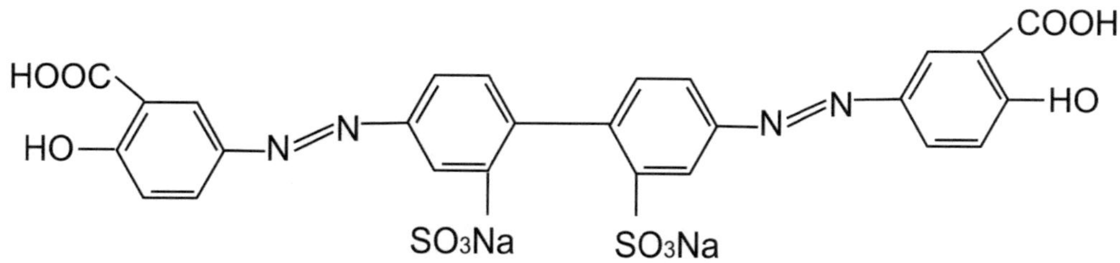

Initial structure

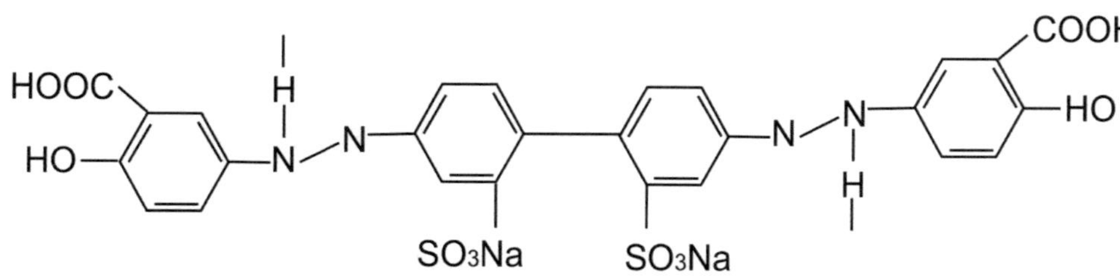

Transformed structure

The dependence of photosensitivity and occurrence of Weigert's effect of azo-dye materials on the humidity is shown in Fig. 3. Curves I and II (Fig. 3, **a**) correspond to 3% and 15% of emulsion humidity respectively.

Thus, the rate of rising and maximum value of Weigert's effect is greater for more humid emulsion layers, the good illustration of which is Fig. 3 b. The parts with higher brightness correspond to higher humidity and anisotropy under the equal energy of exposure.

The obtained results confirm the possibility of existing of one more possible mechanism of Weigert's effect for azo-dye. Under the influence of linearly polarized actinic light of the initial structure of the molecule of azo-dye transforms to structure, where initial — N = N — chromophore is changed in two auxochrome — N - N — and — N - H —. Therefore, powerful absorbtion oscillator — N = N — transforms into two oscillators with weaker absorbtion. In addition the direction of new — N - H — oscillator already is orthogonal to direction of initial oscillator — N = N — (Vasilieva N.V. at al 1978). Thus, the absorption of the given material must decrease along the direction of electric vector of the linearly polarized actinic light and must increase in orthogonal direction. In addition, decrease in absorption along the electric vector of the actinic light must be greater, than increase in absorption in orthogonal direction, because of difference of oscillator's strength mentioned above.

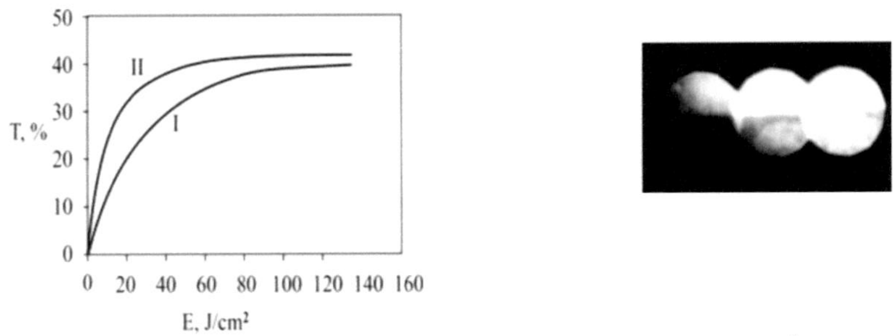

a b

Fig. 3. Humidity dependence of the photosensitivity (**a**) and of the Weigert's effect (**b**) of azo-dye.

2.2 Reversibility of the Weigert's effect in azo-dye colored materials

The results of investigation of multiple recording and erasure of the Weigert's effect in azo-dye, is in good compliance with the mechanism mentioned above. The scheme of experiment is presented in Fig.4. Linearly polarized light of the He-Cd laser 1 (λ= 441.6 nm) is falling on the sample at the Brewster's angle ($\alpha \approx 56^0$). The can rotate in its own plane. Circularly polarized sensing light of the He-Ne (λ= 632.8 nm) laser 3 is falling perpendicularly to the sample, passing through polarizer 4. Polarizer 4 is oriented in such away that the angle between linear polarizations of actinic and probing lights on the plane of the sample is 45^0. The orientation of the analyzer 5 relative to the polarizer 4 is 90^0, so it does not transmit light. Therefore the radiation detector 6 and recorder 7 cannot register the light intensity until the exposure of the investigated sample with the actinic light.

Under the influence of the linearly polarized actinic light the Weigert's effect in azo-dye takes place, and as a result, the linear polarization of the probing wave transforms to an elliptical one. Because of this the analyzer 5 transmits the part of probing light and the recorder registers it. The intensity of this part of the probing wave gives the information of the value of anisotropy (Weigert's effect). Initially, the sample is irradiated up to saturation of Weigert's effect. After this the actinic light is blocked and the sample is rotated by the angle of 90^0 in its plane. Therefore, the linear polarization of the actinic light becomes orthogonal relative to the axis of induced anisotropy, due to which the deletion of anisotropy starts. After the full deletion of the anisotropy, the impact of actinic light is terminated and the sample returns to its initial state for repetition of the recording process.

The multiple recording and deletion of anisotropy showed that fatigue of the investigated material was not observed. The dynamics of 5 series of recording and deletion of anisotropy is represented in Fig.5.

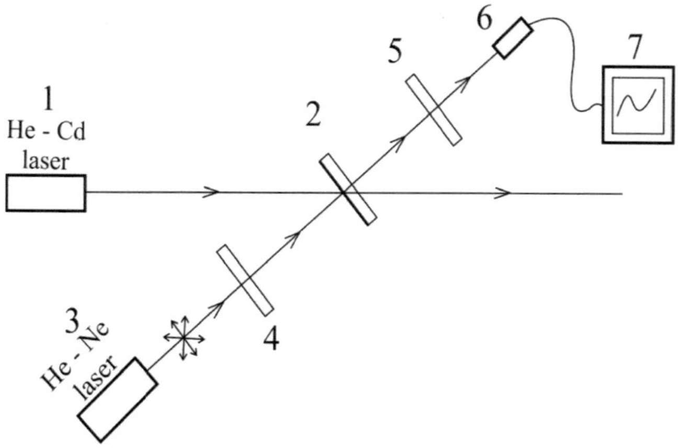

Fig. 4. The scheme of multiple recording and erasure of Weigert's effect

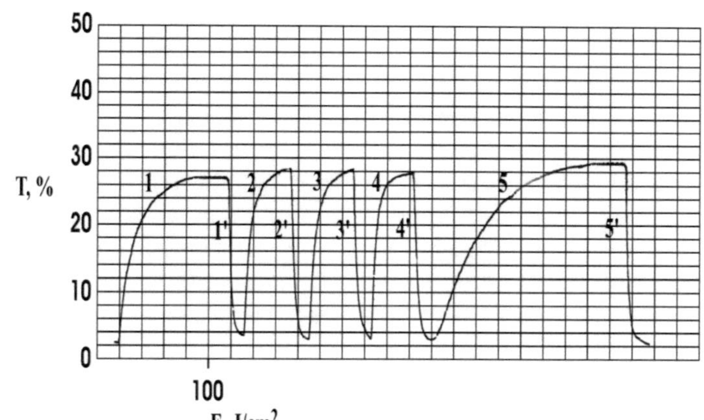

Fig. 5. The dynamics of 5 series of recording and erasure of Weigert's effect

The first four pair of curves 1-1'; 2-2'; 3-3'; 4-4' correspond to the recording and erasure of Weigert's effect, when linear polarization of the actinic light is changing consistently on 90^0. As follows from the obtained results, the first process creation of Weigert's effect requires more than five times as much exposure energy, than the first exposure-erasure process (curves 1-1'). In addition, in the subsequent series of recording and deletion (curves 2-2'; 3-3'; 4-4') the recording process requires far less exposure energy, but for the erasure the energy remains the same as in the first case. The final stage the process of recording and deletion of the Weigert's effect is different than first four (curves 5-5'). In this case the sample does not return to its initial state after the end of the fourth stage of deletion (curve 4') and the process of creation of anisotropy continues by the same polarization of the actinic light (curve 5). In this case the saturation of the Weigert's effect needs 1.8 times more exposure energy than in the initial step of recording (curve 1). Nevertheless, as it follows from curve

5', the speed of the deletion of the Weigert's effect remains the same as in the previous cases (curves 1'; 2'; 3'; 4'). Thus, the recording and deletion of Weigert's effect in the azo-dyes have various character. Particularly, the recording process is characterized by saturation, i.e. Weigert's effect achieves the maximum value and maintains its value during the whole time of the recording. In contrary to that, upon the end of anisotropy deletion process, a new recording process by the same polarization of actinic light starts.

Let the azo-dye material represent the ensemble of statistically oriented absorbing oscillators of - N = N -. For oscillators oriented parallel to the electric vector of actinic light the transformation is maximum and is zero for oscillators oriented at 90^0. Let the electric vector of actinic light be parallel to one of the absorbing oscillators in the initial stage of the repeating recording and erasure (Fig. 6 **a**). Due to of influence of actinic light the oscillator — N = N — transforms in a new — N — H — oscillator oriented orthogonally (Fig.6 **b**). The value of Weigert's effect in this dynamic process achieves its maximum after the full transformation of all oscillators. After the change of the mutual orientation of the sample and the polarization by 90^0 the electric vector of actinic light becomes parallel to the new — N — H — oscillator and begins its transformation (Fig. 6 **c**).

Two opposite processes will take place on this stage: a) formation of new anisotropy, orthogonal to initial one, as a result of second chromophore oscillators — N = N — transformation and b) deletion of the first anisotropy as a result of reconstruction of chromophore oscillators — N = N — which were transformed on the first stage (Fig.6 **d, c**). Thus, the sample reaches the state when the value of the residual anisotropy and the new anisotropy with orthogonal direction become equal to each other and the resulting anisotropy of the sample becomes zero. I.e. material behaves as an isotropic medium, but in reality, here takes place seeming isotropy, when two orthogonal anisotropies compensate each other. Accordingly, to achieve such a neutral state of the material, relatively small energy is needed. After mutual reorientation of the chromophores in the sample and the polarization by 90^0, a new phase of anisotropy formation is determined by two opposing processes again: 1. due to reconstruction of the second chromophore - N = N - in connection with the transformation of their auxochrome oscillators and 2. due to transformation of the first chromophores - N = N - because of emergence of their auxochrome oscillators. Accordingly, the exposure energy, for the achievement of the maximum value of anisotropy, will be smaller than in the original recording process. Subsequent repetition of these processes recording and erasure, gives similar results (curves 2'; 3-3'; 4-4' in Fig.5). As it follows from Fig.5 the stage 5-5' of described process, is differs from previous stages. The matter is that, in this case, the mutual orientation of the investigated sample and polarization of actinic light has not been changed after the achievement of minimal anisotropy. The difference of this process with respect to previous one, is also the result of two opposite processes. Particularly, formation of new anisotropy, orthogonal to previous one, is continuing because of continuation of destruction of the second chromophore oscillator — N = N —, which is oriented parallel to electric vector of actinic light. At the same time, deletion of previous anisotropy because of continuation of the reconstruction of destroyed chromophore oscillator — N = N — which are oriented orthogonally to the electric vector of actinic light (Fig. 6 e) takes place. Consequently, the energy needed for the anisotropy saturation on this stage must be higher than on all previous recording stages (curve 5 of Fig.5). In this case, direction of the final anisotropy is orthogonal to the initial one, obtained at the first stage. Under the next mutual reorientation of the investigated sample and polarization of actinic light by 90⁰, the anisotropy deletion process with the same speed as at the previous stages takes place (curves 1'; 2'; 3'; 4'; 5' of Fig. 5). The anisotropy recording and erasure processes here have different natures. Namely, the formation of anisotropy caused only by destruction of oscillators takes place only at the initial stage of the recording (curve 1 in Fig.5). In all the following stages of multiple recording and deletion of anisotropy up to stage 5-5′, takes place two processes in opposite directions as it was described above (Fig.6). It is interesting that after the first stage of anisotropy formation, it is impossible to return the molecules of the azo dye (absorbing oscillators) into the initial state (Fig. 6 **a**) by irradiation with linearly polarized actinic light.

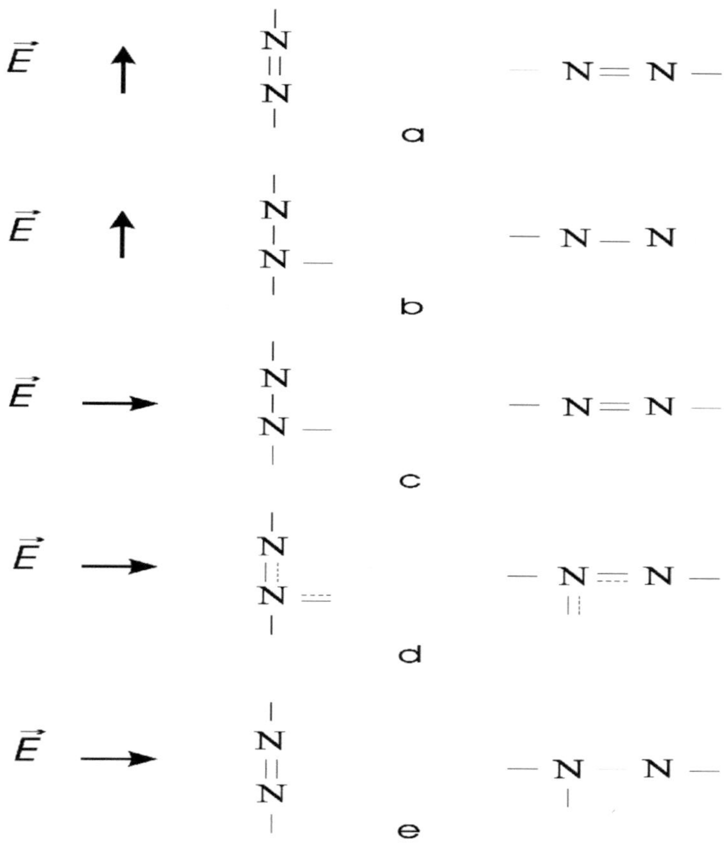

Fig. 6. The scheme of transformation of absorbing oscillators of aso-dye

3. Holographic recording in the general case of linear polarization

An interesting result gives the investigation of holographic recording in the azo-dye material with Weigert's effect in the general case of linear polarization of the object waves. For the simple theoretical calculations the Jones vector-matrix method of was used (Jones, 1941; Kakichashvili, 1974).

Photosensitive material H with Weigert's effect perpendicular to axis OZ is placed into a rectangular system of coordinates (Fig. 7).

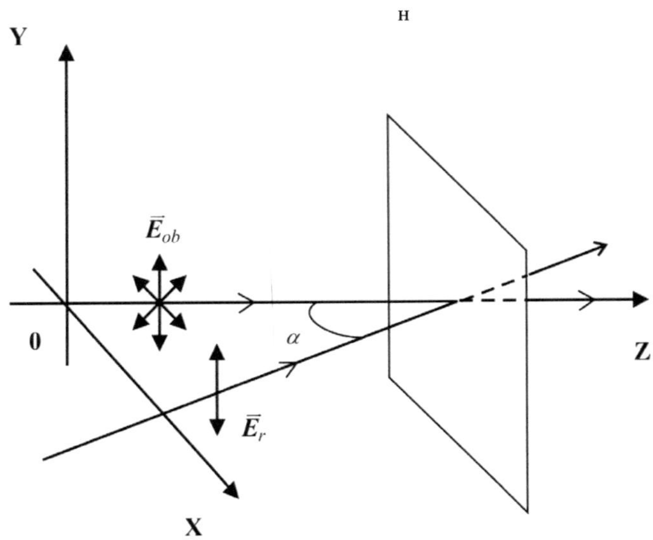

Fig. 7. The scheme of holographic recording

The object wave represents a linearly polarized plane wave of all the possible orientations of linear polarization, which spreads along the OZ axis and falls normally on the surface of the photosensitive medium H

$$\vec{E}_{ob} = \vec{E}_{01}\begin{pmatrix}\cos\theta \\ \pm\sin\theta\end{pmatrix}\exp i(\omega t - kz + \delta) \tag{2}$$

where $\vec{E}_{01}$ is amplitude, θ is the orientation of the electrical field vector in the wave, ω is the frequency of the wave, $k = 2\pi / \lambda$ is the wave number in vacuum, δ is the relative phase.
The reference wave represents linearly polarized plane wave, the electric vector of which oscillates parallel to OY axis, it is located in XZ plane and spreads at a small angle with respect to the object wave (Fig. 7)

$$\vec{E}_r = \vec{E}_{02}\begin{pmatrix}0 \\ 1\end{pmatrix}\exp i(\omega t - kz\cos\alpha) \tag{3}$$

where $\vec{E}_{02}$ is the amplitude of the reference wave; a is the angle of incidence on the plane **H**. Thus, both the object and reference waves are located in **XZ** plane and their electric vectors are situated in the plane of the photosensitive medium **H**.
Let the dependence of the refractive indices of the photosensitive medium on the intensity have linear character (Kakichashvili, 1974):

$$\begin{aligned}\hat{n}_1 &= \hat{n}_0 + aE_1^2 \\ \hat{n}_2 &= \hat{n}_0 + aE_2^2\end{aligned} \tag{4}$$

After application of the known theory of interference pattern calculation (Born, 1970;) and the theory of polarization holography the reconstructed image of the object wave has the form (Kakichashvili, 1974):

$$\vec{E}_{-1} = M_{-1}\vec{E}_r \approx -2ikda\vec{E}_{01}E_{02}^2 \exp(-ikd\hat{n}_0)\begin{pmatrix}\cos\theta \\ \pm\sin\theta\end{pmatrix}\exp i(\omega t - kz + \delta) \quad (5)$$

This is the imaginary image of the object which preserves both the polarizing character and direction of the object wave. Thus, the theory shows the possibility of the holographic recording and adequate reconstruction of linear polarization in general case on the basis of the Weigert's effect.

3.1 Experimental results and discussion

Experimental investigation confirms the theoretical result. The scheme of holographic recording is shown in Fig. 8. The collimated laser beam (λ=441,6 nm), with vertical linear polarization, is divided into two parts with identical intensity by means of a usual glass prism **P** and polarization prism **PP**. The prism P deflects half of the beam to overlap with object wave in the plane of photosensitive medium H without changing its polarization and creates the plane wave, electric vector of which oscillates along OY axis (Fig.7). The object wave is formed by means of a special polarizing prism PP composed of two wedges, one of which represents the optically active crystal of quartz. The second part of the prism represents the wedge of usual glass with analogous geometrical form and the same refractive index. After passing through this prism, the light does not change its spreading direction, but its polarization varies from +90^0 to -90^0 along the cross-section. The analyzer **A** gives a possibility for polarization analysis both of object PP and of its reconstructed image. Fig. 8 **b** shows the distribution of polarization of the reference and object waves along the cross-sections of these waves. Fig. 9 **a** shows the view of the object after passing the analyzer **A** whose orientation changes from +90° up to -90°. Fig. 9 **b** shows the reconstructed image of the object at the same orientations of analyzer, by which the hologram reconstructs the polarization of the object wave with sufficient precision. Fig. 9 **c,d,e** represents the result of holographic recording of the test-object composed of accidentally oriented polarizers.

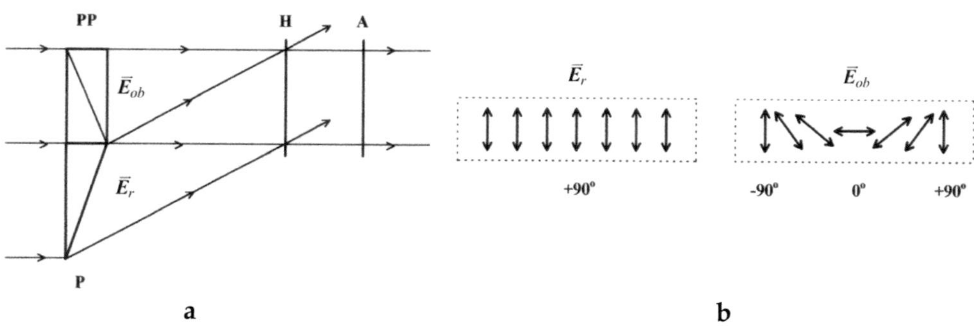

Fig. 8. The experimental scheme of holographic recording **a** and projection pictures of reference and object waves **b**

Fig.9 **c** is a view of the test-object without analyzer, Fig. 9 **d** is the same image through analyzer and Fig.9 **e** is the reconstructed image of the object after the analyzer.

In accordance with Fig. 9 the polarization of the reconstructed images doesn't differ considerably from the polarization of the object wave. Little difference may be the result of the nonlinearity of recording, which was not taken into account in the theoretical calculations. On the other hand it was found that with increasing initial absorption of the photosensitive medium, the proportion of the dichroism of Weigert's effect is much greater and polarization of the reconstructed image is closer to the polarization of the object image. However, the intensity of the reconstructed image decreases strongly in this case, and the conditions of its observation sharply deteriorate.

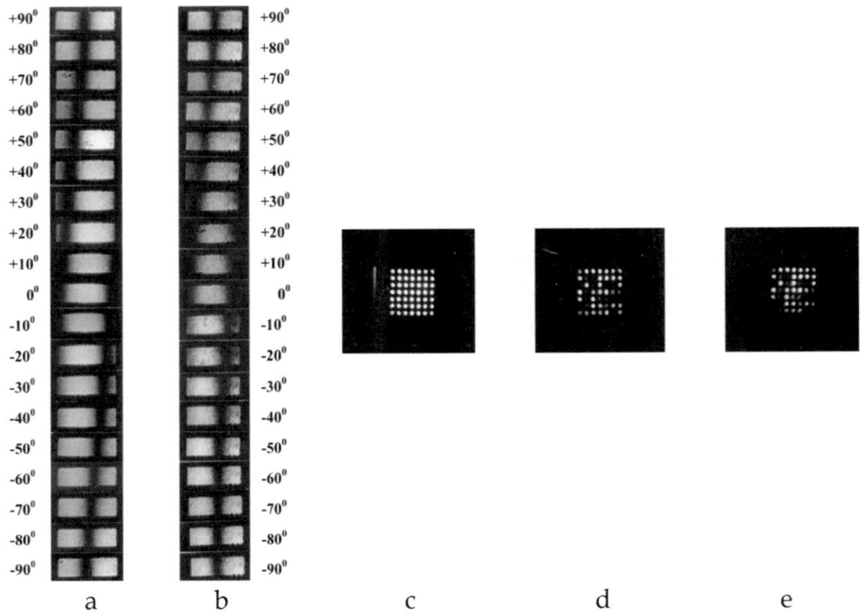

Fig. 9. The views of the objects and their reconstructed images

Thus, it is possible to formulate the following conditions, which are necessary for holographic recording and adequate reconstruction in the general case of linear polarization:

1. Selection of the light-sensitive material's, for example azo-dyes, with a wide range of linear recording.
2. Selection of the optimal absorption of the azo-dye with Weigert's effect.

4. Diffractive optical elements of an anisotropic structure

Application of the Weigert's effect gives interesting results for creation of holographic diffractive optical elements. Particularly, as it is known, the standard holographic diffractive optical elements (diffraction gratings, zone plates, raster systems, holographic mirrors) have isotropic structure. Because of this, they don't influence the polarization state and they have a possibility of transformation and separation only of phase, amplitude and spectrum of the light. In addition, their maximum diffraction efficiency could not be more than 33% for the thin holograms. In contrast, the holographic diffractive optical elements based on the Weigert's effect possess anisotropic structure and have opportunity to influence the state of polarization of light. Diffractive efficiency of these optical elements reaches 100%. For example, the diffractive gratings with efficiency 100% have been already obtained (Shatalin&Kakichashvili, 1987). From this point of view, the most interesting holographic diffraction optical elements represent zone plates (Fresnel lenses) and raster systems of an anisotropic structure.

4.1 Zone plates and raster systems on the basis of Weigert's effect

For creation of the Zone plates (Fresnel lenses) and raster systems the axial holographic recording scheme was used (Fig. 10). The circularly polarized plane wave of actinic light (λ=441,6 nm) is divided by means of calcite lens 1 in two spherical waves with orthogonal linear polarizations and identical intensity. Calcite lens was cut parallel to the optical axis, which allowed the formation of two spherical waves, of different divergence, propagating coaxially. Quarter-wave phase plate 2 converts the linearly orthogonally polarized waves into the circularly orthogonal polarized ones. The converging lens 3 is providing overlapping of separated waves in one area in the plane of the photosensitive material 4. Thus, in the plane of the photosensitive material an interference pattern with the radial distribution of linear polarization is formed. As a photosensitive material the azo-dye colored gelatin was used. Diffractive efficiency of the obtained zone plates reached 80%. Fig. 11 shows the photograph of the zone plate in the polarization microscope for its orientations at an angles 0^0, 45^0 and 90^0 with respect to the crossed polarizers.

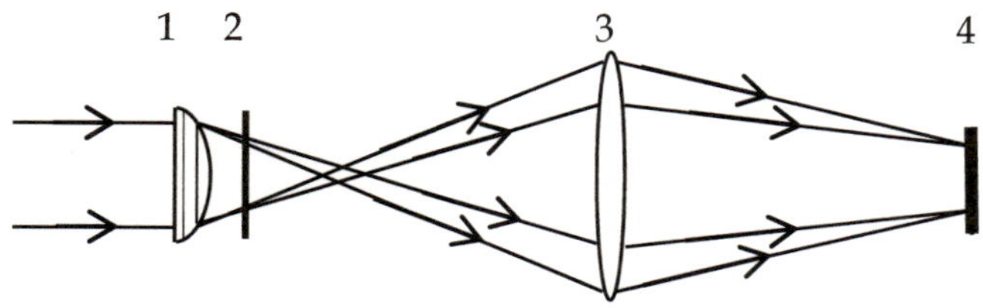

Fig. 10. The scheme of holographic recording of the zone plate

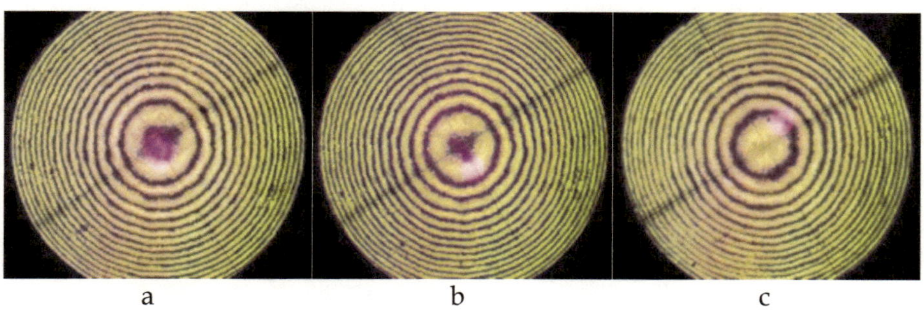

Fig. 11. The image of the zone plate in the polarization microscope

The diffraction from the obtained zone plate is presented in Fig. 12. Fig. 12 **a** corresponds to a zone plate illuminated with a linearly polarized wave. Fig. 12 **b** shows a zone plate illuminated with a circular polarized wave. As it follows from Fig.12 the zone plate behaves as the diverging lens in the first case and as a collecting lens in the second case. For the left circular polarization of the incident wave, this zone plate behaves as a collecting lens and for the right circular polarization it behaves as a diverging lens. Fig. 13 shows the image of the anisotropic complex object obtained with such lenses. The object was illuminated with left- (a) and right- (b) circularly polarized light. As it can be seen from

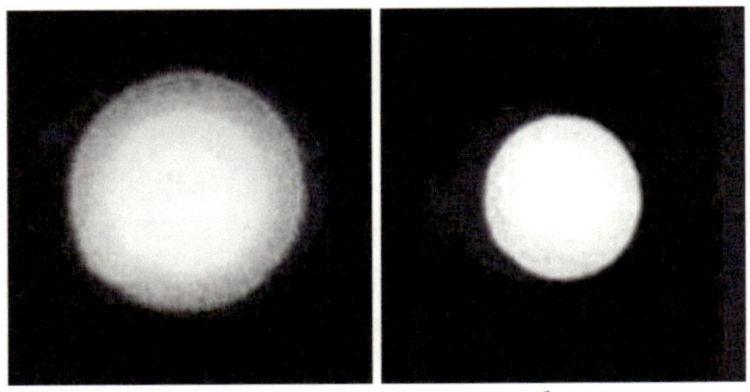

a b

Fig. 12. Diffraction of the linearly polarized (**a**) and circularly polarized (**b**) light on the zone plate

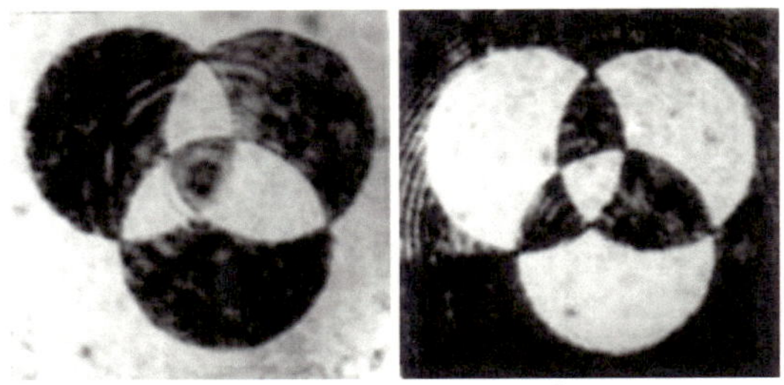

a b

Fig. 13. The image of the anisotropic complex object obtained by means of zone plate.

Fig. 13 these images are complementary. So, the zone plate created with of orthogonal circular polarizations of recording waves automatically performs the subtraction, which is important for solving some problems of microlithography and optical information processing.

Similar technology is used to create raster systems. In this case the mosaic of the holographic microlenses with an anisotropic structure is arranged. Fig.14 shows the image of such raster placed between crossed polarizers and the diffraction pictures on this raster. Fig. 14 a is the image of the raster between crossed polarizaers. Fig. 14 b shows the diffraction of the expanded collimated beam He-Ne laser (λ= 632.8 nm) on this raster and Fig. 14 c shows the diffraction of the unexpanded beam on one of the microlense of the raster.

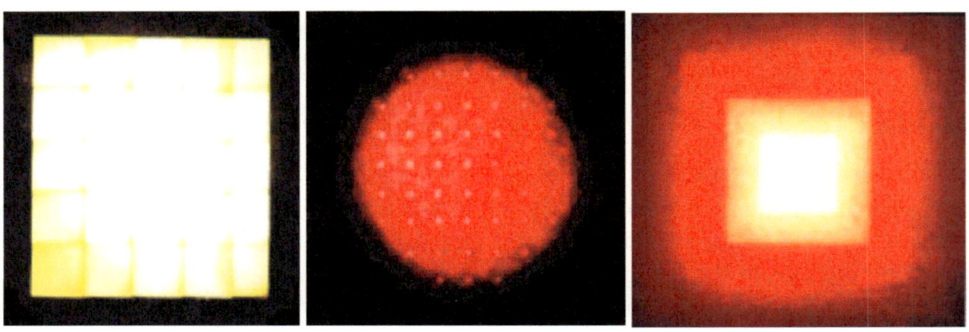

a b c

Fig. 14. The raster's image (**a**) and the diffraction pictures on them (**b,c**)

4.2 Holographic chiral structure on the basis of Weigert's effect

It is known that the method of N - Jones matrices is used to describe the evolution of the polarization state of polarized light as it propagates through the anisotropic optical device consisting of a sequence of anisotropic phase plates, the optical axes are rotated through a certain angle relative to each other (Jones, 1948). Similar systems have been made long ago, in the form of spiral structures. They proved that by the passage of linearly polarized light through the mentioned system, rotation of the polarization plane takes place, analogically to the optical rotator (Dawson&Young, 1960). However, the case when the wavelength of the light is comparable with the period of the spiral structure was not considered in these studies. The physical pictures of the phenomenon, in these extreme cases, are substantially different. As examples the liquid crystals or the sculptured thin films with the chiral structures can be cited (Chilaya et.al., 1997; Laktakia&Messier, 2005). In particular, in these cases, in addition to optical rotation, the selective reflection at wavelengths which are comparable to the period of such structures takes place. It is obvious that there exists a possibility of the realization of similar chiral structures with the help of holographic method based on the Weigert's effect.

A simple theoretical analysis confirmed this assumption. Consider the formation of the interference pattern of two orthogonally circularly polarized plane light waves of equal intensity propagating in opposite directions (counter-propagating beams) (Fig. 15). Let the electric vectors of interfering waves be (Jones, 1941; Kakichashvili, 1974):

$$\vec{E}_+ = \vec{E}_0 \begin{pmatrix} 1 \\ i \end{pmatrix} \exp(i\varpi t - kx + \delta_1) \tag{6}$$

$$\vec{E}_- = \vec{E}_0 \begin{pmatrix} 1 \\ -i \end{pmatrix} \exp(i\varpi t + kx + \delta_2) \tag{7}$$

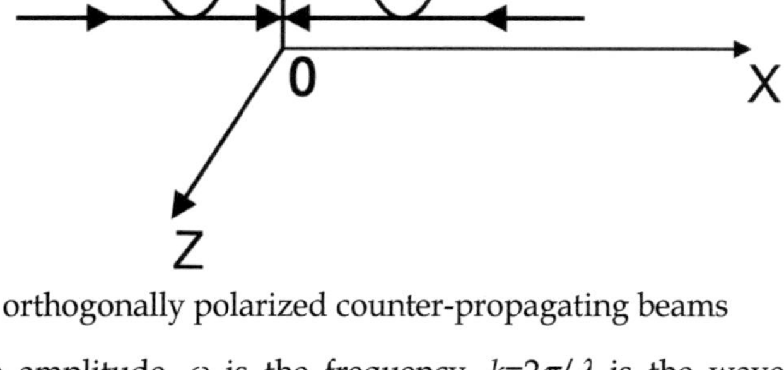

Fig. 15. Circularly orthogonally polarized counter-propagating beams

where $\vec{E}_0$ is the amplitude, ω is the frequency, $k=2\pi/\lambda$ is the wave number, λ is the wavelength, x is the direction of propagation, δ_1, δ_2 are initial phases. After analogous to that in Chapter 3, for the interference pattern of these waves, we will have:

$$\vec{E}_{\Sigma} = 2\begin{pmatrix} \sin[2kx + (\delta_2 - \delta_1)] \\ \cos[2kx + (\delta_2 - \delta_1)] \end{pmatrix} \tag{8}$$

Thus, from (8) follows that the total interference pattern is polarized linearly and its electric field vector describes a helix along the axis OX. Fig. 16 shows the distribution of the vector of electric field in the standing wave in this case.

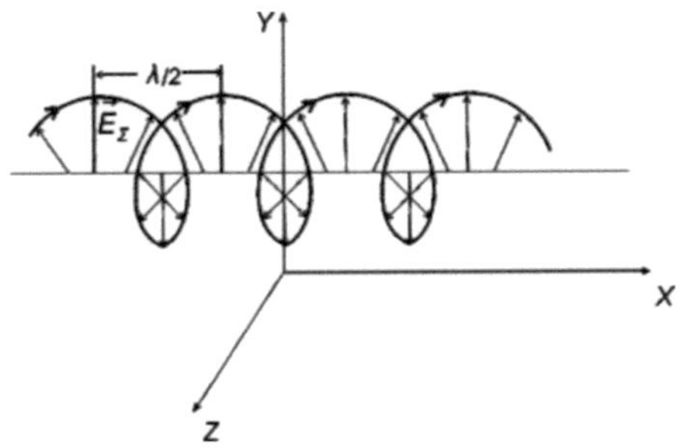

Fig. 16. Vector standing wave with helical structure

Clearly, if such a wave is created in a photosensitive material with the Weigert's effect, the obtained picture of the anisotropy will have a chiral structure, similar to the cholesteric liquid crystal.
If the dependence of the refractive indices of the photosensitive medium on the intensity has linear nature (Kakichashvili, 1974):

$$\begin{aligned} \hat{n}_1 &= \hat{n}_0 + aE_1^2 \\ \hat{n}_2 &= \hat{n}_0 + aE_2^2 \end{aligned} \tag{9}$$

For the matrix of this holographic diffractive optical element, created in this case, we will have (Jones, 1941; Kakichashvili, 1974):

$$M_d = 2ikda \exp - ikd\hat{n}_0 \begin{pmatrix} \cos[2kx + (\delta_2 - \delta_1)] & \sin[2kx + (\delta_2 - \delta_1)] \\ \sin[2kx + (\delta_1 - \delta_2)] & -\cos[2kx + (\delta_2 - \delta_1)] \end{pmatrix} \tag{10}$$

The diffracted wave produced from such structure is:

$$\begin{aligned} \vec{E}_d = M_d \vec{E}_+ = 2ikdaE_0 \exp - ikd\hat{n}_0 \begin{pmatrix} \cos[2kx + (\delta_2 - \delta_1)] & \sin[2kx + (\delta_2 - \delta_1)] \\ \sin[2kx + (\delta_2 - \delta_1)] & -\cos[2kx + (\delta_2 - \delta_1)] \end{pmatrix} \begin{pmatrix} 1 \\ i \end{pmatrix} \times \\ \times \exp i[\omega t - kx + \delta_1] = 2ikdaE_0 \begin{pmatrix} 1 \\ -i \end{pmatrix} \exp i[\omega t + kx + (\delta_2 - kd\hat{n}_0)] \end{aligned} \tag{11}$$

The expression of $\vec{E}_d$ represents the second wave (8) diffracted on the obtained chiral structure. From (11) follows that similar holographic structure behaves like a cholesteric liquid crystal.

The results of analyze were confirmed experimentally by means of holographic recording in counterpropagating beams, two orthogonal circularly polarized waves. The optical scheme of experimental equipment is presented in Fig. 17. Circularly polarized beam of He-Cd-laser (λ= 441.6 nm), divides into two mutually perpendicular linearly polarized beams of equal intensity, with the help of birefringent crystal of the calcite **1**. Crystal was cut in directions parallel to the main crystallographic planes, so that the incident beams splits into two parallel to each other ones, the electric vectors of which oscillate at an angle of ± 45^0 with respect to the drawing plane. The collimator **2 - 3** creates two parallel collimated beams, which are falling into two total internal reflection prisms **4** and **4'**. Between the prisms **4** and **4'** is placed a photosensitive material **5**, for instance, a glass plate, covered with a layer of azo dye embedded in gelatin with a strong Weigert's effect. The $\lambda/4$- phase plate **6** ensures the formation of two opposing waves with the orthogonal circular polarizations propagating in the photosensitive layer. To minimize the influence of Fresnel reflection on the recording process, the space between the prisms **4,4'** was filled with immersion liquid, into which the photosensitive material **5** is placed. In addition the photosensitive material was tilted at a small angle to separate the diffracted light from the Fresnel component. Obtained holographic structure, with the distributed anisotropy, represents the selective mirror, with sufficiently low reflection efficiency (η=1,4%). The dependence of the reflection (diffraction efficiency) on the polarization state, for the same wavelength used for recording (λ=441,6 nm), was investigated. The use of the same wavelength for reading is due to the fact that the step of obtained anisotropic holographic structure is determined by the wavelength of recording and corresponds exactly to it. For other wavelengths the diffraction is absent. Fig.18 illustrates the dependence of diffractive efficiency of the obtained holographic structure on ellipticity of the polarization of the incident wave. As it follows from Fig. 18, at the right circular polarization of the probe light, when the ellipticity is ε=1, the reflection efficiency has the maximal value (η=1,4%). With decreasing the ellipticity to 0, i.e. for linear polarization of the sensing light, the reflection efficiency decreases proportionally to 0,8%. Upon of the change of rotation of the electric vector of the incident wave in opposite proportional decrease in the efficiency of reflection , takes place and goes to zero for the left circular polarization. It is important, that the reflected light remains left-circularly polarized, in all cases. All of these facts confirm that the distribution of the anisotropy, in the obtained holographic structure, has chiral nature. A low reflection efficiency is the result of a high absorption (about 90%) in the photosensitive medium for the wave λ=441,6 nm. Due to this the interference pattern has a strongly damping character inside of the photosensitive medium. Thus, the standing vector wave in reality will have heterogeneous character inside photosensitive medium. In particular, the condition $\vec{E}_+ = \vec{E}_=$ does not hold inside the photosensitive material and the vector of the interference field, in reality, describes an ellipse with ellipticity changing along the standing wave. In this case the anisotropy (Weigert's effect) will be created only because of the difference between of intensities along the big and small axes of the ellipse. Therefore, linearity of the polarization is achieved only for the middle area of the photosensitive medium, in the interference pattern ($d/2$), where the intensity already is very low. Thus, the modulation depth of such anisotropic structure and accordingly its diffractive efficiency will be also low. In practice, the maximum value of the birefringence induced by light in the azo-dye materials reaches $\Delta n \approx 0.01$, which is enough to obtain higher efficiency. If the recording takes place in a wavelength range, where the absorption of the photosensitive material is much lower

(λ=488.0 nm and λ=514.5 nm), the polarization in the interference pattern is near to linear. In addition, the intensity of the interference pattern, in the photosensitive layer, in this case, will be higher, and thus the result of the Weigert's effect will be also larger. Accordingly, the diffraction efficiency of such structure may reach higher values.

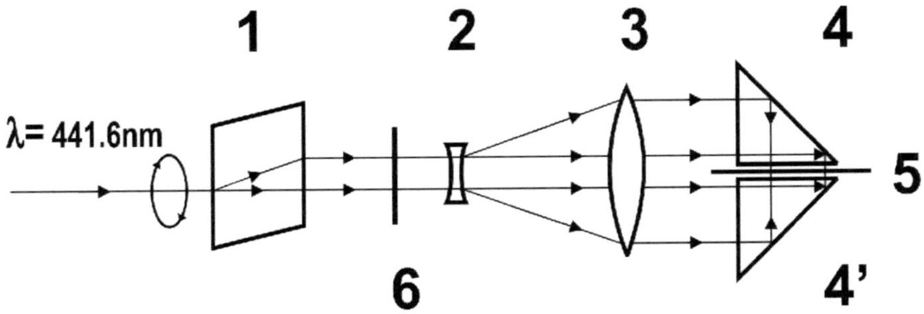

Fig. 17. The scheme of the recording of the holographic chiral structure

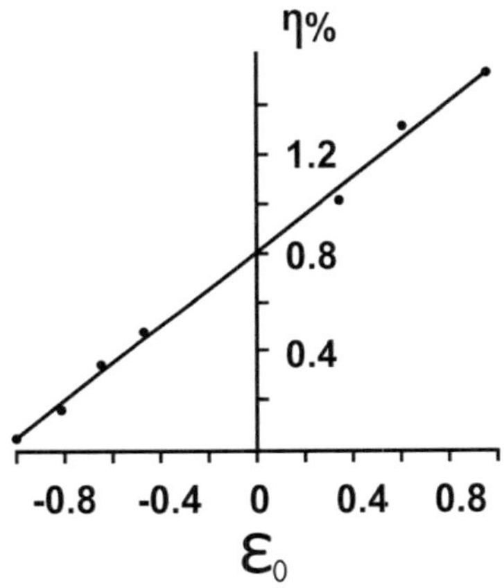

Fig. 18. The diffractive efficiency dependence on the ellipticity of the falling wave.

5. Selfrecording phenomenon

As it was noted, the use of photosensitive materials with the effect of Weigert gives a possibility of holographic recording, under the modulation of interference pattern only with state of polarization. One of the known examples of this is the interference of two waves with orthogonal circular polarizations. The schematic images of interference pattern for equal intensities and different intensities of the interfering waves in this case is shown in Fig. 19 a, b, (Kakichashvili, 1974). The investigations of some authors showed that the holographic recording in similar conditions in azo-dye materials provides 100% diffraction efficiency of thin holograms (Shatalin&Kakichashvili, 1987). In this case the reconstructed wave is polarized also circularly, but orthogonally to the reconstructing wave. It should be noted that one of the important results of the mentioned holographic process is that the diffraction efficiency of holograms increases continuously, reaches a maximum and remains constant on the whole further process of recording. Therefore, overexposure, in this case, does not take place and it is observed in all of the azo-dyes (Todorov at al., 1984;

Naydenova, L. at al., 1998). It is interesting the influence of the diffracted light on the dynamic process of holographic recording. In particular, at sufficiently high diffraction efficiency holograms (η> 10%), the intensity of the diffracted light is close to the intensity of the recording beams. Accordingly, the diffracted wave may create the additional interference pattern together with basic interference pattern and can provide the additional holographic recording. Therefore, the wave diffracted by the hologram can have an influence on the dynamic of holographic recording. Observation of this influence is very difficult in the process of holographic recording. However, it should be more pronounced on the stage of the reconstruction process. In particular, after blocking one of the waves in the recording process, the interaction of the diffracted wave with the second beam will create its own interference pattern, which can create sufficient conditions for the additional process of holographic recording. This process may be named as the self-recording phenomenon.

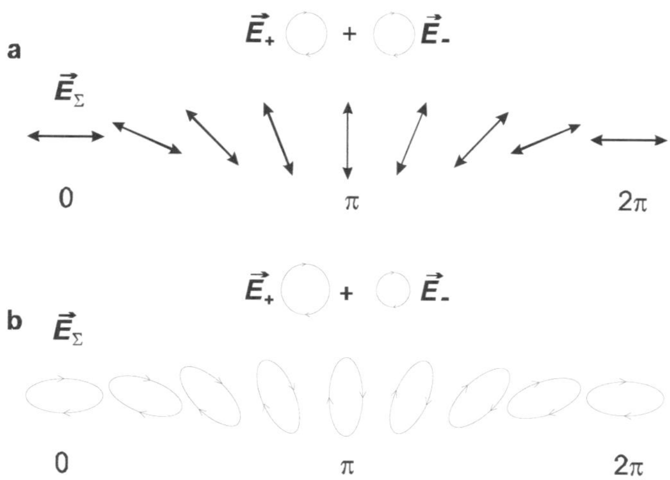

Fig. 19. Interference of two wave with orthogonal circular polarizations.

For the verification of this assumption, holographic recording in coincident beams with orthogonal circular polarization was realized. As the photosensitive material azo-dye colored gelatin was used. The optical scheme of the experimental equipment is presented on the Fig. 20. Passing through $\lambda/4$ - phase plate **1,** the linearly polarized beam of **Ar** laser (λ=488, 0 nm) divides into two orthogonally linearly polarized beams by means ofpolarization Rochon prism **2**. The rotation of $\lambda/4$ - phase plate **1** allows varying the ratio of intensities of these beams. The collecting lens **3**, creates two diverging beams of light propagating parallel to each other. The second $\lambda/4$ - phase plate **4** converts the orthogonal linear polarization of the beams into orthogonal circular polarization. The second collecting lens **5** forms two collimated beams with orthogonal circular polarizations, converging on the plane of the photosensitive medium **6**. Described scheme enables the realization of holographic recording at different ratios of the intensities of the interfering beams. Non-destructive readout was carried out with linearly polarized radiation **7** **He-Ne** laser (λ= 632,8 nm), which is directed perpendicular to the hologram **8**. In the beginning, the holographic recording was carried out, when the intensities of interfering waves were equal. The resulting hologram represents a diffraction grating with anisotropic structure, where the distribution of the anisotropy direction corresponds to the distribution of linear polarization direction in the interference pattern.

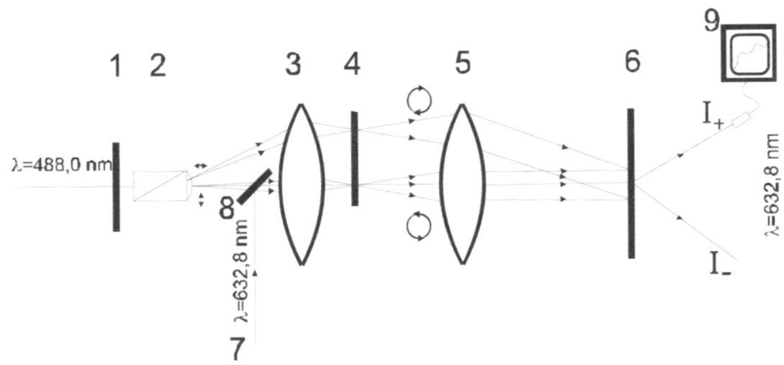

Fig. 20. The scheme of the dynamic holographic recording

The intensities of the ±I orders of the diffracted wave (λ=632,8 nm) were measured by means of recorder **9**, for the control of the dynamic processes of holographic recording and reconstruction. Fig.21 shows the results of investigation of the dynamic processes of holographic recording and reconstruction at equal intensities of the recording waves in the cases of:

I - orthogonal circular polarizations of the recording waves - curve **1** (only recording process);

II - orthogonal circular polarizations of recording waves - curves **2 a,b** (recording - **a**, and reconstruction - **b** by means of actinic light);

III - collinear circular polarizations of recording waves - curves **3 a',b'** (recording - **a'**, and reconstruction - **b'** by means of actinic light).

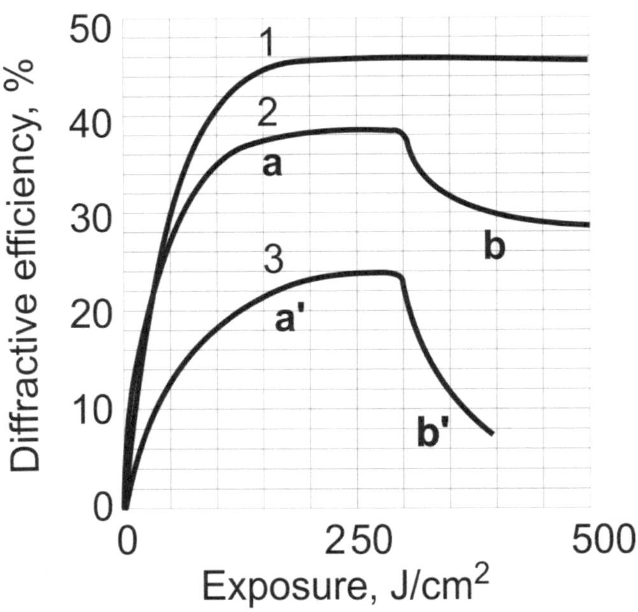

Fig. 21. The dynamic of the holographic recording and reconstruction

The interference pattern, in the cases I and II of holographic recording, is modulated only by polarization and, vice versa, in the case III is modulated only by intensity. Curve **1,** reflects only recording process, and shows that the diffractive efficiency of hologram increases continuously, reaches the maximum value and maintains it during the whole further process of recording. Curve **2** shows that the diffractive efficiency is increasing, analogically as in the case I (curve **2 a**), but decreases after blocking one of the recording beams (curve **2 b**). Similarly, curve **3** shows that the diffractive efficiency is rising in the beginning in the case **III** for the collinear circular polarization of the recording waves (curve **3 a'**) and decreases after interrupting one of the interfering waves similarly to the case **II** (curve **3 b'**). However, despite the similarity, the important difference between the processes II and III, described by the curves **2** and **3**, is obvious. In particular, from Fig. 21 follows that the maximum value of diffractive efficiency is almost twice as much for the orthogonal circular polarization of the recording waves, than for their collinear circular polarizations. It means that the holographic structure with modulation of anisotropy is more efficient than the isotropic structure. From the point of view of the investigated problem, a more important difference between the mentioned processes is that the diffractive efficiency in the reconstruction stage for the anisotropic hologram, produced in orthogonal circular polarizations of the recording waves, decreases much slower than for isotropic hologram, which is recorded in the collinear circular polarizations of the recording waves. The only explanation of such sharp difference between the results of reconstruction can be the difference between the rates of the erasure and the rate of the new holographic recording. Namely, because of great diffractive efficiency of hologram with an anisotropic structure, the intensity of the diffracted part of the actinic light, is closer to the intensity of the falling light, passing through hologram. Accordingly, after blocking of one of the recording beams, the diffracted wave creates a new interference pattern together with the reconstructing wave, which can stimulate the additional process of holographic recording. Consequently, it is clear that the dynamic of the reconstruction process, will depend on the ratios between the process of deletion and the secondary process of the recording. It turned out, that the difference between the intensities of the circularly orthogonally polarized waves of recording, is the most important condition for observation of the process of self-recording (Fig.22). Curve **1** corresponds to the diffraction efficiency of hologram for the ratio of intensities $I_+/ I_-=1/2$ of the recording waves and curve **2** corresponds to the ratio of $I_+/ I_-=1/3$.

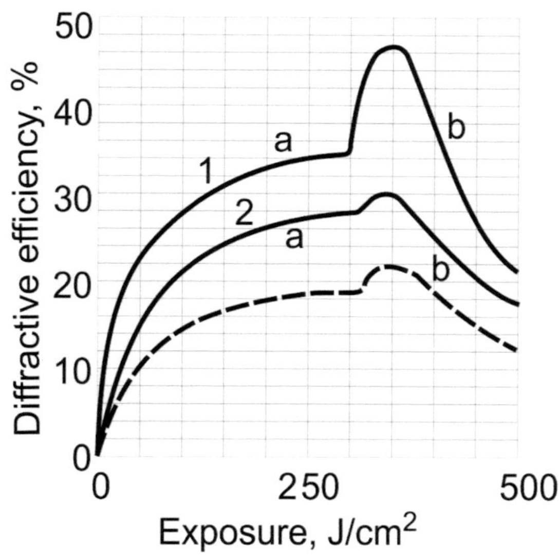

Fig. 22. The dynamic of the holographic recording and reconstruction

The wave with lower intensity was blocked in both cases of holographic process at the moment of reaching the maximum diffractive efficiency (curves *1* and **2 a**). The +I-order of diffracted sensing wave with more high intensity was measured for the observation of the reconstruction process. From Fig.22 follows the instant increase of the diffractive efficiency, which reaches its maximum value in a short interval of time and then decreases in an ordinary way (curves **1** and 2 **b**). At the same time the investigations show that the optimum of the self-recording corresponds to the ratio of $I_+/I_-=1/2$ of the intensities of the recording waves. It must be noted also that the diffraction picture has asymmetrical character in the recording process, i.e. intensities of diffracted ±I-orders of the sensing wave (λ=632,8 nm) are different. The process of self-recording, was also observed during the measurement of the less intensive -I-order of diffraction of the sensing wave (dotted line in Fig.22). It indicates that the observed phenomenon is not a result of the usual redistribution of intensities between ±I - orders of diffracted waves. It is important that the obtained result is not observed at the holographic recording by means of waves of the other polarizations except of orthogonal circular.

Let us consider one of the possible explanations of obtained results. As it was mentioned above, in the case of equal intensities of the circularly orthogonally polarized recording waves, the interference pattern has linear polarization whose orientation changes depending on the phase difference between the interfering waves. Such an interference pattern generates the adequate distribution of anisotropy in the photosensitive medium with Weigert's effect. It is clear that the maximum of the diffraction efficiency of similar hologram corresponds to the maximum of anisotropy (Weigert's effect). Obtained hologram is exposed impact only by one beam with circular polarization after blocking one of the recording waves when its diffractive efficiency is already maximal. The diffracted wave provides the establishment of new interference pattern, which may initiate continuation of the holographic recording process. However, the further increasing in the diffractive efficiency must not be take place, because the value of anisotropy already reached maximum value. Accordingly, the additional holographic recording can only lead to the decrease in speed erasing process (curves **2 a, b** and **3 a'**, b'of Fig.22). Unlike the previous case, if the intensity of the recording waves are different, the interference pattern will be elliptically polarized. The orientation of ellipse varies depending on the phase difference between interfering waves in this case. Because the photosensitive material has anisotropic response, created anisotropic structure is similar to previous one, but maximum value of anisotropy must be lower, as a result of competing effects in orthogonal directions along the ellipse axes. Therefore, the maximum value of the diffractive efficiency of the hologram is also lower in this case. Accordingly, we can assume that such holographic structure should have a resource to further increase the magnitude of the anisotropy and, thus, for the future increase in its diffraction efficiency. Simple experimental investigation of dynamic processes of Weigert's effect in azo dyes, with linear and elliptical polarization of actinic light (λ=488, 0 nm), was carried out to test this assumption. To estimate the created anisotropy, with help of the recorder, transparency of the samples placed between crossed polarizers, at a wavelength of λ= 632,8 nm was measured. Polarization of the actinic light waves were chosen in such away that the direction of the electric vector of linearly polarized light ($\vec{E}_{YL}$) and the direction of the electric vector of the elliptically polarized light ($\vec{E}_{YE}$), oscillating along the main axis of the ellipse coincided and were oriented vertically (OY axes). In addition (Fig. 23), the intensity of the beams were chosen so that the electric vector of linearly polarized light $\vec{E}_{YL}$ (Fig. 23 **a**) and component's of electric vector $\vec{E}_{YE}$ of the elliptically polarized light(Fig. 23 **b**), along the main axis, were equal. So the intensity of the linearly polarized beam, as well as, the intensity of elliptically polarized beam, were equal in the direction of the axis OY. Polarization of the sensing wave (λ= 632,8 nm) is oriented at the angle of 45^0, regarding to axis OY. The results of the investigation are presented in Fig.24. Curve **1** shows the dynamics of increasing of anisotropy in the light-sensitive material for the linear polarization of the actinic light (λ=488, 0 nm). Curve **2 a** corresponds to similar dependence, for the elliptical polarization of actinic light. As follows from these results despite the great intensity, the maximum value of anisotropy, created by elliptically polarized light is significantly lower, than the anisotropy created by linearly polarized light with lower intensity. However, under additional irradiation of the sample (which already was irradiated up to saturation, by means of elliptically polarized light) by means of linearly polarized actinic light, the electric vector of which directed along the main axis of the ellipse of polarization, the value of the anisotropy increases again (curve 2 **b**). Thus, when the interference pattern has elliptical polarization, the anisotropy of holographic structure which is created, has an additional resource to further increase of its anisotropy. After the blocking of the beam of lower intensity, in the process of holographic recording, when the diffractive efficiency is maximum, the diffracted beam creates the interference pattern (with the second recording wave), with polarization close to linear. Therefore, this interference pattern may initiate a further increase in anisotropy of the hologram and accordingly the further increase of its diffraction efficiency, i.e. of the self-recording process.

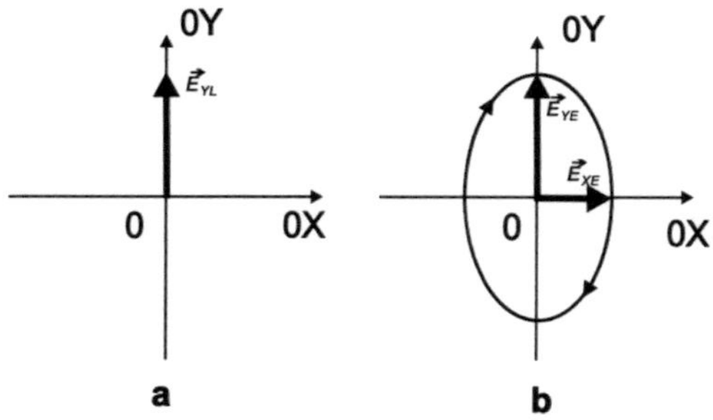

Fig. 23. The schematic picture of creation of the Weigert's effect by the linearly polarized (**a**) and elliptically polarized (**b**) light.

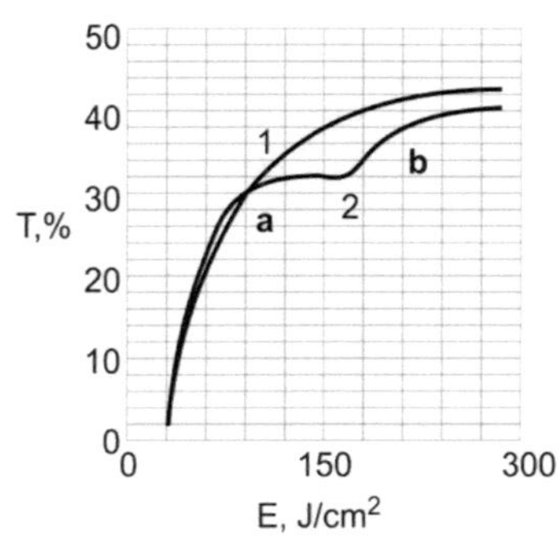

Fig. 24. The dynamic of the Weigert's effect

6. Holographic recording in unpolarized light

After passing of anisotropic or gyrotropic transparent objects, unpolarized light changes its statistical polarization structure. In this case the interaction with the initial reference wave gives interference pattern, which is modulated by intensity , as well as by polarization. Application of photosensitive material with Weigert's effect gives a possibility of holographic registration of above mentioned change of unpolarized light. For the verification of this assumption the corresponding holographic recording was carried out. As the simple anisotropic and gyrotropic objects can be use corresponding phase plates. It is also known that the interference pattern becomes to evenly lit if on the way of one of the interfering unpolarized waves put an anisotropic half-wave or gyrotropic half-wave (rotator at 90^0) phase plate (Langsdorf&Du Bridge, 1931; Vavilov, 1932). However, the polarization analysis shows that the total interference pattern is modulated with the polarization state. The optical scheme of holographic recording is shown in Fig. 25. The collimated beam of the He-Cd (λ=441.6 nm) by means of the Fresnel biprism **1** splits into two beams that overlaping in the plane of the photosensitive material **2**. Azo-dye doped gelatin was used as photosensitive material. On the way of one of the interfering wave the half-wave phase plate 3 was placed. The microlens **4** and the polarizer **5** allow to view the interference pattern and carry out its polarization analysis. In the absence of the half-wave phase plate, the interference pattern is modulated only by intensity, and its bright areas are also unpolarized (Fig. 26 **a**). After placing of the phase plate the modulation of interference pattern by the intensity disappears and it becomes uniformly illuminated (Fig. 26 **b**). However, the polarization analysis shows that the interference pattern is modulated with the polarization.

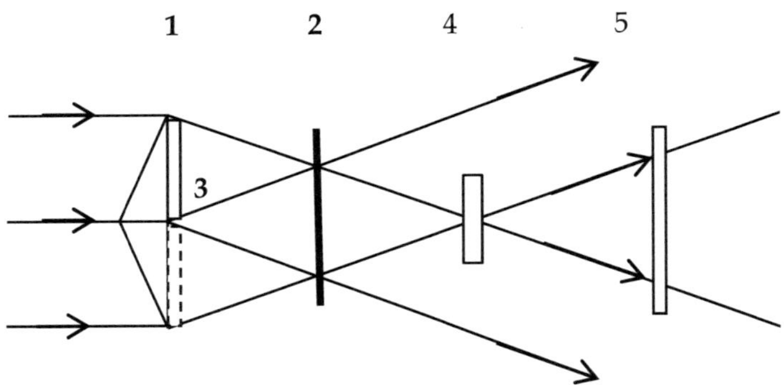

Fig. 25. The scheme of the holographic recording.

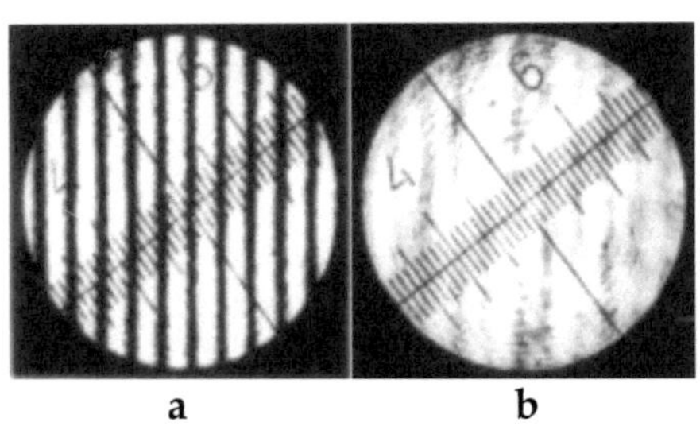

Fig. 26. Interference pattern of unpolarized light without (**a**) and with (**b**) $\lambda/2$ anisotropic phase plate.

Fig. 27 shows the image of the interference pattern in three different orientation of the polarizer. Fig. 27 **a** shows the image of the interference pattern when the polarizer, is oriented along to the fast axis of the phase plate. Fig. 27 **b** shows the image of the interference pattern when the polarizer is oriented at the angle of 45^0 to the fast axis of the phase plate. Fig. 27 **c** shows the image of the interference pattern when the polarizer is oriented at the angle 90^0. From Fig. 27 **a** and **c** follows that there is a shift of the interference fringes as a result of changing the orientation of analyzer by 90^0. This testifies the existence in the interference pattern of two linear polarizations of orthogonal orientation. It is clear that under intermediate orientation of analyzer, the interference pattern will have uniform luminosity, which is confirmed by Fig. 27 **b**. Therefore, the ordinary photosensitive materials without Weigert's effect will not register such interference pattern. Thus, for realization of holographic recording in this case the application of the photosensitive material with Weigert's effect is necessary. The maximum of the diffraction efficiency of created holographic grating reached 10%. It is obviously that the greatest interest in this experiment represents comparison of polarization characteristics of the recorded object wave with that of the reconstructed wave. Investigation showed that reconstructed wave is also unpolarised. From this point of view is interesting to compare the statistical polarization structures of the wave formed with the help of half-wave plate and of its holographic reconstruction. For the comparison, the half-wave phase plate was removed. The object wave was blocked and with the help of micro lens **5** and polarizer **6**, the interference pattern between reconstructed and reference waves after hologram was observed. To obtain of equal intensities of interfering waves, in this case, the intensity of the reference wave was decreased with the help of a neutral light filter. The results of the investigation is presented in Fig. 28 **a,b, c**, which is practically identical to the results represented on Fig. 27 **a,b,c**.

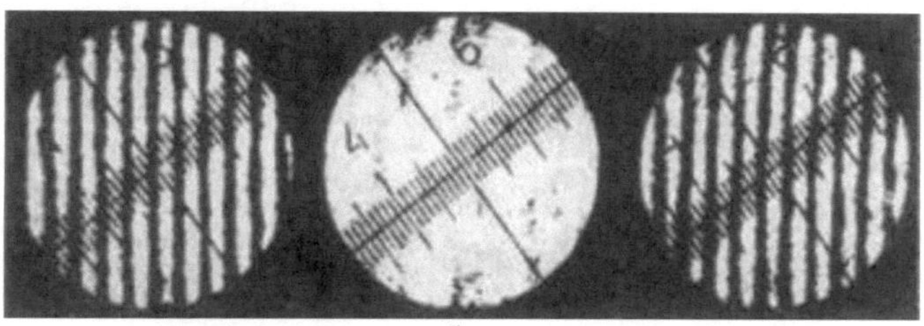

a b c

Fig. 27. The interference pattern with $\lambda/2$ anisotropic phase plate after polarizer

a b c

Fig. 28. The interference with the reconstructed image of the $\lambda/2$ anisotropic phase plate after polarizer

These results confirm that holographic recording in the unpolarized light, on the basis of Weigert's effect, when the object wave is forming with the help of half-wave anisotropic phase plate, gives a possibility of the adequate reconstruction of the changing of statistical polarization structure of unpolarized wave.
As noted above, if the unpolarized object wave is formed with the help of half-wave gyrotropic phase plate the interference pattern contains two mutually orthogonal circular polarization. In this case, when as a recording medium a photosensitive material with Weigert's effect is used, the holographic recording was impossible. Obtained results show that for holographic recording of full information about the wave scattered by the object, in general, Weigert's effect is insufficient and here is needed a photosensitive material with gyrotropic response.
In spite of this, the holographic recording in unpolarized light, extends considerably the possibilities of holography. One of the specific examples of the use of unpolarized light in holography is the solution to some problems of photoelasticity.
The holographic recording for strained transparent object was investigated experimentally, by means of unpolarized light. At the first stage of holographic recording, the photosensitive material was dichromated gelatin without the Weigert's effect. The reconstructed image is shown in Fig. 29 a. This image represents a picture of the isochromes. At the second stage, on the path of the reference wave the half-wave anisotropic phase plate was placed. The reconstructed image of the object, in this case, are represented in Fig 29 **b,c**. Fig. 29 **b** corresponds to holographic recording, when the optical axis of the phase plate is oriented vertically. Fig. 29 **c** corresponds to holographic recording, when the optical axis of the phase plate is oriented at an angle of 45 degrees to the vertical. These images represent the picture of the isochromes and isoclins. From Fig. 29 follows that, the reconstructed image contains some information about the distribution of anisotropy in the object. If on the way of reference wave is placed the half-wave girotopic phase plate (quartz rotator 90^0) the holographic recording does not take place in this photosensitive material.

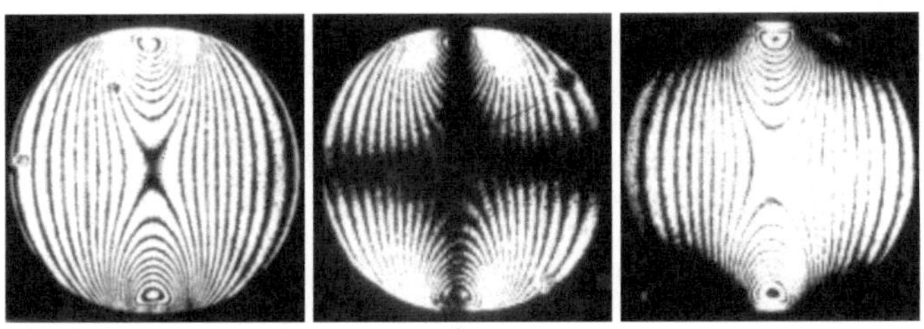

a b c

Fig. 29. The rceonstructed image of the strained transparent object.

In the next stage of holographic recording, as a photosensitive material, was used azo-dye doped gelatin with Weigert's effect. The reconstructed images are represented in Fig. 30 **a,b,c** Fig.30 **a** correspond to the holographic recording without phase plate. In this case, the reconstructed image has practically an uniform illumination. Fig.30 **b** represents the reconstructed image when on the way of the reference wave, in the process of holographic recording, half-wave anisotropic phase plate was placed. Here is clearly observed the pattern of isochromes and isoclins. So, from Fig. 30 **a** and **b** follows that these images represent complementary images of isochromes and isoclins, which are shifted on the one lane relative to each other. The first part of isochrome is the result of holographic recording with the help of unpolarized parts of interference pattern. The second part of isochrome is the result of holographic recording caused by influence of linearly polarized parts of interference pattern. This assumption is confirmed in Fig. 30 **c**. It represents the reconstructed image, when on the way of the reference wave, in the process of holographic recording, was placed a half-wave gyrotropic phase plate (quartz rotator 90^0), when the interference pattern consists of circularly and linearly polarized parts. Fig. 30 **c** shows picture only of the second part of isochromes, which is the result of holographic recording caused by influence of only linearly polarized parts of interference pattern. From the final result follows that, photosensitive material with the Weigert's effect can not register circularly polarized component of the interference pattern. Because of this we may conclude that to fully solve of this task of holography it is necessary to use a photosensitive material with gyrotropic response.

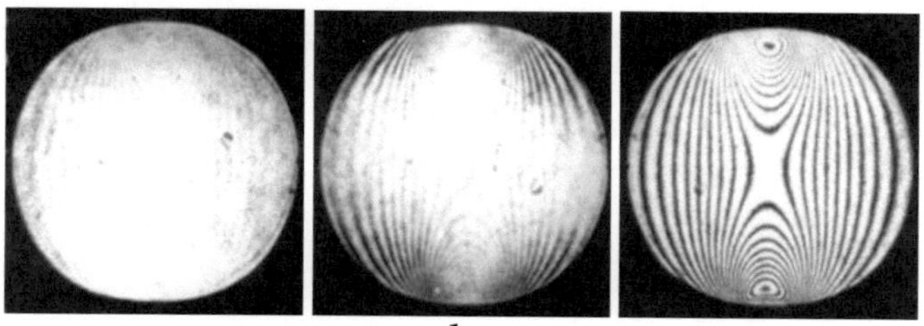

a **b** **c**

Fig. 30. The rceonstructed image of the strained transparent object.

7. Conclusion

From the results described above it can be concluded that the holography based on the Weigert's effect has the following advantages:

7.1 Holographic recording and reconstruction of the polarization in some particular cases.

7.2 Holographic recording and reconstruction of statistical polarization structure of unpolarized light.

7.3 The possibility of creation of the holographic diffractive optical elements with anisotropic structure with advanced features.

7.4 Some new processes (for example selfrecording) in the investigation of dynamic processes of holographic recording based on the Weigert's effect.

7.5 Some new approaches in investigation of mechanisms of Weigert's effect in the Azo-dyes.

8. Acknowledgment

I would like to express my great love to my mother Neburishvili L.K., which helps me in all my life and express my gratitude to the late Professor Kakichashvili Sh.D. with whom I had the privilege to work in the years 1982-1992.

9. References

Gabor, D. (1948). A New Microscope Principle, *Nature*, vol.161, pp. 777-778.

Leith, E.N.; Upatnieks, I. (1962). Reconstructed wave fronts and communication theory. *JOSA*, vol.52, № 11, pp.1123-1130.

Denisiuk, J.N. (1963). On the mapping of the optical properties of the object of the wave field scattered light from them. *Opt. & Spectr.*, vol. 15, № 4, pp.522-532, (Russian).

Born, M., Wolf, E. (1970). *Principles of Optics*, NAUKA, Moscow, (Russian).

Langsdorf, A., Du Bridge, I. (1931). Rotation of polarization in the unpolarized light. *JOSA*, vol. 21, pp.1-4,.

Vavilov, S.I. (1932). Proceedings, About the inside structure of unpolarized light. Ad. Sc. of USSR, ser.7, №9, pp.1451-1458, (Russian).

Weigert, F. (1919). Uber Einen Neuen Effect der Strahling in Lichtempfindlichen Schichten, *Verhandl. Dt. Physik. Ges.*, bd.21, ss. 479-483 (German).

Kondo, T. (1932). Uber den Photoanisotropen Effect (Weigert-effect) an Farbstoffen. *I//Z. Wissenschaftliche Photogr., Photophys. Und Photochem.*, bd.31, H. 6, ss. 153-267 (German).

Zocher, H., Coper K. (1928). Uber die Erzeugung optischer Aktivitat durch Zirkulares Licht. *Z.Phys.Chem.*, Bd.132, S.313-319 (German).

Kakichashvili, Sh. D. (1972). On Polarization Recording of Holograms. *Opt. & Spectr.*, vol.1, No.6, pp. 324-327, (Russian).

Eltsov, A.V. (1982). *Organic Photochromes*, Chimia, Leningrad, p. 285, (Russian).

Vasilieva N.V. at al 1978, *Practical Works In Organic Chemistry*, Prosveshchenie, Moscow, p. 303, 1978 (Russian).

Shatalin, I.D. (1989). To Mechanisms of Photoanisotropy In Photochemical Trans-Cis isomerisation. *Optics & Spectr.*, Vol.66, No.2, pp.362-364. (Russian).

Ebralidze, T.D., Ebralidze, N.A., Bazadze. (2002). M.A., Weigert's effect Mechanism Observed In Dyes. *Appl. Opt.*, Vol.41, No.1, pp.78-79.

Kakichashvili, Sh. D., Shaverdova, V, G. (1979). Photoanisotropy in Mordant Yellow Azo-Dyes. *Jorn. of Sci. and Appl. Phot. and Cinematograph*, Vol. 24, No.5, pp. 342-345, (Russian).

Jones, R.C. (1941). A New Calculus for the Treatment of Optical Systems. J. Opt. Soc. Amer., vol.31, No.7, 488-493.

Kakichashvili, Sh.D. (1974). Method of Polarization Recording of Holograms. *Kvantovaja Electron (Quant. Electr.)*, vol.1, N 6, pp.1435-1441, (Russian).

Shatalin, I.D., Kakichashvili, Sh.D. (1987). Polarization hologramm with 100% diffractive efficiency (polarization cinoform). Pisma v JTF (Sov. Tech. Phys. Lett.), vol.13, No.17, 1051--1055, (Russian).

Todorov at al., 1984, Polarization holography. 2: Polarization holographic gratings in photoanisotropic materials with and without intrinsic birefringence, Appl. Opt. vol. 23, Issue 24, 4588-4591.

Naydenova, L. at al., 1998., Diffraction from polarization holographic gratings with surface relief in side-chain azobenzene polyesters, J. Opt. Soc. Am. B Vol. 15, Issue 4, 1257-1265.

Jones, R.C. (1948). New Method of Calculation of Optical Systems. VIII. Properties of **N** - matrixes. *Journ. Opt. Soc. Amer.*, vol.38, 671-684.

Dawson, F.F., Young, N.O. (1960). Helical Kerr Cell". *Journ. Opt. Soc. Amer.*, vol.50, 170-171.

Chilaya, G. at al (1997). New Pretransitional Liquid Crystal Phase With Anomalous Selective Reflection. *Mol. Materials*, Vol.8, pp. 245-255.

Laktakia A., Messier, R. (2005). Sculptured thin films: nanoengineered morphology and optics. SPIE Press., vol.PM142, ISBN 978-0-8194-5606-9.

7

Holographic Image Storage with a 3-Indoly-Benzylfulgimide/PMMA Film

Neimule Menke[1] and Baoli Yao[2]
[1]*School of Physical Science and Technology, Inner Mongolia Universtiy*
[2]*State Key Laboratory of Transient Optics and Photonics, Xi'an Institute of Optics and Precision Mechanics, Chinese Academy of Sciences*
China

1. Introduction

Fulgide is well-known as a thermally irreversible organic photochromic compound [1,2]. It is one of the important materials in the field of optical signal processing and optical storage. Focusing on the holographic storage applications of 3- indoly-benzylfulgimide/PMMA film, in this chapter include the following themes.

In section 2, first the Fulgide, photochroism and photo-induced anisotropy are introduced. Then the spectra and kinetics of the photochromic and photo-induced anisotropic properties of 3-indoly-benzylfulgimide/PMMA film were studied.

In section 3, the applications of ordinary holography and polarization holography of fulgide film were studied, which were respectively based on the photochromic and photo-induced anisotropy properties. The properties of holographic recording such as diffraction efficiency, spatial resolution and optimal exposure were measured.

In section 4, the holographic optical image storage was realized in a 3-indoly-benzylfulgimide/PMMA film by using different kinds of holographic storage techniques, e.g., transmission-type holographic recording and reflection-type holographic recording, reference beam reconstruction and phase conjugated beam reconstruction, Fraunhofer holographic recording and Fourier-transform holographic recording, different kinds of polarization holographic recording (i.e. parallel linear polarization holographic recording, orthogonal linear polarization holographic recording, parallel circular polarization holographic recording and orthogonal circular polarization holographic recording etc.), and collinear holographic storage technology etc. The diffraction efficiencies and diffractive images' signal-to-noise-ratios (SNR) of different kinds of holograms were compared. The storage density of 2×10^8 bits/cm^2 was obtained in the Fourier-transform holographic data storage by using orthogonal polarization holographic recording, which had a greatly improved signal-to-noise ratio of the diffraction image. And different kinds of multiplexing holographic storage (like polarization multiplexing, circumrotation multiplexing and angle multiplexing) and holographic interferometer were realized in the film.

2. Definition of fulgide, photochroism and photo-induced anisotropy

2.1 Photochroism

"Photochromism"is simply defined as a light-induced reversible change of color. "Photochromism is a reversible transformation of a chemical species induced in one or both directions by absorption of electromagnetic radiation between two forms, A and B, having different absorption spectra".The thermodynamically stable form A is transformed by irradiation into form B. The back reaction can occur thermally or photochemically (like shown in Fig.1) [2,3].

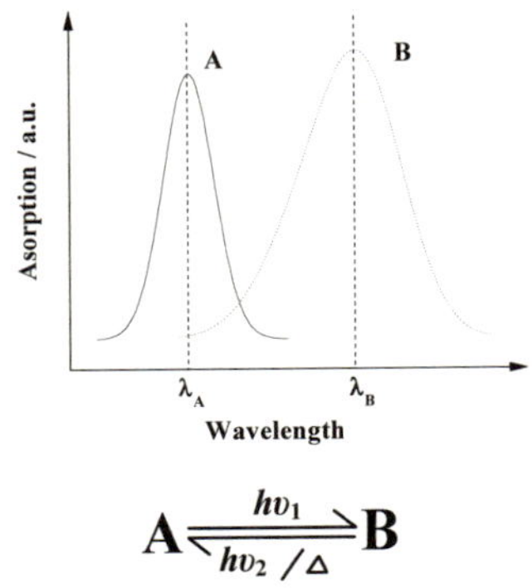

Fig. 1. A simple sketch map of photochromic

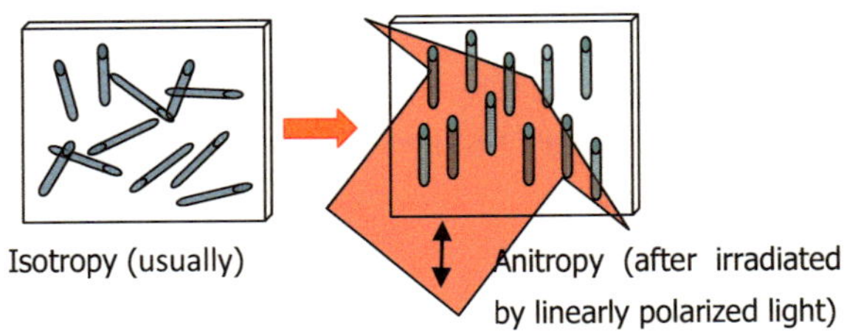

Fig. 2. A simple sketch map of the mechanism of photo-induced anisotropy

2.2 Photo-induced anisotropy

In transparent materials with anisotropic dielectric permittivity, optical anisotropy can be observed, in which include the pleochroism (anisotropy of the material's absorption coefficient) and birefringence (anisotropy of the material's refractive index). **Pleochroism** means that the absorption of the material to the light depends not only on the wavelength of light, but also on the polarization state of light: for uniaxial crystals, it is called as **dichroism**; for biaxial crystals, it is called as **trichroism**. **Birefringence** means that the refractive index of the material depends not only on the wavelength of light, but also on the polarization state of light.

Many crystals have crystal lattice structure themselves, so they are **natural optical anisotropic media**. However, in some amorphous material, under the action of certain external field (such as electromagnetic fields, mechanical forces, etc.), their atoms or molecules will be orientated in certain rules, thus the material will change from isotropic into anisotropic macroscopically, which is called as the **artificial optical anisotropy**. In which, under the irradiation of polarized light, some isotropic materials will turn to be anisotropic, or the degree of anisotropic properties in some materials will change, this phenomenon is called as the **photo-induced anisotropy** (like shown in Fig.2). Which include the photo-induced dichroism (usually photo-induced anisotropic materials show the properties of uniaxial crystal) and photo-induced birefringence.

2.3 Fulgide

2.3.1 Structures of fulgides

Early in 20th century, Stobbe and Eckert, (Leipzig University, Germany) [1,2] found that the condensation products of succinate and aromatic group of aldehydes, ketones have photochromic property. They stated in their article in 1905 that they named the derivatives of 1,3-butadiene-2,3-dicarboxylic acid and its acid anhydride as "fulgenic acid " and "fulgide", respectively (like shown in Fig.3), after the Latin word "fulgere" (glitter or shine), because some derivatives exhibited a variety of characteristic colors by light, and they usually formed shiny crystals.

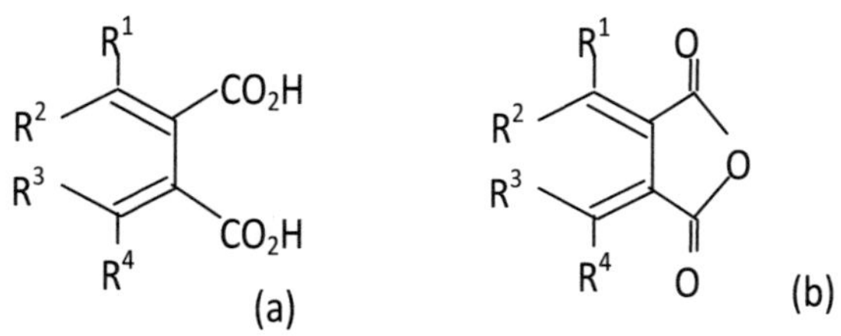

Fig. 3. Chemical molecular formula of the fulgenic acid (a) and fulgide (b)

To be photochromic, fulgides should have at least one aromatic ring or heteroaromatic ring (Ar) on the exo-methylene carbon atom, so that they form a 1,3,5-hexatriene structure that may undergo 6π-electrocyclization.

When three of the four substituents are same, the fulgide compounds have two isomers: cis-isomer and trans-isomer. According to geometrical shape of the double bond connecting the aromatic ring and succinic acid, they are called as "E-form" And "Z-form" respectively (for example, the isomers of phenyl fulgide are shown in Fig.4). In which, the E-form has an "all cis-hexatriene" structural unit, so it can be photocyclized. When two of the four substituents are same, the fulgide compounds have three isomers: (E,E), (E,Z) and (Z,Z). When the four substituents are different with each other, the fulgide compounds have four isomers: (E,E), (E,Z), (Z,E) and (Z,Z).

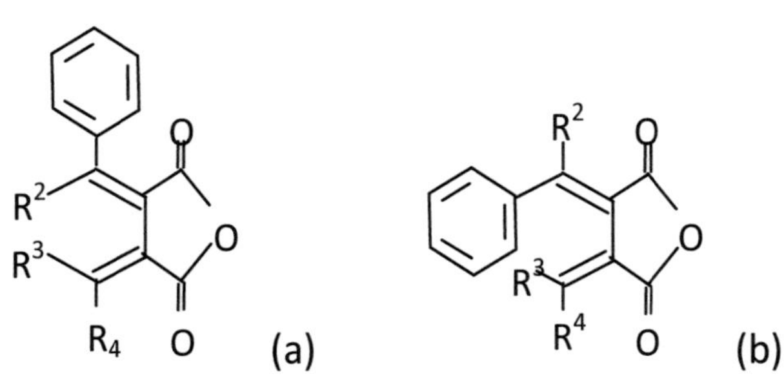

Fig. 4. The two isomers of phenyl-fulgide: (a) Z-form; (b) E-form

2.3.2 The photochroism of fulgides

The photochroism of fulgide occurs between one of the colorless open forms "E-form" and the photocyclized colored form (abbreviated as the C-form), like shown in Fig.5. The coloration mechanism of fulgide is the photochemical 6π-electrocyclization of the hexatriene moiety, is a cyclic reaction consistent with Woodward-Hoffmann selection rules. However, there exits an additional photochemical E-to-Z isomerization, the Z-form is not considered as an important component in the photochromic system. There has been no report that the Z-form cyclizes directly by absorbing on photon to give the C-form. Therefore, E-to-Z isomerization, competing with the photochromic E-to-C isomerization, is an energy-wasting as well as system-complicating process in terms of the "photochroism of fulgides". Using Improvement of the structure, this additional reaction can be suppressed. Usually E- and Z-forms have maximum absorption in the UV region; C-form has maximum absorption in the visible region .

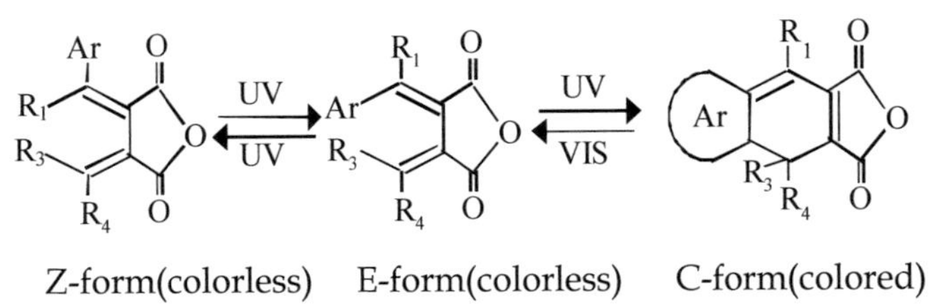

Fig. 5. The photochromic reaction of fulgides

2.3.3 The derivatives of fulgide

The photochromic mechanism of fulgide is the photochemical 6π-electrocyclization process, so the carbonyl group and the aromatic ring are both not indispensable. For example, anhydride part can be instead of other functional groups, like succinimides (fulgimide), butanolides (fulgenolide), dieters (fulgenate) and compounds having an anhydride ring with modified carbonyl group (as shown in Fig.6) [1,2].

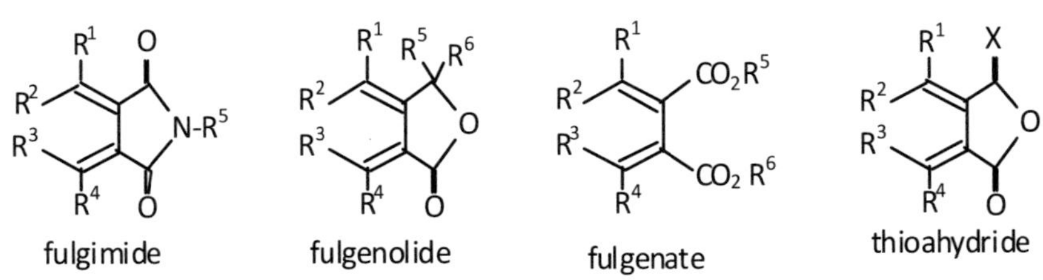

Fig. 6. Chemical molecular formula of the fulgide derivatives

2.3.4 The photo-induced anisotropy of fulgide films

We found that there exists photo-induced anisotropy in fulgide-doped polymeric films [4], which can be used in polarization holography application. When the colored states are irradiated by a linearly polarized 650 nm laser, the films returns to the bleached states and photo-induced anisotropy is produced during this process. The mechanism of photo-induced anisotropy in the fulgide films is the following [4]: anisotropic absorbing molecules of fulgide are immobilized randomly in the PMMA polymeric matrix, which shows isotropic characteristic at the initial state; when the sample is irradiated by linearly polarized light photo-selection of molecules take place. Molecules with long axes parallel to the exiting light polarization direction absorb the light strongly and turn to the other form very quickly, whereas those with long axes perpendicularly orientated to exiting light polarization have low absorption and stay at the initial form. As a result special orientation of two form molecules is induced and the sample shows optical anisotropy properties macroscopically. The angular distributions of molecules in anisotropy inducing progress are shown in Fig.7. So, under the irradiation of circularly polarized light, the sample shows also optical isotropy.

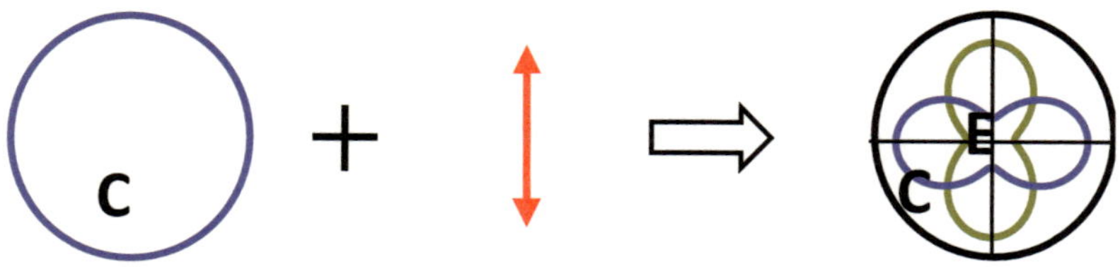

Fig. 7. Angular distributions of molecules in anisotropy inducing progress

2.4 Holographic storage application of fulgide films

Photochromic organic compounds show great potential in the field of rewriteable holographic storage owing to its benefits like lower price, higher signal noise ratio (SNR), higher spatial resolution, higher sensitivity, less toxic, no need to special fixing information after recording, stable in the air or moisture, and easy to be modulated [1,2]. Their disadvantages are lower diffraction efficiency and difficulty to realize non-destructive readout. In which, fulgides are famous as thermally irreversible organic photochromic compounds.

The photochromic and photo-induced anisotropic properties of materials can be used in ordinary and polarization holographic recording respectively. It is known that fulgide-doped polymeric films are photochromic and photo-induced anisotropic, which can be used for both.

For isotropy photochromic materials, only when the object light $\vec{O}$ and reference light $\vec{R}$ have components with same polarization state, there the intensity grating exists and the holograms can be recorded, called as **ordinary hologram**. But for photo-induced anisotropic materials, even if the $\vec{O}$ and $\vec{R}$ have orthogonal polarization states, the holograms also can be recorded, because that at this time although the intensity of the superposed light is a constant, but its polarization state changes with the phase difference between $\vec{O}$ and $\vec{R}$. Photo-induced anisotropic materials can record the polarization state of exciting beam, so the holographic gratings can be recorded, called as **polarization hologram**, in which not only the intensity and phase signals of $\vec{O}$ can be stored (the photo-induced anisotropy is depending on the exposure), but also its polarization state can be stored.

Here we just consider four kinds of polarization holographic storage, including the parallel linearly, parallel circularly, orthogonal linearly and orthogonal circularly polarized holograms. Fig.8 is a simple sketch map showing the space modulation of the distributions and orientations (optical axis) of C-form and E-form molecules of fulgide films in these four kinds of polarization holographic storage under the linearity recording condition when $O=R=A$. The arrows and ellipses on the horizontal line indicate the periodical distributions of intensity magnitudes and polarization states of the recording field corresponding to the different phase differences $\Delta\varphi$ along the x-axis. The rectangle frames under them indicate the distribution and orientation of the molecules under the corresponding light irradiations, where the black and white color indicate the C-form and E-form molecules respectively, and the arrows indicate the direction of induced optical axes, and the star flower (✻) means the sample is isotropic at the place.

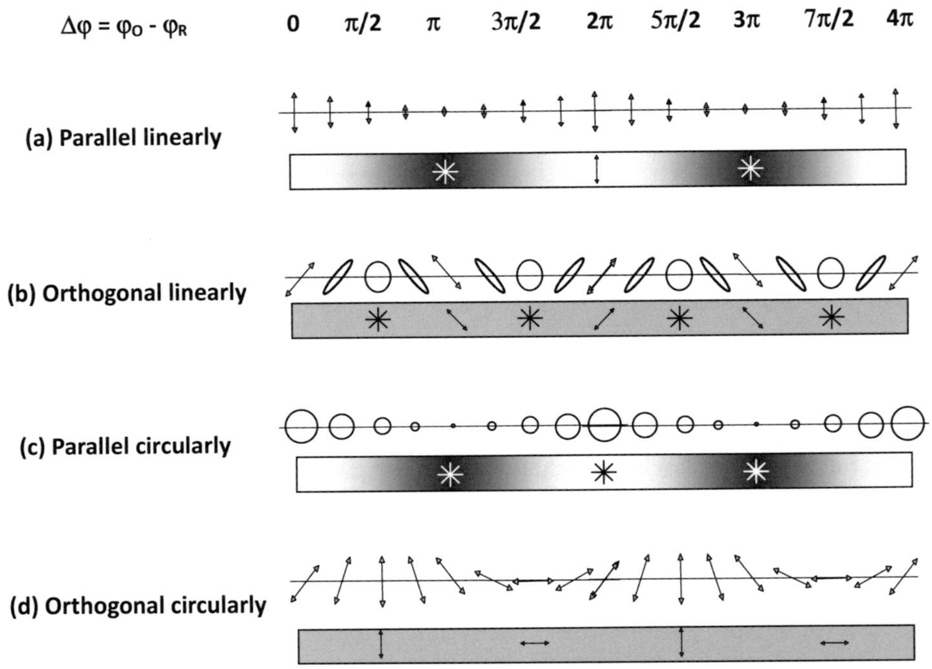

Fig. 8. The distributions of the interference fields in four kinds of polarization holography storage and the corresponding molecular distributions and orientations

From Fig.8, it can be seen that in the parallel circularly polarized hologram just ordinary hologram exits, and in the orthogonal polarized condition, just polarization hologram exits, but in the parallel linearly polarized holography, the ordinary hologram and polarization hologram will be recorded together.

2.5 Photochroism and photo-induced anisotropy of 3-indoly-benzylfulgimide/PMMA film

2.5.1 Material

The fulgide material studied here is 3-indoly-benzylfulgimide, which was synthesized by the Stobbe condensation routine [2]. The target compound of 3mg was dissolved in a 0.lml 10% (by weight) PMMA-cyclohexanone solution. Then the solution was coated on a 1-mm thick K_9 glass plane (∅ 25 mm × 1.5 mm) with a spin coater and dried in air. The thickness of the film is determined to be about 10μm by microscopy of the cross section. The photochromic or photo-induced anisotropic properties of fulgides are due to a reversible photochromic (photoisomerization) reaction that occurs between one of the colorless E-form (bleached state) and the C-form (colored state). These are the two spectrally separated photochromic forms, whose molecular formulas are shown in Fig.9. In this film we also found the photo-induced anisotropy property [4].

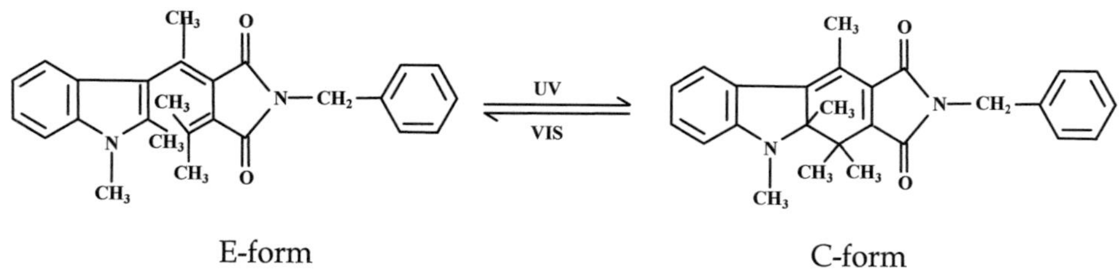

Fig. 9. Molecular formula of the Indolyfulgimide and the photochromic reaction. Left, E-form; right, C-form

2.5.2 Spectra of photochromic and photo-induced anisotropy properties of the sample

The absorption of the C-form ($A_C(\lambda)$) and E-form film ($A_E(\lambda)$) were measured using a UV-VIS-IR spectrophotometer (UV-3101PC, Shimadzu Inc., Japan), which were shown in Fig.10a. And the measurement of the photo-induced dichroism is performed by measuring the transmission spectra of the film for testing light polarized parallel ($T_{//}(\lambda)$) and perpendicular ($T_{\perp}(\lambda)$) to the polarization direction of the exciting beam after the C-form film is excited by the linearly polarized 650 nm laser (shown in Fig.11a).

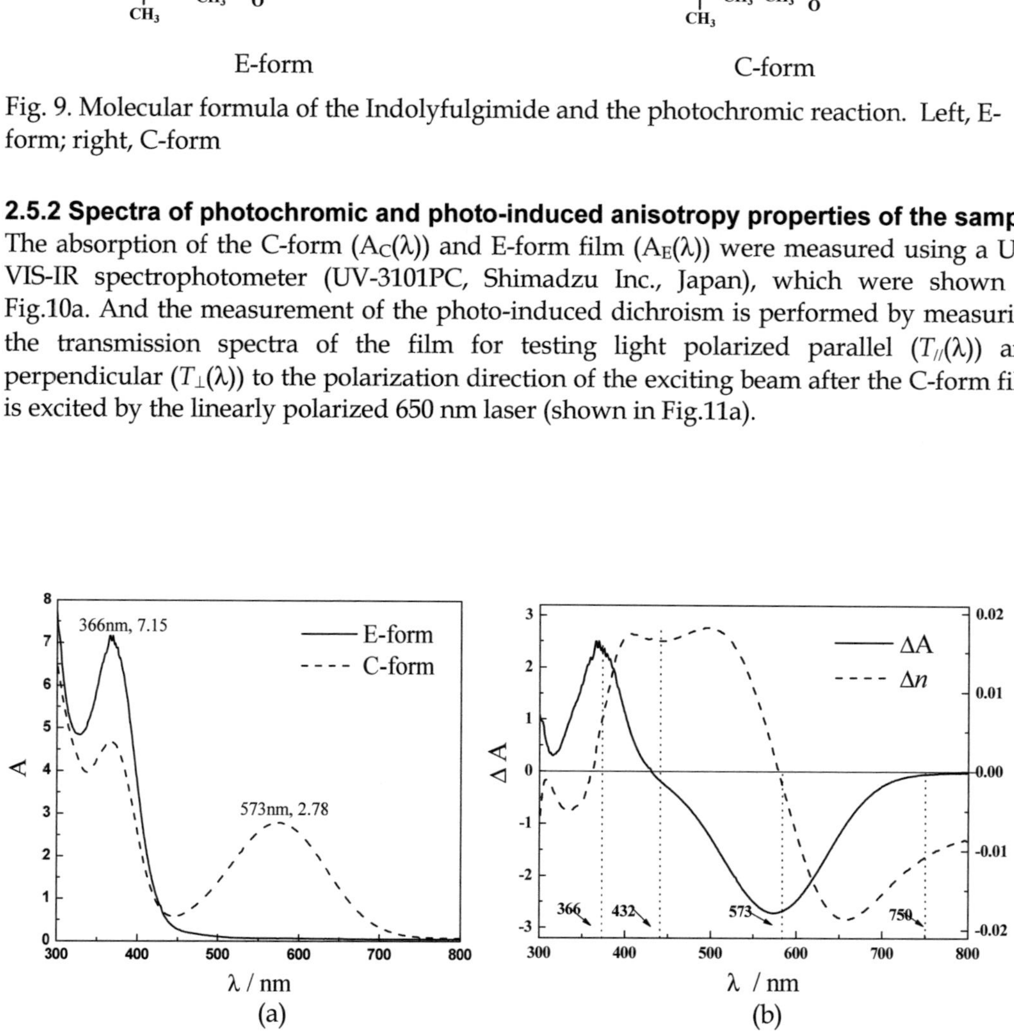

Fig. 10. Spectra of photochromic properties of Indolyfulgimide / PMMA film:(a) Absorption spectra of two forms; (b) Absorption difference spectrum and the corresponding refractive index changing spectrum

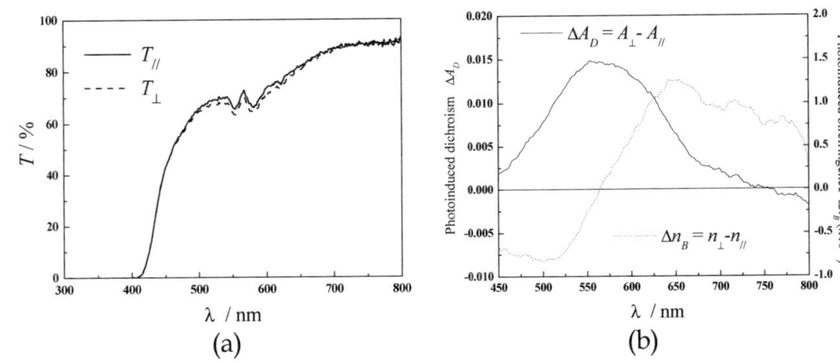

Fig. 11. Spectra of photo-induced anisotropic properties of Indolyfulgimide / PMMA film excited by linearly polarized light: (a) Transmission spectra on directions parallel and perpendicular to exciting beam polarization; (b) Dichroism and birefringence spectra

From Fig.10a and Fig.11a, the photo-induced absorption changing spectrum $\Delta A(\lambda) = A_E(\lambda) - A_C(\lambda)$ and the photo-induced dichroism spectrum $\Delta A_D(\lambda) = A_{\perp}(\lambda) - A_{//}(\lambda) = \lg(T_{//}(\lambda)/T_{\perp}(\lambda))$ were obtained, which are shown as solid lines in Fig.10b and Fig.11b respectively. Assuming that ΔA and ΔA_D are zero outside of the band 300~800nm, the photo-induced refractive index changing spectrum $\Delta n(\lambda) = n_E(\lambda) - n_C(\lambda)$ and the photo-induced birefringence spectrum $\Delta n_B(\lambda) = n_{\perp}(\lambda) - n_{//}(\lambda)$ can be calculated according to the Kramers-Kronig relation [5], where n_E, n_C, $n_{//}$ and $n_{\perp}$ are the refractive indexes of E-form, C-form and of the film excited by linearly polarized light along the photo-induced extraordinary and ordinary axes, respectively. The calculated Δn, Δn_B are plotted as dot lines in Fig.10b and Fig.11b respectively. Kramers-Kronig relation can be satisfied during all the photochromic reaction progress, so at one wavelength λ, $\Delta n(\lambda)$ is proportional to $\Delta A(\lambda)$ and $\Delta n_B(\lambda)$ is proportional to $\Delta A_D(\lambda)$ at different exciting time. From the Fig.10b and Fig.11b, it can be seen that at 633nm in this sample, $\Delta n(633\text{nm})/\Delta A(633\text{nm})=0.00994$ and $\Delta n_B(633\text{nm})/\Delta A_D(633\text{nm})=0.006115$.

2.5.3 Dynamics of photochromic and photo-induced anisotropy properties of the sample

The transmission growing up kinetics of the sample at 633nm were measured on the parallel and perpendicular directions to exciting beam polarization, when the C-form sample was being excited with 314mW/cm^2 and 157mW/cm^2 intensity linearly polarized 633nm He-Ne lasers (I_W) respectively, and an 1mW/cm^2 633nm laser beam is used as the testing beam (I_T), the optical setup and the results are shown in Fig.12 and Fig.13a. From Fig.13a, it can be seen that the photochromic reaction and photo-induced anisotropy progress of fulgide material is an optical cumulating progress, which just depending on the Exposure, so it is enough to consider just one exciting beam intensity condition for the analysis.

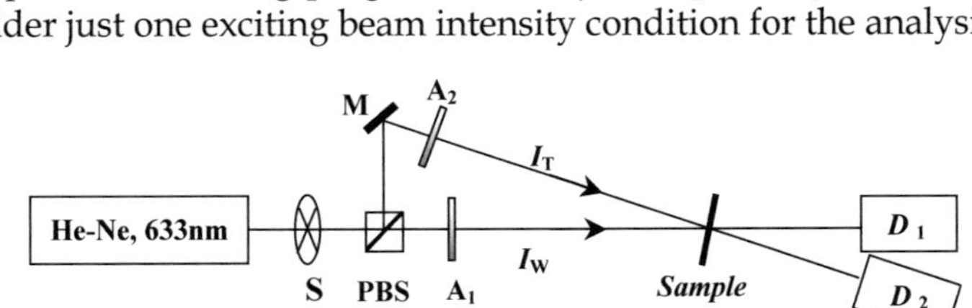

Fig. 12. Schematic of the experimental setup for measuring transmission kinetics.

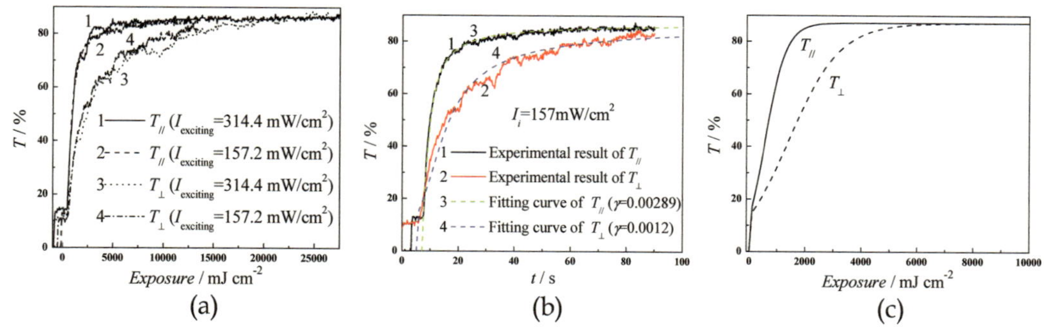

Fig. 13. Photo-induced anisotropic transmission curves of Indolyfulgimide/PMMA film on the directions parallel or perpendicular to exciting beam polarization depending on the exposure or erasing time: (a) Experiment curves measured at different exiting beam intensity; (b) Simulation of experimental results; (c) Calculated curves of uniformity light

For the analysis, we consider a fully bistable Fulgide system, neglecting side reactions like E-Z isomerization or aging effects. Using numerical calculation method [6], where the intensity Gaussian beam distribution of the He-Ne laser beam has been considered, the experimental curves were simulated (shown as the dash lines in Fig.13b) and the best fitting values $\gamma_{//}$=0.00289cm²/mJ, $\gamma_{\perp}$=0.0012cm²/mJ were obtained.. Then the photo-induced anisotropic transmission curves of uniformity light are calculated like shown in the Fig.13c.

3. Holographic recording properties of one kind of indoly-benzylfulgimide/PMMA film

3.1 Kinetics of diffraction efficiency (DE)

3.1.1 Measurement set up

The system configuration for measuring the real-time hologram first order diffraction kinetics of the Fulgide film is schematically illustrated in Fig.14. A He-Ne laser (Melles Griot Inc., USA, 25-LHP-928, 632.8nm, 35mW, vertical linear polarized) is used to generate recording beams (object beam I_O and reference beam I_R) and readout beam (reconstruction beam I_C), and a laser diode LD (Power Technology Inc., USA, IQ2A18, 405nm, 10mW, vertical linear polarized) is used as the auxiliary light source (I_A) and erasing light source (I_E). The He-Ne laser beam is split into the I_O, I_R and I_C after beam splitter BS_1 and polarization beam splitter PBS, in which I_R and I_C are phase conjugated (counter-propagated) beams. The diffracted light I_D of I_C, diffracted by the dynamic holographic grating established by the interference between the I_O and I_R, will be phase conjugated with the I_O, whose power was real time detected by a digital power meter 'D' (United Detector Technology company, USA，11A Photometer / Radiometer, 254~1100nm, Imax=10mW, resolution is 0.01nW) and a digital oscilloscope 'O' (Tektronix company, USA, TDS3032, 300MHz, 2.5GS/s, 1mV) after reflected by the BS_2 (R47%). The I_O and I_R are symmetrically incident on the sample (Fulgide film), whose intersection angle 2θ=16.5°, so the recorded grating is a non-incline grating. Shutter S_1 and S_2 controls the exposure time of red and purple beams. The continuously adjustable attenuators A_1~A_3 are used to adjust the intensities of the waves, in this experiment I_O=I_R=78.6mW/cm² and I_C=0.786mW/cm² (i.e. I_O:I_R:I_C=100:100:1). This insures that the sub-reflection gratings formed by I_O and I_C as well as I_R and I_C can be ignored. The Quarter-wave plates Q_1~Q_4 and the polarizer P are

Polarisation of Object Light	Polarisation of Reference Light	Polarisation of Reconstruction Light	Polarization states of diffracted lights		+1st order diffraction efficiencies	
			+1	-1	Amplitude hologram $\eta_{A,+1}$	Phase hologram $\eta_{P,+1}$
↕	↕	↔	↔	↔	$(\tau_{1om}/2)^2$	$(\tau_{0om}\cdot J_1(\Psi_{1om}))^2$
		↕	↕	↕	$(\tau_{1em}/2)^2$	$(\tau_{0em}\cdot J_1(\Psi_{1em}))^2$
		↺	↺ (ellipse)	↺ (ellipse)	$\frac{(\tau_{1om})^2+(\tau_{1em})^2}{8}$	$\frac{\tau_{0em}^2\cdot J_1(\Psi_{1em})^2+\tau_{0om}^2\cdot J_1(\Psi_{1om})^2}{2}$
		↻	↻ (ellipse)	↻ (ellipse)	$\frac{(\tau_{1om})^2+(\tau_{1em})^2}{8}$	$\frac{\tau_{0em}^2\cdot J_1(\Psi_{1em})^2+\tau_{0om}^2\cdot J_1(\Psi_{1om})^2}{2}$
↺	↺	↔	↔	↔	$(\tau_{1Mm}/2)^2$	$(\tau_{0Mm}\cdot J_1(\Psi_{1Mm}))^2$
		↕	↕	↕	$(\tau_{1Mm}/2)^2$	$(\tau_{0Mm}\cdot J_1(\Psi_{1Mm}))^2$
		↺	↺	↺	$(\tau_{1Mm}/2)^2$	$(\tau_{0Mm}\cdot J_1(\Psi_{1Mm}))^2$
		↻	↻	↻	$(\tau_{1Mm}/2)^2$	$(\tau_{0Mm}\cdot J_1(\Psi_{1Mm}))^2$
↔	↕	↔	↕	↕	$(\tau_{1eo}/2)^2$	$(\tau_{0eo}\cdot J_1(\Psi_{1eo}))^2$
		↕	↔	↔	$(\tau_{1eo}/2)^2$	$(\tau_{0eo}\cdot J_1(\Psi_{1eo}))^2$
		↺	↻	↻	$(\tau_{1eo}/2)^2$	$(\tau_{0eo}\cdot J_1(\Psi_{1eo}))^2$
		↻	↺	↺	$(\tau_{1eo}/2)^2$	$(\tau_{0eo}\cdot J_1(\Psi_{1eo}))^2$
↻	↺	↔	↻	↺	$(\tau_{1eo})^2/2$	$(\tau_{0eo}\cdot \sin\Psi_{1eo})^2/2$
		↕	↻	↺	$(\tau_{1eo})^2/2$	$(\tau_{0eo}\cdot \sin\Psi_{1eo})^2/2$
		↺	↻	—	$(\tau_{1eo})^2$	$(\tau_{0eo}\cdot \sin\Psi_{1eo})^2$
		↻	—	↺	0	0

Table 1. DWPS and DE of different kinds of polarization holograms

used to change the polarization states of the waves, here four different polarization recording: parallel linearly polarization recording, parallel circularly polarization recording, orthogonal linearly polarization recording and orthogonal circularly polarization recording were studied, everyone was constructed by horizontal, vertical, left circular and right circular polarized four kinds of lights like shown in table 1.

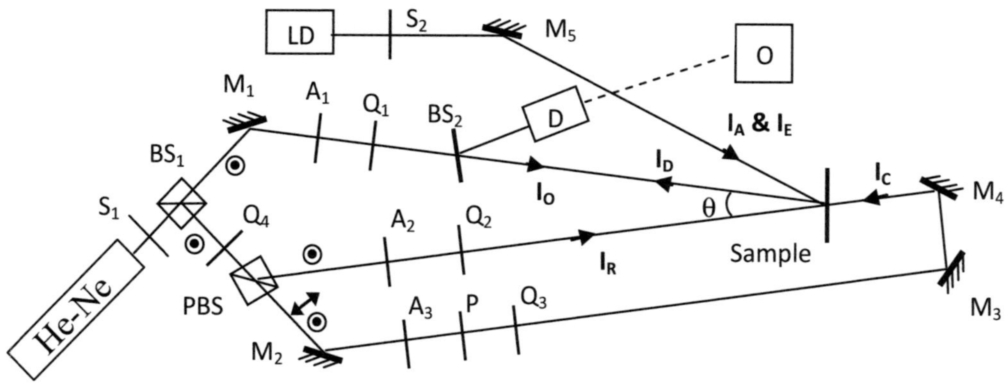

Fig. 14. Schematic of the experimental setup for measuring real-time polarization hologram diffraction kinetics.

3.1.2 Measurement results

In four kinds of polarization recording, and different polarization reading, the diffracted wave polarization states (DWPS) obtained in the experiments are shown in table 1 and the measured kinetic first order diffraction efficiency (DE) curves $\eta_{+1} \sim t$ are shown in Fig.15(a,c,d,e). From them the curves of the conditions when I_C has same polarization state with I_R were compared in Fig.16a. It can be seen that there exist an optimal exposure about $2\times78.6mW/cm^2\times3.75s\approx590\ mJ/cm^2$.

3.1.3 Theoretical analysis

DE dynamic curves at 633nm of different kinds of holograms recorded in the sample can be calculated from the photochromic and photo-induced anisotropic properties of the sample written in section 2.5, by using the DE formulas written in Table 1 [7]. Where the $(\tau_e\text{-}\tau_o)$ indicate the photoinduced anisotropy of the sample under the irradiation of linearly polarized light at some exposure, $(n_e\text{-}n_o)$ indicate the corresponding birefringence, $\tau_M=(\tau_e+\tau_o)/2$ indicate the amplitude transmission of the sample for nonpolarized light or circularly polarized light at this time (at the area of light strips in the interference field), τ_m indicates the amplitude transmission of the sample before the illumination of light (at the area of dark strips in the interference field, isotropy), τ_E and τ_C indicate the amplitude transmission of the E-form and C-form sample respectively, n_M, n_m, n_E and n_C indicate the corresponding refractive index of the sample, $\Psi_i=k_0{\cdot}n_i{\cdot}d$ (i=e, o, M, m, E, C), and it is defined that $\tau_{0ij}=(\tau_i+\tau_j)/2$, $\tau_{1ij}=(\tau_i\text{-}\tau_j)/2$, $\Psi_{0ij}=(\Psi_i+\Psi_j)/2$, $\Psi_{1ij}=(\Psi_i\text{-}\Psi_j)/2$ (i, j = e, o, M, m, E, C).The theoretically calculated diffraction efficiency kinetics curves of parallel linearly polarization hologram are shown in Fig.15b. In parallel circularly polarization recording hologram, the DE curves are same with each other for any kinds of polarized reconstruction light I_C, which is also same with that of parallel linearly polarization hologram for circularly polarized I_C. In orthogonal linearly polarization recording hologram, the DE curves are also same with each other for any kinds of polarized I_C, which is shown in Fig.16b. For orthogonal circularly polarization recording hologram, when I_C has same polarization state with I_R ($\vec{O} \perp \vec{R} // \vec{C}$), the DE curve is shown in Fig.16b; when I_C has orthogonal polarization state with I_R, the DE is zero; when I_C is linearly polarized, the DE is half of that in $\vec{O} \perp \vec{R} // \vec{C}$ condition. The theoretical DE kinetics curves of different kinds of polarization recording holograms for I_C has same polarization state with I_R, are compared in Fig.16b.

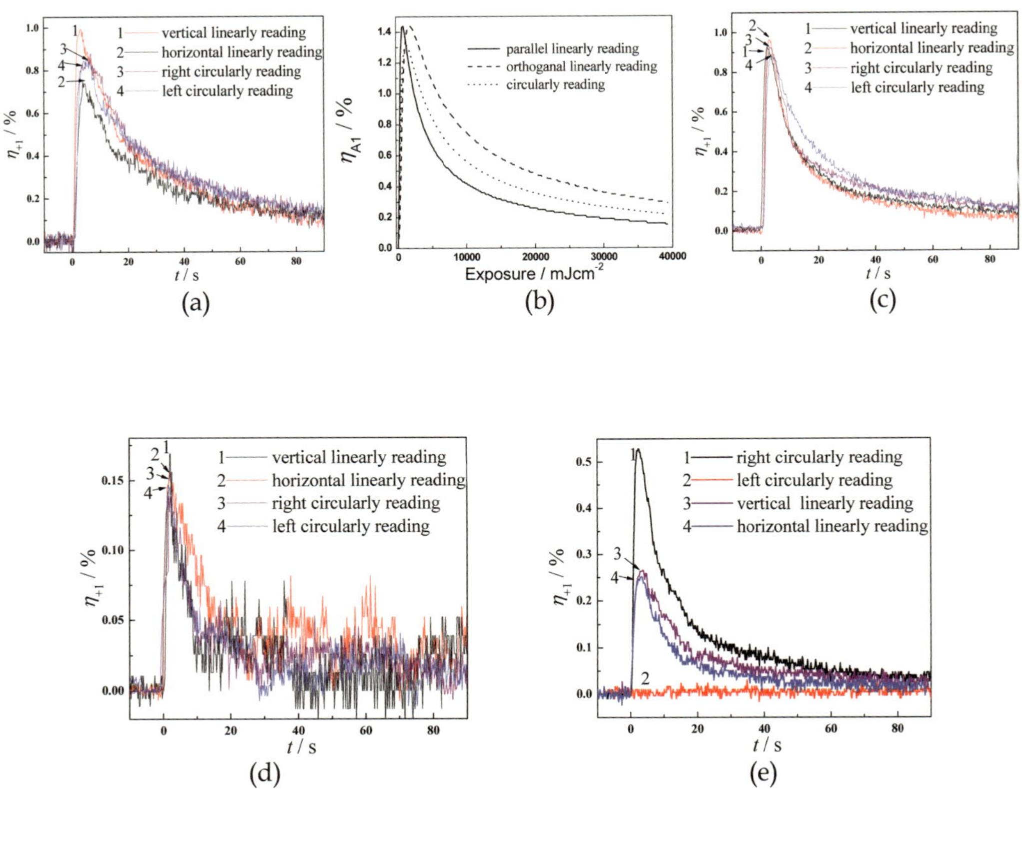

Fig. 15. The diffraction efficiency kinetics curves comparison of different kinds of polarization recording and different kinds of reading in Fulgide film: (a) Experimental results (ER) of parallel linearly polarization recording (b) theoretically calculated curves of parallel linearly polarization hologram (c) ER of Parallel circularly polarization recording; (d) ER of Orthogonal linearly polarization recording; (e) ER of Orthogonal circularly polarization recording

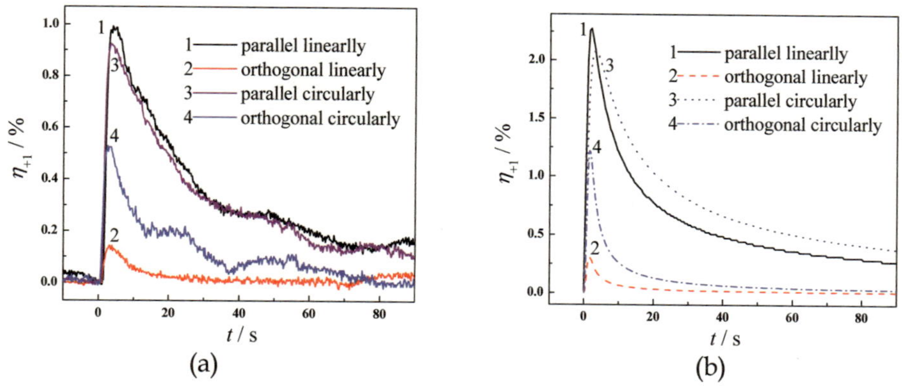

Fig. 16. The diffraction efficiency kinetics curves comparison of different kinds of polarization recording holograms in Fulgide/PMMA film written by Gaussian beams: (a) Experimentally measured results; (b) Theoretically calculated results

It can be seen that the maximum values' ratio of the measured values is basically coincide with the theoretically analyzed one. Only for parallel linearly polarization hologram, in the orthogonal linearly polarization reconstruction condition ($\vec{C} \perp \vec{O} // \vec{R}$), the diffraction efficiency is lower than parallel linearly polarization reconstruction condition ($\vec{C} // \vec{O} // \vec{R}$). It is because that, in calculation the affection of $\vec{C}$ is not considered, which can be proved to be very small when the $I_C=I_O/100=I_R/100$ comparing to the affection of non-linear absorption of the film, whose detail calculation progress will not be given here, where it also can be deduced that the reading beam affection is larger in $\vec{C} \perp \vec{O} // \vec{R}$ condition than in $\vec{C} // \vec{O} // \vec{R}$ condition.

And the theoretical results are larger than the experimental results, and the reaction is quicker (optimal exposures are about 590 mJ/cm^2 and 430mJ/cm^2 respectively in experimental results and theoretical results), this may be caused by: (1) the sample is not homogeneous, the density is different at different area; (2) the incidence angles of beams θ in the experiment are about 8.2°, so the intensities of them on the sample plane will be $\cos\theta \approx$ 0.9898 times of the values used in calculation; (3) the photoreaction rate constants used in calculation is a little bit larger than real ones.

And it can be seen that no matter during the ordinary holograph recording process in photochromic media, or during the polarization holograph recording process in photo-induced anisotropy media, there exits an overshooting peak in the diffraction efficiency, which then decays to a lower permanent level or also to zero. From the theoretical analyses, it can be deduced that is caused by the diminishing of fringe contrast mainly caused by the nonlinear saturation effects of photoisomerization process and photo-induced anisotropy process. In experiment, there also exit the diminishing of fringe contrast caused by a photochemically active readout beam and unequal intensities of object and reference waves. It can be theoretically calculated that the effects of them show very smaller than that of the nonlinear saturation effects, which will not be given here.

3.2 The DE spectra of different kinds of holograms in fulgide film

Suppose that the holographic recording is a linearity recording, from the spectra of ΔA, Δn, ΔA_D and Δn_B, shown in Fig.3b and Fig.4b, using the DE formulas of different kinds of polarization holographies, shown in table 1, the DE spectra of the ordinary holograms (i.e. parallel circularly polarization recording) and polarization holograms (orthogonal linearly and orthogonal circularly polarization recording) can be calculated, which are shown in Fig.17, where the solid line, dot dash lines and dash lines indicate the η_{total}, η_A and η_P respectively. It can be seen that at 450nm and 700nm the diffraction efficiencies are higher, while the absorption is very small, where the recorded information can be read out without any photochromic reaction, i.e. the non-destructive reconstruction can be realized.

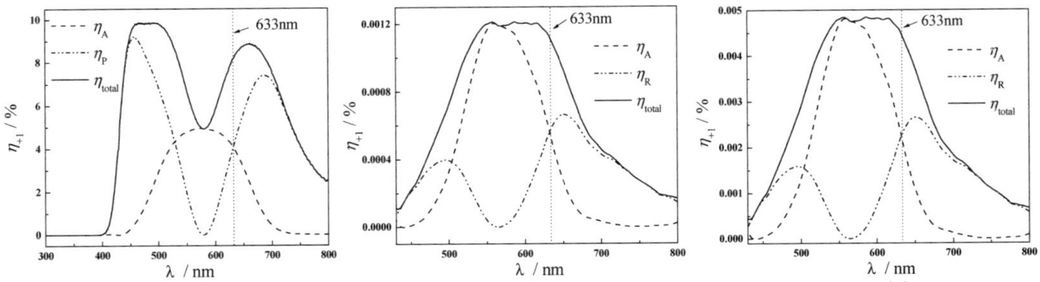

(a) (b) (c)

Fig. 17. Diffraction efficiency spectra of the (a) Ordinary holography; (b) Orthogonal linearly polarization holography; (c) Orthogonal circularly polarization holography

3.3 Effects of auxiliary light and object reference ratio to DE of Fulgide film

No matter during the ordinary holograph recording process in photochromic media, or during the polarization holograph recording process in photo-induced anisotropy media, there exits an overshooting peak in the diffraction efficiency, which then decays to a lower permanent level or also to zero, because of the diminishing of fringe contrast caused by a photochemically active readout beam, unequal intensities of object and reference waves and the nonlinear saturation effects of photoisomerization process and photo-induced anisotropy process. It is known that in ordinary holographic recording, this decreasing process can be eliminated by illuminating the hologram with a uniform control beam that has the effect of molecular back-conversion photochrome [8]. It was found that in polarization holographic recording, this method also can be used [8]. In this section experiments done with an ordinary hologram and a polarization hologram recorded in a 3-indoly-benzylfulgimide/PMMA film at 633 nm have shown that a control beam at 405 nm can increase the stable-state diffraction efficiency, thus, allowing to decrease the rigorous requirements on the recording time, the object reference ratio and the reading beam intensity in the holographic recording. The affections of object reference ratio (ORR) to DE of different holograms recorded in 3-indoly-benzylfulgimide/PMMA film were also measured.

The optical set up shown in Fig.14 is also used in this experiment. Here recording with linearly polarized beams with identical states of polarization (scalar hologram) and recording with circularly polarized beams with orthogonal states of polarization (polarization hologram) were studied. The holograms were reconstructed by light with the same polarization as the reference light.

3.3.1 Measurement of the effect of polarization of the auxiliary light

First the effect of the state of polarization of the auxiliary light was studied. The diffraction efficiencies of two different kinds of holographs were measured when the 405nm auxiliary beam is vertical and horizontal linearly polarized respectively. And the results show that the effect of the polarization of the auxiliary light is very small, the regular is: in the parallel linearly polarized recording the diffraction efficiency is slightly higher when the auxiliary light is parallel to the polarization state of the recording lights than orthogonally polarized; in the orthogonal circularly polarized recording the polarization direction of the auxiliary light nearly have no effects. So we choose the vertical polarization for the LD laser in the below experiments.

3.3.2 Measurement of the effect of the intensity of the auxiliary light

Under the irradiation of different intensity auxiliary lights, the diffraction efficiencies of two kinds of holograms were measured in real time. The kinetics curves of the diffraction efficiencies of parallel linearly polarized recording and the maximum values and stable values contrasts are shown in Fig.18(a,b), and the results in orthogonal circularly polarized recording are shown in Fig.18(c,d).

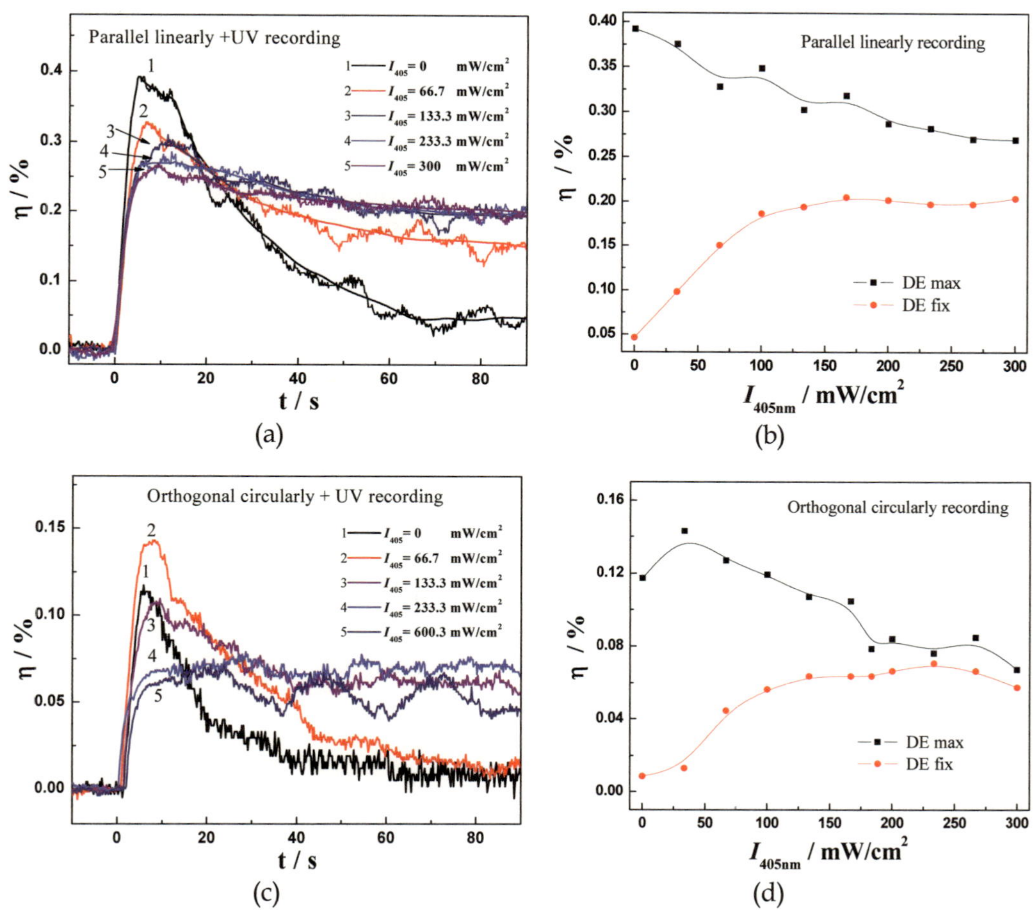

Fig. 18. (a,b): Diffraction efficiencies of parallel linearly polarization holograph under the irradiation of different intensity auxiliary lights: (a) kinetics curves; (b) maximum values and stable values compare. **(c,d):** Diffraction efficiencies of orthogonal circularly polarization holograph under the irradiation of different intensity auxiliary lights: (c) kinetics curves; (d) maximum values and stable values compare

These results show that, under the irradiation of auxiliary light, the diffraction efficiency stable-state values can be increased no matter in ordinary hologram or in polarization hologram, and there exits an optimal intensity, under which the maximum stable-state diffraction efficiency can be obtained. When I_O=I_R=78.6mW/cm^2 and I_C=0.786mW/cm^2, in the parallel linearly recording, the optimal intensity of the purple light, is larger than 300mW/cm^2; in the orthogonal circularly recording, the optimal intensity of the purple light is about 233mW/cm^2. And it can be seen that, when there is no auxiliary light, the diffraction efficiency is depending on the exposure too much, there exits rigorous requirement on the recording time. But if turn on the auxiliary light, the requirement is decreased.

3.3.3 Measurement of the effect of the auxiliary light on the ORR requirement

In this experiment I_O=78.6mW/cm^2, I_C=0.786mW/cm^2 and I_A=267mW/cm^2 were used. In the conditions of with and without auxiliary light, the diffraction efficiencies of two kinds of holograms were measured at different intensities of reference light. The experimental results of parallel linearly polarization hologram and of orthogonal circularly polarization hologram are shown in Fig.19 and Fig.20 respectively. It can be seen that in all kinds of polarization hologram, no matter turn on the auxiliary light or not, the ORR has great effect on the diffraction efficiency. But in ordinary holography, with auxiliary light irradiation, the diffraction efficiency changing in the area near ORR=1:1 is much slower than without auxiliary light irradiation, so the rigorous requirement on ORR can be decreased. In the polarization holography, when the reference beam intensity is a little bit higher than that of object beam, the diffraction efficiency will be maximum, the reason is currently unclear and is a subject of further investigation.

3.4 Dependence of DE on the reading beam incidence angle

The angular selectivity of the sample (*i.e.* the dependence of DE on the reading beam incidence angle) was measured, and the experimental set-up is shown in Fig.21. A 650nm LD is used as the recording and reading light source, and a 405nm LD is used as the erasing light source. Adjustment of coherence was made by creating a Michelson interferometer. The 650nm laser was first adjusted perpendicularly to the face of the beam-splitter cube, and then the transmitted and reflected beams intersect with each other on the sample after reflected by the mirrors. The interference fringes were tuned for optimum contrast by moving one of the mirrors (M2) with a translation stage. The two writing beams are symmetrically incident on the sample (Fulgide film), whose intersection angle $2\theta_B$=10° corresponding to a grating spacing Λ = 3.73 μm, so the recorded grating is a non-incline grating, and the incidence angle of object light (signal light) and reference light are θ_s=-5°and θ_r=5°respectively. The sample is placed on the precision rotary platform (M-062 model, PI company, Germany), and the recording point of the hologram in the sample is on the shaft of platform. So when the platform is rotated, the recording point does not move. The continuously adjustable attenuators A_1~A_2 are used to adjust the intensities of the waves, in this experiment I_O= I_R= 75mW/cm^2 and I_C=0.1875mW/cm^2 (*i.e.* I_O:I_R:I_C =400:400:1). By observing the spots on the screen behind the sample, the best exposure time can be determined (see the next Section). Before reconstruction, the sample was rotated to one direction 20°. In the reconstruction time, the platform is rotated to the opposite direction

40° with a uniform speed. The diffracted light was real time detected by a digital power meter and a digital oscilloscope. Measurement result is shown in Fig.22 as solid line, and the dashed line is its theoretical fitting curve by the formula $y=y_0 \cdot \mathrm{sinc}^2(ax)$.

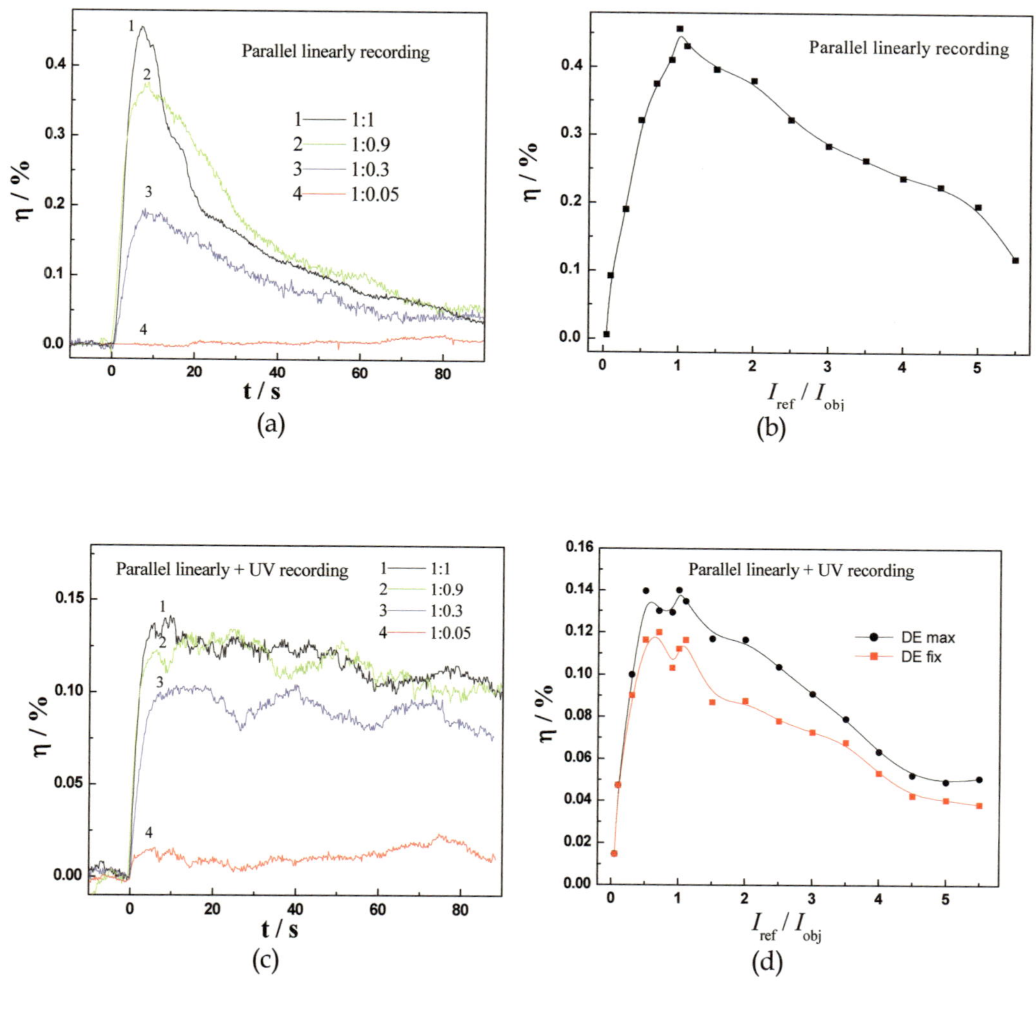

Fig. 19. The diffraction efficiency of parallel linearly polarization holograph recording at different ORR: (a) kinetic curves (*without auxiliary light*); (b) maximum values compare (*without auxiliary light*); (c) kinetic curves(*with auxiliary light*); (b) maximum values and stable values compare (*with auxiliary light*)

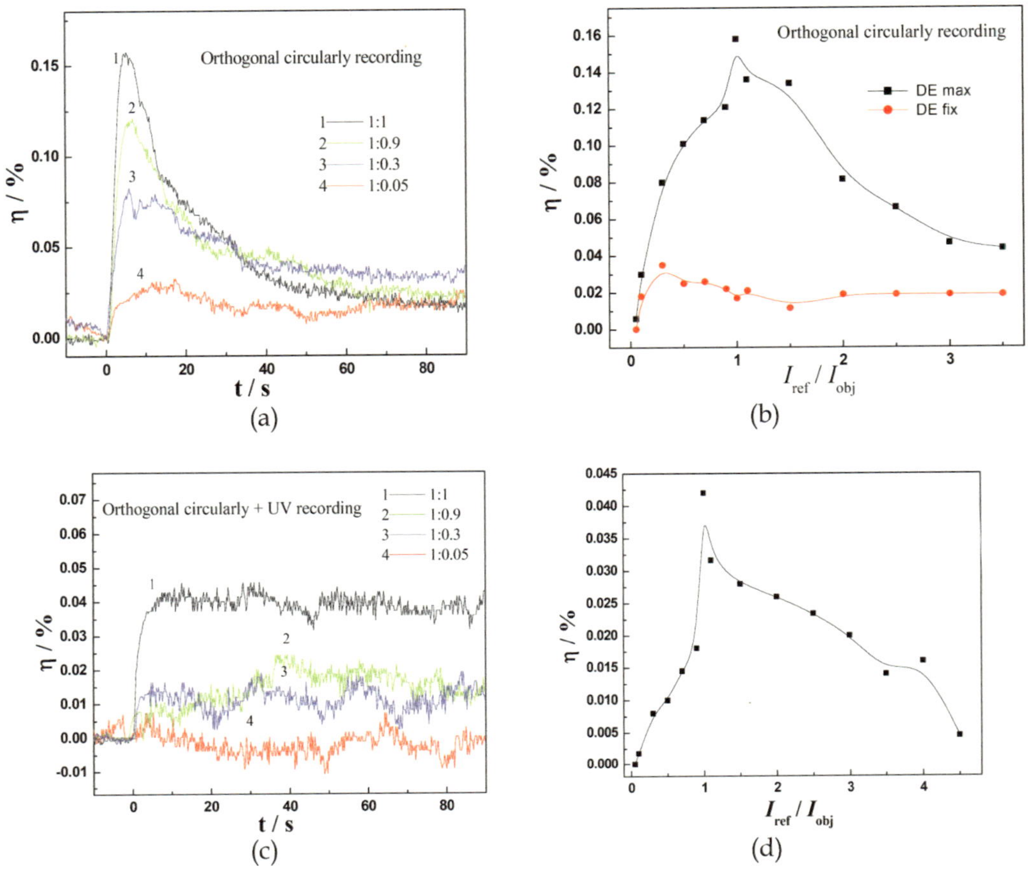

Fig. 20. The diffraction efficiency of orthogonal circularly polarization holograph recording at different ORR: (a) Kinetic curves(*without auxiliary light*); (b) maximum values and stable values compare(*without auxiliary light*); (c) Kinetic curves (*with auxiliary light*); (b) stable values compare (*maximum value basically same with the stable values, with auxiliary light*)

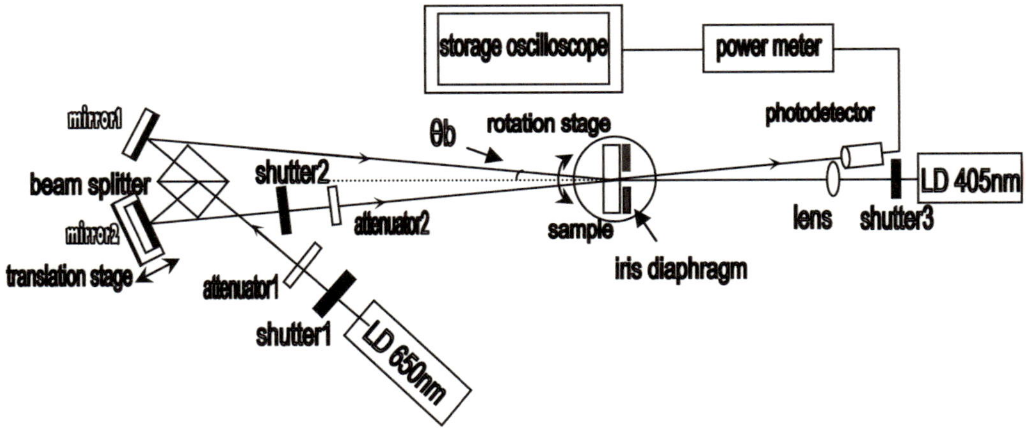

Fig. 21. The measurement set-up of angular selectivity of the sample

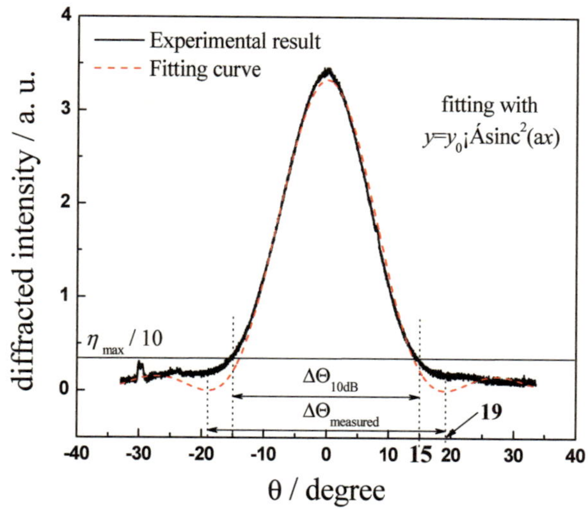

Fig. 22. The curves showing the DE dependence on the reading beam incidence angle

From the Fig.22, it can be seen that the angular selectivity $\Delta\Theta_{measured}$=38°, if it is taken as the angle between the first minimum diffractions at both sides of the fitting curve's peak. But Side lobes do not exist in the experimental curve, so the angular selectivity can be taken as the angular width at the 1/10 of the maximum diffraction efficiency, $\Delta\Theta_{10dB}$=30°.
In theory[10], the angular selectivity of the gratings $\Delta\Theta_{calculated}$ is:

$$\Delta\Theta = \frac{2\sqrt{\pi^2 - \nu^2}\lambda}{\pi n d}\frac{\cos\theta_s}{|\sin(\theta_r - \theta_s)|} \tag{1}$$

Where $\nu = \pi\Delta n d / (\lambda\sqrt{\cos\theta_r \cos\theta_s})$. In this experiment, the λ=650nm, the thickness of the sample is d=10μm, the refractive index of the sample is about $n\approx1.5$, and for this sample at 650nm the refractive index difference between E-form and C-form is $\Delta n \approx 1.84\times10^{-2}$, so it can be calculated that $\Delta\Theta_{calculated}\approx27.3°$.
It can be seen from the experimental value of angular selectivity is greater than the calculated value. One of the reasons is that, for this calculation the incident beam is considered as an infinite plane wave, however, the spot diameter is small in actual storage experiment. According to diffraction theory, the limited size of the beam would inevitably lead to an angle broadening, so the measured curves are broadened.

3.5 The spatial resolution measurement of fulgide film

In the reflection type holographic image storage (where the *Ic* is used as reference beam, shown in section 4.1), the angles between the two recording waves is 173°, so it is proved that this sample can store the gratings with spatial frequency of 6300lines/mm.

3.6 Measurement of the optimum exposure of the sample

From the diffraction dynamic curves, the optimum exposure can be obtained. From Fig.15 it can be seen that he optimum exposure of the Fulgide is 590mJ/cm^2, the similar data also can be obtained from the changing of the diffracted spots' pattern, details please see in Reference [9].

4. Holographic image storage in fulgide film

4.1 Reference beam reconstruction hologram and phase conjugated beam reconstruction hologram (including transmission type hologram and reflection type hologram)

4.1.1 Experiment methods

The experimental setups for non-collinear holographic storage are shown in Fig.23and Fig.24. Fig.23 is the experimental setup of reference beam reconstruction holography. Fig.24 is the experimental setup of conjugated beam reconstruction holography. In every one of these two optical setups, the transmission type holographic recording and reflection type holographic recording both can be realized.

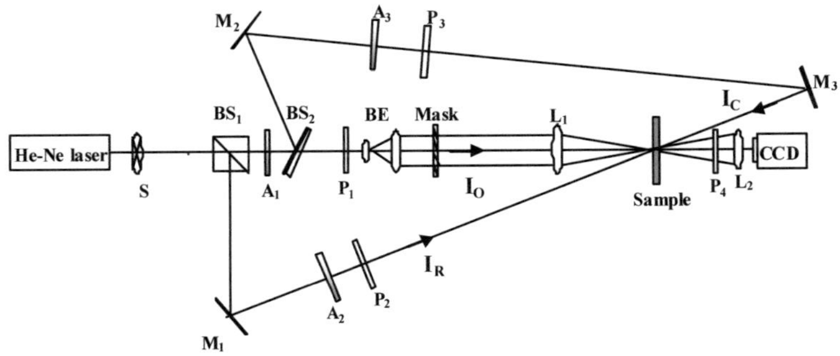

Fig. 23. Experimental setup of reference beam reconstruction holography

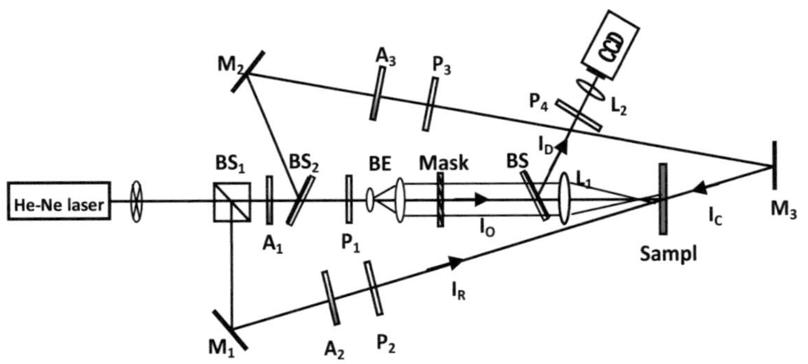

Fig. 24. Experimental setup of conjugated beam reconstruction holography

In Fig.23, the reference beam reconstruction is used. The nonpolarized He-Ne laser (632.8nm, 3mW) is split into three beams by beam splitters BS_1 and BS_2. The three beams are object beam I_O and two phase conjugated beams I_R and I_C. In transmission type holographic recording, reference beam is I_R, which is I_C in reflection type holographic recording. The signal of object beam is loaded by the Mask, which is imaged on the CCD's photosensitive surface by the positive lens L_1 and L_2. The focused I_O beam cross with reference beams after the focus of L_1, where a recording medium, the fulgide/PMMA film, has been placed for recording the hologram. In reconstruction process, using the reference beam as reconstruction beam, diffractive image is captured by the CCD. The shutter S is used to control the exposure time (the best exposure time is 10s in this experiment). The continuously adjustable attenuators A_1, A_2 and A_3 are used to adjust the intensities of I_O, I_R and I_C. Polarizer P_1, P_2 and P_3 are used to adjust the polarization states of I_O, I_R and I_C. Polarizer P_4 in front of CCD is used to filter scattered light. For the ordinary holographic storage the Polarizers are not used. The intensities of object wave and reference wave are both 6mW/cm^2 in experiment. The angles between the two recording waves are 7° and 173°. The intensity of reconstruction beam is 60μW/cm^2. And the diameter of hologram is about 2mm.

In Fig.24, the phase conjugated beam is used as reconstruction beam. I_R and I_C are conjugate with each other. For transmission type holographic recording, reference and reconstruction beams are I_R and I_C respectively, which are exchanged with each other, for reflection type holographic recording experiment. Diffracted beam I_D is conjugated with object beam I_O. After being reflected by BS_3, the diffracted image can be detected by CCD real-timely. Other conditions are same with that of set up shown in Fig.23.

4.1.2 Results and discussions

4.1.2.1 Transmission type hologram and reflection type hologram

In Fig.25, the reconstructed images of parallel linearly polarized transmission type hologram and reflection type hologram (recorded in setup shown in Fig.23) are shown.

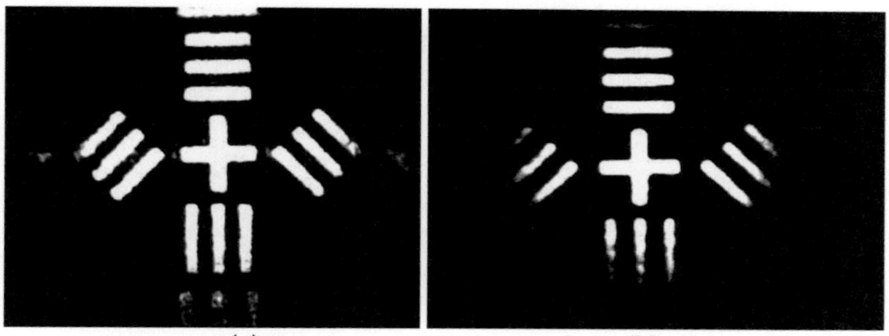

(a) (b)

Fig. 25. The reconstructed images of parallel linear polarized transmission hologram and reflection hologram: (a) reconstructed image of transmission recording hologram; (b) reconstructed image of reflection recording hologram

It can be seen that: compared with transmission-type hologram, reflection-type hologram has higher SNR. This is because that the noise in the reconstructed image of transmission-type hologram is come from the forward scattering, but that of reflection-type holographic recording hologram is come from the backward scattering. Usually the forward scattering is always larger than backward scattering, so the reflection-type hologram has smaller noises. But reflection-type hologram has lower diffraction efficiency.

4.1.2.2 Reference beam reconstruction hologram and phase conjugated beam reconstruction hologram

In Fig.26, the reconstructed images of reference beam reconstruction hologram and conjugated beam reconstruction hologram are shown. It can be seen that: compared with reference beam reconstruction, the phase conjugated beam reconstruction can effectively correct the phase aberration caused by the mis-adjustment of optical setup and the real-time detection of the changing progress of the diffraction image can be realized.

Fig. 26. Reconstructed images of (a) reference beam reconstruction hologram (Phase aberration image) and (b) conjugated beam reconstruction hologram (Corrected image)

4.2 Fraunhofer hologram and Fourier transform hologram

According to the different arrangement of optical setup, holograms can be divided into Fresnel hologram, Fraunhofer hologram, Image plane hologram, Fourier transform hologram and quasi-Fourier transform hologram etc. In which Fraunhofer hologram and Fourier transform hologram are recorded here, the optical setups are shown in Fig.27 and Fig.28.

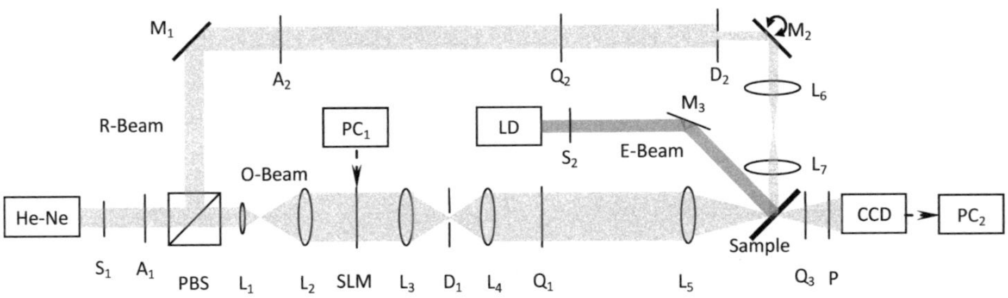

Fig. 27. Fraunhofer angular multiplexing holographic storage experimental setup

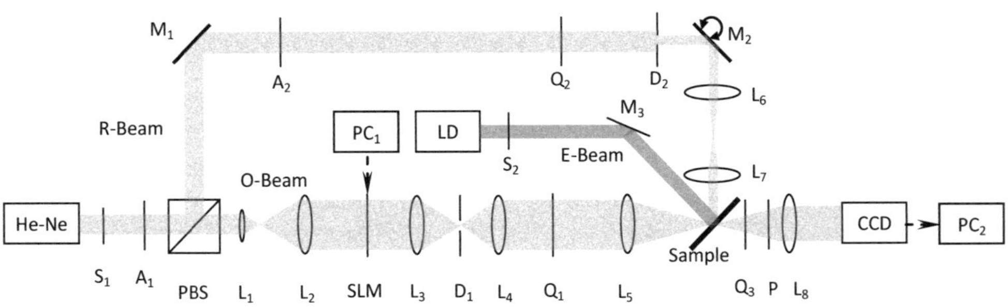

Fig. 28. Fourier transform angular multiplexing holographic storage experimental setup

It can be seen that in two optical setups, there are many common things, difference exist in finally imaging set up in the object beam path. In both setups, a He-Ne laser (633nm, 3mW) is used for recording and read-out beams. A diode laser (405nm, 10mW) is used as an erasing beam. Shutter S_1 and S_2controls the exposure time and erasing time respectively. The continuously adjustable attenuators A_1 andA_2 are used to adjust the intensities of the object, reference, and readout waves. The nonpolarized He–Ne laser is split (1:1) into two beams, a horizontally polarized object wave (O-Beam) and avertically polarized reference wave (R-Beam), by a polarization-sensitive beam splitter (PBS). Lenses L_1 and L_2 in the object beam path comprise a beam expander. Located in the front focal plane of the Fourier transform lens L_3, the spatial light modulator (SLM) was positioned such that the encoded data are loaded from the computer PC_1. The polarization state of the object wave becomes vertical after passing through the SLM. Diaphragm D1 is placed in the spatial frequency spectrum plane of the 4f system composed of L_3 andL_4, which is used to filter high-order diffractive waves, so on the back focal plane of L_4 the object image without grid structure to be recorded can be obtained. In the reference beam path, lenses L_6 and L_7 comprise a 4f system. Rotatable mirror M_2 is located at the focal plane of L_6, which is used to perform angular multiplexing (detail see the section 4.7.2). The angle between the object wave and the normal of the sample and the angle between the reference wave and the normal of the sample are both 45°. Quarter-wave plates Q_1, Q_2, and Q_3 are used to change the polarization states of the recording, read-out, and diffracted waves. Polarizer P is used to filter scattered light.

In the Fourier transform holographic storage experiment, the imaging system is another 4f system composed of lens L_5 and L_8. The object image to be stored located in the front focal plane of L_5 and the CCD is placed in the back focal plane of L_8. The fulgide film is placed on the spectrum plane of the system to record the Fourier transform holograms.

Only a difference of a quadratic phase factor occurs between Fraunhofer diffraction and Fourier transform. So when objects (or images to be stored) placed anywhere around the imaging lens the Fraunhofer hologram can be obtained. Here in the Fraunhofer holographic storage experiment, the image to be stored placed at about two-focal-distance before the imaging lens L_5, so the upside down same size image of the object will be formed at about two-focal-distance after the L_5, where is also the CCD photosensitive surface. Fulgide films were still near the back focal plane of L_5, so the size of the hologram is small, narrow laser beams can be used as reference beam and reconstruction beam.

Compared with Fraunhofer hologram, Fourier-transform hologram has small recording spot, higher storage density and the 4f system has better imaging quality than the single-lens. But because of the small recording spot, the diffracted light is weaker, which can be stronger when stronger reconstruction light is used, however the erasing effect will be sharpen up. The experiment results are shown in Fig.29, here will not be given individually.

4.3 The holograms with different polarization recording waves

In Section 3.1 the diffracted wave polarization states (DWPS) and the diffraction efficiencies (DE) of different polarization holograms recorded in a 3-indoly-benzylfulgimide /PMMA film are given. In this section, different polarization holographic image storages are realized in this film. Experimental setup as shown in Fig.28, four kinds of polarization holograms, like the parallel linearly polarization hologram, parallel circularly polarization hologram, orthogonal linearly polarization hologram and orthogonal circularly polarization hologram, are recorded, the Q_1, Q_2 are used to adjust the polarization states of O-beam and R-beam. The polarization state of scattering noise is similar to that of the original reconstruction light. And in the orthogonal polarization if the reference beam itself is used as the reconstruction light, the DWPS is orthogonal to the polarization state of reconstruction light, so using Q_3 and P the scattering noise can be filtered.

Fig.29 shows the comparison of retrieval diffractive images with different polarization recording waves on the fulgide film, when reconstruction beams has same polarization state with reference beams. It can be seen that these results are same with the measured results shown in Section 3.1. When the intensities of recording wave and readout wave are certain, the parallel linear polarization hologram has the highest DE, the parallel circular polarization hologram followed and the orthogonal circularly polarization hologram has the lowest DE. But in parallel polarization holograms, the scattering noises cannot be filtered, so their signal-to-noise-ratios (SNR) are lower. In orthogonal polarization holograms, the scattering noise can be filtered, so their SNRs are high.

(a) (b) (c) (d)

Fig. 29. Results of holographic storage experiments with different polarization recording waves on Fulgide film: (a) parallel linear polarization recording; (b) parallel circular polarization recording; (c) orthogonal linear polarization recording; (d) orthogonal circular polarization recording

4.4 Fourier transformation orthogonal circular polarization holographic optical data storage

Orthogonal circular polarization hologram has higher DE and high SNR. Fourier transformation hologram has high storage density. So Fourier transformation orthogonal circular polarization recording is chosen as the method of high density holographic data storage in the fulgide film. Optical setup as shown in Fig.28, the intensities of O-beam and R-beam are both about $14mW/cm^2$, optimum exposure time is 10s, and erasing time is smaller than 5s. The encoded binary data images loaded on the SLM are translated by our developed software in PC_1 by reading the data stream of the computer file. The black pixel represents "0" and the white pixel represents "1". The marginal periodically distributed black-white pixels are used as a reference for locating the pixels in the data-decoding process. The diffracted images will be captured by CCD and send to PC_2, which can be successfully decoded and recovered to the original file without any errors. The holograms can be restored after erased by the violet light.

In Fig.30 shows the experimental results. The images are separately stored file, encoded binary monochromic image, retrieval diffractive image, decoded result, retrieved file, measurement of size of holographic image. The retrieval diffractive image that is clear is processed by decoding procedure, and the obtained retrieved file is same as the stored file. In the experiment, data size of each stored holographic page is 81×61 bits, and the size of hologram is 60μm×42μm. So the storage area density of 2×10^8 bits/cm² is obtained. The nonhomogeneity and flaws of the surface of sample, misalignmen of optical elements or uncertainty of adjustment brings out some distortion of the diffractive image and error codes that can be reduced to the minimum by designing reasonable encoding and decoding procedures.

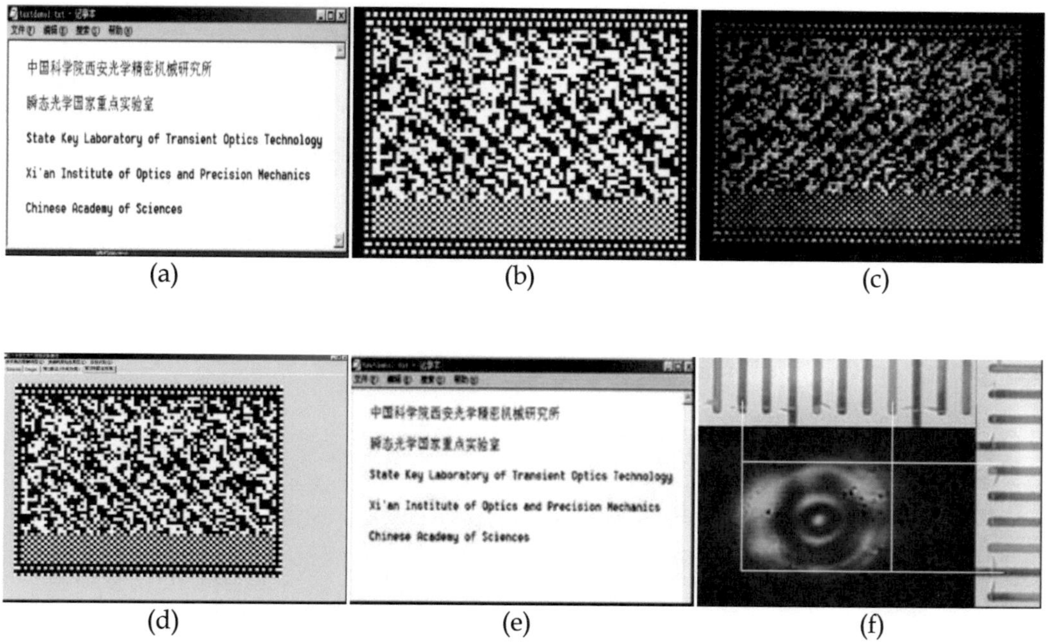

Fig. 30. Results of orthogonal circular polarization holographic optical data storage on BR-D96N film by Fourier transformation holographic method. (a) stored file; (b) encoded binary monochromic image; (c) retrieval diffractive image; (d) decoded result; (e) retrieved file; (f) measurement of size of hologram (one grid of the scale corresponds to 10μm).

4.5 Application of the auxiliary light effect in holographic image storage

The effect of auxiliary light is applied in real image holographic storage. An orthogonal linearly polarization transformation type holographic storage experiment setup was used, which is shown in Fig.31, where the reference light is also used as the reconstruction light and the diffuse reflection objects used as the target. A 633nm, 35mW vertically polarized He-Ne laser is used as the recording and reading light source, which has turned to the elliptically polarized light after a $\lambda/4$ plate, and then divided into two orthogonal polarized object light and reference light through the polarization beam splitter PBS. The polarizer P in front of the CCD was used to filter the scattered noises of reconstruction light. The 405nm LD laser is used as the eraser and auxiliary light source.

The images of the self-diffracted signal at three different instants are presented in Fig.32. The first image (Fig.32a) is taken at $E \approx E_{opt}$, whereas the second one (Fig.32b) corresponds to the case $E >> E_{opt}$ (auxiliary light is absent in both cases). As one can see, the recorded image is essentially lost through saturation, but it is restored at a significant level when the auxiliary control beam is turned on (Fig.32c), which proved that the diffraction efficiency can be increased and stabled when irradiated by an auxiliary light, and the rigorous requirement on the recording exposure is decreased.

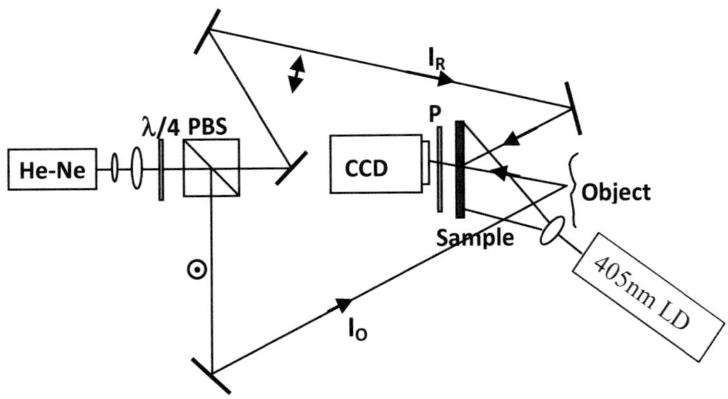

Fig. 31. Diffuse reflection object orthogonal linearly polarization transformation type holographic storage experiment setup

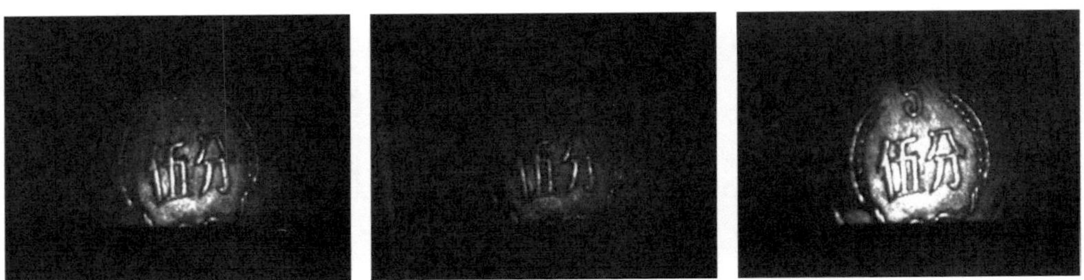

(a) (b) (c)

Fig. 32. The diffracted images of different holograms: (a) at $E \approx E_{opt}$, without auxiliary light; (b) at $E >> E_{opt}$, without auxiliary light; (c) exposed long enough time to be stable after the auxiliary light is turned on

4.6 Collinear hologram

In collinear holographic storage system, the experimental setup for rewritable collinear holographic image storage is shown in Fig.33(a). A He-Ne laser is used as the light source for recording and readout, and an ultraviolet laser diode is used as the light source for erasing. The non-polarized laser beam (632.8nm, 2mW) turns to be vertical polarized light by passing through the polarizer P, after being expanded and collimated by the lens L_1 and L_2. Then the beam is projected to transparent mask SLM (special light modulator). The pattern on the SLM used in recording process is shown in Fig.33(b), in which the center cross is the objective information pattern and the outer semi circles are the reference pattern. So the modulated laser beams by the SLM include both objective light (I_O) and reference light (I_R), whose intensities are respectively 318mW/cm^2 and 382mW/cm^2 (I_O:$I_R \approx$1:1.2). The patterns on the SLM are imaged on the CCD's photosensitive surface by the positive lens L_3 with focal length of 70mm. The I_O and I_C beams are focused and interfere with themselves at the focus of L_3, where a record medium, the BR film, has been placed for recording the hologram. In reconstruction process, only the outer semi circle patterns, as shown in Fig.33(c), is displayed on the SLM, where the object light pattern is covered. So the reconstruction light is same with the original reference light I_C, whose intensity is about 50mW/cm^2. And the retrieved diffractive image is captured by the CCD sensor. The diameter of hologram stored in the media is about 0.2mm, as shown in Fig.34(a). In this system the shutter S is used to control the exposure time (the best exposure time is 13 second in this experiment). The continuously adjustable attenuator A is used to adjust the intensities of I_O and I_C (stronger for recording and weaker for readout). The diagram D, placed in front of the CCD, is used to filter the transmission light of I_C (zero-order diffraction), and to pass only the diffraction light I_D (+1 order diffraction). After erased by ultraviolet light, the holograms can be recorded repeatedly.

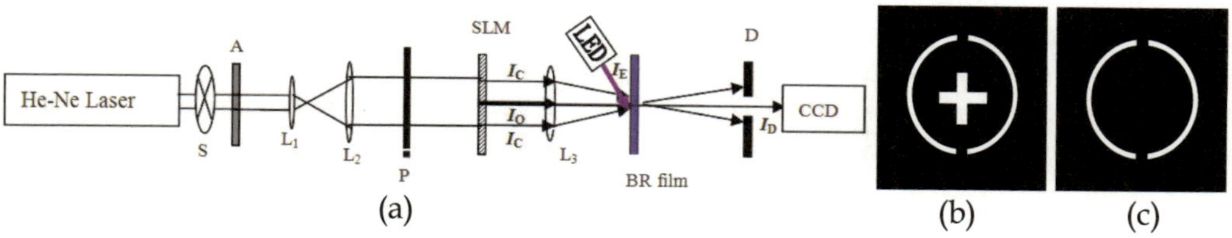

Fig. 33. Optical setup and the patterns displayed on the SLM in rewritable collinear holographic image storage; (a) Optical setup; (b) Recording process pattern; (c) Reading process pattern

(a) (b) (c)

Fig. 34. Size measurement, objective image and reconstructed image of a collinear hologram stored in the media: (a) Size measurement (the scale is 100mm/div); (a) Original objective image; (b) Reconstructed image

In Fig.34(b,c), the results of parallel linear polarized collinear holographic storage experiment on a BR-D96N film are shown. Since the collinear hologram is recorded on the focus of lens, the size of the recording point is very small. So the diffracted light is very weak, which makes the scattering light of the reconstruction light entering into CCD (noises) look too much. In that case, the signal noise ratio (SNR) is very low. But this kind of noise can be filtered by using orthogonal polarization hologram recording technology. And high storage density can be realized in collinear holography because of its small recording area.

Comparing with traditional "two-beam interference recording method", the collinear holographic storage system has simpler optical setup and smaller volume. And in this system the demands for environment are lower, because the objective light and reference light pass through the same one path, in which there are same interferences from the vibration of environment, the change of temperature and the variety of airflow, those effects can be cut down.

4.7 Different kinds of multiplexing holographic storage

High-density high-capacity volume holographic multiplexing is one of the most attractive aspects of holographic data storage. The study on holographic multiplexing storage application of fulgide materials is very limited. Using the optical setups shown in Fig.28, polarization multiplexing, angular multiplexing and rotational multiplexing holographic experiments are carried out initially.

4.7.1 Polarization multiplexing holograms

Polarization multiplexing is based on the photo-induced anisotropic property of the material, at the same location of sample different kinds of polarization holograms are recorded. Todorov et al were first to show that two holographic recording could be stored independently inside the same film when using different combinations for the polarization states of the reference and the object beam during recording. Su et al presented a technique for polarization multiplexing in $LiNbO_3$. Koek et al have presented a technique for simultaneous readout polarization multiplexing in bacteriorhodopsin.

Here different kinds of polarization multiplexing holographic storage were realized in fulgide film, including linearly polarization multiplexing and circularly polarization multiplexing, optical setup as shown in Fig.28. In circularly polarization multiplexing experiments, at the same location of the sample parallel circularly polarization hologram and orthogonal circularly polarization hologram were recorded. By adjusting the Q_3 and P, the diffraction images of parallel circularly polarization hologram and of orthogonal circularly polarization hologram can be obtained individually and together, like shown in Fig.35. So it can be seen that the storage density can be doubled by using polarization multiplexing.

(a) (b) (c)

Fig. 35. The reconstruction images in polarization multiplexing holographic storage in fulgide film: reconstruction image (a) of parallel circular polarization hologram; (b) of orthogonal circular polarization hologram; (c) of both holograms

4.7.2 Angular multiplexing holograms

Using the optical setup shown in Fig.28, the angular multiplexing holographic storage is realized. In the reference beam path, lenses L_6 and L_7 comprise a 4f system. Rotatable mirror M_2 and sample are located at the front focal plane of L_6 and at the back focal plane of L_7. When the mirror M_2 rotates within a certain range round its shaft, the exposure position of the reference beam on the sample does not move with the rotation of M_2, but only the incident angle will change, so angular multiplexing can be performed. The angle between the object wave and the normal of the sample and the angle between the centre line of reference wave and the normal of the sample are both 45°. In the angular multiplexing, the angle between two reference beams corresponding to two neighboring holograms should be greater than the minimum horizontal selection angle $\Delta\Theta$[10] :

$$\Delta\Theta = \frac{2\sqrt{\pi^2 - \nu^2}\lambda}{\pi n d} \frac{\cos\theta_s}{|\sin(\theta_r - \theta_s)|} \tag{2}$$

Where $\nu = \pi\Delta n d / (\lambda\sqrt{\cos\theta_r \cos\theta_s})$. In this experiment, the λ=633nm, the thickness of the sample is d=10μm, the refractive index of the sample is about $n\approx1.5$, and for our sample at 633nm the refractive index difference between E-form and C-form is $\Delta n\approx1.7\times10^{-2}$, θ_r=45°, θ_s=-45°, so it can be calculated that $\Delta\Theta\approx3.16°$. Because limited by the size of lens L_6 and L_7, the rotation range of M_2 is $\Delta\alpha_{max}$=8°, so the corresponding reference beam maximum multiplexing angle range is $\Delta\theta_{max}$=16°. Therefore, in this experiment, the angle between two reference beams corresponding to two neighboring holograms is chosen as $\Delta\theta$=4°. Five images were multiplexing recorded respectively with exposure time: 20s, 18s, 16s, 14s and 12s.

Reading order of diffracted images is reverse to the recording order. The read out time of each image is 0.2s. The results are shown in Fig.36. It can be seen that, no crosstalk exist between five images and the images' qualities are good. So the storage density can be increased 4 times.

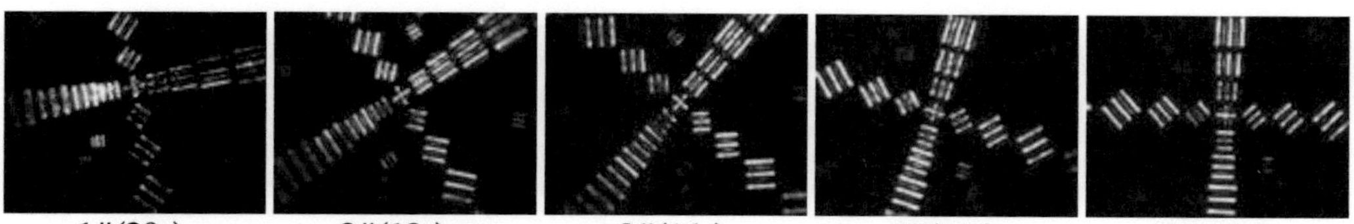

1#(20s) 2#(18s) 3#(16s) 4#(14s) 5#(12s)

Fig. 36. The experimental results of angular multiplexing holograms

Because each time when a new hologram is recorded, the hologram recorded before will be erased. In multiple holograms recording, the erase to the first image is largest and the last recorded image is not affected. If each hologram is recorded for the same time, diffraction efficiency of the first hologram will be lowest, and the diffraction efficiency of final hologram will be highest. Therefore, in order to obtain same diffraction efficiency for the holograms, the exposure time should be reduced with the increasing of the recording order of holograms. The simple diagram is shown in Fig.37.

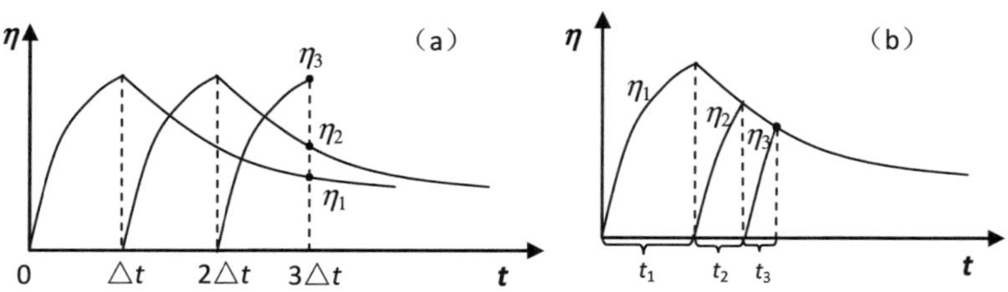

Fig. 37. A simple diagram to explain the exposure time of each hologram in holographic multiplexing: (A) condition with equivalent recording time; (b) condition with decreased recording time

4.7.3 Circumrotation multiplexing holograms

Circumrotation multiplexing is a special angular multiplexing method [10]. In basic circumrotation multiplexing light path of two recording beams unchanged. Every time when a new hologram is recorded, recording medium is rotated at an angle around an axis perpendicular to sample's surface. Fig.38a is a diagram of circumrotation multiplexing holograms, i.e. a diagram of the grating vector directions. The x-y plane is material's surface, and z axis is the rotational axis. To avoid the merger of gratings, the rotation angle of the medium must meet one of the following two conditions: (1) which is more than vertical selection angle; (2) the unnecessary reconstructed images can be filtered by the aperture of detector. The phase conjugated beam reconstruction hologram storage system shown in Fig.24 was used for the circumrotation multiplexing holograms storage experiment. According to the experimental condition, rotation angle is chosen as 10°, after 180° of rotation, 18 images are recorded at the same position of the sample. In Fig.38b, the reconstructed images of the holograms recorded at 0° and 170°are given. It can be seen that no crosstalk exist between the holograms, but reconstructed images' quality is poor.

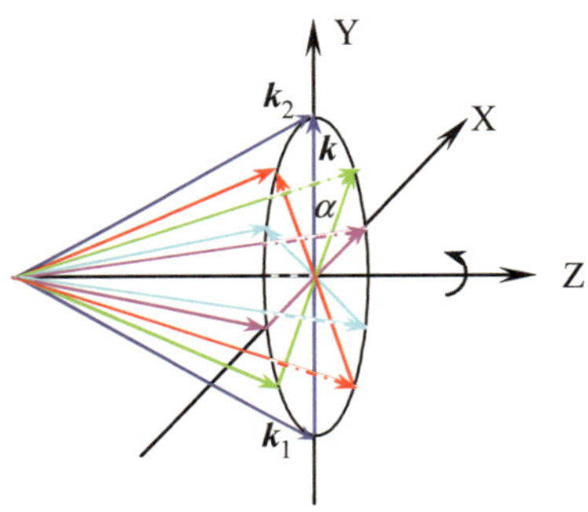

(a)

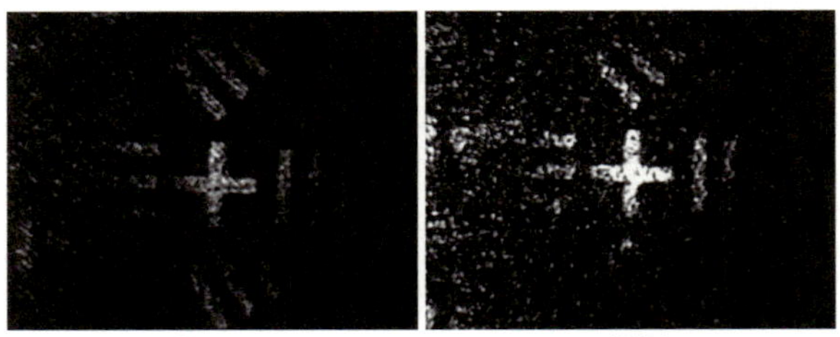

0° 170°

(b)

Fig. 38. Circumrotation multiplexing holograms storage experiment: (a) schematic diagram; (b) experimental results

4.8 Holographic interferometer based on fulgide film

Holographic interferometer is an important aspect of holographic application. At present no reports about the application of fulgide films on holographic interferometer is found. In the section, using the current experimental setups the holographic interferometer application experiment of fulgide film are carried out, including double exposure method and single exposure method.

The experimental setup shown in Fig.28 is used for holographic interferometer application experiment of fulgide film. The recording medium is placed near the frequency-plane. The process of double exposure experiment is as follows: Two holograms are recorded early or late at the same place of the sample by using same setup. When reconstructed by original reference beam, two object waves will be reconstructed together and interference with each other, whose interference fringe can be captured by the CCD. The process of real-time holographic interferometer (single exposure method) experiment is as follows: A hologram of original object beam (OB1) is recorded first. Then the object wave surface to be tested (OB2, whose intensity is decreased to similar with the diffraction light intensity) and the reference beam irradiate the hologram at the same time. The interference fringe between the diffracted wave of OB1 and transmitted wave of OB2 can be captured by the CCD.

According to the theoretical analysis and experimental results, it can be understood that the experimental results of double exposure interferometer and real-time interferometer are same, and their processing methods are same too. So here just four experimental results of double exposure interferometer are given, like shown in Fig.39.

Using data processing, the thickness variety values of measured objects at any point on the X-Y surface can be got. In Fig.40 shows the thickness variety of measured objects got by processing the experiment results shown in Fig.39(a,b). And the tilt angles of the optical wedge and of axicon are calculated as 1.83°and 1° respectively. Form Fig.39(c,d) it can be calculated that the rotation angle of optical wedge is 7.91°and the movement of axicon is 1.86 mm [11].

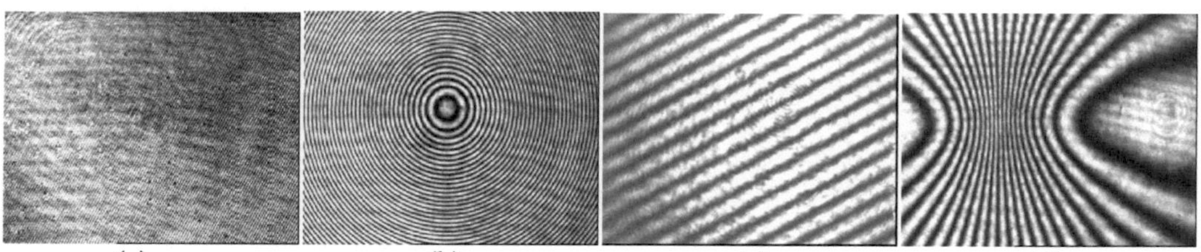

(a) (b) (c) (d)

Fig. 39. Experimental results of double exposure interferometer: (a) no object(air) in the first exposure and the wedge is the object to be tested in the second exposure; (b) no object(air) in the first exposure and the axicon is the object to be tested in the second exposure; (c) the wedge is the object, which is rotated slightly in two exposure; (d) the axicon is the object, which is moved slightly toward right-or-left in two exposure

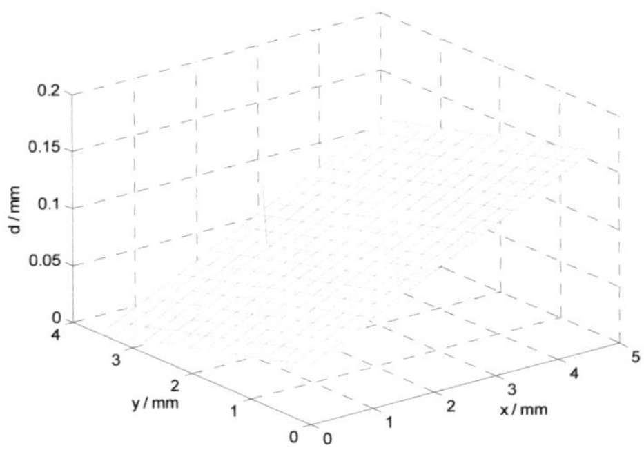

(a)

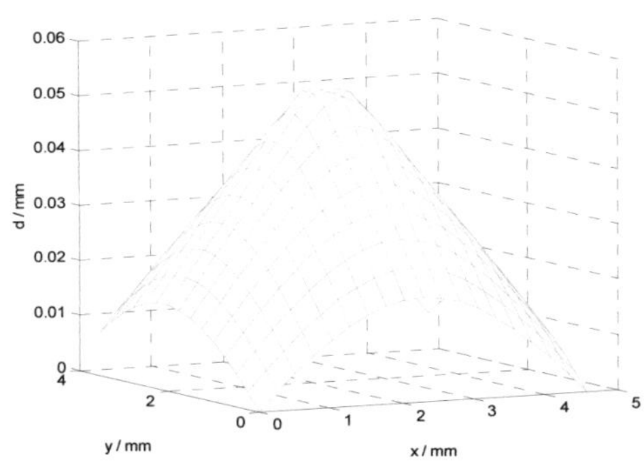

(b)

Fig. 40. (a) thick variety of optical wedge, (b) thick variety of axicon

4.8.1 Application of polarization multiplexing technique in holographic interferometer

The object to be measured is moving particle. The parallel linearly polarization hologram and orthogonal linear polarization hologram are recorded respectively, and then the two images can be reconstructed individually owing to polarization multiplexing technique, in addition the recorded order of the two holograms is known, so the direction and velocity of the object's movement can be got. But we cannot distinguish the recorded order of the two holograms if they are both recorded by parallel linearly polarization holographic technology. Fig.41 shows the experimental results (because for real moving particles the dispersion of light is too large, the transparent film is used as the measured object.)

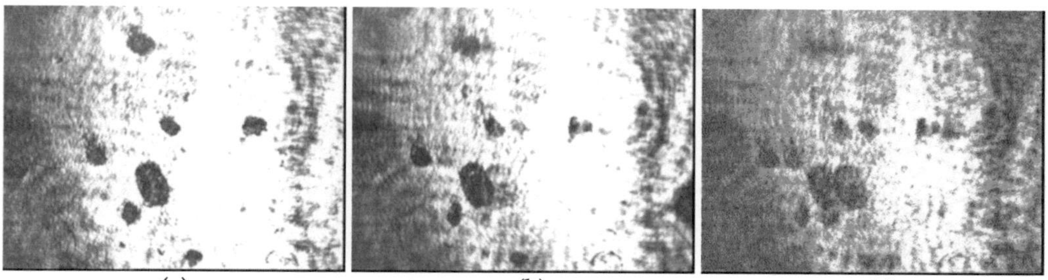

(a) (b) (c)

Fig. 41. The double exposure experimental results to record movement track of the particles at different times (a) 1st image recorded as parallel linearly polarization hologram; (b) 2nd image recorded as orthogonal linearly polarization hologram; (c) reconstructed image obtained when the polarization P is rotated to 30°

5. Conclusion

The holographic storage applications of 3-indoly-benzylfulgimide/PMMA film were studied in detail including the ordinary holography and polarization holography, which are respectively based on the photochromic and photoinduced anisotropy properties. The properties of holographic recording such as diffraction efficiency, spatial resolution and optimal exposure were measured; especially the diffraction efficiency spectra and dynamic curves of different kinds of polarization holographic recording were theoretically analyzed and experimentally measured.

The holographic optical image storage was realized in the fulgide films by using different kinds of holographic storage techniques. The experimental results show that: compared with transmission-type holographic recording, reflection-type holographic recording hologram has lower diffraction efficiency and higher SNR; compared with reference beam reconstruction, the phase conjugated beam reconstruction can effectively correct the phase aberration caused by the mis-adjustment of optical setup; compared with Fraunhofer holograms, Fourier-transform holograms have lower diffraction efficiency and higher storage density; compared with parallel polarization holograms, in orthogonal polarization holograms the scattering noise can be filtered, so one can obtain high SNR, in which the orthogonal circularly polarization hologram also has high diffraction efficiency, so it is the best polarization recording method; compared with traditional non-collinear holography, the collinear holographic storage system has simpler optical setup and smaller volume, lower environmental demand and higher storage density. The storage density of 2×10^8 bits/cm^2 was obtained in the Fourier-transform holographic data storage by using orthogonal polarization holographic recording, which had a greatly improved signal-to-noise ratio of the diffraction image.

Different kinds of multiplexing holographic storage, like polarization multiplexing, circumrotation multiplexing and angle multiplexing, were also realized in fulgide film, where 2 images, 5 images and 18 images were stored at the same position of the film, and diffracted without crosstalk with each other. And the application of the fulgide films in holographic interferometry was also studied initially.

6. References

[1] Yokoyama, Y. (2000). Fulgides for Memories and Switches. Chem. Rev., Vol.100, No.5, pp. 1717-1739

[2] Fan, M. (1997). Photon Storage Principles and Photochromic Materials. Progress in Chemistry, Vol.9, No.2, pp. 170-178 (in Chinese)

[3] Bouas-Laurent, H. & Dürr H. (2001). Organic Photochromism. Pure and Applied Chemistry. Vol.73, pp. 639–665

[4] Menke, N.; Yao, B.; Wang, Y.; Zheng, Y.; Lei, M.; Chen, G.; Chen, Y.; Fan, M. & Li, T. (2006). Optical Image Processing Using The Photoinduced Anisotropy of Pyrrylfulgide, J. Opt. Soc. Am. A, Vol.23, No.2, pp. 267-271

[5] Menke, N.; Yao, B.; Wang, Y.; Dong, W.; Lei, M.; Chen, Y.; Fan, M. & Li, T. (2008). Spectral Relationship of Photoinduced Refractive Index and Absorption Changes in Fulgide Films, Journal of Modern Optic, Vol.55, No.6, pp. 1003-1011

[6] Du, J.; Menke, N.; Yao, B.; Wang, Y. & Chen, Y. (2011). Photoreaction constants of fulgide films at different wavelengths, Proceedings of International Conference on Remote Sensing, Environment and Transportation Engineering 2011, IEEE Catalog Number: CFP1104M-PRT Volume 7, pp.6148-6154 (June 24-26, 2011, Nanjing, China)

[7] Menke, N.; Yao, B.; Wang, Y. & Chen, Y. (2011). Polarization holography in 3-indoly-benzylfulgimide/PMMA film. Journal of Atomic, Molecular, and Optical Physics Volume2011, ArticleID509507, doi:10.1155/2011/509507, pp.1-20

[8] Cheng, M.; Menke, N.; Yao, B.; Wang, Y. & Chen, Y. (2011). Improvement of the diffraction efficiency of holographic gratings in Fulgide films by auxiliary light, Proceedings of International Conference on Remote Sensing, Environment and Transportation Engineering 2011, IEEE Catalog Number: CFP1104M-PRT, Volume 7, pp.6047-6052 (June 24-26, 2011, Nanjing, China)

[9] Menke, N.; Yao, B.; Wang, Y.; Zheng, Y.; Lei, M.; Chen, G.; Chen, Y.; Fan, M.; Han, Y. & Meng, X. (2003). Holographic Recording Characteristics of a Rewritable Fulgide/PMMA Film, Acta Photonica Sinica, Vol.32, No.7, pp. 819-822 (in Chinese)

[10] Tao., S. (December 1998). Optical Holographic Storage (First Edition). Beijing University Press, Beijing, pp.267-278

[11] Ji, K.; Menke, N.; Menke, N.; Yao, B.; Wang, Y. & Chen, Y. (2011). Holographic interfermetry based on fulgide film, Proceedings of International Conference on Electronics and Optoelectroonics2011, IEEE Catalog Number: CFP1137N-PRT, Volume 3, pp.V3-276-V3-280 (July 29-31, 2011, Dalian, China)

8

Three-Dimensional Vector Holograms in Photoreactive Anisotropic Media

Tomoyuki Sasaki[1], Akira Emoto[2], Kenta Miura[1], Osamu Hanaizumi[1], Nobuhiro Kawatsuki[3] and Hiroshi Ono[2]

[1]*Department of Electronic Engineering, Graduate School of Engineering, Gunma University,*
[2]*Department of Electric Engineering, Nagaoka University of Technology,*
[3]*Department of Materials Science and Chemistry, Graduate School of Engineering, University of Hyogo,*
Japan

1. Introduction

Polarization gratings, in which optical anisotropy is periodically modulated, are very attractive from the point of view of their interesting optical properties, including polarization selectivity of the diffraction efficiency and polarization conversion in the diffraction process (Nikolova et al., 1984). These properties make polarization gratings useful for numerous optical applications related to polarization discrimination (Davis et al., 2001; Asatryan et al., 2004), control (Nikolova et al., 1997; Hasman et al., 2002), and measurement (Gori, 1999; Provenzano et al., 2006). Polarization gratings can be fabricated by holographic exposure with polarized interference light in photoanisotropic media such as azobenzene-containing polymer (azopolymer) films (Todorov et al., 1984; Ebralidze et al., 1992; Huang and Wagner, 1993; Samui, 2008). Since azobenzene molecules reorient in accordance with polarization of light resulting from trans-cis-trans photoisomerization reactions, periodic structures are formed as spatial distribution of the molecular orientation by holographic recording using vectrial light (i.e., by vector holography). In addition, surface relief gratings are obtained by holographic recording in azopolymers (Rochon et al., 1995; Kim et al., 1995; Ramanujam et al., 1996; Naydenova et al., 1998; Labarthet et al., 1998, 1999). Nikolova et al. have investigated the diffraction properties of various types of vector holograms recorded in side-chain azobenzene polyesters (Nikolova et al., 1996). Birabassov and Galstian have analyzed the relationship of polarization states between holographic recording and reconstruction beams with the use of azopolymers (Birabassov et al., 2001). It has been also demonstrated that vector holograms realize recording and reconstruction of polarized optical images (Ono et al., 2009a, 2009b).

In a previous paper, we presented three-dimensional (3D) vector holography, which is a novel concept for holographic recording using vectrial light (Sasaki et al., 2008). Common vector holograms are recorded in initially isotropic media such as amorphous azopolymers. In contrast, 3D vector holograms are recoded in initially anisotropic media. Since the propagation velocity of light in anisotropic media varies in accordance with the polarization state, standing waves with a multidimensionally varying state of polarization were obtained in the region of overlap. We recorded anisotropic gratings in a liquid crystalline medium by the 3D vector holography, and investigated their diffraction properties. As a result, it was demonstrated that the 3D vector holography enables formation of higher-order periodical modulation of optical anisotropy. In addition, we observed that the multidimensional distribution exhibited interesting diffraction properties.

The present chapter reported the principle of 3D vector holography in detail, and investigated optical characteristics of 3D vector holograms recorded in a photoreactive anisotropic medium. Interference of two polarized light in anisotropic media was described by employing the extended Jones matrix method. Diffraction properties of the holograms, recorded in the model medium, were observed and the obtained results were analyzed through the use of the finite-difference time-domain (FDTD) method. By comparing the experimental and calculated results, we discussed the adequacy of the formation mechanism of the 3D vector holograms. In addition, the diffraction efficiency was simulated for various recording conditions to reveal the relationship between the multidimensional structure and the optical behaviour.

2. Principle of three-dimensional vector holography

We consider interference of mutually coherent polarized light in uniaxial anisotropic media. As illustrated in Fig. 1, the xz -plane is the incident plane and the z -axis is taken normal to the film plane. Assuming that the two recording beams are plane waves and that the amplitude of their incident angles is small, the electric field of interference light is described using the Jones vector as

$$\mathbf{E}(x, z) = \mathbf{J}_1(z)\mathbf{E}_1 \exp(ik_1x) + \mathbf{J}_2(z)\mathbf{E}_2 \exp(ik_2x), \tag{1}$$

where $\mathbf{J}_1$ and $\mathbf{J}_2$ are the extended Jones matrixes of the medium for the two beams, $\mathbf{E}_1$ and $\mathbf{E}_2$ are the Jones vectors of the two beams, $k_1 = (2\pi/\lambda)\sin\theta_1 \equiv k\sin\theta_1$, $k_2 = k\sin\theta_2$, θ_1 and θ_2 are the incident angles of the two beams, and λ is the wavelength in a vacuum. Here, to simplify the analysis, reflection at the air-medium boundaries is neglected. When the ordinary and extraordinary indexes and the direction of the optical axis $\mathbf{c}_0$ are homogeneous in the medium, the extended Jones matrix is given by (Lien, 1990, 1997)

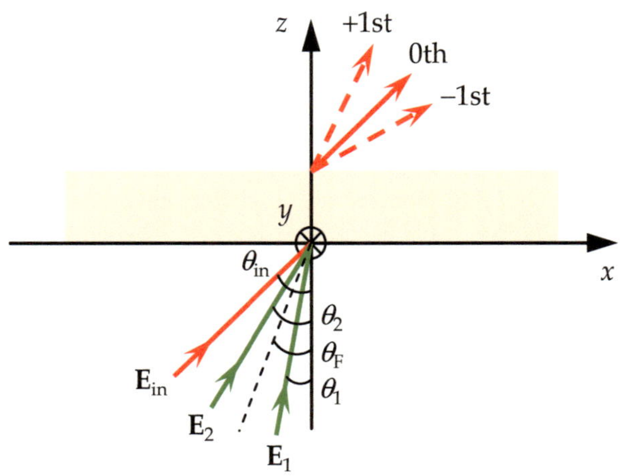

Fig. 1. Schematic illustration of arrangement for holographic recording and reconstruction. $\mathbf{E}_{in}$ is the electric field vector of reconstruction light

$$\mathbf{J}_m = \mathbf{S}_m \mathbf{G}_m \mathbf{S}_m^{-1}, \tag{2}$$

where

$$\mathbf{G}_m = \begin{bmatrix} \exp(ik_{z1}z) & 0 \\ 0 & \exp(ik_{z2}z) \end{bmatrix}, (0 \le z \le d), \tag{3}$$

$$\mathbf{S}_m = \begin{bmatrix} 1 & c_2 \\ c_1 & 1 \end{bmatrix}, \tag{4}$$

$$k_{z1}/k = \sqrt{n_o^2 - (k_m/k)^2}, \tag{5}$$

$$\frac{k_{z2}}{k} = -\frac{\varepsilon_{xz}}{\varepsilon_{zz}}\frac{k_m}{k} + \frac{n_o n_e}{\varepsilon_{zz}}\sqrt{\varepsilon_{zz} - \left(1 - \frac{n_e^2 - n_o^2}{n_e^2}\cos^2\theta_0 \sin^2\phi_0\right)\left(\frac{k_m}{k}\right)^2}, \tag{6}$$

$$c_1 = \frac{[(k_m/k)^2 - \varepsilon_{zz}]\varepsilon_{yx} + [(k_m/k)(k_{z1}/k) + \varepsilon_{zx}]\varepsilon_{yz}}{[(k_{z1}/k)^2 + (k_m/k)^2 - \varepsilon_{yy}][(k_m/k)^2 - \varepsilon_{zz}] - \varepsilon_{yz}\varepsilon_{zy}}, \tag{7}$$

$$c_2 = \frac{[(k_m/k)^2 - \varepsilon_{zz}]\varepsilon_{xy} + [(k_m/k)(k_{z2}/k) + \varepsilon_{xz}]\varepsilon_{zy}}{[(k_{z2}/k)^2 - \varepsilon_{xx}][(k_m/k)^2 - \varepsilon_{zz}] - [(k_m/k)(k_{z2}/k) + \varepsilon_{zx}][(k_m/k)(k_{z2}/k) + \varepsilon_{xz}]}. \tag{8}$$

Here, $m = 1, 2$, d is the film thickness, n_o and n_e are the ordinary and extraordinary indexes, θ_0 is the angle between the xy-plane and $\mathbf{c}_0$, and ϕ_0 is the angle between the x-axis and the projection of $\mathbf{c}_0$ onto the xy-plane. The dielectric tensor of the medium can be written as

$$\boldsymbol{\varepsilon} = \begin{bmatrix} \varepsilon_{xx} & \varepsilon_{xy} & \varepsilon_{xz} \\ \varepsilon_{yx} & \varepsilon_{yy} & \varepsilon_{yz} \\ \varepsilon_{zx} & \varepsilon_{zy} & \varepsilon_{zz} \end{bmatrix}, \tag{9}$$

$$\varepsilon_{xx} = n_o^2 + (n_e^2 - n_o^2)\cos^2\theta_0 \cos^2\phi_0, \tag{10}$$

$$\varepsilon_{xy} = \varepsilon_{yx} = (n_e^2 - n_o^2)\cos^2\theta_0 \sin\phi_0 \cos\phi_0, \tag{11}$$

$$\varepsilon_{xz} = \varepsilon_{zx} = (n_e^2 - n_o^2)\sin\theta_0 \cos\theta_0 \cos\phi_0, \tag{12}$$

$$\varepsilon_{yy} = n_o^2 + (n_e^2 - n_o^2)\cos^2\theta_0 \sin^2\phi_0, \tag{13}$$

$$\varepsilon_{yz} = \varepsilon_{zy} = (n_e^2 - n_o^2)\sin\theta_0 \cos\theta_0 \sin\phi_0, \tag{14}$$

$$\varepsilon_{zz} = n_o^2 + \left(n_e^2 - n_o^2\right)\sin^2\theta_0 . \tag{15}$$

By setting

$$\mathbf{E}(x, z) \equiv \begin{bmatrix} A_x \exp(i\delta_x) \\ A_y \exp(i\delta_y) \end{bmatrix} \equiv \begin{bmatrix} E'_x \\ E'_y \end{bmatrix}, \tag{16}$$

we obtain the polarization distribution in interference light with the use of the Stokes parameters as (Kliger, 1990)

$$\begin{bmatrix} S_0 \\ S_1 \\ S_2 \\ S_3 \end{bmatrix} = \begin{bmatrix} A_x^2 + A_y^2 \\ A_x^2 - A_y^2 \\ 2A_xA_y\cos\left(\delta_y - \delta_x\right) \\ 2A_xA_y\sin\left(\delta_y - \delta_x\right) \end{bmatrix} . \tag{17}$$

Figure 2 illustrates the polarization distribution in the interference light calculated by varying n_o and n_e. The paremeters used in this calculation are as follows: $d = 10$ µm, $\theta_0 = 0$, $\phi_0 = \pi/2$, $\lambda = 532$ nm, $\theta_1 = -\theta_2 = 1.5°$ (i.e., $(\theta_1 + \theta_2)/2 \equiv \theta_F = 0$), $\mathbf{E}_1 = (0, 1)$ (i.e., $\mathbf{E}_1$ is *s*-polarized light), and $\mathbf{E}_2 = (1, 0)$ (i.e., $\mathbf{E}_2$ is *p*-polarized light). The one cycle of the interference pattern for the x-direction is given as $\Lambda_x = \lambda/\left|\sin\theta_1 - \sin\theta_2\right| \cong 10$ µm. As seen in Fig. 2, the polarization distribution of the interference light varies depending on values of n_o and n_e. When $n_e - n_o \equiv \Delta n = 0$, namely, the medium is isotorpic, the polarization state is modulated only for the x-direction. However, that is modulated for the x- and z-dirctions in the case of $\Delta n \neq 0$. This is because the refractive indexes for the *s*- and *p*-polarized components differ in the anisotorpic medium since the optical axis is parallel to the incident plane. These calculated results clearly indicate that multidimensional periodic distribution of optical anisotorpy can be formed by the holographic recording. Therefore, we call these holograms, recorded in anisotropic media by vectrical light, 3D vector holograms.

Figure 3 illustrates interference patterns calculated by varing polarization states of the two beams in the case of $n_o = 1.52$ and $n_e = 1.75$. The other paremeters used in this calculation are the same as those mentioned above. In this chapter, linearly polarized (LP) light with the azimuth angle of ±45° for the x-axis are described as $q_\pm$-polarized, respectively. In addition, right- and left-handed circularly polarized light are denoted by *r* and *l*, respectively. The results shown in Fig. 3 demonstrate that various interference patterns are formed depending on polarization states of the two beams. When both the beams are *s*-polarized, the intensity was modulated one-dimensionally and the polarization state was remained in the medium. In contrast, the polarization state was two-dimensionally modulated when both the beams are q_+- or *r*-polarized. This is because *s*-polarized light propagates with no change of polarization state since the electric field vector is parallel to the optical axis $\mathbf{c}_0$. Figure 4 shows interference patterns calculated by varying ϕ_0 (i.e., the azimuth angle of $\mathbf{c}_0$). In this calculation, the two beams are *s*- and *p*-polarized and the other parameters are the same as those mentioned above. As seen in Fig. 4, the interference pattern also depended on ϕ_0. This fact suggests that the periodic structure, recorded by the 3D vector holography, is affected by the initial alignment of the medium.

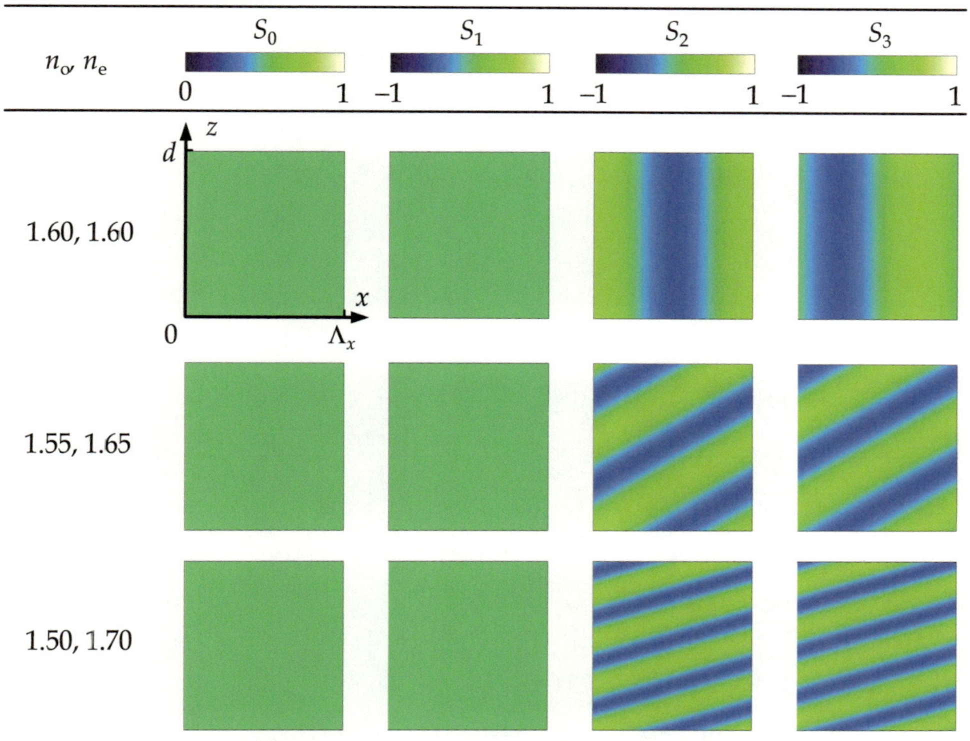

Fig. 2. Spatial distribution of the Stokes parameters in interference light calculated by varying the refractive indexes

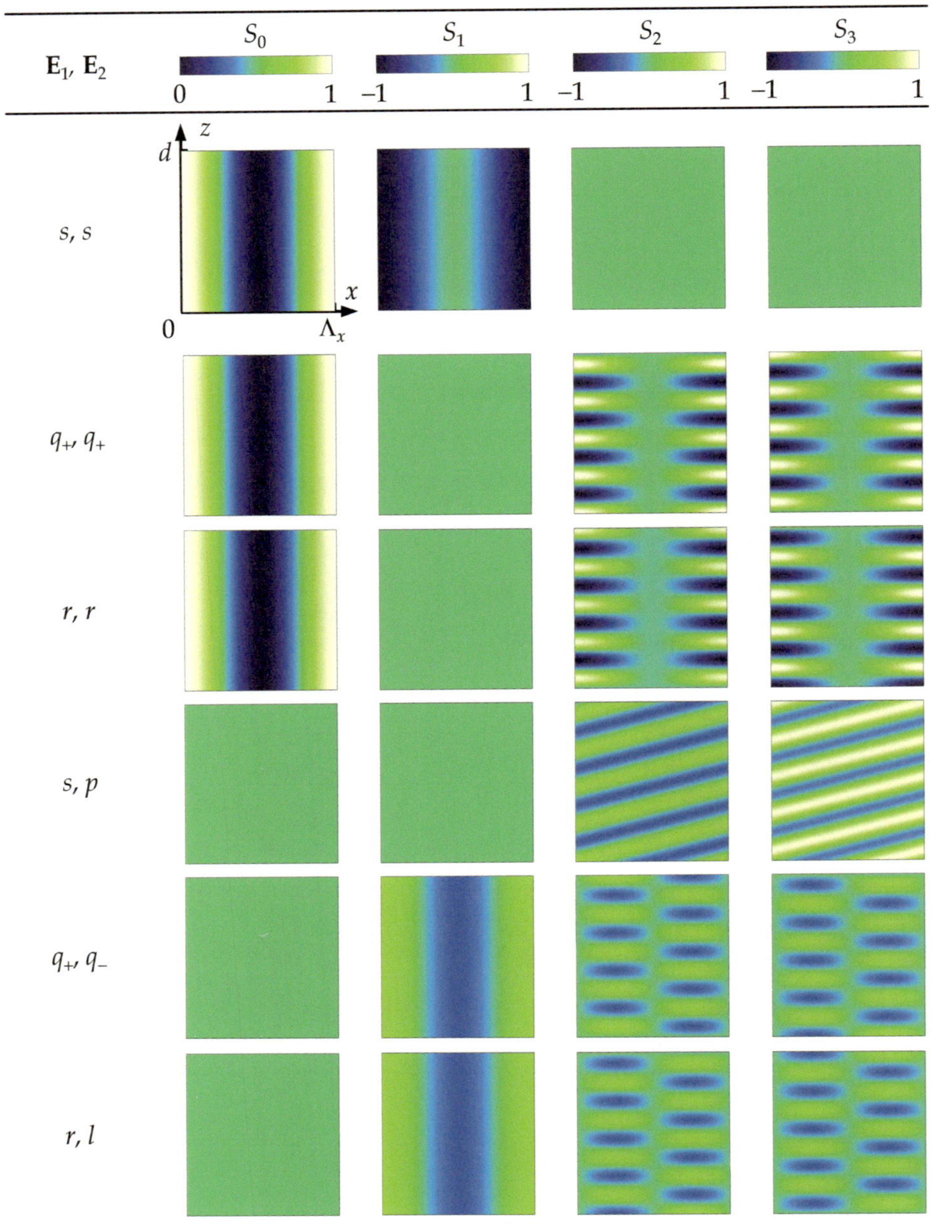

Fig. 3. Spatial distribution of the Stokes parameters in interference light calculated by varying polarization states of two recording beams

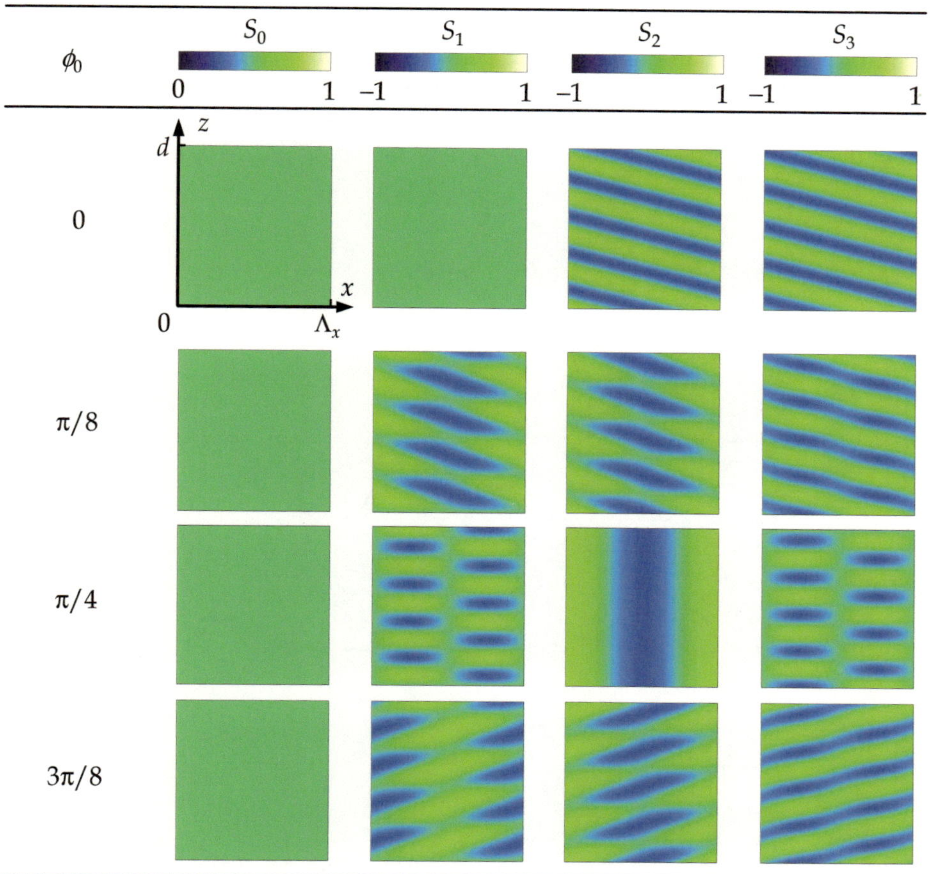

Fig. 4. Spatial distribution of the Stokes parameters in interference light calculated by varying the direction of the optical axis

In conclusion, we confirmed theoretically that 3D vector holography realizes fabrication of various multidimensional anisotropic structures owing to polarization states of recording beams, optical anisotorpy of the medium, and alignment of the optical axis.

3. Photoreactive anisotropic medium with uniaxial anisotropy

In order to confirm the principle described in section 2, a grating was recorded in a model medium by 3D vector holography and its diffraction properties were observed. We prepared an azo-dye doped liquid crystalline material as the photoreactive anisotropic medium by mixing nematic mixture E7, a side-chain liquid crystalline polymer (SLCP), and azo-dye 4-[N-(2-hydroxyethyl)-N-ethyl]amino-4'-nitroazobenzene, more commonly known as disperse red 1 (DR1), with a weight ratio of E7 : SLCP : DR1 = 59 : 40 : 1. Here, E7 and DR1 were obtained from BDH-Merck and Ardrich, respectively. SLCP was synthesized using a poly(methyl methacrylate) backbone comprising 4-cyanophenyl benzoate side groups. The number and weight averages of SLCP used in this study were 11,700 and 32,800, respectively. The chemical structures of the three components were illustrated in Fig. 5. They were stirred at around 100°C until a homogeneous solution was obtained. The mixture was sandwiched between two rubbed poly(vinyl alcohol)-coated glass substrates with 10 μm thick spacers in order to obtain a homogeneously aligned sample. By using a polarizing optical microscope, we observed that the mixture exhibited the nematic phase at room temperature. Namely, it was confirmed that the azo-dye doped liquid crystalline material with homogeneous alignment was a uniaxial anisotropic medium.

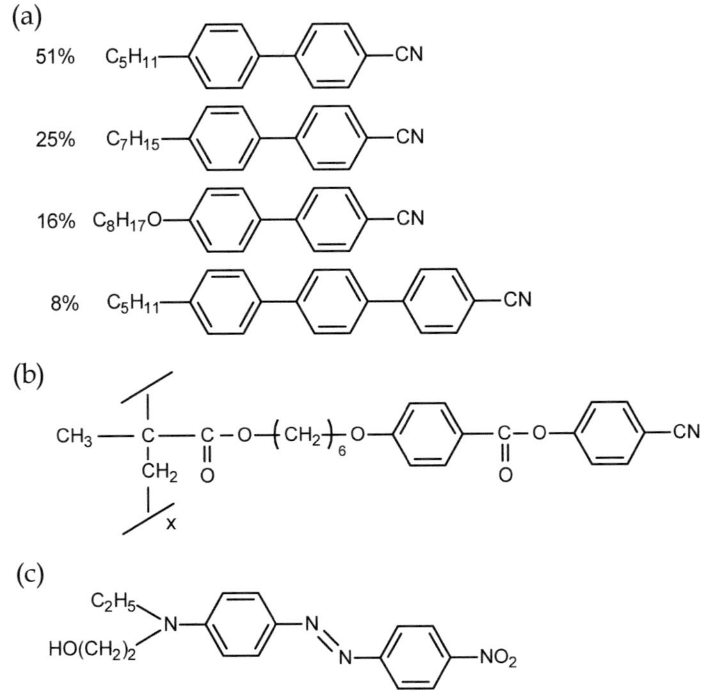

Fig. 5. Chemical structures of (a) E7, (b) SLCP, and (c) DR1

We consider the molecular reorientation induced by irradiation with polarized light. In order to discuss the polarization state of interference light, we assume that its propagation direction is in the xz-plane and that the angle between the z-axis and the propagation direction is given by

$$\Theta_F = \sin^{-1}\left[(\sin\theta_F)/\bar{n}\right], \tag{18}$$

where $\bar{n}$ is the average refractive index of the medium. Here, we transform the coordinate system from xyz into $x'y'z'$ as follow

$$\begin{bmatrix} x' \\ y' \\ z' \end{bmatrix} = \begin{bmatrix} \cos\Theta_F & 0 & \sin\Theta_F \\ 0 & 1 & 0 \\ -\sin\Theta_F & 0 & \cos\Theta_F \end{bmatrix} \begin{bmatrix} x \\ y \\ z \end{bmatrix} \equiv \mathbf{R}^{-1}(\Theta_F) \begin{bmatrix} x \\ y \\ z \end{bmatrix}. \tag{19}$$

In the $x'y'z'$-coordinate system, the orientation direction (i.e., the optical axis) in the initial state is written as

$$\mathbf{c}_0' = \mathbf{R}^{-1}(\Theta_{\mathrm{F}})\ \mathbf{c}_0 , \tag{20}$$

where $\mathbf{c}_0$ is the normalized vector of the initial optical axis in the xyz-coordinate system. Here, $\mathbf{c}_0'$ can be described as

$$\mathbf{c}_0' = \left(\cos\theta_0' \cos\phi_0',\ \cos\theta_0' \sin\phi_0',\ \sin\theta_0'\right), \tag{21}$$

where θ_0' is the angle between the $x'y'$-plane and $\mathbf{c}_0'$ and ϕ_0' is the angle between the x'-axis and the projection of $\mathbf{c}_0'$ onto the $x'y'$-plane. The reoriented optical axis in the $x'y'z'$-coordinate system is written as

$$\mathbf{c}_{\mathrm{re}}' = \left(\cos(\theta_0' + \theta_{\mathrm{re}}')\cos(\phi_0' + \phi_{\mathrm{re}}'),\ \cos(\theta_0' + \theta_{\mathrm{re}}')\sin(\phi_0' + \phi_{\mathrm{re}}'),\ \sin(\theta_0' + \theta_{\mathrm{re}}')\right), \tag{22}$$

where θ_{re}' is the photoinduced tilt angle and ϕ_{re}' is the photoinduced azimuth angle. For simplicity, we assume that the amplitudes of the photoinduced angles are small and are proportional to the intensity of light. In addition, it has been revealed that the director of the azo-dye doped liquid crystalline material tilts to the perpendicular direction for the polarization azimuth. Therefore, using Equations 16 and 17, the two photoinduced angles are written as

$$\theta_{\mathrm{re}}' = C_\theta \left|\mathbf{c}_0' \cdot \mathbf{E}_{\mathrm{n}}\right| S_0 , \tag{23}$$

$$\phi_{\mathrm{re}}' = C_\phi \left|\mathbf{c}_0' \cdot \mathbf{E}_{\mathrm{n}}\right| \left[S_1 \sin(2\phi_0') - S_2 \cos(2\phi_0')\right], \tag{24}$$

where C_θ and C_ϕ are constants, which represent sensitivity to light intensity, and $\mathbf{E}_{\mathrm{n}}$ is the normalized electric field vector of interference light in the $x'y'z'$-coordinate system, namely,

$$\mathbf{E}_{\mathrm{n}} = \frac{1}{\sqrt{\left|E_x'\right|^2 + \left|E_y'\right|^2}} \begin{bmatrix} E_x' \\ E_y' \\ 0 \end{bmatrix}. \tag{25}$$

In the xyz-coordinate system, the reoriented optical axis is given by

$$\mathbf{c}_{\mathrm{re}} = \mathbf{R}(\Theta_{\mathrm{F}})\ \mathbf{c}_{\mathrm{re}}' \equiv \left(c_x,\ c_y,\ c_z\right) = \left(\cos\theta\cos\phi,\ \cos\theta\sin\phi,\ \sin\theta\right), \tag{26}$$

where θ is the angle between the xy-plane and $\mathbf{c}_{\mathrm{re}}$ and ϕ is the angle between the x-axis and the projection of $\mathbf{c}_{\mathrm{re}}$ onto the xy-plane. From Equation 26, we obtain

$$\theta(x, z) = \sin^{-1} c_z , \tag{27}$$

$$\phi(x, z) = \sin^{-1}\left(c_y / \cos\theta\right), \tag{28}$$

Therefore, the photoinduced tilt and azimuth angles in the xyz-coordinate system are obtained as

$$\theta_{\mathrm{re}}(x, z) = \theta - \theta_0 , \tag{29}$$

$$\phi_{\mathrm{re}}(x, z) = \phi - \phi_0 . \tag{30}$$

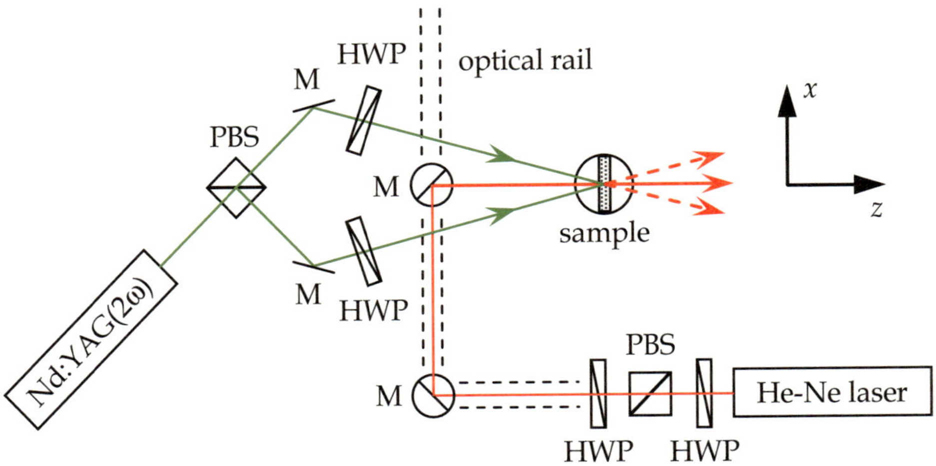

Fig. 6. Schematic illustration of the experimental setup (HWP, half-wave plate; PBS, polarizing beam splitter; M, mirror). Circles represent rotary stage in the incident plane

4. Observed diffraction properties

Figure 6 schematically illustrates the experimental setup for the holographic recording and the reconstruction. A frequency doubled, Nd-doped yttrium aluminum garnet laser with an operating wavelength of 532 nm was used for recording the hologram. This is because the azo-dye effectively reacts to irradiation with green light. The LP recording beam was divided into *s*- and *p*- polarizations using a polarizing beam splitter, and the two beams intersected on the sample as illustrated in Fig. 6. The intensity of the two recording beams was 900 mW/cm^2, respectively. The reconstruction properties were probed by an LP He-Ne laser with a wavelength of 633 nm. The intensity of diffracted beams was measured by a photodiode detector while varying the incident angle of the probe beam. All measurements were conducted at room temperature.

Figure 7 presents the measured positive and negative first-order diffraction efficiencies for $\theta_1 = -\theta_2 = 1.5°$ (i.e., $\theta_F = 0$), $\theta_0 = 0$, and $\phi_0 = \pi/2$. The ±1st-order diffraction efficiencies $\eta_\pm$ were defined as the ratio of the intensity of the diffracted beam to the total intensity of the transmitted light. As seen in Fig. 7, strong diffraction appeared in the ±1st-order when the probe beam was *p*- and *s*-polarized, respectively. When the probe beams was q_+-polarized, relatively strong diffraction was observed in both the orders. The results shown in Fig. 7 clearly indicate that the diffraction efficiency depends on the incident angle of the probe beam θ_{in}. It is also important to remember that the angular dependence is asymmetry for $\theta_{in} = 0$ even though $\theta_F = 0$. Figure 8 illustrates observed polarization states of the diffracted beams in the case of $\theta_{in} = 0$. This result implies that the grating diffracts *p*- and *s*-polarized components of incident light as *s*- and *p*-polarized light, respectively. Here, the *p*- and *s*-polarizations were defined for the xz -plane in this chapter. This property with polarization discrimination is useful for applying to optical elements such as a polarizing beam splitter with a thin thickness.

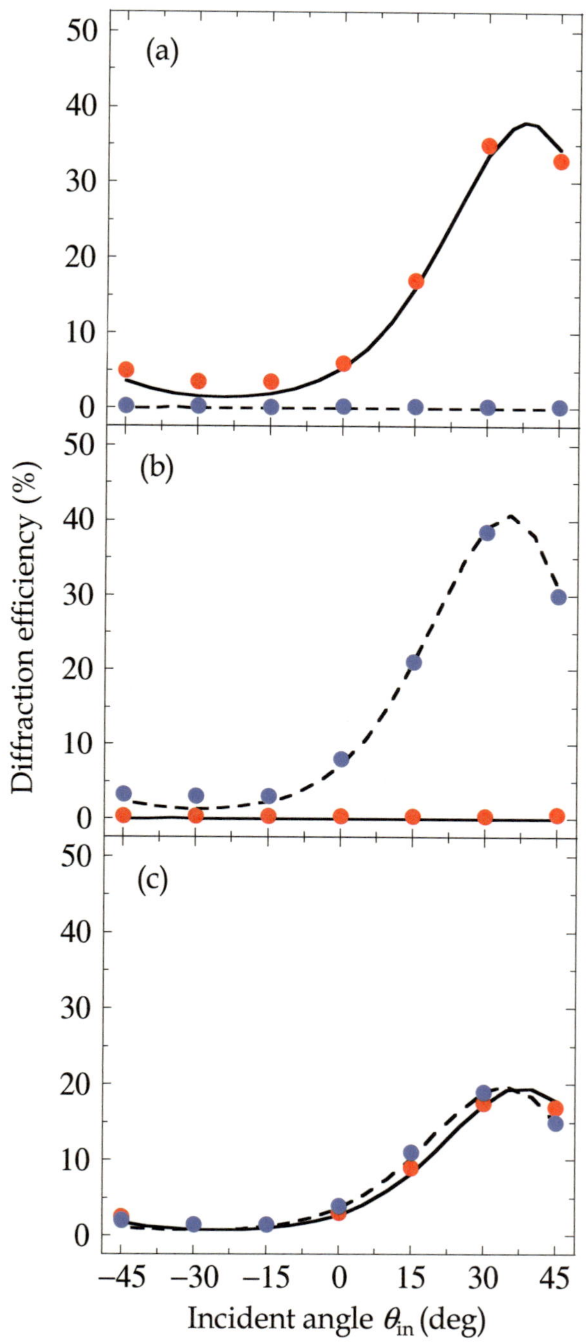

Fig. 7. Angular dispersion of the diffraction efficiency for (a) *p*-, (b) *s*-, and (c) q_+-polarized probe beam. Red and blue plots represent the measured positive and negative first-order diffraction efficiencies. Solid and dashed lines represent the calculated positive and negative first-order diffraction efficiencies

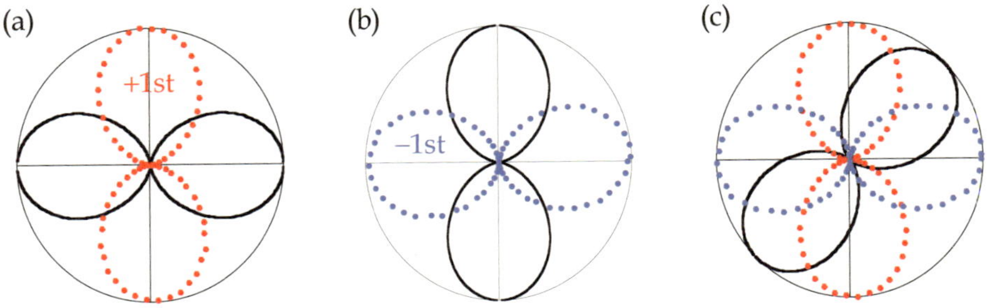

Fig. 8. Polar plots of polarization states for incident and diffraction beams. The incident beam, indicated as solid lines, is (a) *p*-, (b) *s*-, and (c) q_+-polarized. Red and blue plots represent measured data for positive and negative first-order diffraction beams

5. Theoretical characterization

5.1 FDTD method

In order to characterize the observed results mentioned above, optical behaviour of the periodical distribution of the dielectric tensor, which was described in section 3, were calculated by employing the FDTD method. This method imposes no limitations on the dimension of the object and the incident conditions since Maxwell's equations are directly solved numerically by the calculus of finite differences in the algorithm (Yee, 1966; Teflova & Gedne, 2005). The FDTD method can analyze anisotropic media such as liquid crystals (Kriezis & Elston, 1999, 2000; Titus et al., 2001; Scharf, 2007). In addition, it has been reported that the FDTD method is suitable for characterizing anisotropic gratings recorded by vector holography (Oh & Escuti, 2006, 2007; Ono et al. 2008a, 2008b).

Figure 9 schematically illustrates the model for the FDTD calculation in this study. We set the calculation conditions as follows: the wavelength of the incident light is 633 nm, the size of the analytical area (x -direction × z -direction) is 162 μm × 15 μm, the grid spacing is 20 nm × 20 nm, and the time step is 0.047 fs. As the boundary condition, the perfectly matched layer (PML) with the thickness of 1 μm was used (Teflova & Gedne, 2005). The glass substrates and the alignment layers were ignored and antireflection (AR) layers were installed at the air-hologram boundaries for reducing the effect of multiple interference (Southwell, 1983). The diffraction properties in the far field were calculated by Fourier transforming each component of the electric field at the output plane.

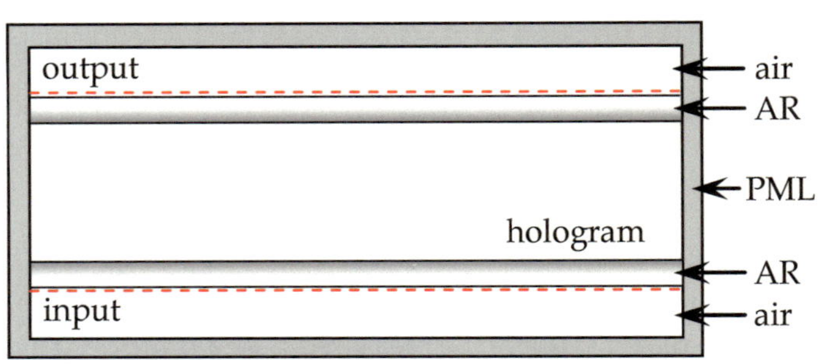

Fig. 9. Schematic layout of the analytical area for the FDTD calculation (AR, anti-reflection coatings; PML, perfectly matched layer)

5.2 Analysis of observed diffraction properties

The calculated diffraction efficiencies for the experimental conditions, described in section 4, were shown in Fig. 7. The parameters used this calculation are as follows: $n_o = 1.52$, $n_e = 1.75$, $C_\theta = 4.9\times10^{-2}$ cm²/W, and $C_\phi = 4.8\times10^{-2}$ cm²/W. As seen in Fig. 7, the observed diffraction efficiencies were well explained with the use of the theoretical model and the FDTD calculation. This fact suggests that the 3D vector hologram is actually recorded in the azo-dye doped liquid crystalline material. The calculated distribution of the photoinduced tilt and azimuth angles were illustrated in Fig. 10. It is clearly known that a slanted grating is formed based on the spatially varying state of polarization. Figure 11 illustrates the calculated distribution of electric fields in and around the hologram when the incident beam is *p*- and *s*-polarized. Here, the AR layers were not implemented in this calculation to accentuate the air-hologram boundaries. For the *p*-polarization, the *s*-component was gradually generated in the hologram and propagated to the positive first-order direction. In contrast, the *p*-component was gradually generated in the hologram and propagated to the negative first-order direction when the incident light was *s*-polarized. The calculated results are consistent with the observed property of polarization conversion shown in Fig. 8. This fact also demonstrates the accuracy of the theoretical model.

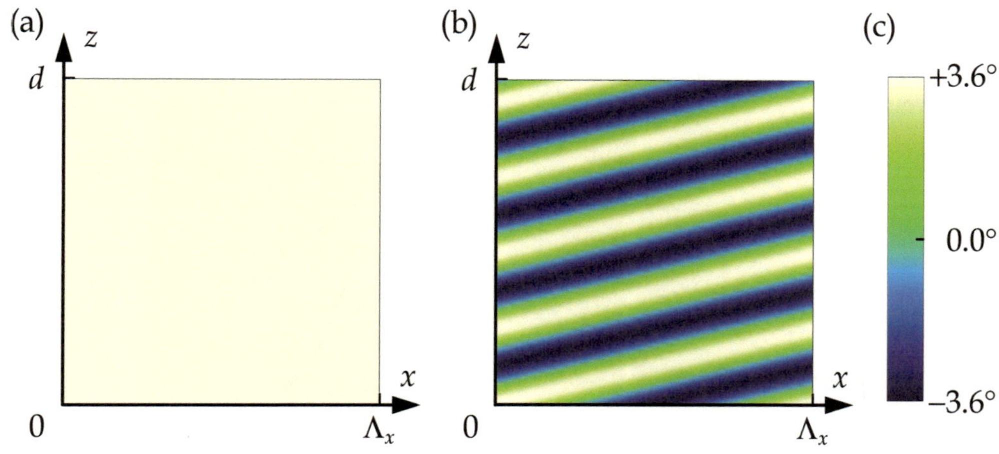

Fig. 10. Calculated distribution of (a) θ_{re} and (b) ϕ_{re} and (c) their value

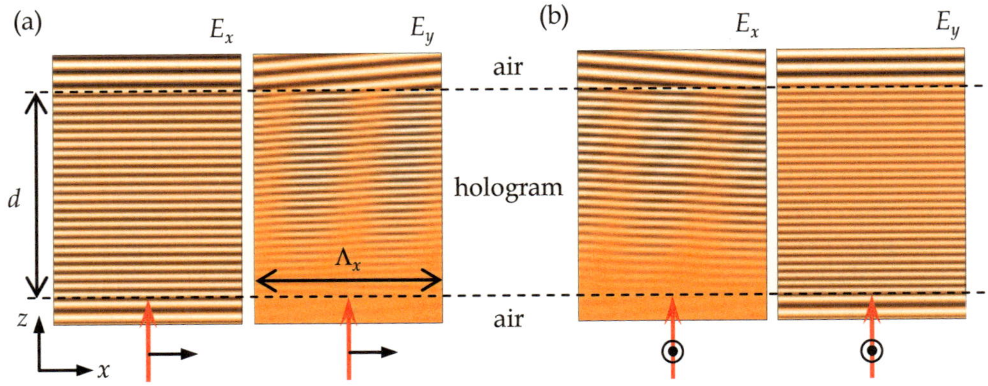

Fig. 11. Calculated electric fields for the (a) *p*- and (b) *s*-polarized probe beam. E_x and E_y are the real part of the x - and y -components for the electric field vector

5.3 Simulated diffraction properties

This section investigates the relationship between various parameters of the 3D vector hologram and its diffraction properties through the theoretical model. Figure 12 presents the calculated results of the grating structure for varying Λ_x. The grating pitch in the grating Λ and the slanted angle Θ_S, which is the angle between the *x*-axis and the grating vector, are given by

$$\Lambda = \Lambda_z \cos\Theta_S = \Lambda_z \cos\left(\tan^{-1}\frac{\Lambda_z}{\Lambda_x}\right), \tag{31}$$

where Λ_z is the grating pitch for the *z*-axis. In this case, Λ and Λ_z are determined by the distribution of S_2 and S_3 as described in Equations 23 and 24. We confirmed that Λ and Λ_z decreased with decreasing Λ_x. It should also be pointed out that $\Lambda_z \cong \Lambda_x/\Delta n = 2.31$ μm for relatively large Λ_x (namely, for $|\theta_1| = |\theta_2| << 1$).

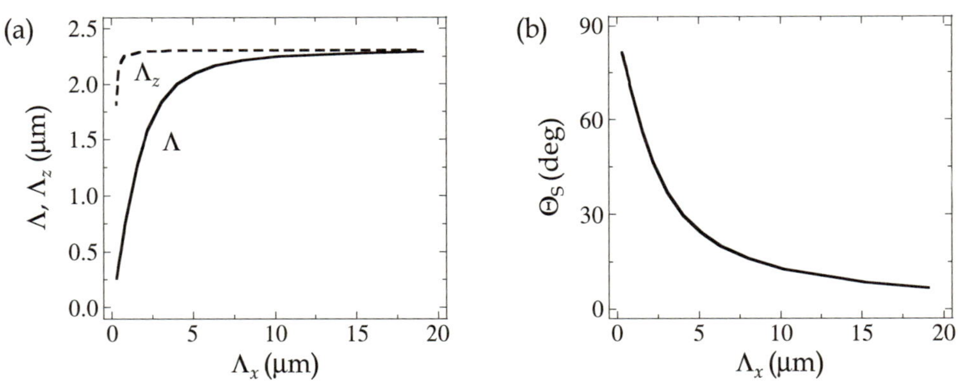

Fig. 12. Dependence of (a) Λ, Λ_z, and (b) Θ_S on Λ_x

Figure 13 illustrates angular dispersion of $\eta_{\pm 1}$ calculated for $\Lambda_x = 2,\ 4,\ 8$ μm when the probe beam is *p*- and *s*-polarized. The maximum values of the diffraction efficiency

$\eta_{\pm 1}(\theta_{in}) \equiv \eta_{\pm 1}(\theta_{max})$ were almost the same for respective Λ_x. However, θ_{max} varied by Λ_x as seen in Fig. 13. In addition, θ_{max} was remarkably influenced by the polarization state of the probe beam. For example, θ_{max} at $\Lambda_x = 2\ \mu m$ were approximately 21° and 2° when the probe beam was *p*- and *s*-polarized. We attribute the reason for the discrepancy to a change of the Bragg angle owing to the difference of the refractive index for the two polarizations. This is one of a characteristic of the 3D vector hologram.

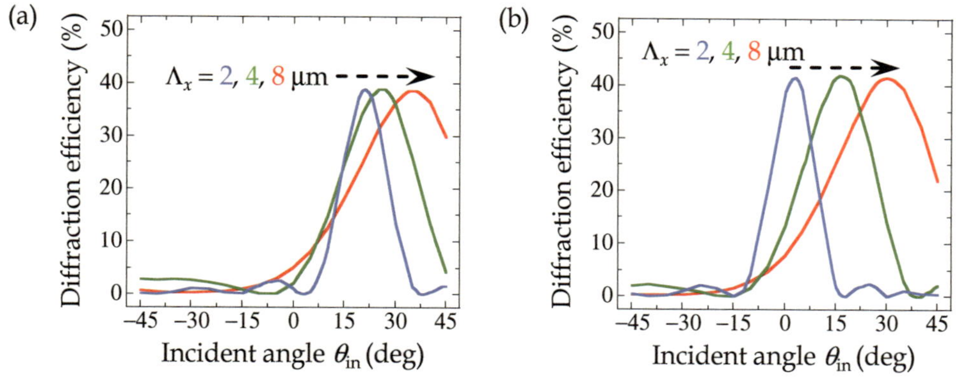

Fig. 13. Angular dependence of the diffraction efficiency calculated by varying Λ_x. (a) η_+ for *p*-polarization. (b) η_- for *s*-polarization

Figure 14 presents $\eta_{+1}(\theta_{in})$ for the *p*-polarized probe beam calculated by varying the film thickness d. It is clearly shown in Fig. 14 that the angular dependence of the diffraction efficiency decreases with increasing Λ_x (i.e., Λ) and decreasing d. This is understood by considering the difference of thick and thin gratings, similarly common diffraction gratings in isotropic media (Kogelnik, 1969; Hariharan, 1996). Figure 15 illustrates dependence of η_{+1} at $\theta_{in} = \theta_{max} = 24.7°$ and η_{-1} at $\theta_{in} = \theta_{max} = -12.5°$ on d in the case of $\Lambda_x = 1\ \mu m$. As seen in Fig. 15, η_{+1} for the *p*-polarization and η_{-1} for the *s*-polarization reached about 100% around $d = 22\ \mu m$ and then decreased with increasing d, respectively. These curves were well fitted sinusoidal functions similarly transmission type thick gratings in anisotropic media. In contrast, it was confirmed that η_{+1} for the *s*-polarization and η_{-1} for the *p*-polarization were nearly equal to zero, respectively.

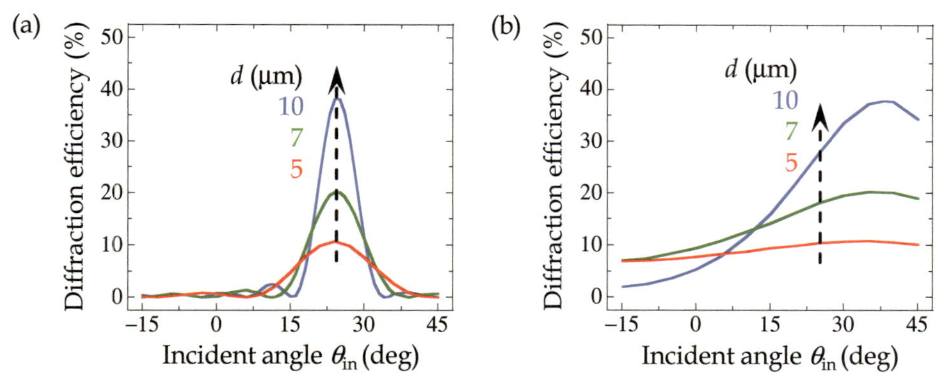

Fig. 14. Angular dependence of η_+ for *p*-polarization at (a) $\Lambda_x = 1\ \mu m$ and (b) $\Lambda_x = 10\ \mu m$

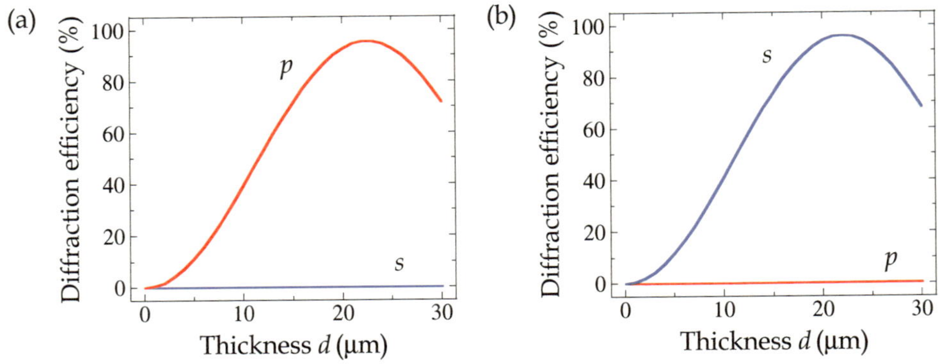

Fig. 15. Dependence of (a) η_+ at $\theta_{in} = 24.7°$ and (b) η_- at $\theta_{in} = -12.5°$ on the film thickness. Red and blue lines are the calculated results for *p*- and *s*-polarizations

6. Conclusions

In this chapter, the principle of 3D vector holography has been described. We considered interference of two polarized plane waves with coherency in anisotropic media, and formulated a type of 3D vector hologram as spatial distribution of the dielectric tensor based on the polarization state of interference light. Optical behaviour of the periodic distribution was simulated by employing the FDTD method and the result was indeed consistent with the observed diffraction properties of the holographic grating, recorded in an azo-dye doped liquid crystalline medium with uniaxial optical anisotropy. Hence, we confirmed that various periodic structures, consisted of spatial distribution of optical anisotropy, could be formed by means of the 3D vector holography. Furthermore, detailed diffraction properties of the holographic grating were studied with the use of the FDTD method. As a result, it was revealed that the 3D vector hologram exhibit various useful characteristics including high diffraction efficiencies, angular dispersion, polarization dependence, and polarization conversion. We believe that the 3D vector holography realizes fabrication of higher-order structures of optical anisotropy and that these structures can be applied to diffractive optical elements with multiple functions, 3D optical memories, anisotropic photonic crystals and so on.

7. References

Asatryan, K. E., Frédérick, S., Galstian, T., & Vallée, R. (2004). Recording of polarization holograms in photodarkened amorphous chalcogenide films. *Appl. Phys. Lett.*, Vol. 84, No. 10, (March 2004), pp. 1626-1628, ISSN 0003-6951

Birabassov, R. & Galstian, T. V. (2001). Analysis of the recording and reconstruction of the polarization state of light. *J. Opt. Soc. Am. B*, Vol. 18, No. 10, (October 2001), pp. 1423-1427, ISSN 0740-3224

Ebralidze, T. D. & Ebralidze, N. A. (1992). Hologram record by means of film anisotropy photoinduction. *Appl. Opt.*, Vol. 31, No. 23, (August 1992), pp. 4720-4724, ISSN 0003-6935

Hariharan, P. (1996). *Optical holography: principles, techniques, and applications.* Cambridge Univ. Press, ISBN 0-521-43348-7, New York

Huang, T. & Wagner, K. H. (1993). Holographic diffraction in photoanisotropic organic materials. *J. Opt. Soc. Am. A*, Vol. 10, No. 2, (February 1993), pp. 306-315, ISSN 0740-3232

Kim, D. Y., Tripathy, S. K., Li, L., & Kumar, J. (1995). Laser-induced holographic surface relief gratings on nonlinear optical polymer films, *Appl. Phys. Lett.*, Vol. 66, No. 10, (March 1995), pp. 1166-1168, ISSN 0003-6951

Kliger, D. S., Lewis, J. W., & Randall, C. E. (1990). *Polarized light in optics and spectroscopy.* Academic press, ISBN 0-12-414975-8, California

Kogelnik, H. (1969). Coupled wave theory for thick hologram gratings. *Bell Syst. Tech. J.*, Vol. 48, No. 8, pp. 2909-2947, ISSN 0005-8580

Kriezis, E. E. & Elston, S. J. (1999). Finite-difference time domain method for light wave propagation within liquid crystal devices. *Opt. Commun.*, Vol. 165, No. 1-3, (July 1999), pp. 99-105, ISSN 0030-4018

Kriezis, E. E. & Elston, S. J. (2000). Light wave propagation in liquid crystal displays by the 2-D finite-difference time-domain method. *Opt. Commun.*, Vol. 177, No. 1-6, (April 2000), pp. 69-77, ISSN 0030-4018

Labarthet, F. L., Buffeteau, T., & Sourisseau, C. (1998). Analyses of the diffraction efficiencies, birefringence, and surface relief gratings on azobenzene-containing polymer films. *J. Phys. Chem. B*, Vol. 102, No. 15, (April 1998), pp. 2654-2662, ISSN 1089-5647

Labarthet, F. L., Buffeteau, T., & Sourisseau, C. (1999). Azopolymer holographic diffraction gratings: time dependent analyses of the diffraction efficiency, birefringence, and surface modulation induced two linearly polarized interfering beams. *J. Phys. Chem. B*, Vol. 103, No. 32, (August 1999), pp. 6690-6699, ISSN 1089-5647

Lien, A. (1990). Extended Jones matrix representation for the twisted nematic liquid-crystal display at oblique incidence. *Appl. Phys. Lett.*, Vol. 57, No. 26, (December 1990), pp. 2767-2769, ISSN 0003-6951

Lien, A. (1997). A detailed derivation of extended Jones matrix representation for twisted nematic liquid crystal displays. *Liq. Cryst.*, Vol. 22, No. 2, (February 1997), pp. 171-175, ISSN 0267-8292

Naydenova, I., Nikolova, L., Todorov, T. Holme, N. C. R., Ramanujam, P. S., & Hvilsted, S. (1998). Diffraction from polarization holographic gratings with surface relief in side-chain azobenzene polyesters, *J. Opt. Soc. Am. B*, Vol. 15, No. 4, (April 1998), pp. 1257-1265, ISSN 0740-3224

Nikolova, L. & Todorov, T. (1984). Diffraction efficiency and selectivity of polarization holographic recording. *Opt. Acta*, Vol. 31, No. 5, (May 1984), pp. 579-588, ISSN 0030-3909

Nikolova, L., Todorov, M., Ivanov, M., Andruzzi, F., Hvilsted, S., & Ramanujam, P. S. (1996). Polarization holographic gratings in side-chain azobenzene polyesters with linear and circular photoanisotropy. *Appl. Opt.*, Vol. 35, No. 20, (July 1996), pp. 3835-3840, ISSN 0003-6935

Nikolova, L., Todorov, M., Ivanov, M., Andruzzi, F., Hvilsted, S., & Ramanujam, P. S. (1997). Photoinduced circular anisotropy in side-chain azobenzene polyesters. *Opt. Matter.*, Vol. 8, No. 4, (November 1997), pp. 255-258, ISSN 0925-3467

Oh, C. & Escuti, M. J. (2006). Time-domain analysis of periodic anisotropic media at oblique incidence: an efficient FDTD implementation. *Opt. Express*, Vol. 14, No. 24, (November 2006), pp. 11870-11884, ISSN 1094-4087

Oh, C. & Escuti, M. J. (2007). Numerical analysis of polarization gratings using the finite-difference time-domain method. *Phys. Rev. A*, Vol. 76, No. 043815, (October 2007), pp. 1-8, ISSN 1050-2947

Ono, H., Sekiguchi, T., Emoto, A., & Kawatsuki, N. (2008). Light wave propagation in polarization holograms formed in photoreactive polymer liquid crystals. *Jpn. J. Appl. Phys.*, Vol. 47, No. 5, (May 2008), pp. 3559-3563, ISSN 0021-4922

Ono, H., Sekiguchi, T., Emoto, A., Shioda, T., & Kawatsuki, N. (2008). Light wave propagation and Bragg diffraction in thick polarization gratings. *Jpn. J. Appl. Phys.*, Vol. 47, No. 10, (October 2008), pp. 7963-7967, ISSN 0021-4922

Ono, H., Wakabayashi, H., Sasaki, T., Emoto, A., Shioda, T., & Kawatsuki, N. (2009). Vector holograms using radially polarized light. *Appl. Phys. Lett.*, Vol. 94, No. 071114, (February 2009), pp. 1-3, ISSN 0003-6951

Ono, H., Suzuki, K., Sasaki, T., Iwato, T., Emoto, A., Shioda, T., & Kawatsuki, N. (2009). Reconstruction of polarized optical images in two- and three-dimensional vector holograms. *J. Appl. Phys.*, Vol. 106, No. 083109, (October 2009,) pp. 1-7, ISSN 0021-8979

Provenzano, C., Cipparrone, G., & Mazzulla, A. (2006). Photopolarimeter based on two gratings recorded in thin organic films. *Appl. Opt.*, Vol. 45, No. 17, (June 2006), pp. 3929-3933, ISSN 0003-6935

Ramanujam, P. S., Holme, N. C. R., & Hvilsted, S. (1996). Atomic force and optical near-field microscopic investigations of polarization holographic gratings in a liquid crystalline azobenzene side-chain polyester, *Appl. Phys. Lett.*, Vol. 68, No. 10, (March 1996), pp. 1329-1331, ISSN 0003-6951

Rochon, P., Batalla, E., & Natansohn, A. (1995). Optically induced surface gratings on azoaromatic polymer films, *Appl. Phys. Lett.*, Vol. 66, No. 2, (January 1995), pp. 136-138, ISSN 0003-6951

Samui, A. B. (2008). Holographic recording medium. *Recent Patents on Mater. Sci.*, Vol. 1, No. 3, pp. 79-94, ISSN 1874-4648.

Sasaki, T., Ono, H., & Kawatsuki, N. (2008). Three-dimensional vector holograms in anisotropic photoreactive liquid-crystal composites. *Appl. Opt.*, Vol. 47, No. 13, (April 2008), pp. 2192-2200, ISSN 0003-6935

Scharf, T. (2007). *Polarized light in liquid crystals and polymers.* Wiley, ISBN 978-0-471-74064-3, New Jersey

Taflove, A. & Gedne, S. C. (2005). *Computational electrodynamics: the finite-difference time-domain method.* Artech House, ISBN 1-58053-832-0, Massachusetts

Titus, C. M., Kelly, R., Gartland, E. C., Shiyanovskii, S. V., Anderson, J. A., & Bos, P. J. (2001). Asymmetric transmissive behaviour of liquid-crystal diffraction gratings. *Opt. Lett.*, Vol. 26, No. 15, (August 2001), pp. 1188-1190, ISSN 0146-9592

Todorov, T., Nikolova, L., & Tomova, N. (1984). Polarization holography. 2: Polarization holographic gratings in photoanisotropic materials with and without intrinsic birefringence. *Appl. Opt.*, Vol. 23, No. 24, (December 1984), pp. 4588-4591, ISSN 0003-6935

Yee, K. S. (1966). Numerical solution of initial boundary value problems involving Maxwell's equations in isotropic media. *IEEE Trans. Antennas Propag.*, Vol. 14, No. 3, pp. 302-307, ISSN 0018-926X

Part 2

Holographic Data Storage

9

Diffraction Property of Collinear Holographic Storage System

Yeh-Wei Yu and Ching-Cherng Sun
National Central University
Taiwan

1. Introduction

1.1 The collinear volume holographic storage system

The collinear storage system was proposed by Optware [1,2]. It is a coaxially aligned optical structure for signal and reference beams, which are encoded simultaneously by the same spatial light modulator (SLM) and the two beams interfere with each other in the recording medium through a single objective lens. The system has been proven a capability of large storage capacity, high transfer rate, short access time, and besides, it is compatible with existing disc storage systems such as CDs and DVDs [3-4]. Recent report of the collinear Volume holographic storage (VHS) has performed a storage density as high as 270 Gbits/inch^2 [5]. Many advantages are proposed, including uniform shift selectivity in both radial and tangential directions, a fairly large wavelength shift and a fairly large tilt tolerance [6].

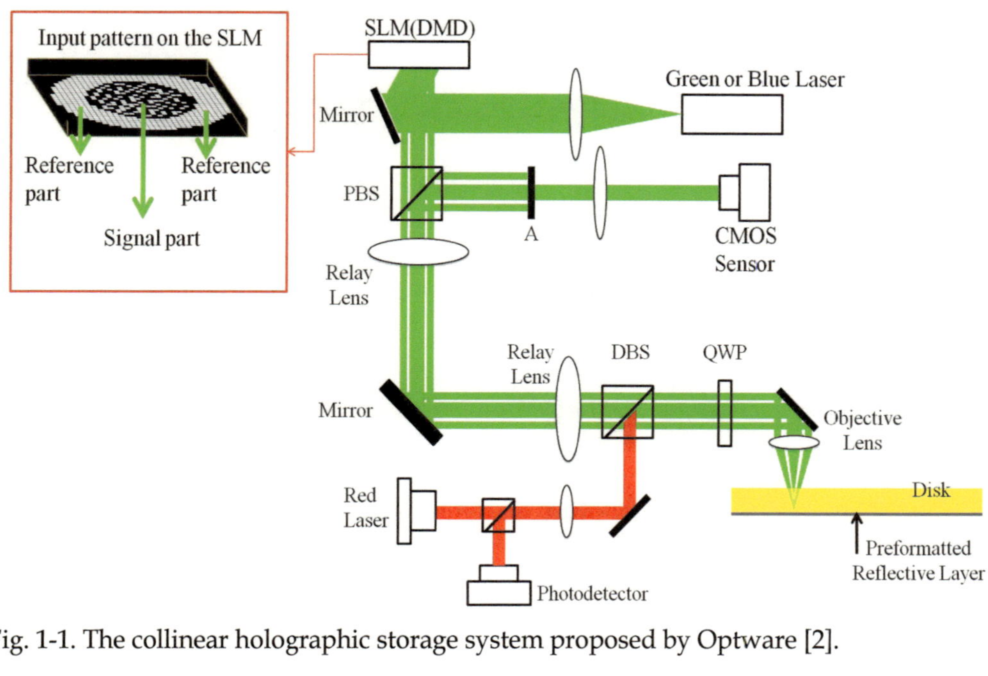

Fig. 1-1. The collinear holographic storage system proposed by Optware [2].

The collinear storage system can be illustrated in Fig. 1-1. Two laser sources are requested in the system, where a red laser is used for the position servor and the green or blue laser is used for data writing and reading. In the writing process, the SLM is imaged on the front focal plane of the objective lens. The central region of the SLM is used to modulate the signal, and the surrounding region of the SLM is used to modulate the reference. After passing through the objective lens, both signal and reference are focused on the reflecting surface of the disk. The interference fringe caused by signal and reference are then recorded by the disk. In the reading process, only the region of the reference beam on the SLM is turn on. The reading beam is then imaged on the front focal plane of the objective lens through the relay lens, and is focused on the reflecting plane of the disc by the objective lens. Then the readout light is diffracted. The reading beams and the diffracted signal are reflected by the reflective layer and re-image on the front focal plane of the objective lens. The combination of quarter wave plate (QWP) and polarization beam splitter (PBS) functions as an isolator, which routes the reflecting diffractive signal and reflecting reading beam to the CMOS sensor. Since the element A blocks reading beams, only diffracted signal can be imaged on the CMOS sensor.

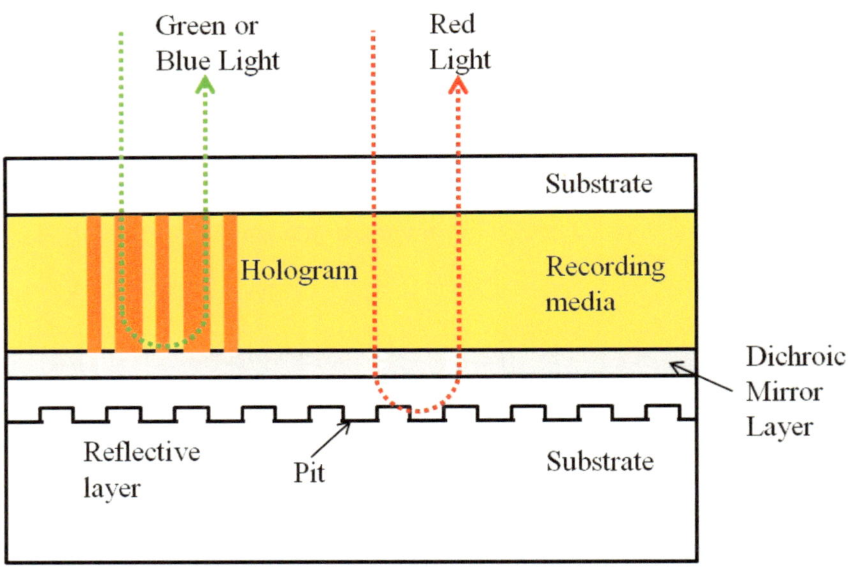

Fig. 1-2. The diagram of the structure of the disc.

The structure of the disc is shown in Fig. 1-2. The dichroic mirror layer reflects the green or blue beam. The red beam propagates through the dichroic mirror layer, and reaches the reflective layer on the substrate. Pits located on the reflective layer induce constructive or destructive interference of the refelected beam. The reflected beam is routed to the photodector, on which the lightness and the darkness can be decoded to do servo-positioning. This system can be simplified as an equivalent model. The two lenses in Fig. 1-3 stand for the objective lens in Fig. 1-2. The collinear system is generally a reflecting architecture, which causes both transparent grating and reflective grating to exist inside the holographic disk. Since the existence of reflective grating decreases the signal to noise ratio (SNR) in the retrieved signal, many techniques are proposed to remove it [7-9]. Accordingly, only the transparent grating needs to be considered. As shown in Fig. 1-3, the mirror on the back surface of the holographic disk in Fig. 1-2 is replaced by a disc of double-thickness in a modeled transmission algorithm. The two lenses form a 4f system: SLM locates at the front focal plane of the first lens, the disc is located at the back focal plane of the first lens, and the CCD is placed at the back focal plane of the second lens. Fig. 1-3(a) shows the effective model of the writing process. The reference beams from the outer ring of the SLM and the signal beams from the inner circle of the SLM are transformed to frequency domain by the lens and then recorded by the lens. Fig. 1-3(b) shows the effective model of the reading process. Only the light from outer ring can pass through as a reading beam, which is transformed by the lens into frequency domain. Subsequently, the diffracted signal beam and the un-diffracted reading beam are transformed together by the second lens into spatial domain. The aperture blocks the un-diffracted reading beam and leaves only the diffracted signal beam to image on the CCD.

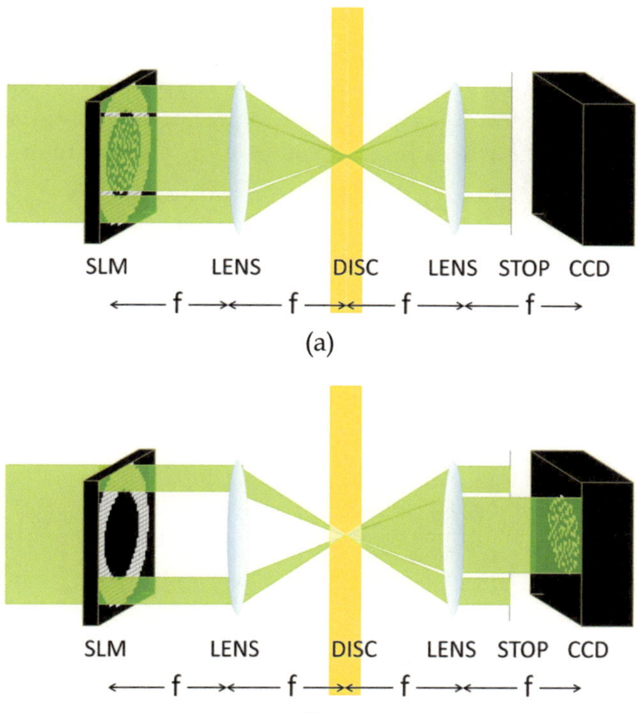

Fig. 1-3. The simplified model of the coaxial holographic storage system.

1.2 Basic theorem

1.2.1 Two models for the collinear system

In general, the analysis of the VHS system is based on the following two models. The first is to calculate the point spread function (PSF) through Bragg mismatch [10,11]; the second is to apply Fresnel transform to simulate the light field in the recording disc and through Born's approach or the method called "volume hologram being an integrator of the lights emitted from elementary light sources" (VOHIL) to calculate the diffraction pattern on the CCD plane [12-15]. The procedures of the approaches are described in detaisl from Refs. 10 to Refs.15. For sake of completeness, brief introductions are given here.

For the first model, light from each pixel (*i,j*) on the SLM can be treated as a point light source, it is transferred to a plane wave by the FT lens and is incident on the disk. Thus, the wave vector k_{ij} of the plane wave can be expressed as

$$k_{ij} = \left(-\frac{2\pi}{\lambda}\frac{x_i}{f}, -\frac{2\pi}{\lambda}\frac{y_j}{f}, \left[\left(n\frac{2\pi}{\lambda} \right)^2 - \left(\frac{2\pi}{\lambda}\frac{x_i}{f} \right)^2 - \left(\frac{2\pi}{\lambda}\frac{y_j}{f} \right)^2 \right]^{1/2} \right), \quad (1\text{-}1)$$

where, (x_i, y_j) is the position of the pixel (i, j). The plane waves from the pixel of information and reference pattern interfere with each other, and reference pattern interfere with each other and is recorded by the media. In the reading process, plane waves from the reference pixels are diffracted by all existing gratings. The diffracted plane waves from the grating can be calculated by coupled mode theory. Considering diffraction of a plane wave from a pixel (k', l') by a grating $K_{klij}=k_{ij}-k_{kl}$, which is written by the interference between the waves from a data pixel (i, j) and a reference pixel (k,l), the Bragg mismatch is obtained as

$$\Delta k_z = \left(k_{k'l'_z} + K_{klij_z} \right) - \left[(nk_0)^2 - \left(k_{(k'+i-k),(l'+j-l)_x} \right)^2 - \left(k_{(k'+i-k),(l'+j-l)_y} \right)^2 \right]^{1/2}. \quad (1\text{-}2)$$

Thus, the light intensity detected by the CCD is a summation of each pair of plane waves.

For the second model, in the writing process, the optical field in the recording disk can be calculated by Fresnel transform

$$U_{disk}(u,v,\Delta z) = \frac{\exp\left[jk(2f+\Delta z) \right]}{j\lambda^3 f^3} FFT2\left\{ U_{SLM}(x,y) \exp\left[-j\frac{\pi \Delta z}{\lambda f^2}\left(x^2+y^2 \right) \right] \right\}, \quad (1\text{-}3)$$

where, FFT2 is the operator of Fast Fourier Transform in Matlab, which transfers (x, y) into (u/λf, v/λf), U_{SLM} is the signal modulated by the SLM, x and y are the lateral coordinates of the SLM, u and v are the lateral coordinates of the recording medium, λ is the wave length, f is the focal length of the lens, and Δz is the distance deviated from the front focal plane of the second FT lens, within the volume of the recording medium. The corresponding intensity distribution is $|U_{disk}|^2$, thus generates an index variation proportional to the intensity distribution. The phase distribution of the medium is then obtained from the index variation.

In the reading process, the optical field of the reading beam can also be calculated by Fresnel transform. The diffracting optical field is obtained by multiplying the recorded phase distribution by the reading beam. Considering the short propagation distance insides the disk, the diffractive optical field propagating to the front focal plane of the second FT lens is calculated by angular spectrum technique

$$U_{df}(u,v,\Delta z)=FFT2\left\{A\left(f_u,f_v,\Delta z\right)\exp\left(\Delta z\cdot 2\pi\sqrt{\frac{1}{\lambda^2}-f_u^2-f_v^2}\right)\right\}, \quad (1\text{-}4)$$

where, U_{df} is the diffractive optical field propagating to the front focal plane of the second FT lens. A is the angular spectrum of the optical field. Therefor, light propagating through the second FT lens and imaging on the CCD can be calculated directly by FFT method. By repeating the above procedures for all layers and summing the calculation results, the final diffractive optical field can be get. Due to the facility of computer calculation, the proposed model can simulate light distribution at any position along the propagation path.

1.2.2 VOHIL theorem

Volume Hologram Being an Integrator of the Lights Emitted from Elementary Light Sources (VOHIL) is a powerful model in calculating the relative diffracting ratio under weak coupling condition [15]. It is used to simplify the optical model. Fig. 1-4 shows an example for calculation by VOHIL theorem. In the writing proces, signal (E_1) and reference (E_2) interfer inside the recording media. The optical field of E_1 and E_2 are expressed

$$E_i=A_i\exp(j\phi_i),\ i=1,2, \quad (1\text{-}5)$$

where A_j stands for amplitude and ϕ_j stands for phase. The interference fringe is

$$I=\left|A_1\right|^2+\left|A_2\right|^2+A_1^*A_2\exp\left[j(\phi_2-\phi_1)\right]+A_1A_2^*\exp\left[j(\phi_1-\phi_2)\right]. \quad (1\text{-}6)$$

Express the reading beam as

$$E_r=A_r\exp\left(j\phi_3\right), \quad (1\text{-}7)$$

which is used to probe the media. The diffracted light is

$$D=(\left|A_1\right|^2+\left|A_2\right|^2)A_r+A_1^*A_2A_r\exp\left[j(\phi_2-\phi_1+\phi_3)\right]+A_1A_2^*A_r\exp\left[j(\phi_1-\phi_2+\phi_3)\right]. \quad (1\text{-}8)$$

To detect the diffracted signal on the path of the signal, only the last term need to be considered. So, the diffracting light from each position is

$$D\propto A_rA_2^*A_1\exp\left[j\phi_1(x_i)\right]\exp\left[j\Delta\phi(x_i)\right], \quad (1\text{-}9)$$

where $\Delta\phi(x_i)$ stands for the phase difference at each position, and is

$$\Delta\phi(x_i)=\phi_3(x_i)-\phi_2(x_i), \quad (1\text{-}10)$$

the final diffracting light is the integration along the wave propagating direction and can be expressed

$$D\propto\int_{-\frac{t}{2}}^{\frac{t}{2}}A_rA_2^*A_1\exp\left(j\phi_1\right)\exp\left[j\left(\phi_3-\phi_2\right)\right]dx, \quad (1\text{-}11)$$

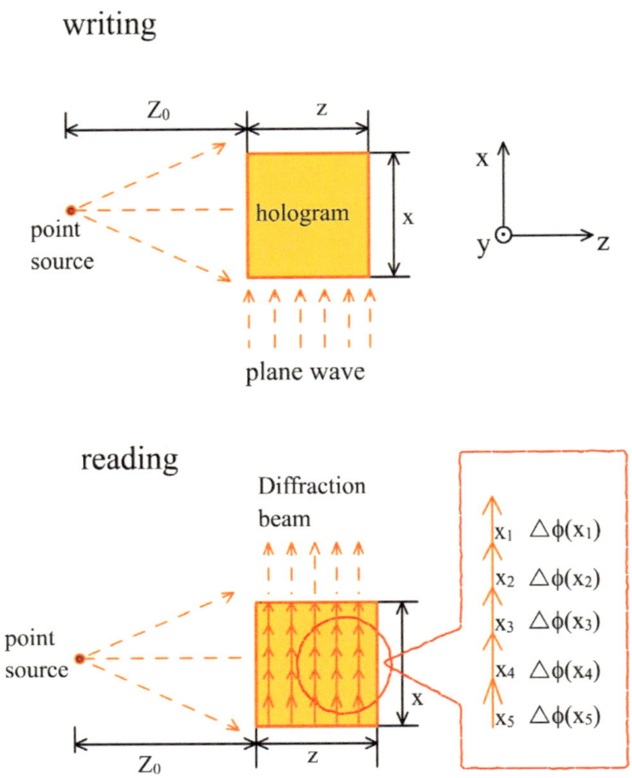

Fig. 1-4. The diagram of VOHIL

where t is the thickness of the recording media along the wave propagating direction. If $\phi_3(x)$ is equal to $\phi_2(x)$, the signal can be perfectly diffracted because of phase match condition. If $\phi_3(x)$ is not equal to $\phi_2(x)$, the phase difference $\Delta\phi(x)$ will be the function of x, the diffractive intensity will be decreased because of phase mismatch. As long as the reading beam is different from the reference in wavelength, incident angle, light source position or other parameters, the diffractive light will be suppressed by phase mismatch. Therefore, Eq. (1-11) can be used to calculate angular selectivity, shift selectivity, wavelength selectivity, temperature tolerance and so on.

In the following paragraph, solutions of collinear system based on VOHIL model are derived out, which clearly point out the physical concepts of the system. Thus, improvement of the system can be proposed based on the physical concepts.

1.3 Point spread function of the system

For volume holographic storage system, both the storage capacity and transfer rate are proportional to the capacity per page. Point spread function (PSF) of the system determines how much data can be stored in one page. If the reference pattern is not well designed, point spread function (PSF) of the collinear system blurs seriously, and thus limits the storage capacity of the system. The radial line (RL) amplitude modulation in reference beam was then proposed to improve the PSF [4]. Even so, 80% of energy is wasted due to the opaque part in the reference region. The random binary phase (RBP) modulation is proposed to solve this problem, which is shown to improve the PSF without wasting energy in reference modulation [16, 17]. In this section, a novel technique called lens array phase (LAP) modulation is further proposed to improve the PSF.. Thus, the storage capacity and the data transfer rate can be enhanced at the same time.

The effective optical model of collinear system is shown in Fig. 1-3. The optical field of the reference (A_R), signal (A_S) and reading light (A_P) in the disc can be expressed

$$A_R(u,v,\Delta z)=\frac{\exp\left[jk(2f+\Delta z)\right]}{j\lambda f}\Im\left\{U_R(x,y)\exp\left[-j\frac{\pi\Delta z}{\lambda f^2}\left(x^2+y^2\right)\right]\right\}, \tag{1-12}$$

$$A_S(u,v,\Delta z)=\frac{\exp\left[jk(2f+\Delta z)\right]}{j\lambda f}\Im\left\{U_S(x,y)\exp\left[-j\frac{\pi\Delta z}{\lambda f^2}\left(x^2+y^2\right)\right]\right\}, \tag{1-13}$$

$$A_P(u,v,\Delta z)=\frac{\exp\left[jk(2f+\Delta z)\right]}{j\lambda f}\Im\left\{U_P(x,y)\exp\left[-j\frac{\pi\Delta z}{\lambda f^2}\left(x^2+y^2\right)\right]\right\}, \tag{1-14}$$

where U_R, U_S and U_P is the optical field of the reference, signal and probing beams on the SLM, respectively; *x* and *y* are the lateral coordinates of the SLM; *u* and *v* are the lateral coordinates of the recording medium; $\Im$ stands for Fourier transform which transfers (x, y) into (u/λf, v/λf); k is the wave vector of the light source; λ is the wavelength; f is the focal length of the lens, and Δz is the distance deviated from the focal plane within the volume of the recording medium. The interference of the signal beam and the reference beam can be expressed as

$$I=\left|A_R\right|^2+\left|A_S\right|^2+A_RA_S^*+A_R^*A_S\,. \tag{1-15}$$

In the reading process, reading beam A_p is used to probe the hologram, and the diffraction is described

$$D=A_p\left(\left|A_R\right|^2+\left|A_S\right|^2\right)+A_pA_RA_S^*+A_pA_R^*A_S\,. \tag{1-16}$$

In the collinear volume holographic storage system, the dc term is blocked, and the conjugation term is suppressed by deconstruction. So, only the last term need to be considered. Therefore, the optical field diffracted from different layers at a specific depth in the holographic disc is expressed as

$$U(u,v,\Delta z)=\frac{\exp\left[jk(2f+\Delta z)\right]}{j\lambda^3f^3}\left.\left\{\begin{array}{l}\Im\left\{U_P(x,y)\exp\left[-j\frac{\pi\Delta z}{\lambda f^2}\left(x^2+y^2\right)\right]\right\}\\ \Im^*\left\{U_R(x,y)\exp\left[-j\frac{\pi\Delta z}{\lambda f^2}\left(x^2+y^2\right)\right]\right\}\\ \Im\left\{U_S(x,y)\exp\left[-j\frac{\pi\Delta z}{\lambda f^2}\left(x^2+y^2\right)\right]\right\}\end{array}\right\}\right|_{\substack{x\to\frac{u}{\lambda f}\\ y\to\frac{v}{\lambda f}}}. \tag{1-17}$$

The solution for the diffracted optical field in the CCD plane is

$$U_{\det}(\xi,\eta)=\frac{\exp(jk4f)}{(\lambda f)^2}\int_{-T}^{T}\exp\left[\frac{j\pi\Delta z}{\lambda f^2}\left(\xi^2+\eta^2\right)\right]\left\{\begin{aligned}&\left\{U_P(-\xi,-\eta)\exp\left[-j\frac{\pi\Delta z}{\lambda f^2}\left(\xi^2+\eta^2\right)\right]\right\}\\&\otimes\left\{U_R^{*}(\xi,\eta)\exp\left[j\frac{\pi\Delta z}{\lambda f^2}\left(\xi^2+\eta^2\right)\right]\right\}\\&\otimes\left\{U_S(-\xi,-\eta)\exp\left[-j\frac{\pi\Delta z}{\lambda f^2}\left(\xi^2+\eta^2\right)\right]\right\}\end{aligned}\right\}d\Delta z, \quad (1\text{-}18)$$

where ⊗ stands for the convolution, ξ and η are the lateral coordinates on the CCD plane, and T is the thickness of the disk. Considering that the probing beam is always the same as the reference beam, and using a point source as the input signal, the PSF can be obtained by calculating the diffracted field on the CCD plane.

$$PSF(\xi,\eta)=\frac{\exp(jk4f)}{(\lambda f)^2}\int_{-T}^{T}\exp\left[\frac{j\pi\Delta z}{\lambda f^2}\left(\xi^2+\eta^2\right)\right]ACOR\{U_R(-\xi,-\eta)\exp[-j\frac{\pi\Delta z}{\lambda f^2}(\xi^2+\eta^2)]\}d\Delta z, \quad (1\text{-}19)$$

where *ACOR* stands for the autocorrelation. Equation (1-19) shows that the autocorrelation of the reference pattern times a defocusing phase term is dominant in the PSF function.

Four different reference patterns are shown in Fig. 1-5. Eq. (1-19) is applied to simulate the PSF based on the reference patterns, and autocorrelation theorem is used to speed up the simulation [18]

$$ACOR\{O(x,y)\}=IFFT2\{FFT2[O(x,y)]FFT2[O^*(-x,-y)]\}, \quad (1\text{-}20)$$

where, FFT2 and IFFT2 is the operator of Fast Fourier Transform in Matlab and Inverse Fast Fourier Transform in Matlab, respectively. In the calculation, the thickness is 0.6mm, the wavelength is 532nm and the focal length is 5mm. In order to simplify the effect of the refraction index diference between the air and the holographic disc, the effective focal length is set as 7.5 mm instead of 5 mm, and the effective wavelength is 532 nm/1.5 instead of 532 nm The simulation result of the PSF for the corresponding reference patterns is shown in Fig. 1-6, where λ=532nm, T=0.6mm, pixel size is 13μmx13μm, fill factor of each pixel is 71.6% and the size of the SLM is 4.4mmx4.4mm. The simulation results shows the PSF of the horizontal-line reference pattern is much wider than that of the vertical-line reference pattern along x direction. Similarly, the PSF of the vertical-line reference pattern is also much wider than that of the horizontal-line reference pattern along y direction. Amoung these reference patterns, the radial-line (RL) pattern proposed by Shimura et al. indeed performs a narrower PSF. It is because the RL amplitude modulation avoids the Bragg degeneracy from different pixels in the reference region.

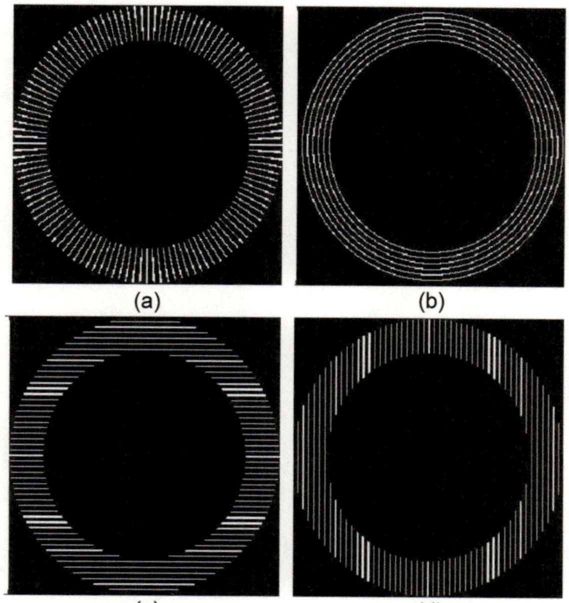

Fig. 1-5. Four different reference patterns. (a) the radial-line (RL), (b) the multi-ring, (c) the horizontal lines and (d) the vertical lines [14].

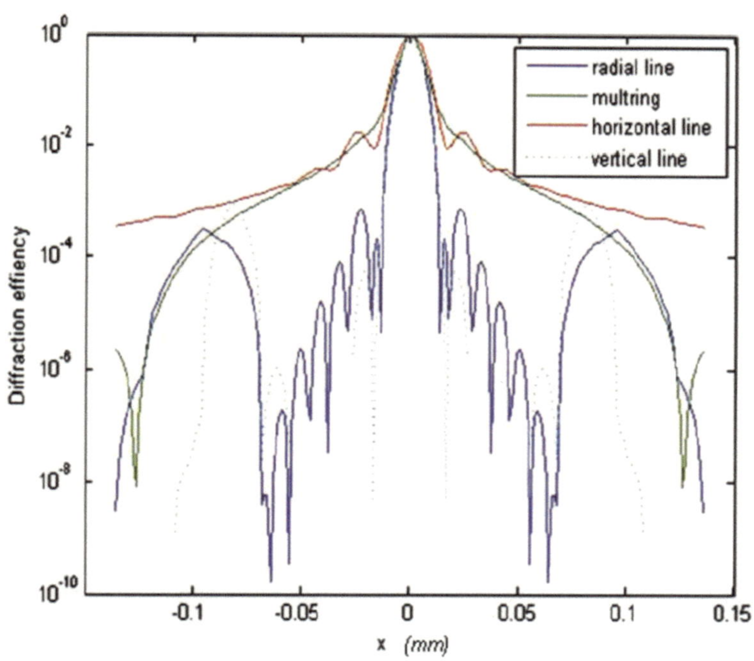

Fig. 1-6. Calculation of intensity of the PSF for the four different reference patterns in Fig. 1-5 [14].

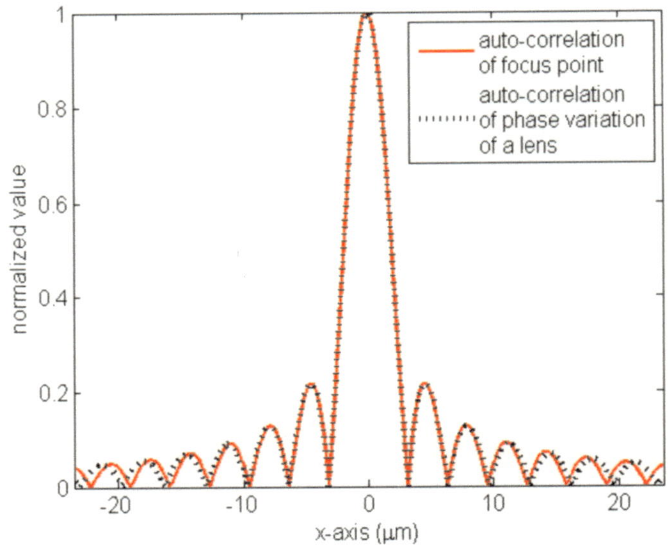

Fig. 1-7. A comparison of the simulation result of the autocorrelation of the phase just behind the lens and at the focal point.

The physical concept described by Eq. (1-19) reveals that the path to achieve the narrowest PSF is through the reference modulation with the narrowest autocorrelation width. Since the autocorrelation of a point source is the sharpest, it looks like the best choice for reference modulation. However, it is not a good policy in the view point of energy. So, a lens array was proposed to approach the effect of point source. The phase modulation of a lens can be described as

$$L(x,y)=\exp\left[-j\pi\left(x^2+y^2\right)/\lambda f_L\right]P(x,y), \tag{1-21}$$

where $P(x,y)$ is the lens pupil, f_L is focal length of the lens. Fig. 1-7 shows a comparison of the autocorrelation of the phase just behind the lens and at the focal point. In the simulation, the focal length f_L is 3.5 mm and the dimensions are 588.24μm×588.24μm. The two curves seem to overlap with each other. Accordingly, lens array phase (LAP) modulation can work as good as points array, and is expressed as

$$LAP(x,y)=\sum_{s=-N}^{N}\sum_{t=-N}^{N}\exp\left[-j\pi\frac{(x-sw)^2+(y-tw)^2}{\lambda f_L}\right]rect\left(\frac{x-sw}{w}\right)rect\left(\frac{y-tw}{w}\right), \tag{1-22}$$

where w is the width of each single lens. Here, the role of the defocusing phase term inside autocorrelation in Eq. (1-19) is ignored, because the phase variation of it is ignorable when comparing with the lens array. Since the width of focus point is related to diffraction limit, the lens array with a higher numeric aperture in each unit lens can provide a tighter PSF. Figure 1-8 shows three kinds of reference modulation. Fig. 1-8 (a) is the RL modulation with 120 radial lines, where the line width is equal to 13.68μm, Fig. 1-8 (b) is the RBP modulation with pixel pitch equal to 13.68μm and Fig. 1-8 (c) is the LAP modulation with the focal length and area of each single lens equal to 3.5mm and 588.24μm×588.24μm, respectivly. The parameters are choosen with acceptable cut-off frequency and narrowest focus point.

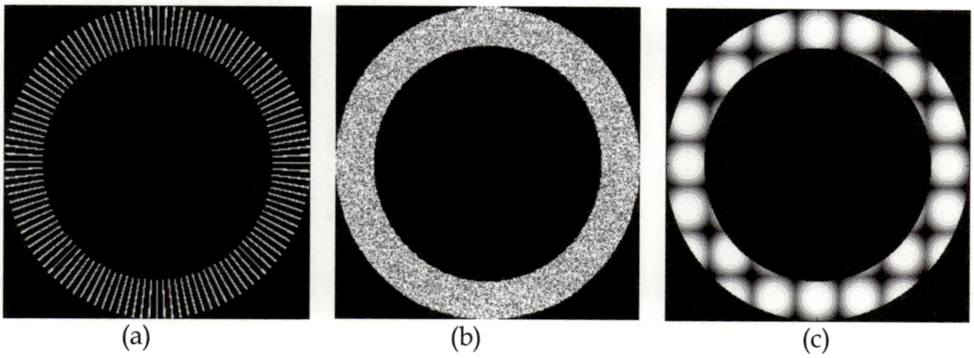

Fig. 1-8. Three kinds of reference modulation: (a) RL modulation; (b) RBP modulation; (c) LAP modulation [19].

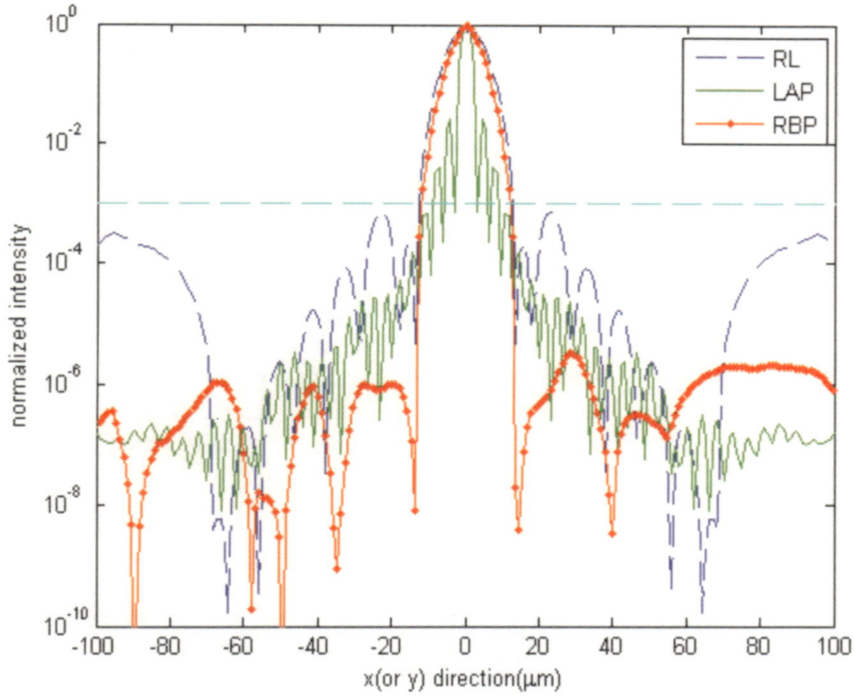

Fig. 1-9. PSF for different reference patterns [19].

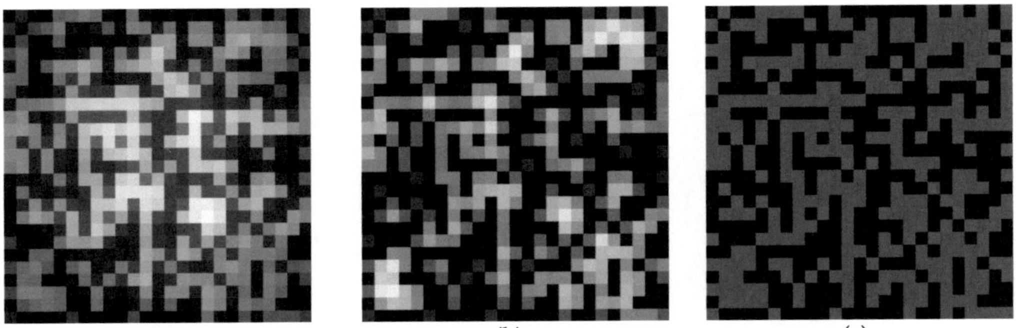
(a) (b) (c)

Fig. 1-10. Simulation result of signal detected by the CCD for (a) RL amplitude modulation, (b) RBP modulation and (c) LAP modulation [19].

Fig. 1-9 is the simulation result of PSF based on Eq. (1-19). It shows that the LAP modulation can tighten the PSF effectively. Fig. 1-10 is the simulated result of the diffracted signal detected by the CCD. The simulation use Eq. (1-18). It shows that the LAP modulation results in the clearest diffracted image with the best readout quality.
The signal to noise ratio (SNR) of the readout pattern is defined as

$$SNR = \frac{(m_1 - m_0)}{\sqrt{\sigma_1^2 + \sigma_0^2}}, \tag{1-23}$$

where σ_0 and σ_1 are the standard deviations of signal 0 and signal 1, respectively ; m_0 and m_1 are the mean values of signal 0 and signal 1, respectively. Eq. (1-23) is used to judge the quality of the diffracted signal. Calculated SNRs for the reference beams with the RL, RBP and LAP modulation are 2.3, 8.1 and 63.2, respectively. The lens array reference with short focal length is shown much helpful in enhancing the SNR of the readout signal of a collinear VHS.

1.4 Shift selectivity of the system

To keep increasing the storage capacity of the collinear VHS, calculation methods for the diffraction property, such as point spread function and shifting selectivity, are highly demanded. It is shown that point spread function is a key issue to calculate the storage capacity of each stored page, and that shifting selectivity is related to the total number of stored pages [20]. In this section, the physics concept of the shifting selectivity for a pixel signal is introduced. Two models of shifting selectivity are proposed. One considering a pixel signal, which is important for precisely describing the shift selectivity. Another model is simplified by considering a point signal as the signal.
The simplified model is based on three assumptions:

1. The shift selectivity is the same for signal point at any position. Therefore, only the center point in the SLM is turned on as signal.
2. In the reading process, the conjugating image point on the CCD plane is dominant. So, only the diffracting intensity at this point is considered.
3. It has been shown that the shift selectivity is independent of the thickness of the disc [21]. Thus, whatever the original thickness of the disk is, the thickness of the disc can always be set as zero.

The shift selectivity of the system is named as "point shift selectivity". The schematic diagram of the storage system for the theoretical model is shown in Fig. 1-3. In the writing process, considering only the center point of the signal, the signal propagating to the disk is a plane wave along z axis, and the grating is expressed

$$R_g^* S_g = \Im\{U_R\}^*, \tag{1-24}$$

where R_g and S_g is the optical field of reference and signal respectively in the medium of the disc. U_R is the optical field of reference in the input plane. The operation $\Im\{\}$ denotes Fourier transform. When the disk shifts, the grating can be expressed

$$G = \Im\{U_R\}^* \otimes \delta(u-\Delta u)\delta(v-\Delta v), \tag{1-25}$$

where (u,v) is the coordinate of the disk, Δu and Δv is the shifting of the disk in the u direction and v direction respectively. When a probing beam is used to read the disk, the diffracted light can be expressed

$$U_d = \Im\{U_P\}\Big|_{\substack{u/(\lambda f)\\ v/(\lambda f)}} \Im\{U_R\}^*\Big|_{\substack{(u-\Delta u)/(\lambda f)\\ (v-\Delta v)/(\lambda f)}}, \tag{1-26}$$

where U_P is the optical field of the probing beam in the input plane. Let the probing beam be the same as the reference beam, the diffracted optical field at the conjugating image point on the CCD plane can be expressed

$$U(\Delta u,\Delta v) = \iint \Im\{U_P\}\Big|_{\substack{u/(\lambda f)\\ v/(\lambda f)}} \Im\{U_R\}^*\Big|_{\substack{(u-\Delta u)/(\lambda f)\\ (v-\Delta v)/(\lambda f)}} \exp\left[-\frac{2\pi j}{\lambda f}(u\xi+v\eta)\right] du dv \Bigg|_{\substack{\xi=0\\ \eta=0}}, \tag{1-27}$$

$$= \Im\left\{|U_R|^2\right\}\Big|_{\substack{\Delta u/(\lambda f)\\ \Delta v/(\lambda f)}}$$

where (ξ,η) is the coordinate of the CCD plane. Thus, the diffracted intensity depending on disk deviation can simply be expressed

$$I(\Delta u,\Delta v) = \left| \Im\left\{|U_R|^2\right\}\Big|_{\substack{\Delta u/(\lambda f)\\ \Delta v/(\lambda f)}} \right|^2. \tag{1-28}$$

Thus, the point shifting selectivity equals to square of Fourier transform of intensity distribution of the reference pattern.

Fig. 1-11 shows five different reference patterns. Fig. 1-11 (a) is a ring pattern with phase modulation of $\exp[j2\pi\cos(y/\Lambda)]$. Fig. 1-11 (b) is a ring pattern with amplitude modulation of $\cos(y/\Lambda)$. Where $\Lambda/2\pi$ is the period of the cosine function and is equal to 273.6μm. Fig. 1-11 (c) is a ring pattern with random binary amplitude modulation, and the pixel size is 13.68μm×13.68μm. Fig. 1-11 (d) is random phase modulation in a ring pattern, where the modulation pitch is 13.68μm×13.68μm. Fig. 1-11 (e) is a ring pattern without any modulation inside. In the simulation, the wavelength is 408nm; the focal length of the objective lens is 4mm.

(a) $\exp(j2\pi\cos(y/\Lambda))$ (b) $\cos(y/\Lambda)$ (c) random amplitude

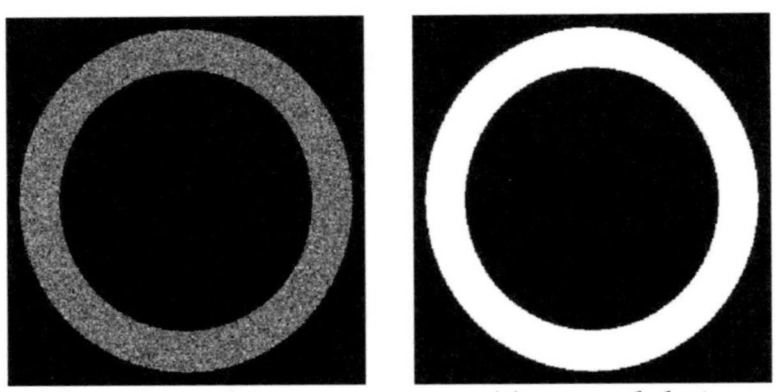

(d) random phase (e) no modulation

Fig. 1-11. The reference patterns used in the simulation [22].

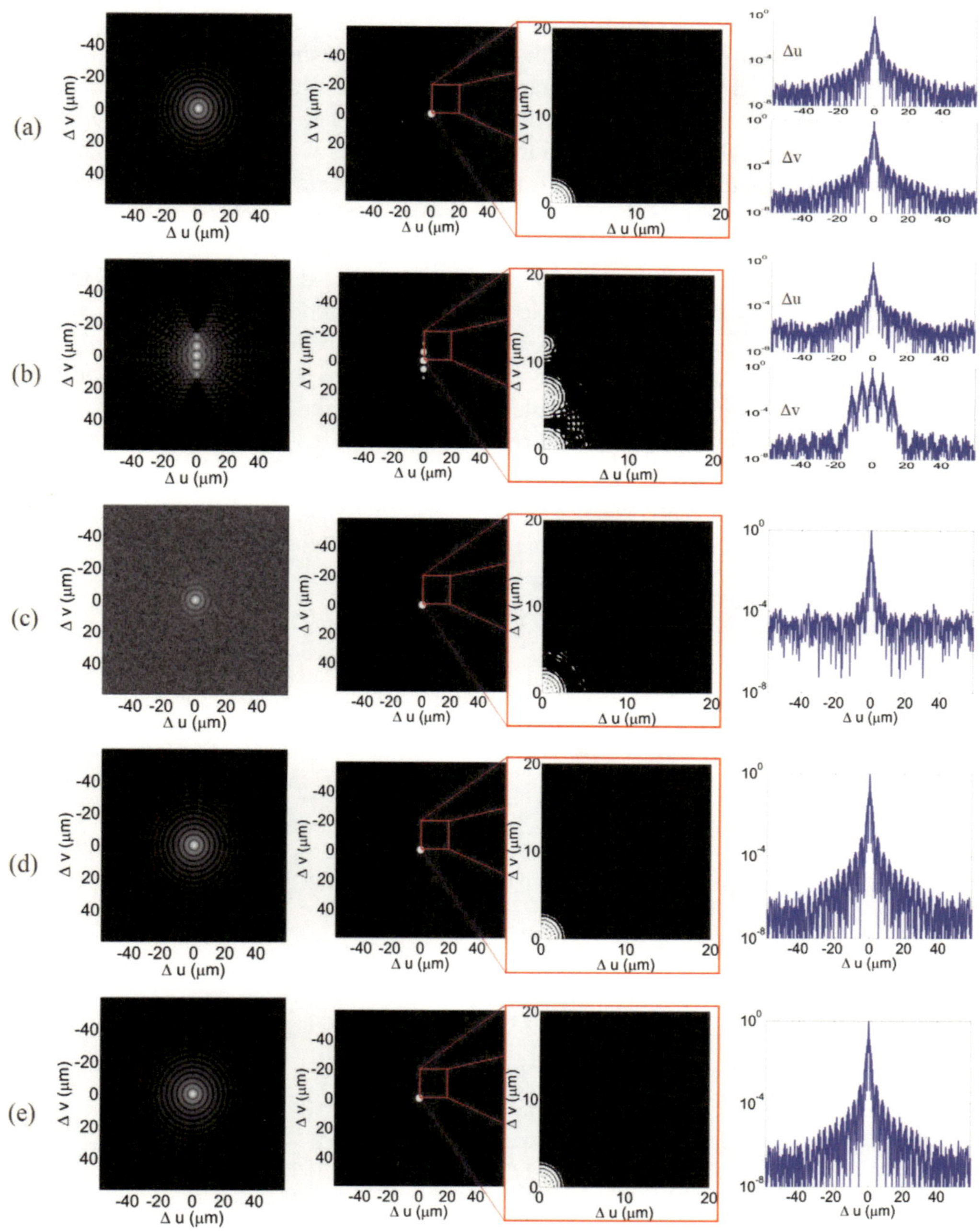

Fig. 1-12. The simulation result of the point shifting selectivity, based on Eq. (1-28), for a point located at the center of the signal plane corresponding to the reference patterns in Fig. 1-11. The figures from left to right are followed with 2-D shifting selectivity, 2-D shifting selectivity after thresholding and its enlarged pattern, 1-D shifting selectivity [22].

Fig. 1-12 (a)~(e) show a simulation of point shifting selectivity corresponding to the reference patterns shown in Fig. 1-11 (a)~(e). The simulation uses Eq. (1-28) and Fast Fourier Transform in Matlab for fast calculation. Figures in the first columns are 2-D shift selectivity, where the horizontal and vertical axes correspond to the shifting along Δu and Δv, respectively. The grey level in each point represents the diffraction intensity (in log scale) collected by the corresponding image point of the CCD. To emphasize cross-talk caused by the 2-D shifting selectivity, a threshold value of 10^{-3} is used in the second column. Each white point stands for the intensity larger than the threshold value, and black point stands for the intensity below the value. Figures in the last column show the 1-D shift selectivity. It is shown that the point shift selectivity shown in Fig. 1-12 (a), Fig. 1-12 (d) and Fig. 1-12 (e) are all the same. Obviously, the point shift selectivity cannot tell the effect of the phase modulated reference patterns. Even so, the simple relationship still provides a simple prediction rule for designing reference pattern.

For a precise calculation, a more complicated model should be adopted. The model named "pixel shift selectivity" does not use the three assumptions in point shift selectivity. It considers the response of one pixel to simulate the real condition. Inside the disk, the optical field of the reference (A_R), signal (A_S) and reading light (A_P) have been described in Eq. (1-12), Eq. (1-13) and Eq. (1-14). When the disk shifts, based on VOHIL model, the optical field diffracted from specific layer at a depth (Δz) can be expressed

$$U(u-\Delta u, v-\Delta v, \Delta z) = \frac{\exp\left[jk(2f+\Delta z)\right]}{j\lambda^3 f^3} \Im\left\{U_P(x,y)\exp\left[-j\frac{\pi\Delta z}{\lambda f^2}\left(x^2+y^2\right)\right]\right\} \times \left\{\delta(u-\Delta u, v-\Delta v) \otimes \left\{\begin{array}{l} \Im^*\left\{U_R(x,y)\exp\left[-j\frac{\pi\Delta z}{\lambda f^2}\left(x^2+y^2\right)\right]\right\} \\ \Im\left\{U_S(x,y)\exp\left[-j\frac{\pi\Delta z}{\lambda f^2}\left(x^2+y^2\right)\right]\right\} \end{array}\right\}\right\}. \tag{1-29}$$

The optical field is thus propagating to the CCD plane and is integrated along Δz.

$$U_{det}(\xi,\eta,\Delta u,\Delta v) = \frac{2T\exp(4jkf)}{(\lambda f)^2} \int_{-\infty}^{\infty}\int_{-\infty}^{\infty} \Im\left\{\begin{array}{l} U_p(\xi_2-\xi,\eta_2-\eta)U_R^*(\xi_2+\xi_1,\eta_2+\eta_1) \\ \mathrm{sinc}\left[-2T\frac{\xi_2(\xi_1+\xi)+\eta_2(\eta_1+\eta)}{\lambda f^2}\right] \end{array}\right\}\Bigg|_{\frac{\Delta u}{\lambda f}, \frac{\Delta v}{\lambda f}} U_S(\xi_1,\eta_1)d\xi_1 d\eta_1, \tag{1-30}$$

where, T is the thickness of the recording media, (ξ, η) is the coordinate on the output CCD plane, (ξ_1, η_1) is the coordinate on the input SLM plane, and (ξ_2, η_2) is the parameter caused by convolution. Considering a pixel in the center of the SLM as the signal and integrating the intensity of the whole pixel in the center of the CCD, the pixel shift selectivity is

$$I_{pixel}(\Delta u,\Delta v)=$$

$$\iint_{pixel}\left|\iint_{pixel}\Im\left\{\begin{matrix}U_P(\xi_2-\xi,\eta_2-\eta)U_R^*(\xi_2+\xi_1,\eta_2+\eta_1)\\ \operatorname{sinc}\left[-2T\dfrac{\xi_2(\xi_1+\xi)+\eta_2(\eta_1+\eta)}{\lambda f^2}\right]\end{matrix}\right\}\Bigg|_{\frac{\Delta u}{\lambda f}\ \frac{\Delta v}{\lambda f}} d\xi_1 d\eta_1\right|^2 d\xi d\eta\ . \tag{1-31}$$

Eq. (1-31) shows that the shifting selectivity is nearly the autocorrelation of the Fourier transform of the reference pattern, and is independent of thickness as long as (ξ, η) is close to ($-\xi_1$,$-\eta_1$). This condition can always be satisfied when the diffracting signal locates near the conjugate image position. Thus, the broader reference pattern will cause narrower Fourier transform pattern. Fig. 1-13 shows the 2-D pixel shifting selectivity. It is find the first two patterns perform relative higher shifting selectivity in one direction. The reason is the Fourier transform pattern of the reference pattern is narrower in the direction. For the ring patterns in random phase or amplitude modulation, there seems no obvious difference. Among all these patterns, the ring pattern without any modulation shows the best shifting selectivity because the Fourier transform pattern of the last one is the narrowest.

The simulation result in Fig. 1-13 shows that the reference pattern is important to the shifting selectivity. But, as mentioned in Section 1.3, the reference pattern is also important to the PSF. Fig. 1-14 shows the simulated 1-D PSF for the five patterns shown in Fig. 1-11. Actually, to design an appropriate reference pattern to simultaneously obtain narrower PSF and higher shifting selectivity is one of the most important issues in a collinear system.

1.5 Summary

In this chapter, the collinear holographic storage system is studied. An effective model is used to analyze the system. Based on Fresnel transform and the VOHIL model, paraxial solutions to describe the diffraction characteristic of the collinear holographic system are carried out. The solutions are used to figure out the the physics insight of the system. In Eq. (1-19), it shows that the PSF is related to the auto-correlation function of the reference pattern times a defocusing phase term. Accordingly, a lens array used as the reference is introduced. From the simulated diffracted signal on the CCD plane, the SNR for LAP modulation can be obtained as high as 63.2, which is much larger than 2.3 in RL reference and 8.1 in RBP reference. Besides, the discussed physics concept is important in designing a more effective VHS system when other system characteristics such like system tolerance are taken into consideration.

Equation (1-28) based on a simplified model shows point shifting selectivity is related to Fourier transform of intensity distribution of the reference pattern. It provides a simple prediction rule for designing some reference patterns to improve the shift selectivity of the system. The equation shows that the broader the reference pattern is, the higher shifting selectivity we get. To obtain an accurate calculation, Eq. (1-31) is derived out without using the three assumptions in point shift selectivity, and the result is called pixel shift selectivity. It

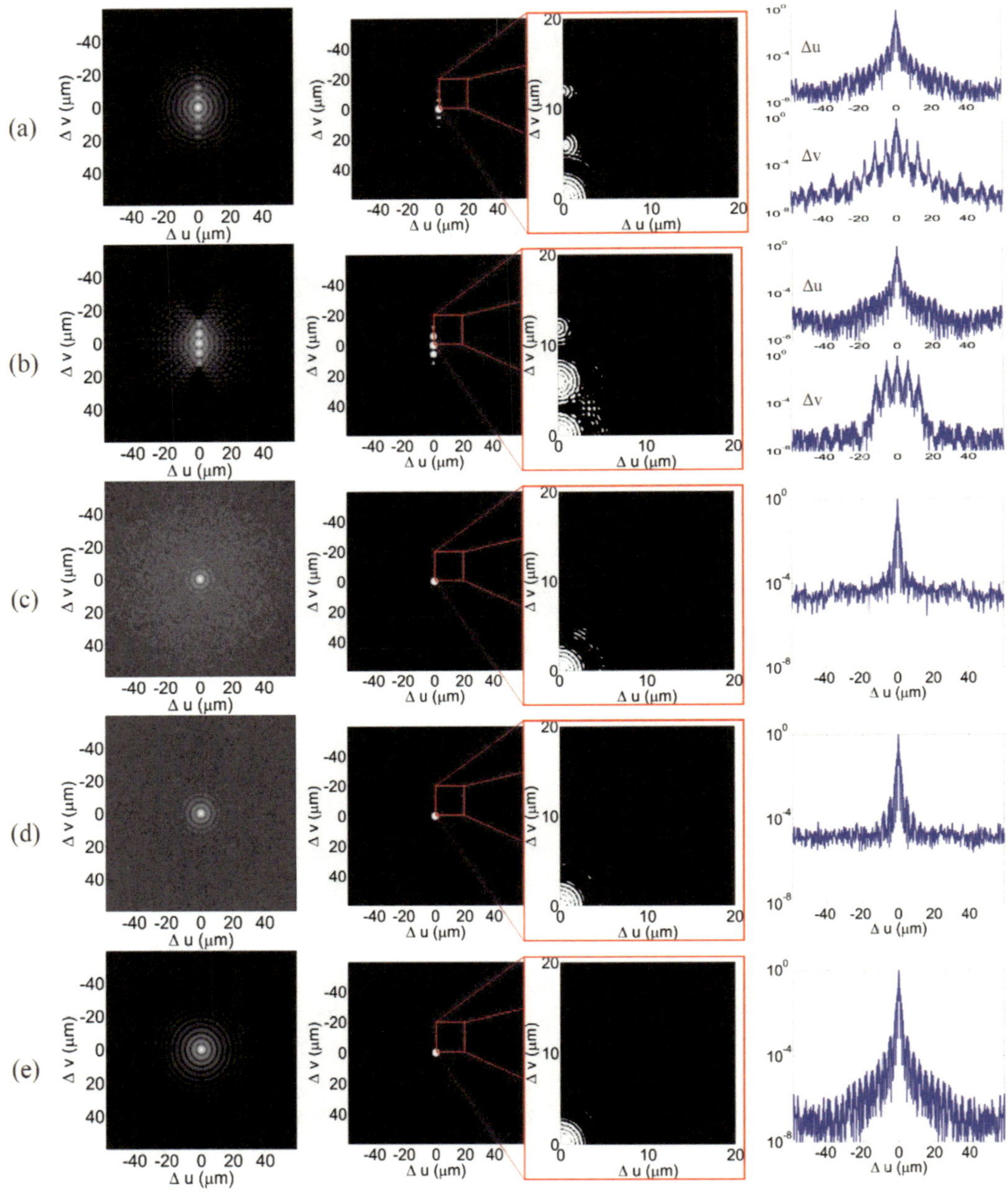

Fig. 1-13. The simulation result of the shifting selectivity for a pixel located at the center of the signal plane corresponding to the reference patterns in Fig. 11. The figures from left to right are followed with 2-D shifting selectivity, 2-D shifting selectivity after thresholding and its enlarged pattern, 1-D shifting selectivity [22].

shows that the shifting selectivity is nearly an autocorrelation of the Fourier transform of the reference pattern, and is independent of thickness of the recording medium when the diffracting signal is located near the conjugate image position. The simulation results shows both amplitude and phase modulation make no obvious change in shifting selectivity, so shifting selectivity is not affected through advanced design of reference pattern. The solution will be useful in the design of a reference pattern to perform high-quality readout in the collinear holographic storage system.

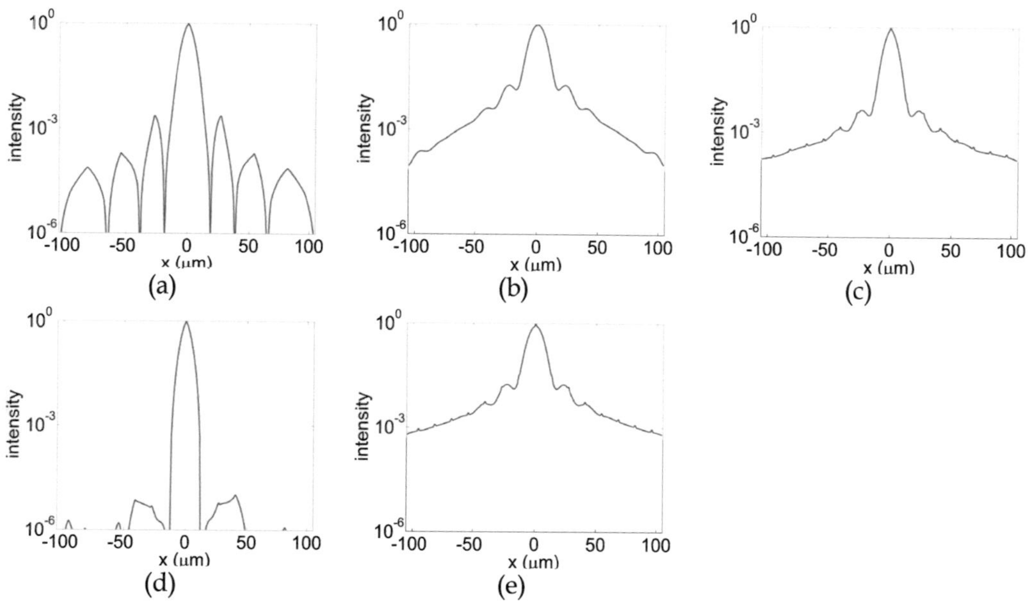

Fig. 1-14. The simulation result of the impulse response for a point located at the center of the signal plane corresponding to the reference patterns in Fig. 11 [22].

2. References

[1] H. Horimai, “Collinear Holography”, The 5th Pacific Rim Conference on Lasers and Electro-Optics, Proceedings 1, Taiwan, (2003).

[2] H. Horimai and L. Jun, “A Novel Collinear Optical Setup for Holographic Data Storage System”, Proc. of SPIE 5380, 297-303 (2004).

[3] H. Horimai and X. Tan, “Advanced collinear holography,” Opt. Rev. 12, 90–92 (2005).

[4] H. Horimai, X. D. Tan, and J. Li, “Collinear holography,” Appl. Opt. 44, 2575–2579 (2005).

[5] K. Tanaka, H. Mori, M. Hara, K. Hirooka, A. Fukumoto, and K. Watanabe, “High density recording of 270 Gbits/inch2 in a coaxial holographic storage system,” Tech. Digest of ISOM 2007, MO-D-03.

[6] H. Horimai and X. Tan, “Holographic versatile disc system,” Proc. of SPIE 5939, 593901(2005).

[7] M. Toishi, “Holographic recording and reconstructing apparatus and holographic recording and reconstructing method,” United States Patent Application Publication, US 20060176532 A1 (2006).

[8] B. King, K. Anderson and K. Curtis, "System and method for reflective holographic storage with associated multiplexing technique," United States Patent, US 6721076 B2 (2004).

[9] H. Horimai, Y. Sakane and K. Kimura, "Optical information-recording medium, optical information recording apparatus and optical information reproducing apparatus including optical information-recording medium and method for manufacturing polarization changing layer," United States Patent, US2004/0165518 A1 (2004).

[10] T. Shimura, S. Ichimura, R. Fujimura, K. Kuroda, X. D. Tan and H. Horimai, "Analysis of a collinear holographic storage system: introduction of pixel spread function," Opt. Letters 31, 1208–1210 (2006).

[11] X. Tan and H. Horimai, "Collinear™ technology for holographic versatile disc (HVD™) system," Proc. of SPIE 6343, 63432w (2006).

[12] H. Horimai and X. Tan, "Advanced collinear holography," Opt. Rev. 12, 90–92 (2005).

[13] S. R. Lambourdiere, A. Fukumoto, K. Tanaka, and K. Watanabe, "Simulation of holographic data storage for the optical collinear system," Jpn. J. Appl. Phys. 45, 1246–1252 (2006).

[14] C. C. Sun, Y. W. Yu, S. C. Hsieh, T. C. Teng and M. F. Tsai "Point spread function of a collinear holographic storage system," Optics Express 15, 18111-18118 (2007).

[15] C. C. Sun, "A simplified model for diffraction analysis of volume holograms," Opt. Eng. 42, 1184–1185 (2003).

[16] K. Tanaka, M. Hara, K. Tokuyama, K. Hirooka, K. Ishioka, A. Fukumoto and K. Watanabe, "Improved performance in coaxial holographic data recording," Optics Express 15, 16196-16209 (2007).

[17] C. B. Burckhardt, "Use of a random phase mask for the recording Fourier transform holograms of data masks," Appl. Opt. 9, 695–700 (1970).

[18] J. W. Goodman, *Introduction to Fourier Optics* (McGraw-Hill, 2002).

[19] Y. W. Yu, C. Y. Cheng, and C. C. Sun, "Increase of signal-to-noise ratio of a collinear holographic storage system with reference modulated by a ring lens array," Opt. Lett. 35, 1130-1132 (2010).

[20] T.Shimura, Y. Ashizuka, M. Terada, R. Fujimura, and K. Kuroda, "What Limits the Storage Density of the Collinear Holographic Memory," Tech. Digest of 32 ODS2007,TuD1.

[21] T. Shimura, S. Ichimura, Y. Ashizuka, R. Fujimura, K. Kuroda, X. D. Tan, and H. Horimai, "Shift selectivity of the collinear holographic storage system," Proc. of 35 SPIE 6282, 62820s (2006).

[22] Y. W. Yu, T. C. Teng, S. C. Hsieh, C. Y. Cheng, and C. C. Sun, "Shifting selectivity of collinear volume holographic storage," Opt. Comm.283, 3895-3900 (2010).

10

Theory of Polychromatic Reconstruction for Volume Holographic Memory

Ryushi Fujimura, Tsutomu Shimura and Kazuo Kuroda
Institute of Industrial Science, The University of Tokyo
Japan

1. Introduction

In volume holographic memory (van Heerden, 1963), the information is stored as a volume hologram and retrieved through the holographic reconstruction process by illuminating the hologram with a readout probe beam whose wavelength, incident angle, and wavefront should be identical to those of the reference beam used in the recording process. This requirement stems from the fact that diffraction from the volume hologram is restricted by Bragg's law. While such a restriction is responsible for the large storage density of volume holographic memories, it also causes some obstacles for implementing practical memory systems. For example, in rewritable recording media, like photorefractive materials, illumination with a readout probe beam will rewrite the recorded hologram, destroying the stored information. Even in a photopolymer, some of the storage capacity will be wasted during the readout if some monomers still exist in the readout volume. These issues are obviously caused by the destructive probe beam having the ability to expose the recording medium in a similar manner to the recording beam.

To avoid such a problem, several nondestructive readout methods have been proposed so far (Gulanyan et al., 1979; Petrov et al., 1979; Külich, 1987), where the readout is performed at a longer wavelength, outside the sensitive spectral region of the recording material. These methods can successfully reconstruct the stored image at a wavelength different from the recording one, but most of these methods may not be practical for holographic memory systems because the multiplexing capability is considerably lowered. For example, anisotropic diffraction (Petrov et al., 1979) requires a specific recording configuration and thus limits the number of multiplexed pages. A spherical probe beam method (Külich, 1987) tends to produce severe crosstalk noise from other multiplexed pages, which demands a large angular separation between two adjacent multiplexed holograms, resulting in a small storage density.

Recently, we proposed another way to reconstruct an image at a different wavelength (Fujimura et al., 2007). Our method, which we call polychromatic reconstruction (PCR), utilizes a spectrally broad light source for the probe beam, as shown in Fig. 1. Each angular spectral component of the recorded gratings can be Bragg-matched with one particular wavelength within the broadband spectrum of the probe beam. Thus, the whole image can be reconstructed from the volume hologram even though the probe wavelength is very different from the recording one. On the other hand, analogous to the spherical probe beam method, the large spectral width of the polychromatic probe beam causes deterioration of the angular selectivity and results in considerable lowering of the multiplexing capability. However, unlike the case of the spherical probe beam, such a drawback can be overcome by using a selective detection method (Fujimura et al., 2010). If the proper optical component, such as a wavelength filter or grating, is additionally inserted in the imaging system, we can detect the signal image alone even though the crosstalk-noise waves are diffracted from other multiplexed pages. Therefore, PCR with the selective detection method is a promising way to achieve nondestructive readout in volume holographic memories without sacrificing the multiplexing capability.

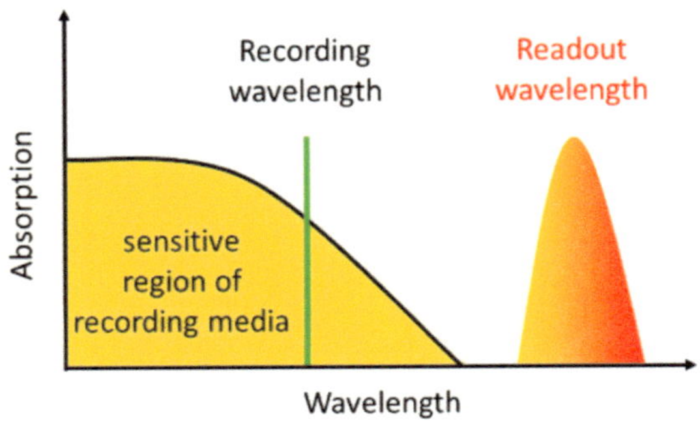

Fig. 1. Concept of the nondestructive readout in the PCR method.

When we implement the PCR method in a holographic memory system, it is important to know the properties of the holographic reconstruction process with the polychromatic light. Indeed, the PCR method shows several notable features as compared with conventional monochromatic reconstruction. For example, the reconstructed image has a wavelength distribution that linearly shifts along the grating vector, and image magnification occurs in a direction perpendicular to the incident plane. In fact, knowledge of the required bandwidth of the polychromatic probe beam is essential to design a practical memory system.

In this chapter, we develop the theory of holographic reconstruction with polychromatic light, especially from the viewpoint of its application to volume holographic memory. Based on the plane-wave expansion model, we will derive expressions for the required bandwidth, the distortion of the reconstructed image, the optimum recording configuration, the diffraction efficiency, the inter- and intra-page crosstalk noise, and the theoretical limit of the storage density. The obtained expressions show unique features of the PCR method and are very informative for constructing actual holographic memory systems utilizing the PCR method.

2. Basic principle of the image reconstruction

2.1 Reconstruction from a plane-wave hologram

First we consider the diffraction from a plane-wave hologram that is formed by signal and reference plane waves. When the grating recorded at a wavelength λ_w is read out with a probe wavelength λ_p, the Bragg condition is given by

$$k_p\left(\mathbf{e}_d-\mathbf{e}_p\right)=k_w\left(\mathbf{e}_s-\mathbf{e}_r\right)\equiv k_w\mathbf{G}, \tag{1}$$

where $\mathbf{e}_s$, $\mathbf{e}_r$, $\mathbf{e}_p$, and $\mathbf{e}_d$ are the unit direction vectors of the signal, reference, probe, and diffracted plane waves, respectively; $k_q = 2\pi n_q/\lambda_q$ ($q = w, p$) is the wave number at the wavelength λ_q and the refractive index n_q; and $\mathbf{G}$ is the grating vector normalized by the recording wave number k_w. In order to obtain diffraction from the grating $\mathbf{G}$, the unit direction vector $\mathbf{e}_p$ and the probe wavelength λ_p should be properly chosen so as to satisfy Eq. (1). Such a restriction can be derived from Eq. (1) and the relation $|\mathbf{e}_d|^2 = 1$, which is written as

$$\left(\mathbf{e}_p + \frac{\mu}{2}\mathbf{G}\right)\cdot\mathbf{G} = 0, \tag{2}$$

where μ is the ratio of the wave numbers of the recording and probe beams, i.e., $\mu = k_w/k_p$. Hereafter, for simplicity, we assume that wavelength dispersion of the refractive index can be neglected, i.e., $n_w \approx n_p \approx n$, and thus, μ can be approximated by the ratio of wavelengths. If $\mathbf{e}_p$ and λ_p satisfy Eq. (2), the grating will reproduce the plane wave with the unit direction vector $\mathbf{e}_d$ expressed as

$$\mathbf{e}_d = \mathbf{e}_p + \mu\mathbf{G}. \tag{3}$$

Generally, there are many combinations of $\mathbf{e}_p$ and λ_p that satisfy Eq. (2). Thus, it is possible to obtain diffraction from a plane wave hologram even when the readout probe wave vector is not identical to the reference wave vector in the recording process, i.e., $\lambda_p \neq \lambda_w$ or $\mathbf{e}_p \neq \mathbf{e}_r$. For example, when the probe wavelength λ_p is given, the incident angle of the probe beam should be adjusted so that the vector $\mathbf{e}_p + \mu\mathbf{G}/2$ lies on the plane normal to the grating $\mathbf{G}$, as shown in Fig. 2. Note that there is a maximum value for the allowed λ_p because the relation $\mu|\mathbf{G}|/2 < 1$ should hold, as is seen from Fig. 2. In this case, the trace of $\mathbf{e}_p$ forms a circle, which is regarded as the Bragg degeneracy.

On the contrary, when $\mathbf{e}_p$ is given, the probe wavelength λ_p should satisfy the following relation:

$$\mu = -\frac{2\mathbf{e}_p \cdot \mathbf{G}}{|\mathbf{G}|^2}. \tag{4}$$

Note that, in this case, the wavelength ratio μ is uniquely determined by the set of ($\mathbf{e}_p$, $\mathbf{G}$).

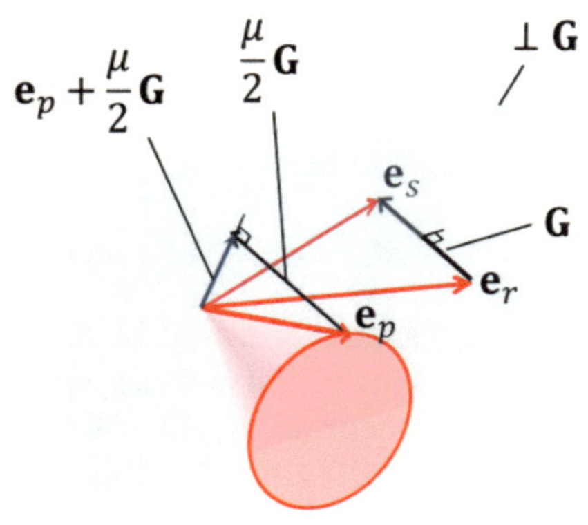

Fig. 2. The possible choice of the incident angle of the probe beam to reconstruct the plane wave hologram represented by **G**.

2.2 Reconstruction of the image information

A volume holographic memory usually stores two-dimensional image information. Therefore, the signal beam consists of many plane waves and creates various grating vectors after interfering with the reference plane wave. We express the i-th signal component $\mathbf{e}_{si}$ as $\mathbf{e}_{si} = \mathbf{e}_{s0} + \boldsymbol{\delta}\mathbf{e}_{si}$, where $\mathbf{e}_{s0}$ is the central direction vector of the diverged signal beam, and $\boldsymbol{\delta}\mathbf{e}_{si}$ is the deviation vector from $\mathbf{e}_{s0}$. Similarly, the corresponding diffracted component $\mathbf{e}_{di}$ is expressed as $\mathbf{e}_{di} = \mathbf{e}_{d0} + \boldsymbol{\delta}\mathbf{e}_{di}$. In this case, the Bragg condition for each angular spectral component is given by

$$k_p\left(\mathbf{e}_{di} - \mathbf{e}_p\right) = k_w\left(\mathbf{e}_{si} - \mathbf{e}_r\right) \equiv k_w\mathbf{G}_i, \tag{5}$$

where $\mathbf{G}_i$ is the normalized grating vector for i-th component. In order to reproduce the whole image, the Bragg condition of Eq. (5) should be satisfied at all components i. If we assume that Eq. (5) is satisfied at the central signal component ($i = 0$) and that the divergence angle of the signal beam is sufficiently small, i.e., the relation $|\boldsymbol{\delta}\mathbf{e}_{si}|^2 << 1$ holds, then Eq. (2) can be rewritten as

$$\left(\mathbf{e}_p + \frac{\mu}{2}\mathbf{G}_i\right)\cdot\mathbf{G}_i \approx \mathbf{e}_{d0}\cdot\boldsymbol{\delta}\mathbf{e}_{si}. \tag{6}$$

In order to satisfy the Bragg condition at any component i, the probe wave vector should be identical to the reference wave vector (i.e., $\lambda_p = \lambda_w$ and $\mathbf{e}_p = \mathbf{e}_r$) so that the relation $\mathbf{e}_{d0} = \mathbf{e}_{s0}$ holds. Otherwise, the obtained diffracted waves are limited to those from the components satisfying the relation $\mathbf{e}_{d0}\cdot\boldsymbol{\delta}\mathbf{e}_{sj} = 0$. Therefore, it is usually considered that the image information cannot be completely reproduced when the probe wavelength is different from the recording one.

However, there are two possible ways to satisfy Eq. (5) for all components even when $\lambda_p \neq \lambda_w$. One is Külich's approach (Külich, 1987), where a probe beam having adequate angular divergence is used instead of the plane wave. In this method, for each signal component i, there is a probe plane wave component $\mathbf{e}_{pi} = \mathbf{e}_{p0} + \boldsymbol{\delta}\mathbf{e}_{pi}$ that satisfies

$$\left(\mathbf{e}_{pi} + \frac{\mu}{2}\mathbf{G}_i\right)\cdot\mathbf{G}_i \approx \mathbf{e}_{d0}\cdot\left(\boldsymbol{\delta}\mathbf{e}_{si} + \frac{1}{\mu}\boldsymbol{\delta}\mathbf{e}_{pi}\right) = 0, \tag{7}$$

where we assume that the divergence angles of the probe and signal beams are sufficiently small, and thus, the relations $|\boldsymbol{\delta}\mathbf{e}_{si}|^2 << 1$, $|\boldsymbol{\delta}\mathbf{e}_{pi}|^2 << 1$, and $\boldsymbol{\delta}\mathbf{e}_{pi}\cdot\boldsymbol{\delta}\mathbf{e}_{si} << 1$ hold. Then, the diffracted wave is reproduced as described by the following relation:

$$\mathbf{e}_{di} = \mathbf{e}_{pi} + \mu\mathbf{G}_i = \mathbf{e}_{d0} + \boldsymbol{\delta}\mathbf{e}_{pi} + \mu\boldsymbol{\delta}\mathbf{e}_{si}. \tag{8}$$

Note that, in this method, special care should be taken about the diffraction due to the Bragg degeneracy. If several probe components simultaneously satisfy Eq. (7) for one particular signal component, its grating component will produce several diffracted waves with different direction vectors, as represented by Eq. (8). This will degrade the quality of the reconstructed image. To avoid such a situation, the angular spectral components of the probe beam should exist on only one particular plane. For example, it is preferable that the probe beam should be expanded by using a cylindrical lens, not a spherical lens.

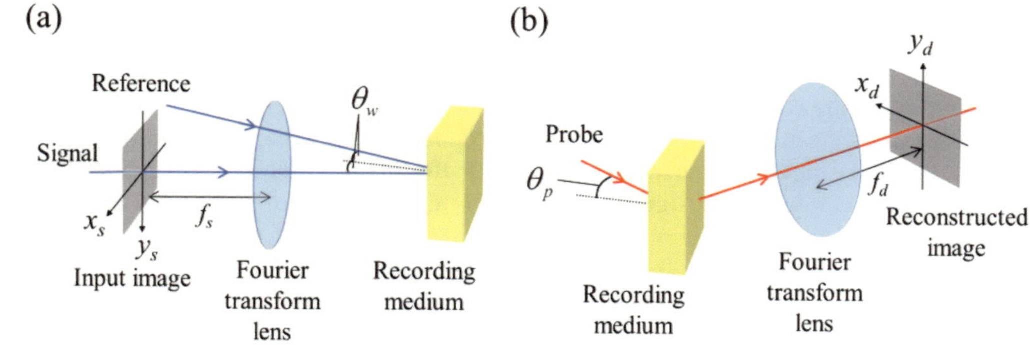

Fig. 3. Schematic diagram of (a) the recording and (b) the reconstruction schemes in the PCR method. θ_w is the internal half crossing angle of $\mathbf{e}_{s0}$ and $\mathbf{e}_r$; θ_p is the internal incident angle of $\mathbf{e}_p$; and f_s and f_d are the focal lengths of the Fourier transform lenses in the recording and reconstruction processes, respectively.

Another way to satisfy Eq. (5) is our approach, the PCR method. This method utilizes a broadband probe beam instead of monochromatic light. The recording and readout schemes of the PCR method are illustrated in Fig. 3. In the recording process, a monochromatic signal beam bearing the image information passes through a Fourier transform lens and records a Fourier hologram in the usual way. In the readout process, the recorded hologram is readout by using a spectrally broad but spatially coherent light source, such as a super luminescent diode (SLD). All grating components satisfy the Bragg condition because the probe beam includes a spectral component that satisfies the following relation for each component i:

$$\mu_i \equiv \frac{k_w}{k_{pi}} \approx \frac{\lambda_{pi}}{\lambda_w} = -\frac{2\mathbf{e}_p \cdot \mathbf{G}_i}{|\mathbf{G}_i|^2}, \tag{9}$$

where k_{pi} and λ_{pi} are the Bragg-matched wavenumber and wavelength for a grating component $\mathbf{G}_i$, respectively. Note that image degradation due to Bragg degeneracy will not occur in the PCR method because the Bragg-matched wavelength λ_{pi} is uniquely determined by the set of ($\mathbf{e}_p$, $\mathbf{G}_i$). In this case, the diffracted waves are reproduced with the direction vector

$$\mathbf{e}_{di} = \mathbf{e}_p + \mu_i \mathbf{G}_i . \tag{10}$$

The characteristic feature of the PCR method is that the wavelength of each diffracted plane wave is different at different grating components $\mathbf{G}_i$. Therefore, the PCR method can be applied only to Fourier holograms, where one grating vector corresponds to one particular point on the object plane. Even though all diffracted waves are obtained in an image hologram and a Fresnel hologram with polychromatic light, the image cannot be reconstructed since waves with different wavelengths cannot construct a point image.

An example of the reconstructed image in the PCR method is presented in Fig. 4, where we calculated the wavelength ratio μ_i and the direction vector $\mathbf{e}_{di}$ for each grating component $\mathbf{G}_i$ by using Eqs. (9) and (10), neglecting off-Bragg diffraction. In this simulation, we assumed that the input image, which was an outline character "A" with dimensions 1 cm × 1 cm, shown in Fig. 4(a), was recorded at $\lambda_w = 532$ nm and was reconstructed with a polychromatic probe beam with a central wavelength λ_{p0} of 815 nm. From the figure, we can see that the reconstructed image was formed with spectral components ranging from 795 nm to 835 nm, and image magnification occurred in the y_d-direction. Such features are considered as a consequence of using the polychromatic light for the holographic reconstruction. In the following section, we will develop a theory of holographic reconstruction with polychromatic light and investigate characteristic features of the PCR method especially in holographic memory systems.

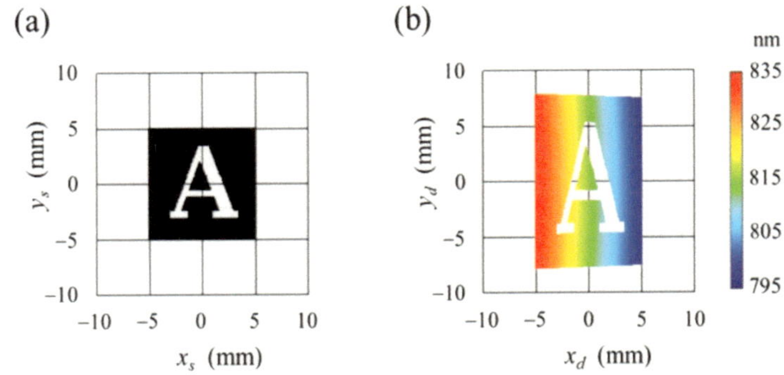

Fig. 4. Simulated results of image reconstruction by PCR. (a) Input image and (b) the reconstructed image. The color in (b) represents the Bragg-matched wavelength of each diffracted wave. The calculation parameters are as follows: $\lambda_w = 532$ nm; $\theta_w = 30°$; $\theta_p = 50°$; $n = 1$; and $f_s = f_d = 100$ mm.

3. Theory of holographic reconstruction with polychromatic light

3.1 Definition of the coordinate system

In this section, we introduce a coordinate system that allows for a more quantitative discussion of the PCR method. A Cartesian coordinate system is defined here using unit direction vectors $\mathbf{e}_{s0}$ and $\mathbf{e}_r$, as shown in Fig. 5(a), whose normal bases are given by

$$\mathbf{e}_x = \frac{\mathbf{e}_{s0} - \mathbf{e}_r}{|\mathbf{e}_{s0} - \mathbf{e}_r|}, \quad \mathbf{e}_y = -\frac{\mathbf{e}_{s0} \times \mathbf{e}_r}{|\mathbf{e}_{s0} \times \mathbf{e}_r|}, \quad \mathbf{e}_z = \frac{\mathbf{e}_{s0} + \mathbf{e}_r}{|\mathbf{e}_{s0} + \mathbf{e}_r|}, \quad (\mathbf{e}_{s0} \neq \pm \mathbf{e}_r) \tag{11}$$

We introduce an elevation angle α_k and azimuthal angle β_k to specify a unit direction vector $\mathbf{e}_k$, as is depicted in Fig. 5(b). In this case the Cartesian components of $\mathbf{e}_k$ are written as

$$\mathbf{e}_k = \begin{bmatrix} e_{kx} \\ e_{ky} \\ e_{kz} \end{bmatrix} = \begin{bmatrix} \cos\alpha_k \sin\beta_k \\ \sin\alpha_k \\ \cos\alpha_k \cos\beta_k \end{bmatrix}. \qquad (k = si, r, p, di) \tag{12}$$

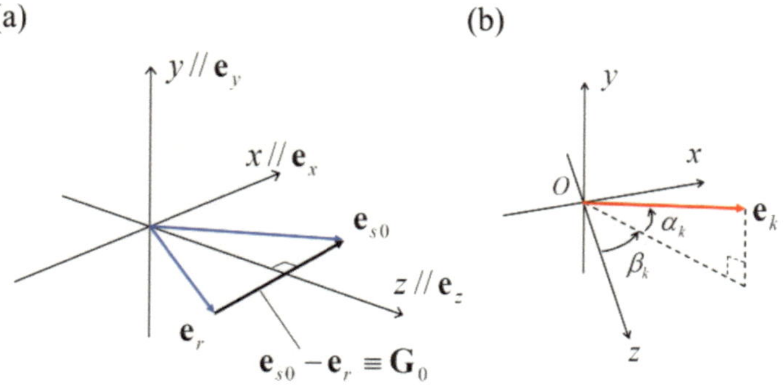

Fig. 5. Definition of (a) the Cartesian coordinate system and (b) the elevation angle α_k and the azimuthal angle β_k of the unit direction vector $\mathbf{e}_k$.

From the symmetry of the coordinate system and the Bragg condition of Eq. (5), the relations $\alpha_{s0} = 0$, $\alpha_r = 0$, $\beta_r = -\beta_{s0}$, and $\beta_{d0} = -\beta_p$ should hold. Thus, the direction vectors are written as

$$\mathbf{e}_{s0} = \begin{bmatrix} \sin\beta_{s0} \\ 0 \\ \cos\beta_{s0} \end{bmatrix}, \ \mathbf{e}_r = \begin{bmatrix} -\sin\beta_{s0} \\ 0 \\ \cos\beta_{s0} \end{bmatrix}, \ \mathbf{e}_p = \begin{bmatrix} \cos\alpha_p \sin\beta_p \\ \sin\alpha_p \\ \cos\alpha_p \cos\beta_p \end{bmatrix}, \ \mathbf{e}_{d0} = \begin{bmatrix} -\cos\alpha_p \sin\beta_p \\ \sin\alpha_p \\ \cos\alpha_p \cos\beta_p \end{bmatrix},$$
$$\mathbf{G}_0 \equiv \mathbf{e}_{s0} - \mathbf{e}_r = \begin{bmatrix} 2\sin\beta_{s0} \\ 0 \\ 0 \end{bmatrix}. \qquad (0 < \beta_{s0} < \pi/2) \tag{13}$$

Note that if the grating component $\mathbf{G}_0$ and the wavelength ratio μ_0 are given, then $\mathbf{e}_p$ and $\mathbf{e}_{d0}$ have only one free parameter. This is because, from Eq. (5), α_p and β_p should satisfy

$$\cos\alpha_p \sin\beta_p = -\mu_0 \sin\beta_{s0}. \tag{14}$$

Furthermore, we express α_{si}, β_{si}, α_{di}, and β_{di}, as $\alpha_{si} = \alpha_{s0} + \delta\alpha_{si}$, $\beta_{si} = \beta_{s0} + \delta\beta_{si}$, $\alpha_{di} = \alpha_{d0} + \delta\alpha_{di}$, and $\beta_{di} = \beta_{d0} + \delta\beta_{di}$, respectively, and hereafter for simplicity, we assume that the divergence angles of the signal and diffracted beams are sufficiently small (i.e., $\delta\alpha_{si}$, $\delta\beta_{si}$, $\delta\alpha_{di}$, $\delta\beta_{di} << 1$) and the relations $|\boldsymbol{\delta}\mathbf{e}_{si}|^2$, $|\boldsymbol{\delta}\mathbf{e}_{di}|^2 << 1$ hold at all components i. Then, the deviation vectors $\boldsymbol{\delta}\mathbf{e}_{si}$ and $\boldsymbol{\delta}\mathbf{e}_{di}$ are approximated by

$$\boldsymbol{\delta}\mathbf{e}_{si} \equiv \mathbf{e}_{si} - \mathbf{e}_{s0} \approx \delta\alpha_{si} \begin{bmatrix} 0 \\ 1 \\ 0 \end{bmatrix} + \delta\beta_{si} \begin{bmatrix} \cos\beta_{s0} \\ 0 \\ -\sin\beta_{s0} \end{bmatrix} \equiv \delta\alpha_{si}\mathbf{e}_{s\alpha} + \delta\beta_{si}\mathbf{e}_{s\beta},$$
$$\boldsymbol{\delta}\mathbf{e}_{di} \equiv \mathbf{e}_{di} - \mathbf{e}_{d0} \approx \delta\alpha_{di} \begin{bmatrix} \sin\alpha_p \sin\beta_p \\ \cos\alpha_p \\ -\sin\alpha_p \cos\beta_p \end{bmatrix} + \cos\alpha_p \delta\beta_{di} \begin{bmatrix} \cos\beta_p \\ 0 \\ \sin\beta_p \end{bmatrix}$$
$$\equiv \delta\alpha_{di}\mathbf{e}_{d\alpha} + \cos\alpha_p \delta\beta_{di}\mathbf{e}_{d\beta}, \tag{15}$$

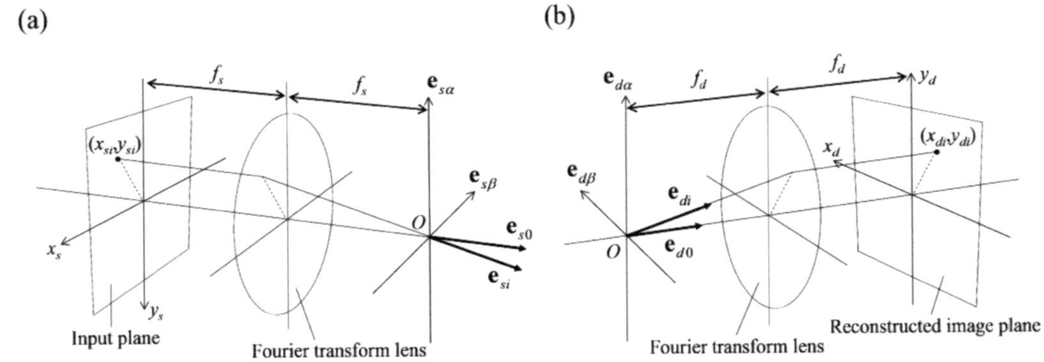

Fig. 6. The relation between the angular spectral component and the imaging position in the Fourier hologram. (a) The recording process and (b) the reconstruction process.

Note that the unit direction vectors $\mathbf{e}_{s\alpha}$ and $\mathbf{e}_{s\beta}$ correspond to the normal bases at the Fourier plane of the input image, as shown in Fig. 6(a). Thus, the position on the input plane (x_{si}, y_{si}) and the set of the deviation angles ($\delta\alpha_{si}$, $\delta\beta_{si}$) are related via

$$\begin{pmatrix} \delta\beta_{si} \\ \delta\alpha_{si} \end{pmatrix} = \frac{1}{f_s} \begin{pmatrix} x_{si} \\ y_{si} \end{pmatrix}, \tag{16}$$

where f_s is the focal length of the Fourier transform lens in the recording process. Similarly, the relation between the position on the reconstructed image plane (x_{di}, y_{di}) and the set of the deviation angles ($\delta\alpha_{di}$, $\delta\beta_{di}$) is given by

$$\begin{pmatrix} x_{di} \\ y_{di} \end{pmatrix} = f_d \begin{pmatrix} \cos\alpha_p \delta\beta_{di} \\ \delta\alpha_{di} \end{pmatrix}, \tag{17}$$

where f_d is the focal length of the Fourier transform lens in the reconstruction process. Furthermore, hereafter we assumed that the dimensions of the input image are in the ranges $-x_{sMax} \le x_{si} \le x_{sMax}$ and $-y_{sMax} \le y_{si} \le y_{sMax}$. Therefore, the maximum deviation angles of the signal beam, $\delta\alpha_{sMax}$ and $\delta\beta_{sMax}$, are

$$\begin{pmatrix} \delta\beta_{sMax} \\ \delta\alpha_{sMax} \end{pmatrix} = \frac{1}{f_s} \begin{pmatrix} x_{sMax} \\ y_{sMax} \end{pmatrix}. \tag{18}$$

3.2 Spectral width required for the reconstruction

In order to reconstruct the image information, an adequate spectral width is needed for the probe beam to satisfy the Bragg condition at every grating component. In this section, we will theoretically estimate such a spectral width. When we rewrite Eq. (9) using the deviation vectors $\boldsymbol{\delta}\mathbf{e}_{si}$, the difference between μ_i and μ_0 is given by

$$\delta\mu_i \equiv \mu_i - \mu_0 \approx -\frac{2\mathbf{e}_{d0} \cdot \boldsymbol{\delta}\mathbf{e}_{si}}{|\mathbf{G}_0|^2}, \tag{19}$$

where we assume that the relation $|\boldsymbol{\delta}\mathbf{e}_{si}|^2 << |\mathbf{G}_0|^2$ holds. Equation (19) indicates that the Bragg-matched wavelength does not depend only on the amplitude $|\boldsymbol{\delta}\mathbf{e}_{si}|$, but is determined by the projection of $\boldsymbol{\delta}\mathbf{e}_{si}$ on $\mathbf{e}_{d0}$. If we rewrite Eq. (19) using the Cartesian components in Eqs. (13) and (15) and the relation of Eq. (14), we obtain

$$\begin{aligned}\delta\mu_i &= -\frac{\sin\alpha_p\delta\alpha_{si} - \cos\alpha_p\sin\left(\beta_{s0}+\beta_p\right)\delta\beta_{si}}{2\sin^2\beta_{s0}} \\ &= \frac{\mu_0}{2}\left[\frac{\tan\alpha_p}{\sin\beta_{s0}\sin\beta_p}\delta\alpha_{si} - \left(\frac{1}{\tan\beta_{s0}}+\frac{1}{\tan\beta_p}\right)\delta\beta_{si}\right].\end{aligned} \tag{20}$$

Therefore, the spectral width required for the full image reconstruction ($\Delta\lambda_{BM}$) is given by

$$\Delta\lambda_{BM} = \lambda_{p0}\left[\frac{\tan|\alpha_p|}{\sin\beta_{s0}\sin\left(-\beta_p\right)}\frac{\delta y_{sMax}}{f_s} + \left(\frac{1}{\tan\beta_{s0}}+\frac{1}{\tan\beta_p}\right)\frac{\delta x_{sMax}}{f_s}\right]. \tag{21}$$

The Bragg-matched wavelength generally depends on both deviation angles $\delta\alpha_{si}$ and $\delta\beta_{si}$. When $\mathbf{e}_p$ lies in the xz-plane (i.e., $\alpha_p = 0°$), however, $\delta\mu_i$ depends only on the azimuthal deviation angle $\delta\beta_{si}$. To see these features more clearly, the spatial distributions of the Bragg-matched wavelength are illustrated in Fig. 7. In this simulation, we calculated the reconstructed image with the Bragg-matched wavelength assuming that the grating recorded at 532 nm was read out at around 815 nm. As is predicted from Eq. (21), the spectral width required for the probe beam has a minimum at $\alpha_p = 0°$. Furthermore, the distortion of the reconstructed image is also smallest at $\alpha_p = 0°$, which will be treated further in Section 3.3.

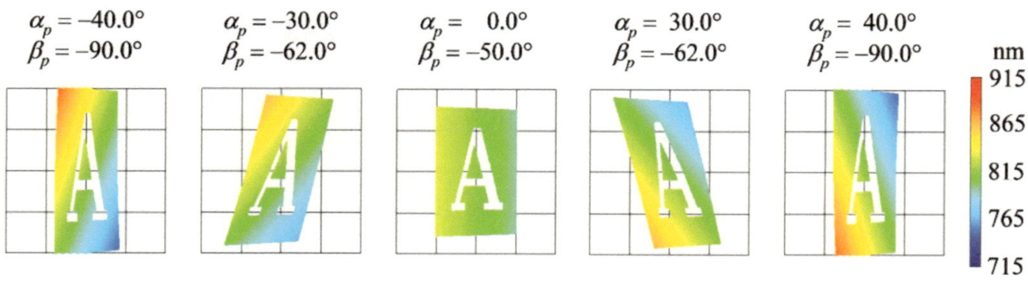

Fig. 7. The reconstructed images in the PCR method at several probe elevation angles α_p. The input image was the same as that in Fig. 4. The calculation parameters were as follows: $\lambda_w = 532$ nm; $\lambda_{p0} = 815$ nm; $\beta_{s0} = 30°$; $n = 1$; $f_s = f_d = 100$ mm; and β_p is determined from Eq. (14) at each α_p.

Rewriting Eq. (20) using the relation of Eq. (14) at the condition $\alpha_p = 0°$ yields

$$\delta\mu_i = -\frac{\mu_0}{2}\left\{\frac{1}{\tan\beta_{s0}} + \frac{1}{\tan\left[\arcsin\left(-\mu_0\sin\beta_{s0}\right)\right]}\right\}\delta\beta_{si}. \tag{22}$$

There is an optimum recording angle β_{s0Opt} that minimizes the required spectral width for a given μ_0, as shown in Fig. 8, which is given by

$$\beta_{s0Opt} = \arctan\left(\frac{1}{\mu_0}\right), \quad (\mu_0 \neq 1) \tag{23}$$

Then, Eq. (22) can be simplified to

$$\delta\mu_i = -\frac{1}{2}\left(\mu_0^{\ 2} - 1\right)\delta\beta_{si}. \tag{24}$$

In this case, the required spectral width $\Delta\lambda_{BM}$ reaches a minimum and is expressed as

$$\Delta\lambda_{BMMin} = \lambda_{p0}\left(\mu_0 - \frac{1}{\mu_0}\right)\frac{\delta x_{sMax}}{f_s}. \tag{25}$$

However, in most cases, the allowed deviation angle $\delta\beta_{sMax}$ is determined by the spectral width $\Delta\lambda_p$ of a given probe light source. Assuming that $\alpha_p = 0°$, from Eqs. (21), we obtain

$$\delta\beta_{s\max} = \left(\frac{1}{\tan\beta_{s0}} + \frac{1}{\tan\beta_p}\right)^{-1}\frac{\Delta\lambda_p}{\lambda_{p0}} \tag{26}$$

At the optimum recording angle, $\delta\beta_{sMax}$ reaches a maximum

$$\delta\beta_{sMax} = \left(\mu_0 - \frac{1}{\mu_0}\right)^{-1}\frac{\Delta\lambda_p}{\lambda_{p0}} \tag{27}$$

For example, if we have a probe light source whose central wavelength λ_{p0} is 815 nm and whose full spectral width $\Delta\lambda_p$ is 50 nm, and the recording wavelength λ_w is 532 nm, then the optimum recording angle β_{s0Opt} is about 33°, and the allowed deviation angle $\delta\beta_{sMax}$ is about 4.0°.

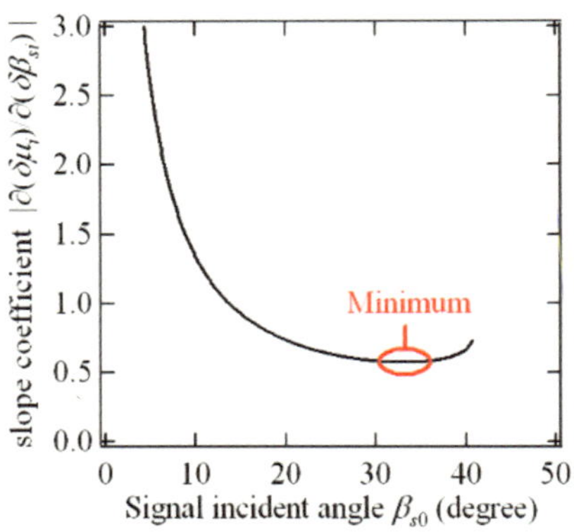

Fig. 8. The slope coefficient of $\delta\beta_{si}$ in Eq. (22) is plotted as a function of the signal incident angle β_{s0}. We assumed that $\lambda_w = 532$ nm and $\lambda_{p0} = 815$ nm.

So far, assuming the small deviation angles of the signal beam, we derived the expression for the Bragg-matched wavelength and found that the wavelength shift is simply proportional to the deviation angles $\delta\alpha_{si}$ and $\delta\beta_{si}$. However, for large deviation angles, their relation cannot be expressed by Eq. (20) and becomes nonlinear. Examples for the reconstructed images at large deviation angles are shown in Fig. 9. In this simulation, the focal length of the Fourier transform lenses was varied from 50 mm to 10 mm while the input image size was kept constant, which resulted in an increase of the maximum deviation angle from 5.7° to 30°. A larger angular spectrum of the signal beam results in greater deviation from the linear relationship and requires a probe beam of much broader spectrum. Moreover, considerable distortion of the reconstructed image occurs at large deviation angle.

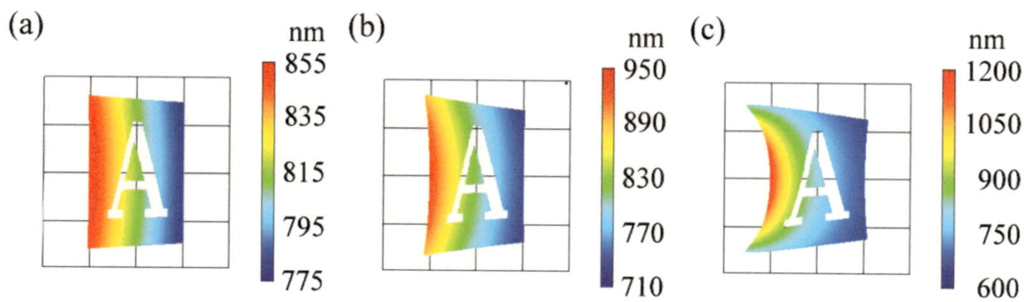

Fig. 9. Influence of large deviation angles $\delta\alpha_{si}$ and $\delta\beta_{si}$ on the reconstructed image. The input image was the same as that in Fig. 4. The focal lengths of the Fourier transform lenses are (a) $f_s = f_d = 50$ mm, (b) $f_s = f_d = 20$ mm, (c) $f_s = f_d = 10$ mm. The calculation parameters were as follows: $\lambda_w = 532$ nm; $\lambda_{p0} = 815$ nm; $\beta_{s0} = 30°$; and $n = 1$;

3.3 Distortion of the reconstructed image

In a thin hologram, it is well-known that magnification or reduction of the reconstructed image occurs during readout at a wavelength different from the recording one (Champagne, 1967). The PCR method, in contrast, produces a directionally-stretched image depending on the readout configuration, as shown in Fig. 7. Such distortion is likely to cause errors in retrieving data from volume holograms, but if the property of the distortion is known and predictable, the stored information can be completely recovered after the image processing of the distorted image. In this section, we will derive an expression for the distortion of the reconstructed image in the PCR method.

From Eq. (10), the deviation vector $\boldsymbol{\delta}\mathbf{e}_{di}$ is approximated by

$$\boldsymbol{\delta}\mathbf{e}_{di} \equiv \mathbf{e}_{di} - \mathbf{e}_{d0} \approx \mu_0 \boldsymbol{\delta}\mathbf{e}_{si} + \delta\mu_i \mathbf{G}_0 , \tag{28}$$

where we neglect the product of $\boldsymbol{\delta}\mathbf{e}_{si}$ and $\delta\mu_0$. From Eqs. (13), (15), and (20), the Cartesian components of $\boldsymbol{\delta}\mathbf{e}_{di}$ are expressed as

$$\boldsymbol{\delta}\mathbf{e}_{di} = \mu_0 \begin{bmatrix} \cos\beta_{s0}\delta\beta_{si} \\ \delta\alpha_{si} \\ -\sin\beta_{s0}\delta\beta_{si} \end{bmatrix} - \frac{\sin\alpha_p \delta\alpha_{si} - \cos\alpha_p \sin\left(\beta_{s0} + \beta_p\right)\delta\beta_{si}}{2\sin^2\beta_{s0}} \begin{bmatrix} 2\sin\beta_{s0} \\ 0 \\ 0 \end{bmatrix}. \tag{29}$$

On the other hand, $\boldsymbol{\delta}\mathbf{e}_{di}$ is also expressed by Eq. (15) using $\mathbf{e}_{d\alpha}$ and $\mathbf{e}_{d\beta}$. Rewriting Eq. (29) using normal bases of $\mathbf{e}_{d\alpha}$ and $\mathbf{e}_{d\beta}$ yields

$$\boldsymbol{\delta}\mathbf{e}_{di} = \left(\frac{\mu_0}{\cos\alpha_p} \delta\alpha_{si} \right) \mathbf{e}_{d\alpha} + \left(\mu_0 \frac{\tan\alpha_p}{\tan\beta_p} \delta\alpha_{si} + \cos\alpha_p \delta\beta_{si} \right) \mathbf{e}_{d\beta} . \tag{30}$$

Comparing the coefficients in (30) with those in Eq. (15), we obtain the relation between the deviation angles ($\delta\alpha_{si}$, $\delta\beta_{si}$) and ($\delta\alpha_{di}$, $\delta\beta_{di}$):

$$\begin{pmatrix} \cos\alpha_p \delta\beta_{di} \\ \delta\alpha_{di} \end{pmatrix} = \begin{pmatrix} \cos\alpha_p & \mu_0 \dfrac{\tan\alpha_p}{\tan\beta_p} \\ 0 & \dfrac{\mu_0}{\cos\alpha_p} \end{pmatrix} \begin{pmatrix} \delta\beta_{si} \\ \delta\alpha_{si} \end{pmatrix} . \tag{31}$$

From Eqs. (16) and (17), we can finally derive the expression for a transfer matrix between the object and reconstructed image plane, that is,

$$\begin{pmatrix} x_{di} \\ y_{di} \end{pmatrix} = \mathbf{T} \begin{pmatrix} x_{si} \\ y_{si} \end{pmatrix}, \quad \mathbf{T} \equiv \frac{f_d}{f_s} \begin{pmatrix} \cos\alpha_p & \mu_0 \dfrac{\tan\alpha_p}{\tan\beta_p} \\ 0 & \dfrac{\mu_0}{\cos\alpha_p} \end{pmatrix} . \tag{32}$$

When $\alpha_p = 0°$ or $\beta_p = -90°$, non-diagonal components in the transfer matrix become zero and the image can be reconstructed without any tilting. However, the distortion becomes minimal at $\alpha_p = 0°$, as was seen in Fig. 7.

3.4 Multiplexing with crystal rotation

In this section, we investigate the multiplexing capability in the PCR method. While many multiplexing methods have been proposed so far, we adopt peristrophic multiplexing (Curtis et al., 1994) and rotation of the crystal angle (hereafter, the latter will be called crystal angle multiplexing). The multiplexing configuration considered here is shown in Fig. 10. Note that other multiplexing methods are also possible in principle, but most of them are probably not suitable for the PCR method because they require a complicated system to read a target page. For example, ordinary angle multiplexing, which varies the incident angle of the reference beam, will also require moving the imaging system. This is because $\mathbf{e}_{d0}$ will be pointed in a different direction at each multiplexed page, as expressed by Eq. (10). The situation is similar in the case of the wavelength multiplexing method. Peristrophic multiplexing and crystal angle multiplexing, in contrast, do not require any additional movement other than the crystal rotation. In addition, the imaging properties are almost unchanged with each multiplexed page because the recording angle is constant. Thus, these multiplexing methods are suitable for holographic memory systems employing PCR.

In the peristrophic and crystal angle multiplexing methods, recording of another page is performed after crystal rotation by the proper angle. In order to retrieve the stored information without crosstalk, the rotation angle should be sufficiently large so that the other multiplexed holograms cannot produce noise diffracted waves that would disturb the detection of the target signal image. On the other hand, the theoretical limit of the storage density is inversely proportional to the angular separation between the adjacent multiplexed holograms. Therefore, the angular selectivity (i.e., how small the angular separation can be made) is an important figure of merit determining the total storage density of the system. In the following sections, we will derive an expression of the minimum rotation angle for each multiplexing method.

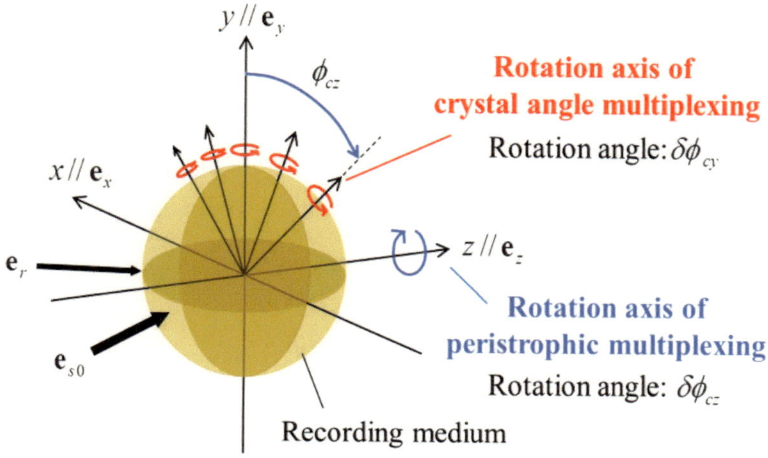

Fig. 10. Configuration of crystal angle multiplexing combined with peristrophic rotation. Peristrophic multiplexing and crystal angle multiplexing are performed by rotating the crystal around the z-axis and an axis inclined at ϕ_{cz} from the y-axis, respectively.

3.4.1 Peristrophic multiplexing

When the crystal is rotated by a small angle $\delta\phi_{cz}$ around the z-axis (peristrophic rotation), the displacement vector of the grating component ($\mathbf{\delta e}_{cz}$) is approximated by

$$\mathbf{\delta e}_{cz} \equiv \begin{pmatrix} 1 & -\delta\phi_{cz} & 0 \\ \delta\phi_{cz} & 1 & 0 \\ 0 & 0 & 1 \end{pmatrix} \mathbf{G}_i - \mathbf{G}_i \approx \begin{bmatrix} 0 \\ 2\sin\beta_{s0}\delta\phi_{cz} \\ 0 \end{bmatrix}. \tag{33}$$

Note that we neglected the product of small quantities. Due to the small rotation of the grating vectors, the Bragg-matched wavelength and the unit direction vector of each diffracted wave are also changed according to Eqs. (9) and (10). The shift of the wavelength ratio after the peristrophic rotation ($\delta\mu_{cz}$) is given by

$$\begin{aligned} \delta\mu_{cz} &\equiv -\frac{2\mathbf{e}_p \cdot (\mathbf{G}_i + \mathbf{\delta e}_{cz})}{|\mathbf{G}_i + \mathbf{\delta e}_{cz}|^2} - \left(-\frac{2\mathbf{e}_p \cdot \mathbf{G}_i}{|\mathbf{G}_i|^2} \right) \\ &\approx -\frac{2\mathbf{e}_{d0} \cdot \mathbf{\delta e}_{cz}}{|\mathbf{G}_0|^2} \\ &= -\frac{\sin\alpha_p}{\sin\beta_{s0}} \delta\phi_{cz}, \end{aligned} \tag{34}$$

where last equality is obtained by substituting Eqs. (13) and (33). The displacement vector of $\mathbf{e}_{di}$ after the peristrophic rotation ($\boldsymbol{\delta}\mathbf{e}_{dcz}$) is expressed as

$$\begin{aligned}\boldsymbol{\delta}\mathbf{e}_{dcz} &\equiv \left[\mathbf{e}_p + (\mu_i + \delta\mu_{cz})(\mathbf{G}_i + \boldsymbol{\delta}\mathbf{e}_{cz})\right] - (\mathbf{e}_p + \mu_i \mathbf{G}_i) \\ &\approx \mu_0 \boldsymbol{\delta}\mathbf{e}_{cz} + \delta\mu_{cz}\mathbf{G}_0 \\ &= -2\delta\phi_{cz}\begin{bmatrix}\sin\alpha_p \\ \cos\alpha_p \sin\beta_p \\ 0\end{bmatrix},\end{aligned} \tag{35}$$

where we use Eqs. (13), (14), (33), and (34) to derive the last equality. If we rewrite Eq. (35) using normal bases of $\mathbf{e}_{d\alpha}$ and $\mathbf{e}_{d\beta}$ and compare the coefficients with those in Eq. (15), similarly to Section 3.3, we obtain the shift amount of the imaging position due to the peristrophic rotation:

$$\begin{pmatrix}\delta x_{dicz} \\ \delta y_{dicz}\end{pmatrix} = 2\mu_0 \sin\beta_{s0} f_d \begin{pmatrix}\dfrac{\tan\alpha_p}{\tan\beta_p} \\ \dfrac{1}{\cos\alpha_p}\end{pmatrix}\delta\phi_{cz}. \tag{36}$$

For simplicity, we assume that $\alpha_p = 0$. Then, from Eqs. (20), (32), (34), and (36), the Bragg-matched wavelength λ_{BMcz} can be expressed as a function of the imaging position (x_d, y_d) and the rotation angle $\delta\phi_{cz}$:

$$\lambda_{BMcz}(x_d, y_d, \delta\phi_{cz}) \approx \lambda_{p0} - \frac{\lambda_{p0}}{2}\left(\frac{1}{\tan\beta_{s0}} + \frac{1}{\tan\beta_p}\right)\frac{x_d}{f_d}$$
$$\left(\begin{array}{c} -x_{sMax}\dfrac{f_d}{f_s} \le x_d \le x_{sMax}\dfrac{f_d}{f_s} \\ -\mu_0 y_{sMax}\dfrac{f_d}{f_s} + 2\mu_0 \sin\beta_{s0} f_d \delta\phi_{cz} \le y_d \le \mu_0 y_{sMax}\dfrac{f_d}{f_s} + 2\mu_0 \sin\beta_{s0} f_d \delta\phi_{cz}\end{array}\right). \tag{37}$$

Note that the inequality in Eq. (37) represents the location of the reconstructed image. The peristrophic rotation causes the reconstructed image to shift in the y_d direction while keeping the spatial distribution of the Bragg-matched wavelength unchanged. In order to record another page without crosstalk, the rotation angle should be large enough to shift the noise image away from the signal imaging area. Thus, from the inequality in Eq. (37), the minimum angular separation in peristrophic multiplexing ($\delta\phi_{czMin}$) is expressed as

$$\delta\phi_{czMin} = \frac{y_{sMax}}{\sin\beta_{s0} f_s}. \tag{38}$$

This equation does not include the wavelength ratio μ_0. Therefore, the readout at a longer wavelength does not affect the angular selectivity in the peristrophic multiplexing, even though the reconstructed image was magnified by a factor of μ_0 in the y_d direction. This is because the amount of spatial shift is also increased by a factor of μ_0, as shown in Eq. (36).

In order to see these behaviours more clearly, in Fig. 11, we show the simulated results of the reconstructed image obtained while varying the peristrophic rotation angle $\delta\phi_{cz}$. We also present results for the conventional monochromatic readout for comparison. Note that the off-Bragg diffraction was also neglected in the monochromatic case, but here, in order to see the influence of the peristrophic rotation on the Bragg condition, the probe light was assumed to be quasi-monochromatic with a small spectral width of 1 nm, instead of taking the finite hologram dimension into account. Although a part of the image faded away due to the Bragg selectivity in the quasi-monochromatic case, the rotation angles required for multiplexing are almost the same in both cases. Therefore, the use of polychromatic light in the PCR method has only a small influence on the multiplexing capability in the peristrophic multiplexing method, and the PCR method could employ peristrophic multiplexing without any adverse influence on the multiplexing capability.

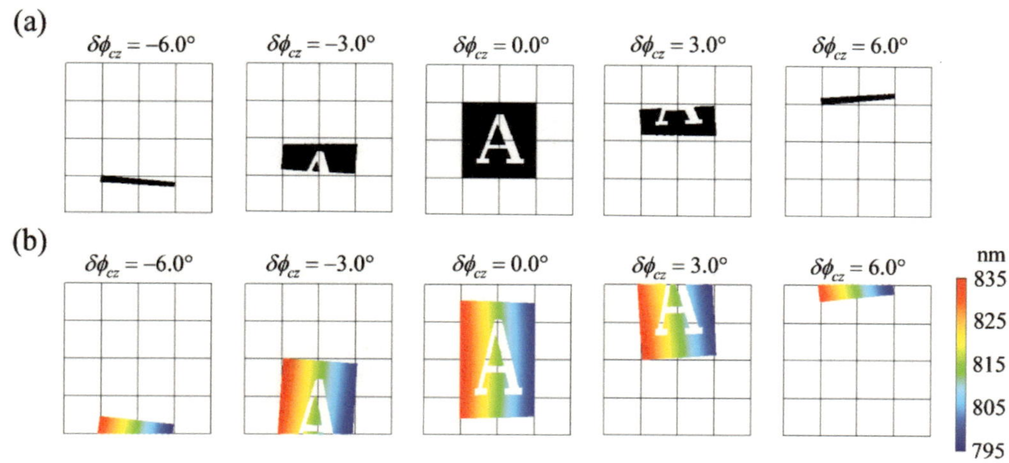

Fig. 11. Influence of the peristrophic rotation on the reconstructed image obtained with (a) conventional quasi-monochromatic light whose central wavelength is 532 nm and whose spectral width is 1 nm, and (b) polychromatic light whose central wavelength is 815 nm and whose spectral width is 40 nm. $\delta\phi_{cz}$ is the peristrophic rotation angle.

3.4.2 Crystal angle multiplexing with peristrophic rotation

Next, let us consider the angular selectivity in the crystal angle multiplexing, where the crystal is rotated around an axis inclined at ϕ_{cz}, as shown in Fig. 10. Note that the inclination angle ϕ_{cz} results from the crystal rotation due to the peristrophic multiplexing. In a similar way to the previous section, the displacement vector $\mathbf{G}_i$ due to a small rotation in the crystal angle multiplexing ($\boldsymbol{\delta}\mathbf{e}_{cy}$) is approximated by

$$\boldsymbol{\delta}\mathbf{e}_{cy} \equiv \begin{pmatrix} 1 & 0 & \cos\phi_{cz}\delta\phi_{cy} \\ 0 & 1 & \sin\phi_{cz}\delta\phi_{cy} \\ -\cos\phi_{cz}\delta\phi_{cy} & -\sin\phi_{cz}\delta\phi_{cy} & 1 \end{pmatrix} \mathbf{G}_i - \mathbf{G}_i \approx \begin{bmatrix} 0 \\ 0 \\ -2\sin\beta_{s0}\cos\phi_{cz}\delta\phi_{cy} \end{bmatrix}. \quad (39)$$

In this case, the shift of the wavelength ratio ($\delta\mu_{cy}$) and the displacement vector of $\mathbf{e}_{di}$ after the crystal rotation ($\boldsymbol{\delta}\mathbf{e}_{dcy}$) are expressed as

$$\delta\mu_{cy} \approx -\frac{2\mathbf{e}_{d0}\cdot\boldsymbol{\delta}\mathbf{e}_{cy}}{\left|\mathbf{G}_0\right|^2} = \frac{\cos\alpha_p\cos\beta_p}{\sin\beta_{s0}}\cos\phi_{cz}\delta\phi_{cy} \tag{40}$$

and

$$\boldsymbol{\delta}\mathbf{e}_{dcy} \approx \mu_0\boldsymbol{\delta}\mathbf{e}_{cy} + \delta\mu_{cy}\mathbf{G}_0 = 2\cos\alpha_p\cos\varphi_{cz}\delta\phi_{cy}\begin{bmatrix}\cos\beta_p\\ 0\\ \sin\beta_p\end{bmatrix}, \tag{41}$$

respectively. Then, the imaging shift due to the rotation of the crystal angle multiplexing is written as

$$\begin{pmatrix}\delta x_{dicy}\\ \delta y_{dicy}\end{pmatrix} = f_d\begin{pmatrix}2\cos\alpha_p\cos\phi_{cz}\delta\phi_{cy}\\ 0\end{pmatrix}. \tag{42}$$

Thus, the reconstructed image will shift in the x_d direction regardless of α_p. Assuming that $\alpha_p = 0$, we obtain the Bragg-matched wavelength λ_{BMcy} as a function of the imaging position (x_d, y_d) and the rotation angles $\delta\phi_{cy}$ and ϕ_{cz}:

$$\lambda_{BMcy}\left(x_d, y_d, \delta\phi_{cy}, \phi_{cz}\right) \approx \lambda_{p0} - \frac{\lambda_{p0}}{2}\left(\frac{1}{\tan\beta_{s0}} + \frac{1}{\tan\beta_p}\right)\frac{x_d}{f_d} + \frac{\lambda_{p0}}{\tan\beta_{s0}}\cos\phi_{cz}\delta\phi_{cy}$$

$$\begin{pmatrix}-x_{sMax}\dfrac{f_d}{f_s} + 2f_d\cos\phi_{cz}\delta\phi_{cy} \le x_d \le x_{sMax}\dfrac{f_d}{f_s} + 2f_d\cos\phi_{cz}\delta\phi_{cy}\\ -\mu_0 y_{sMax}\dfrac{f_d}{f_s} \le y_d \le \mu_0 y_{sMax}\dfrac{f_d}{f_s}\end{pmatrix}. \tag{43}$$

In contrast to peristrophic multiplexing, the crystal angle multiplexing leads to a change in the Bragg-matched wavelength as well as a shift of the imaging location. The Bragg-matched wavelength in Eq. (43) is plotted at several rotation angles $\delta\phi_{cy}$ in Fig. 12. The dotted part of the Bragg-matched line in the figure corresponds to the portion of the image that is not reconstructed by the probe beam because the Bragg-matched wavelength is beyond the probe spectral band. Such behaviour can be clearly seen in Fig. 13.

When the solid part of the Bragg-matched line lies inside the signal imaging area, the noise diffraction will spatially overlap with the signal image and cause severe crosstalk. Therefore, the angular separation between adjacent holograms should be sufficiently large so that the solid part will be shifted away from the signal imaging area. If we assume that the central wavelength of the probe beam is λ_{p0} and its full spectral width is $\Delta\lambda_p$, the minimum angular separation ($\delta\phi_{cyMin}$) is given by

$$\delta\phi_{cyMin} = \min\left[\frac{x_{sMax}}{\cos\phi_{cz}f_s}, \frac{\tan\beta_{s0}}{2\cos\phi_{cz}\lambda_{p0}}\left(\Delta\lambda_p + \lambda_{p0}\left(\frac{1}{\tan\beta_{s0}} + \frac{1}{\tan\beta_p}\right)\frac{x_{sMax}}{f_s}\right)\right], \tag{44}$$

where min[a, b] is a minimum function yielding the smaller value of a and b. Note that the former parameter in the minimum function is the rotation angle required for separating the noise and signal images spatially, and the latter corresponds to that required for spectral separation. The above equation implies that an unnecessarily large spectral width increases the minimum angular separation needlessly. Thus, if we take the spectral width required for the full image reconstruction, $\Delta\lambda_{BM}$ in Eq. (25), as the probe spectral width $\Delta\lambda_p$, then $\delta\phi_{cyMin}$ is given by

$$\delta\phi_{cyMin} = \left(1 + \frac{\tan\beta_{s0}}{\tan\beta_p}\right)\frac{x_{sMax}}{\cos\phi_{cz} f_s}. \tag{45}$$

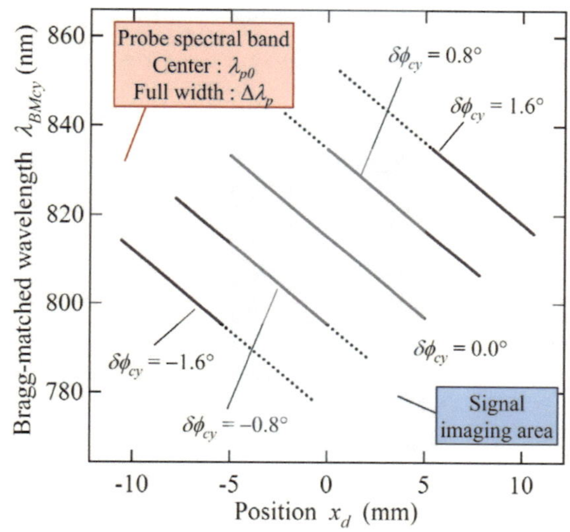

Fig. 12. The imaging location x_d and the Bragg-matched wavelength λ_{BMcy} at each rotation angle $\delta\phi_{cy}$. The reconstructed image field is limited by the probe spectral band $\Delta\lambda_p$. The dotted part of the Bragg-matched line corresponds to the portion of the image that will not be reconstructed by the probe beam. The peristrophic rotation angle ϕ_{cz} is assumed to be zero.

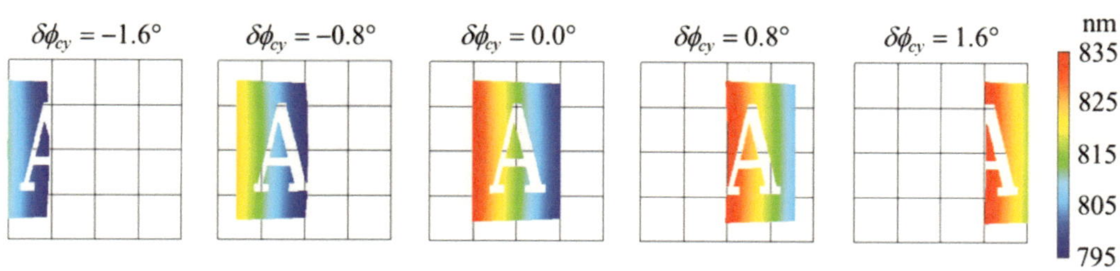

Fig. 13. The reconstructed images after the crystal rotation around the y-axis. $\delta\phi_{cy}$ is the rotation angle. The polychromatic light with a central wavelength of 815 nm and a spectral width of 40 nm was used.

Note that when the wavelength ratio μ_0 is unity (i.e., the readout wavelength is the same as the recording one), the required angular separation $\delta\phi_{cyMin}$ becomes zero. This is because we assumed that the hologram dimension is large enough to neglect off-Bragg diffraction. In other words, if we employ the PCR method, the angular separation specified by Eq. (45) is always required, regardless of the hologram dimension.

3.5 Off-Brag diffraction

So far, we considered only the diffracted wave satisfying the Bragg condition. In reality, however, owing to the finite hologram dimension, the grating produces non-Bragg-matched diffracted waves, referred to as off-Bragg diffraction. In this section, we consider the off-Brag diffraction in the PCR method and derive expressions for crosstalk noise.

We assume that a set of ($\mathbf{G}_i$, $\mathbf{e}_p$, $\mathbf{e}_{di}$) satisfies the Bragg condition at λ_w and λ_{pi} and consider the situation where a grating component $\mathbf{G}_i + \boldsymbol{\delta}\mathbf{e}_{Goff}$ produces a diffracted wave whose wavelength and unit direction vector are $\lambda_{pi} + \delta\lambda_{off}$ and $\mathbf{e}_{di}$, respectively. In this case, the off-Bragg vector $\boldsymbol{\Delta}\mathbf{K}$ can be written as

$$\boldsymbol{\Delta}\mathbf{K} \equiv k_w\left(\mathbf{G}_i + \boldsymbol{\delta}\mathbf{e}_{Goff}\right) - \frac{2\pi n_p}{\lambda_{pi} + \delta\lambda_{off}}\left(\mathbf{e}_{di} - \mathbf{e}_p\right) \approx k_w\left(\boldsymbol{\delta}\mathbf{e}_{Goff} + \frac{\delta\lambda_{off}}{\lambda_{p0}}\mathbf{G}_0\right), \tag{46}$$

where we also assumed that the relation $\delta\lambda_{off} << \lambda_{pi}$ holds. The grating deviation vector $\boldsymbol{\delta}\mathbf{e}_{Goff}$ can be divided into two classes according to the origin. One is related to the grating that is formed by a different signal component $\mathbf{e}_{si} + \boldsymbol{\delta}\mathbf{e}_{soff}$. The other corresponds to the grating recorded in a different multiplexed page. Thus, $\boldsymbol{\delta}\mathbf{e}_{Goff}$ can be expressed as

$$\boldsymbol{\delta}\mathbf{e}_{Goff} = \boldsymbol{\delta}\mathbf{e}_{soff} + \boldsymbol{\delta}\mathbf{e}_{cy} + \boldsymbol{\delta}\mathbf{e}_{cz}, \tag{47}$$

where $\boldsymbol{\delta}\mathbf{e}_{cz}$ and $\boldsymbol{\delta}\mathbf{e}_{cy}$ are the displacement vectors due to the crystal rotation, defined by Eqs. (33) and (39), respectively. Then the Cartesian components of Eq. (46) can be written as

$$\begin{bmatrix} \Delta K_x \\ \Delta K_y \\ \Delta K_z \end{bmatrix} = k_w \begin{pmatrix} 0 & \cos\beta_{s0} & 0 & 0 & 2\sin\beta_{s0} \\ 1 & 0 & 0 & 2\sin\beta_{s0} & 0 \\ 0 & -\sin\beta_{s0} & -2\cos\phi_{cz}\sin\beta_{s0} & 0 & 0 \end{pmatrix} \begin{bmatrix} \delta\alpha_{soff} \\ \delta\beta_{soff} \\ \delta\phi_{cy} \\ \delta\phi_{cz} \\ \delta\lambda_{off}/\lambda_{p0} \end{bmatrix}, \tag{48}$$

where $\delta\alpha_{soff}$ and $\delta\beta_{soff}$ are the off-Bragg deviation angles that are related to $\boldsymbol{\delta}\mathbf{e}_{soff}$ through Eq. (15). If we assume that the modulation of the grating is weak enough to validate the Born approximation, the intensity of the off-Bragg diffracted wave can be expressed as (Barbastathis & Psaltis, 2000)

$$I_{off} \propto \left[\operatorname{sinc}\left(\frac{\Delta K_x L_x}{2}\right)\cdot\operatorname{sinc}\left(\frac{\Delta K_y L_y}{2}\right)\cdot\operatorname{sinc}\left(\frac{\Delta K_z L_z}{2}\right)\right]^2 I_p\left(\lambda_{pi} + \delta\lambda_{off}\right)d\lambda, \tag{49}$$

where L_x, L_y, and L_z are the hologram dimensions in the x, y, and z directions, respectively;

$$\mathrm{sinc}(x) \equiv \frac{\sin(x)}{x} \tag{50}$$

is the sinc function;

$$I_p(\lambda) = \frac{I_{p0}}{\Delta\lambda_p}\mathrm{rect}\left(\frac{\lambda - \lambda_{p0}}{\Delta\lambda_p}\right), \tag{51}$$

is the spectral intensity of the probe beam whose central wavelength, full spectral width, and intensity are λ_{p0}, $\Delta\lambda_p$, and I_{p0}, respectively; and rect(x) is the rectangle function, defined as

$$\mathrm{rect}(x) = \begin{cases} 1 & |x| < 1/2 \\ 1/2 & |x| = 1/2 \\ 0 & |x| > 1/2 \end{cases}. \tag{52}$$

Equation (49) indicates that the off-Bragg diffraction will occur at **ΔK** around the origin. The diffracted intensity will become zero when the relation

$$|\Delta K_i| = \frac{2\pi}{L_i} \quad (i = x, y, z) \tag{53}$$

holds at any component of the off-Bragg vector **ΔK**. Thus, hereafter, we assume that Eq. (53) represents an upper limit of $|\Delta K_i|$ in which the grating $\mathbf{G}_i + \boldsymbol{\delta}\mathbf{e}_{Goff}$ can produce off-Bragg diffraction.

3.5.1 Properties of signal diffraction

Before considering the crosstalk noise, let us first discuss the prpperties of signal diffraction from a target grating $\mathbf{G}_i$. In this case, we can set $\delta\alpha_{soff} = \delta\beta_{soff} = \delta\phi_{cy} = \delta\phi_{cz} = 0°$. Then the off-Bragg vector in Eq. (48) can be simplified to

$$\begin{bmatrix} \Delta K_x \\ \Delta K_y \\ \Delta K_z \end{bmatrix} = k_w \begin{pmatrix} 2\sin\beta_{s0}\,\delta\lambda_{off}/\lambda_{p0} \\ 0 \\ 0 \end{pmatrix}. \tag{54}$$

From Eq. (49), the intensity of the signal diffracted wave at wavelength $\lambda_{pi} + \delta\lambda_{off}$ is written as

$$I_{off} \propto \mathrm{sinc}^2\left(\frac{\pi\delta\lambda_{off}}{\delta\lambda_{offMax}}\right) I_p\left(\lambda_{pi} + \delta\lambda_{off}\right) d\lambda, \tag{55}$$

where

$$\delta\lambda_{offMax} = \frac{\lambda_w \lambda_{p0}}{2n_w \sin\beta_{s0} L_x} \tag{56}$$

is the half spectral width of the signal diffracted wave. Therefore, a target grating $\mathbf{G}_i$ produces the signal diffraction with a certain spectral width which depends on the hologram dimension L_x. The total intensity of the signal diffracted wave (I_{dif}) can be calculated by integrating Eq. (55). The result is given by

$$I_{dif} \propto \int_{-\infty}^{\infty} \mathrm{sinc}^2\left(\frac{\pi\delta\lambda_{off}}{\delta\lambda_{offMax}}\right) I_p\left(\lambda_{pi} + \delta\lambda_{off}\right) d\left(\delta\lambda_{off}\right) \approx \frac{\delta\lambda_{offMax}}{\Delta\lambda_p} I_{p0}, \tag{57}$$

where we assumed that the probe spectral width $\Delta\lambda_p$ is much larger than $\delta\lambda_{offMax}$, meaning that $I_p(\lambda)$ is considered to be a constant over the region where the integrand is appreciable, and we used the relation

$$\int_{-\infty}^{\infty} \mathrm{sinc}^2\left(x\right) dx = \pi. \tag{58}$$

Note that if $\Delta\lambda_p \to 0$, then $I_{dif} \to I_{p0}$. Thus, the diffraction efficiency in the PCR method is a factor of $\delta\lambda_{offMax}/\Delta\lambda_p$ smaller than that in monochromatic readout. This is because the grating component $\mathbf{G}_i$ can diffract only the limited spectral component around λ_{pi}. Such a reduction in the diffraction efficiency is an unavoidable drawback of the PCR method.

3.5.2 Intra-page crosstalk noise

In this section, we will consider the intra-page crosstalk noise which is derived from the unwanted diffraction coming from the same page but a different grating component, $\mathbf{G}_i + \delta\mathbf{e}_{soff}$. In this case, we can set $\delta\phi_{cy} = \delta\phi_{cz} = 0°$, and then the off-Bragg vector in Eq. (48) can be rewritten as

$$\begin{bmatrix} \Delta K_x \\ \Delta K_y \\ \Delta K_z \end{bmatrix} = k_w \begin{bmatrix} \cos\beta_{s0}\delta\beta_{soff} + 2\sin\beta_{s0}\,\delta\lambda_{off}/\lambda_{p0} \\ \delta\alpha_{soff} \\ -\sin\beta_{s0}\delta\beta_{soff} \end{bmatrix}. \tag{59}$$

The off-Bragg diffraction will occur only when a set of ($\delta\alpha_{soff}$, $\delta\lambda_{off}$) satisfies all of the following inequalities:

$$-\frac{\lambda_w}{n_w \cos\beta_{s0} L_x} \le \delta\beta_{soff} + 2\tan\beta_{s0}\frac{\delta\lambda_{off}}{\lambda_{p0}} \le \frac{\lambda_w}{n_w \cos\beta_{s0} L_x}, \tag{60}$$

$$-\frac{\lambda_w}{n_w L_y} \le \delta\alpha_{soff} \le \frac{\lambda_w}{n_w L_y}, \tag{61}$$

$$-\frac{\lambda_w}{n_w \sin\beta_{s0} L_z} \le \delta\beta_{soff} \le \frac{\lambda_w}{n_w \sin\beta_{s0} L_z}. \tag{62}$$

On the other hand, the variable ranges of $\delta\alpha_{soff}$, $\delta\beta_{soff}$, and $\delta\lambda_{off}$ are also restricted by the input image size or the probe spectrum, which are given by

$$-\delta\alpha_{sMax} \le \delta\alpha_{si} + \delta\alpha_{soff} \le \delta\alpha_{sMax}$$
$$\Leftrightarrow -\delta\alpha_{si} - \delta\alpha_{sMax} \le \delta\alpha_{soff} \le -\delta\alpha_{si} + \delta\alpha_{sMax} \tag{63}$$

$$-\delta\beta_{sMax} \le \delta\beta_{si} + \delta\beta_{soff} \le \delta\beta_{sMax}$$
$$\Leftrightarrow -\delta\beta_{si} - \delta\beta_{sMax} \le \delta\beta_{soff} \le -\delta\beta_{si} + \delta\beta_{sMax} \tag{64}$$

$$\lambda_{p0} - \frac{\Delta\lambda_p}{2} \le \lambda_{p0} + \delta\lambda_{pi} + \delta\lambda_{off} \le \lambda_{p0} + \frac{\Delta\lambda_p}{2}$$
$$\Leftrightarrow -\delta\lambda_{pi} - \frac{\Delta\lambda_p}{2} \le \delta\lambda_{off} \le -\delta\lambda_{pi} + \frac{\Delta\lambda_p}{2}, \tag{65}$$

where $\delta\lambda_{pi}$ is the difference between λ_{pi} and λ_{p0}. Note that the above ranges depend on the targeting position (x_{si}, y_{si}) within the input image.

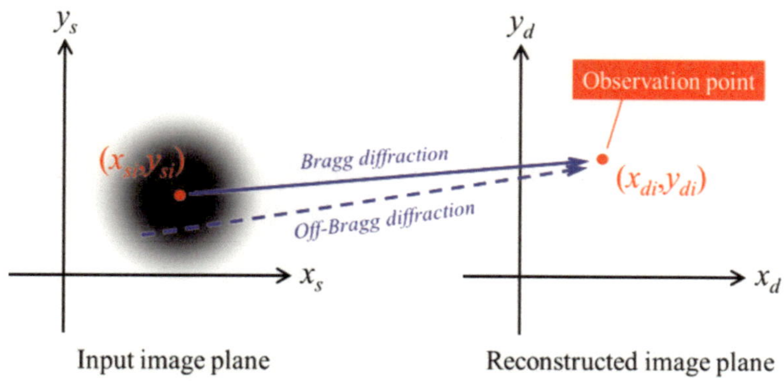

Fig. 14. Schematic diagram of the intra-page crosstalk noise. The off-Bragg diffracted waves coming from other input positions will reach the same position as the signal diffracted wave.

Our goal in this section is to find an upper limit of the off-Bragg deviation angles ($\delta\alpha_{soff}$, $\delta\beta_{soff}$). Such maximum angles specify an input image area suffering from intra-page crosstalk noise, as shown in Fig. 14. First, we will consider the conventional monochromatic case because it can be understood more easily. Since the probe beam is monochromatic, we can set $\delta\lambda_{off} = 0$. Then, the maximum off-Bragg deviation angles $\delta\alpha_{soffMax(mono)}$ and $\delta\beta_{soffMax(mono)}$ are obtained from Eqs. (60), (61), and (62):

$$\delta\alpha_{soffMax(mono)} = \frac{\lambda_w}{n_w L_y} \tag{66}$$

$$\delta\beta_{soffMax(mono)} = \min\left[\frac{\lambda_w}{n_w \cos\beta_{s0} L_x}, \frac{\lambda_w}{n_w \sin\beta_{s0} L_z}\right]. \tag{67}$$

Note that $\delta\alpha_{soffMax(mono)}$ and $\delta\beta_{soffMax(mono)}$ are usually much smaller than $\delta\alpha_{sMax}$ and $\delta\beta_{sMax}$ and satisfy Eqs. (63) and (64), respectively.

Next, we move to the polychromatic case. In this case, the variable range of $\delta\lambda_{off}$ is appreciable owing to the large spectral width $\Delta\lambda_p$, as expressed in Eq. (65). On the other hand, in order to satisfy Eqs. (60) and (62) at the same $\delta\beta_{soff}$, there should exist a common region in the two inequalities; that is,

$$\begin{cases} -\dfrac{\lambda_w}{n_w \sin\beta_{s0} L_z} \leq -2\tan\beta_{s0}\dfrac{\delta\lambda_{off}}{\lambda_{p0}} + \dfrac{\lambda_w}{n_w \cos\beta_{s0} L_x} \\ -2\tan\beta_{s0}\dfrac{\delta\lambda_{off}}{\lambda_{p0}} - \dfrac{\lambda_w}{n_w \cos\beta_{s0} L_x} \leq \dfrac{\lambda_w}{n_w \sin\beta_{s0} L_z} \end{cases}$$
$$\Leftrightarrow \left|\frac{\delta\lambda_{off}}{\lambda_{p0}}\right| \leq \frac{\lambda_w}{2n_w \tan\beta_{s0}}\left(\frac{1}{\cos\beta_{s0}L_x} + \frac{1}{\sin\beta_{s0}L_z}\right). \tag{68}$$

The off-Bragg diffraction will occur at $\lambda_{pi} + \delta\lambda_{off}$ only when $\delta\lambda_{off}$ satisfies both Eqs. (65) and (68). However, because $\Delta\lambda_p$ is sufficiently large, every $\delta\lambda_{off}$ within Eq. (68) will always satisfy Eq. (65). Therefore, Eq. (68) could be considered a sufficient condition for $\delta\lambda_{off}$. Similarly, if we choose a proper $\delta\lambda_{off}$ from Eq. (68), every $\delta\beta_{soff}$ within Eq. (62) will always satisfy Eq. (60). Thus, Eq. (62) is considered to be a sufficient condition for $\delta\beta_{soff}$. Consequently, the variable ranges of $\delta\alpha_{soff}$ and $\delta\beta_{soff}$ are determined by Eqs. (61) and (62), respectively. Thus the maximum off-Bragg deviation angles $\delta\alpha_{soffMax(PCR)}$ and $\delta\beta_{soffMax(PCR)}$ are given by

$$\delta\alpha_{soffMax(PCR)} = \frac{\lambda_w}{n_w L_y}, \tag{69}$$

$$\delta\beta_{soffMax(PCR)} = \frac{\lambda_w}{n_w \sin\beta_{s0} L_z}. \tag{70}$$

The above equations imply that the intra-page crosstalk noise in the PCR method does not depend on the hologram dimension L_x. This feature of the PCR method is clearly shown in Fig. 15, where the normalized diffracted intensity is plotted as a function of the position (x_s, y_s) at several values of L_x. While the input image area suffering from the crosstalk noise shrinks with increasing L_x in the conventional monochromatic case, that in the PCR method is unchanged.

3.5.3 Inter-page crosstalk noise

In this section, we will consider the inter-page crosstalk noise which is caused by the unwanted diffraction coming from a grating component in another multiplexed page, i.e., $\mathbf{G}_i + \boldsymbol{\delta}\mathbf{e}_{soff} + \boldsymbol{\delta}\mathbf{e}_{cz} + \boldsymbol{\delta}\mathbf{e}_{cy}$. Similar to the previous section, we obtain the set of inequalities given by

$$-2\tan\beta_{s0}\frac{\delta\lambda_{off}}{\lambda_{p0}} - \frac{\lambda_w}{n_w \cos\beta_{s0} L_x} \leq \delta\beta_{soff} \leq -2\tan\beta_{s0}\frac{\delta\lambda_{off}}{\lambda_{p0}} + \frac{\lambda_w}{n_w \cos\beta_{s0} L_x}, \tag{71}$$

$$\frac{\delta\alpha_{soff}}{2\sin\beta_{s0}} - \frac{\lambda_w}{2n_w \sin\beta_{s0} L_y} \leq \delta\phi_{cz} \leq \frac{\delta\alpha_{soff}}{2\sin\beta_{s0}} + \frac{\lambda_w}{2n_w \sin\beta_{s0} L_y}, \tag{72}$$

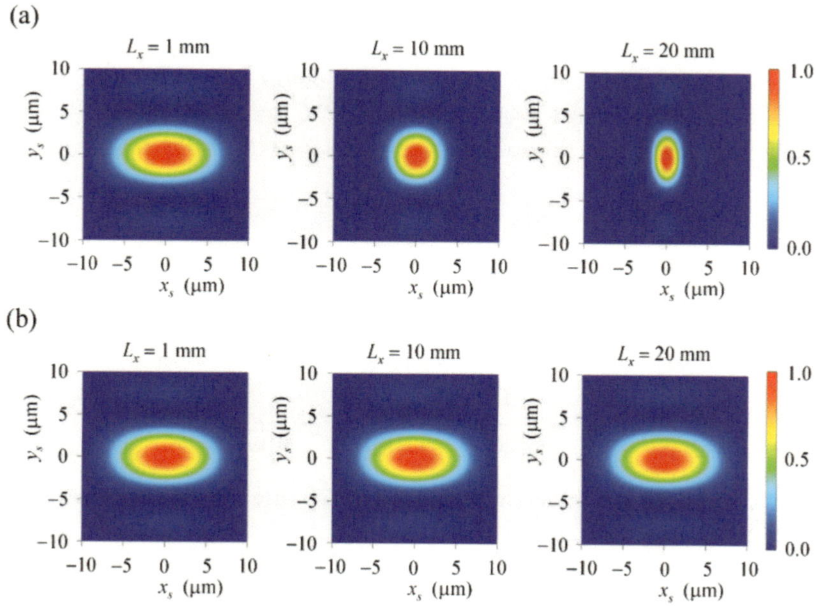

Fig. 15. The input image area suffering from intra-page crosstalk noise in (a) the conventional monochromatic case ($\lambda_w = \lambda_{p0} = 532$ nm, $\Delta\lambda_p = 0$ nm) and (b) the PCR method ($\lambda_w = 532$ nm, $\lambda_{p0} = 815$ nm, $\Delta\lambda_p = 50$ nm). The calculation parameters are as follows: $\beta_{s0} = 30°$; $n_w = 1$; $f_s = 100$ mm; and $L_y = L_z = 10$ mm.

$$-\frac{\delta\beta_{soff}}{2\cos\phi_{cz}} - \frac{\lambda_w}{2n_w\cos\phi_{cz}\sin\beta_{s0}L_z} \le \delta\phi_{cy} \le -\frac{\delta\beta_{soff}}{2\cos\phi_{cz}} + \frac{\lambda_w}{2n_w\cos\phi_{cz}\sin\beta_{s0}L_z}. \tag{73}$$

Note again that the off-Bragg diffraction will occur only when the set of ($\delta\alpha_{soff}$, $\delta\beta_{soff}$, $\delta\phi_{cy}$, $\delta\phi_{cz}$, $\delta\lambda_{off}$) satisfies all of the above inequalities. First, let us consider Eq. (72) to find the rotation angle $\delta\phi_{cz}$ that is large enough to avoid inter-page crosstalk noise. Since the variable range of $\delta\alpha_{soff}$ is expressed as Eq. (63), the off-Bragg diffraction will occur whenever $\delta\phi_{cz}$ satisfies the following inequality:

$$-\frac{\delta\alpha_{sMax} + \delta\alpha_{si}}{2\sin\beta_{s0}} - \frac{\lambda_w}{2n_w\sin\beta_{s0}L_y} \le \delta\phi_{cz} \le \frac{\delta\alpha_{sMax} - \delta\alpha_{si}}{2\sin\beta_{s0}} + \frac{\lambda_w}{2n_w\sin\beta_{s0}L_y}. \tag{74}$$

The above inequality is an expression that is applicable at one particular position (x_{si}, y_{si}) within the input image. In order to avoid crosstalk at every position, the peristrophic rotation angles $\delta\phi_{cz}$ should be set larger than $\delta\phi_{czMin(off)}$; that is,

$$\delta\phi_{czMin(off)} = \frac{\delta\alpha_{sMax}}{\sin\beta_{s0}} + \frac{\lambda_w}{2n_w\sin\beta_{s0}L_y}, \tag{75}$$

where we used the relation $-\delta\alpha_{sMax} \le \delta\alpha_{si} \le \delta\alpha_{sMax}$. The right side in Eq. (75) consists of two parts: one is the minimum peristrophic rotation angle $\delta\phi_{czMin}$ obtained in Section 3.4.1; and the other is the contribution from the off-Bragg diffraction. However, the latter is usually much smaller than the former, and thus the influence of the off-Bragg diffraction can be neglected in the peristrophic multiplexing. Moreover we should also note that Eq. (75) is applicable in both the monochromatic case and the PCR method. This is consistent with the result in Fig. 11.

Next, we will focus on Eqs. (71) and (73) to find the rotation angle $\delta\phi_{cy}$ that is large enough to avoid crosstalk noise. Before considering the case of PCR, we will again treat the conventional monochromatic readout first. In this case, we can set $\delta\lambda_{off} = 0$. Thus, Eq. (71) is simplified to

$$\left|\delta\beta_{soff}\right| \leq \frac{\lambda_w}{n_w \cos\beta_{s0} L_x}. \tag{76}$$

Note that every $\delta\beta_{soff}$ within Eq. (76) always satisfies Eqs. (64). From Eqs. (73) and (76), the off-Bragg diffraction will occur whenever $\delta\phi_{cy}$ satisfies the following inequality:

$$\left|\delta\phi_{cy}\right| \leq \frac{\lambda_w}{2n_w \cos\phi_{cz}} \left(\frac{1}{\cos\beta_{s0} L_x} + \frac{1}{\sin\beta_{s0} L_z} \right). \tag{77}$$

Conversely, in order to avoid inter-page crosstalk noise, the crystal rotation angle should be set larger than $\delta\phi_{cyMin(mono)}$; that is,

$$\delta\phi_{cyMin(mono)} = \frac{\lambda_w}{2n_w \cos\phi_{cz}} \left(\frac{1}{\cos\beta_{s0} L_x} + \frac{1}{\sin\beta_{s0} L_z} \right). \tag{78}$$

Now, let us return to the case of the PCR method. Due to the wide spectral width $\Delta\lambda_p$, the restriction of $\delta\beta_{soff}$ in Eq. (71) is considerably relaxed as compared with the monochromatic case; that is,

$$\begin{cases} \tan\beta_{s0} \dfrac{\left(2\delta\lambda_{pi} - \Delta\lambda_p\right)}{\lambda_{p0}} - \dfrac{\lambda_w}{n_w \cos\beta_{s0} L_x} \leq \delta\beta_{soff} \\ \delta\beta_{soff} \leq \tan\beta_{s0} \dfrac{\left(2\delta\lambda_{pi} + \Delta\lambda_p\right)}{\lambda_{p0}} + \dfrac{\lambda_w}{n_w \cos\beta_{s0} L_x} \end{cases}. \tag{79}$$

On the other hand, the variable range of $\delta\beta_{soff}$ is also limited by the input image size, as is expressed in Eq. (64). Since the off-Bragg diffraction will occur only when $\delta\beta_{soff}$ satisfies both Eqs. (64) and (79), the sufficient condition for $\delta\beta_{soff}$ to produce the off-Bragg diffraction is given by

$$\begin{cases} \max\left[\tan\beta_{s0} \dfrac{\left(2\delta\lambda_{pi} - \Delta\lambda_p\right)}{\lambda_{p0}} - \dfrac{\lambda_w}{n_w \cos\beta_{s0} L_x}, -\delta\beta_{si} - \delta\beta_{sMax} \right] \leq \delta\beta_{soff} \\ \delta\beta_{soff} \leq \min\left[\tan\beta_{s0} \dfrac{\left(2\delta\lambda_{pi} + \Delta\lambda_p\right)}{\lambda_{p0}} + \dfrac{\lambda_w}{n_w \cos\beta_{s0} L_x}, -\delta\beta_{si} + \delta\beta_{sMax} \right] \end{cases}, \tag{80}$$

where max[a, b] is a maximum function yielding the larger value between a and b. If we assume that $\alpha_p = 0$ and take the spectral width required for the full image reconstruction $\Delta\lambda_{BM}$ of Eq. (25) as the probe spectral width $\Delta\lambda_p$, then Eq. (80) can be modified to

$$\begin{cases} -\left(1+\dfrac{\tan\beta_{s0}}{\tan\beta_p}\right)(\delta\beta_{si}+\delta\beta_{sMax})-\dfrac{\lambda_w}{n_w\cos\beta_{s0}L_x}\le\delta\beta_{soff} \\ \delta\beta_{soff}\le-\left(1+\dfrac{\tan\beta_{s0}}{\tan\beta_p}\right)(\delta\beta_{si}-\delta\beta_{sMax})+\dfrac{\lambda_w}{n_w\cos\beta_{s0}L_x} \end{cases}, \tag{81}$$

where we used the relation of Eq. (20) in the derivation. From Eqs. (73) and (81), the off-Bragg diffraction will occur whenever $\delta\phi_{cy}$ satisfies the following inequality:

$$\begin{cases} \left(1+\dfrac{\tan\beta_{s0}}{\tan\beta_p}\right)\dfrac{(\delta\beta_{si}-\delta\beta_{sMax})}{2\cos\phi_{cz}}-\delta\phi_{cyMin(mono)}\le\delta\phi_{cy} \\ \delta\phi_{cy}\le\left(1+\dfrac{\tan\beta_{s0}}{\tan\beta_p}\right)\dfrac{(\delta\beta_{si}+\delta\beta_{sMax})}{2\cos\phi_{cz}}+\delta\phi_{cyMin(mono)} \end{cases}. \tag{82}$$

The above inequality is an expression that is applicable at one particular position (x_{si}, y_{si}) within the input image. In order to avoid crosstalk at every position, the crystal rotation angles $\delta\phi_{cy}$ should be set larger than $\delta\phi_{cyMin(PCR)}$; that is,

$$\delta\phi_{cyMin(PCR)}=\left(1+\frac{\tan\beta_{s0}}{\tan\beta_p}\right)\frac{\delta\beta_{sMax}}{\cos\phi_{cz}}+\delta\phi_{cyMin(mono)}, \tag{83}$$

where we used the relation $-\delta\beta_{sMax}\le\delta\beta_{si}\le\delta\beta_{sMax}$. The right side in Eq. (83) consists of the minimum crystal rotation angle $\delta\phi_{cyMin}$ obtained in Section 3.4.2 and the contribution from the off-Bragg diffraction $\delta\phi_{cyMin(mono)}$. Similarly to the peristrophic rotation, the latter is usually much smaller than the former, and thus the influence of the off-Bragg diffraction can be ignored also in the crystal angle multiplexing.

3.6 Storage capacity

Since a single point on the input image corresponds to one particular plane wave in the recording medium, a Fourier hologram stores the information in such a way that one data bit is stored at one localized point in frequency space, and thus, nonlocally in real space. However, due to the off-Bragg diffraction, one data point in frequency space will reproduce an image with a finite dimension in the reconstructed image plane. In order to distinguish two different bits, corresponding data points in frequency space must maintain a certain interval large enough to avoid crosstalk. Therefore, one data bit in real space actually occupies a finite volume in frequency space. If such a volume is denoted by V_{K1bit}, the storage capacity of the holographic memory (N_{bit}) is expressed as

$$N_{bit}=\frac{V_{KG}}{V_{K1bit}}, \tag{84}$$

where V_{KG} is the volume in frequency space which the grating vector can attain through multiplexing. Note that, if the readout wavelength is different from the recording one, as is in the present case, there exists a grating vector that cannot satisfy the Bragg condition with that probe beam. Such a grating vector should be removed from V_{KG}, because the data stored there cannot be read out. Therefore, the theoretical limit of the storage capacity is ultimately determined by the readout wavelength not the recording one.

Although V_{K1bit} and V_{KG} are generally different in different multiplexing methods, we will consider here the crystal angle multiplexing combined with the peristrophic rotation discussed in Section 3.4. Furthermore, for simplicity, we assume $\alpha_p = 0°$ in this section. First, let us consider V_{KG} after the crystal angle multiplexing around the axis inclined at ϕ_{cz}. The matrix that rotates at ϕ_{cz} around the z-axis, and subsequently rotates at ϕ_{cy} around the y-axis, is expressed as

$$\mathbf{R}\left(\phi_{cy},\phi_{cz}\right)=\begin{pmatrix} \cos\phi_{cy}\cos\phi_{cz} & \cos\phi_{cy}\sin\phi_{cz} & \sin\phi_{cy} \\ -\sin\phi_{cz} & \cos\phi_{cz} & 0 \\ -\sin\phi_{cy}\cos\phi_{cz} & -\sin\phi_{cy}\sin\phi_{cz} & \cos\phi_{cy} \end{pmatrix}. \tag{85}$$

The grating vector $k_w(\mathbf{G}_0 + \boldsymbol{\delta}\mathbf{e}_{si})$ operated on by that matrix is written as

$$\begin{aligned}\mathbf{R}\left(\phi_{cy},\phi_{cz}\right)k_w\left(\mathbf{G}_0+\boldsymbol{\delta}\mathbf{e}_{si}\right)&=2k_w\sin\beta_{s0}\begin{bmatrix}\cos\phi_{cy}\cos\phi_{cz}\\ -\sin\phi_{cz}\\ -\sin\phi_{cy}\cos\phi_{cz}\end{bmatrix}+k_w\begin{bmatrix}\cos\phi_{cy}\sin\phi_{cz}\\ \cos\phi_{cz}\\ -\sin\phi_{cy}\sin\phi_{cz}\end{bmatrix}\delta\alpha_{si}\\ &+k_w\begin{bmatrix}\cos\beta_{s0}\cos\phi_{cy}\cos\phi_{cz}-\sin\beta_{s0}\sin\phi_{cy}\\ -\cos\beta_{s0}\sin\phi_{cz}\\ -\cos\beta_{s0}\sin\phi_{cy}\cos\phi_{cz}-\sin\beta_{s0}\cos\phi_{cy}\end{bmatrix}\delta\beta_{si}.\end{aligned} \tag{86}$$

If above grating is further rotated around the y-axis by a small angle $\delta\phi_{cy}$, we obtain the differential grating vector ($\boldsymbol{\delta}\mathbf{K}_G$) resulting from small shifts of $\delta\alpha_s$, $\delta\beta_s$, and $\delta\phi_{cy}$:

$$\begin{aligned}&\boldsymbol{\delta}\mathbf{K}_G=k_w\left(\delta\alpha_s\boldsymbol{\delta}\mathbf{e}_{K\alpha s}+\delta\beta_s\boldsymbol{\delta}\mathbf{e}_{K\beta s}+\delta\varphi_{cy}\boldsymbol{\delta}\mathbf{e}_{Kcy}\right),\\ &\boldsymbol{\delta}\mathbf{e}_{K\alpha s}=\begin{bmatrix}\cos\phi_{cy}\sin\phi_{cz}\\ \cos\phi_{cz}\\ -\sin\phi_{cy}\sin\phi_{cz}\end{bmatrix},\ \boldsymbol{\delta}\mathbf{e}_{K\beta s}=\begin{bmatrix}\cos\beta_{s0}\cos\phi_{cy}\cos\phi_{cz}-\sin\beta_{s0}\sin\phi_{cy}\\ -\cos\beta_{s0}\sin\phi_{cz}\\ -\cos\beta_{s0}\sin\phi_{cy}\cos\phi_{cz}-\sin\beta_{s0}\cos\phi_{cy}\end{bmatrix},\\ &\boldsymbol{\delta}\mathbf{e}_{Kcy}=-2\sin\beta_{s0}\begin{bmatrix}\sin\phi_{cy}\cos\phi_{cz}\\ 0\\ \cos\phi_{cy}\cos\phi_{cz}\end{bmatrix}.\end{aligned} \tag{87}$$

Thus, the differential volume in frequency space is given by

$$\begin{aligned}dV_K&=k_w^{\ 3}\left[\left(\boldsymbol{\delta}\mathbf{e}_{K\alpha s}\times\boldsymbol{\delta}\mathbf{e}_{K\beta s}\right)\cdot\boldsymbol{\delta}\mathbf{e}_{Kcy}\right]\delta\alpha_s\delta\beta_s\delta\phi_{cy}\\ &=k_w^{\ 3}\cos\phi_{cz}\sin 2\beta_{s0}\delta\alpha_s\delta\beta_s\delta\phi_{cy}.\end{aligned} \tag{88}$$

When we assumed that $\delta\alpha_s$, $\delta\beta_s$, and $\delta\phi_{cy}$ are in the ranges $-\delta\alpha_{sMax} < \delta\alpha_s < \delta\alpha_{sMax}$, $-\delta\beta_{sMax} < \delta\beta_s < \delta\beta_{sMax}$, and $-\delta\phi_{cyMax} < \delta\phi_{cy} < \delta\phi_{cyMax}$, respectively, then V_{KG} after crystal angle multiplexing around the axis inclined at ϕ_{cz}, which is denoted by $V_{KG\_\phi cz}$, can be written as

$$
\begin{aligned}
V_{KG\_\phi cz}(\phi_{cz}) &= \int dV_K \\
&= k_w{}^3 \cos\phi_{cz} \sin 2\beta_{s0} \int_{-\delta\phi_{cyMax}}^{\delta\phi_{cyMax}} d\delta\phi_{cy} \int_{-\delta\alpha_{s\max}}^{\delta\alpha_{s\max}} d\delta\alpha_s \int_{-\delta\beta_{s\max}}^{\delta\beta_{s\max}} d\delta\beta_s \\
&= 8k_w{}^3 \cos\phi_{cz} \sin 2\beta_{s0} \delta\alpha_{s\max} \delta\beta_{s\max} \delta\phi_{cyMax} .
\end{aligned}
\tag{89}
$$

Next, let us consider this in combination with peristrophic multiplexing. In this case, crystal angle multiplexing is performed at each inclined angle ϕ_{czn}; that is,

$$
\phi_{czn} = n\delta\phi_{czMin} = n\frac{\delta\alpha_{s\max}}{\sin\beta_{s0}} \quad (n = -n_{\max}, \cdots -2, -1, 0, 1, 2, \cdots n_{\max}),
\tag{90}
$$

where $\delta\phi_{czMin}$ is the minimum angular separation in peristrophic multiplexing of Eq. (38); n is an integer; and n_{max} is the maximum integer that satisfies the relation

$$
\phi_{czMax} \geq n_{\max} \frac{\delta\alpha_{s\max}}{\sin\beta_{s0}},
\tag{91}
$$

where ϕ_{czMax} is a maximum peristrophic rotation angle. If we take the summation of $\cos\phi_{czn}$ over all n, we obtain

$$
\begin{aligned}
\sum_{n=-n_{\max}}^{n_{\max}} \cos\phi_{czn} &= \sum_{n=-n_{\max}}^{n_{\max}} \cos\left(n\frac{\delta\alpha_{s\max}}{\sin\beta_{s0}} \right) \\
&= \frac{\cos\left(\dfrac{\delta\alpha_{s\max}}{2\sin\beta_{s0}}\right)}{\sin\left(\dfrac{\delta\alpha_{s\max}}{2\sin\beta_{s0}}\right)} \sin\left[\frac{\delta\alpha_{s\max}}{\sin\beta_{s0}}(n_{\max}+1)\right] - \cos\left[\frac{\delta\alpha_{s\max}}{\sin\beta_{s0}}(n_{\max}+1)\right] \\
&\sim \frac{2\sin\beta_{s0}}{\delta\alpha_{s\max}} \sin\phi_{czMax} - \cos\phi_{czMax} ,
\end{aligned}
\tag{92}
$$

where we assumed that $\delta\alpha_{sMax} << 1$ and the relation

$$
\phi_{czMax} \sim \frac{\delta\alpha_{s\max}}{\sin\beta_{s0}}(n_{\max}+1)
\tag{93}
$$

approximately holds. Then, V_{KG} after crystal angle multiplexing in combination with peristrophic multiplexing, which is denoted by V_{KG_total}, is expressed as

$$
\begin{aligned}
V_{KG\_total} &= \sum_{n=-n_{\max}}^{n_{\max}} V_{KG\_\phi cz}(\phi_{czn}) \\
&= 8k_w{}^3 \sin 2\beta_{s0} \delta\beta_{s\max} \delta\phi_{cyMax} \left(2\sin\beta_{s0} \sin\phi_{czMax} - \delta\alpha_{s\max} \cos\phi_{czMax} \right)
\end{aligned}
\tag{94}
$$

In particular, if we set $\delta\phi_{cyMax} = \phi_{czMax} = \pi/2$, we obtain

$$
V_{KG\_total} = 8\pi k_w{}^3 \sin\beta_{s0} \sin 2\beta_{s0} \delta\beta_{s\max} .
\tag{95}
$$

We find that V_{KG_total} increases with increasing maximum deviation angle $\delta\beta_{sMax}$. Note that V_{KG} is, of course, independent of the readout scheme, and thus, Eq. (95) can be applied to both the monochromatic case and the PCR method.

Next, we move on to V_{K1bit}. Let $\delta\alpha_{s1bit}$, $\delta\beta_{s1bit}$, and $\delta\phi_{cy1page}$ denote the angular separations needed to distinguish two different bits or pages. Then V_{K1bit} can be written as

$$V_{K1bit} = k_w^{\ 3} \cos\phi_{cz} \sin 2\beta_{s0} \delta\alpha_{s1bit} \delta\beta_{s1bit} \delta\phi_{cy1page} \,. \tag{96}$$

In the monochromatic case, we can adopt $\delta\alpha_{soffMax(mono)}$ in Eq. (66), $\delta\beta_{soffMax(mono)}$ in Eq. (67), and $\delta\phi_{cyMin(mono)}$ in Eq. (78) as $\delta\alpha_{s1bit}$, $\delta\beta_{s1bit}$, and $\delta\phi_{cy1page}$, respectively:

$$\delta\alpha_{s1bit(mono)} = \frac{\lambda_w}{n_w L_y} \tag{97}$$

$$\delta\beta_{s1bit(mono)} = \frac{\lambda_w}{n_w} \min\left[\frac{1}{\cos\beta_{s0} L_x}, \frac{1}{\sin\beta_{s0} L_z} \right] \tag{98}$$

$$\delta\phi_{cy1page(mono)} = \frac{\lambda_w}{2n_w \cos\phi_{cz}} \left(\frac{1}{\cos\beta_{s0} L_x} + \frac{1}{\sin\beta_{s0} L_z} \right) \tag{99}$$

Then, V_{K1bit} for the monochromatic case is given by

$$V_{K1bit(mono)} = \frac{8\pi^3}{L_x L_y L_z} \left(1 + \min\left[\xi, 1/\xi \right] \right), \tag{100}$$

where

$$\xi \equiv \frac{\tan\beta_{s0} L_z}{L_x}. \tag{101}$$

On the other hand, in the case of the PCR method, we similarly adopt $\delta\alpha_{soffMax(PCR)}$ in Eq. (69), $\delta\beta_{soffMax(PCR)}$ in Eq. (70), and $\delta\phi_{cyMin(PCR)}$ in Eq. (83) as $\delta\alpha_{s1bit}$, $\delta\beta_{s1bit}$, and $\delta\phi_{cy1page}$, respectively:

$$\delta\alpha_{s1bit(PCR)} = \frac{\lambda_w}{n_w L_y} \tag{102}$$

$$\delta\beta_{s1bit(PCR)} = \frac{\lambda_w}{n_w \sin\beta_{s0} L_z} \tag{103}$$

$$\begin{aligned} \delta\phi_{cy1page(PCR)} &= \left(1 + \frac{\tan\beta_{s0}}{\tan\beta_p} \right) \frac{\delta\beta_{sMax}}{\cos\phi_{cz}} + \delta\phi_{cy1page(mono)} \\ &\approx \left(1 + \frac{\tan\beta_{s0}}{\tan\beta_p} \right) \frac{\delta\beta_{sMax}}{\cos\phi_{cz}}, \end{aligned} \tag{104}$$

where we assume that the second term at the right side of Eq. (104) is much smaller than the first term and can be neglected. Then V_{K1bit} for the PCR method is given by:

$$V_{K1bit(PCR)} = \frac{16\pi^3 n_w \sin\beta_{s0}}{\lambda_w L_y L_z}\left(\frac{1}{\tan\beta_{s0}} + \frac{1}{\tan\beta_p}\right)\delta\beta_{sMax} \,. \tag{105}$$

Equation (105) does not include L_x but depends on $\delta\beta_{sMax}$ instead. Due to this feature, V_{K1bit} for the PCR method is much larger than that for monochromatic readout. If we rewrite Eq. (105) using $\Delta\lambda_{BM}$ in Eq. (21) and $\delta\lambda_{offMax}$ in Eq. (56), we obtain

$$V_{K1bit(PCR)} = \frac{8\pi^3}{L_x L_y L_z}\frac{\Delta\lambda_{BM}}{\delta\lambda_{offMax}} \tag{106}$$

Therefore, V_{K1bit} for the PCR method is a factor of $\Delta\lambda_{BM} / \delta\lambda_{offMax}$ larger than that in the monochromatic readout. The resultant storage capacity in each readout method is

$$N_{bit(mono)} = \frac{16\pi n_w^{\ 3} L_x L_y L_z}{\lambda_w^{\ 3}}\frac{\sin^3\beta_{s0}\tan\beta_{s0}}{1+\min\left[\xi, 1/\xi\right]}\delta\beta_{sMax} \tag{107}$$

$$N_{bit(PCR)} = \frac{4\pi n_w^{\ 2} L_y L_z}{\lambda_w^{\ 2}}\frac{\sin 2\beta_{s0}}{\left(\dfrac{1}{\tan\beta_{s0}} + \dfrac{1}{\tan\beta_p}\right)} \tag{108}$$

For example, if $\lambda_w = 532$ nm, $\lambda_{p0} = 815$ nm, $\beta_{s0} = 30°$, $n_w = 1$, $\delta\beta_{sMax} = 4.0°$, and $L_x = L_y = L_z = 10$ mm, then $\beta_p = -50°$. The resultant storage capacities $N_{bit(mono)}$ and $N_{bit(PCR)}$ are 3.2 Tbit and 4.3 Gbit, respectively. Therefore, the storage capacity in the PCR method decreases by more than two orders of magnitude.

4. Selective detection method

In the previous section, we see that the PCR method causes a significant decrease in the storage capacity. However, such a problem can be overcome if we employ the selective detection method together with PCR (Fujimura et al., 2010). The method is based on the selective detection of a target signal image that is submerged in noise waves. By inserting a suitable wavelength separator into the reconstructed image plane, we can retrieve the stored information without crosstalk even if the angular separation is not large enough to suppress the noise diffraction. In this section, we will explain how to remove the crosstalk noise, and we derive theoretical limit of the storage capacity after the improvement.

4.1 Basic concept and principle

As seen in Fig. 12, the Bragg-matched wavelength at a certain imaging position (x_d, y_d) will change after the crystal rotation of $\delta\phi_{cy}$. From Eq. (43), the amount of shift of the Bragg-matched wavelength $\delta\lambda_{cy\_xfix}$ is expressed as

$$\delta\lambda_{cy\_xfix} = \frac{\lambda_{p0}}{\tan\beta_{s0}}\cos\phi_{cz}\delta\phi_{cy} \,. \tag{109}$$

If we take advantage of this difference of the Bragg-matched wavelength, it should be possible to detect the signal image alone, even if the noise images overlap with the target signal image. For example, let us consider inserting a special wavelength filter into the reconstructed image plane as shown in Fig. 16, whose transmittance T_{LVF} is represented by

$$T_{LVF}(x_d,\lambda) = \mathrm{rect}\left(\frac{\lambda - \lambda_{BMcy}(x_d, y_d, 0, \phi_{cz})}{\Delta\lambda_{LVF}}\right), \tag{110}$$

where λ is the wavelength; $\Delta\lambda_{LVF}$ is the full width of the transmitting band; and rect(x) is the rectangle function defined in Eq. (52). Note that the transmission spectrum of this filter depends on the illuminated location on the filter; that is, the optical waves passing through different spatial positions will undergo different spectral filtering by this filter. Such a wavelength filter is known as a linear variable filter (LVF) since the spectral shift of the transmission band is proportional to the spatial shift of the illuminated position.

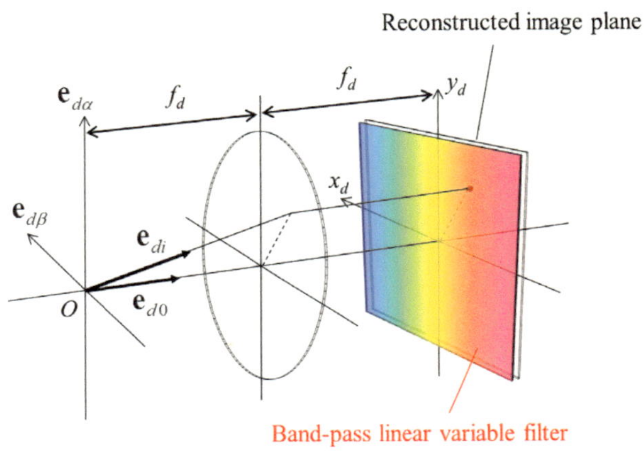

Fig. 16. Configuration for the selective detection method. A special wavelength filter, like a band-pass linear variable filter, is inserted at the reconstructed image plane.

The principle of the selective detection is shown in Fig. 17. Since the transmitting wavelength of the LVF coincides with the spectral dispersion of the target signal image, every diffracted wave that constitutes the signal image can go through the LVF and will be detected by the imager. On the contrary, the LVF will reject the noise diffracted wave whose wavelength lies outside the transmission band of the LVF.

4.2 Theoretical description of the selective detection method

In this section, we will see the influence of the LVF on the properties of the crosstalk noise, and we formulate the achievable storage capacity in the selective detection method. As was seen in Section 3.5.3, the off-Bragg diffraction will occur only when the set of ($\delta\alpha_{soff}$, $\delta\beta_{soff}$, $\delta\phi_{cy}$, $\delta\phi_{cz}$, $\delta\lambda_{off}$) satisfies all of inequalities in Eqs. (71), (72), and (73). Now, due to the transmission band of the LVF, the variable range of $\delta\lambda_{off}$ should be modified to

$$-\frac{\Delta\lambda_{LVF}}{2} \leq \delta\lambda_{off} \leq \frac{\Delta\lambda_{LVF}}{2}. \tag{111}$$

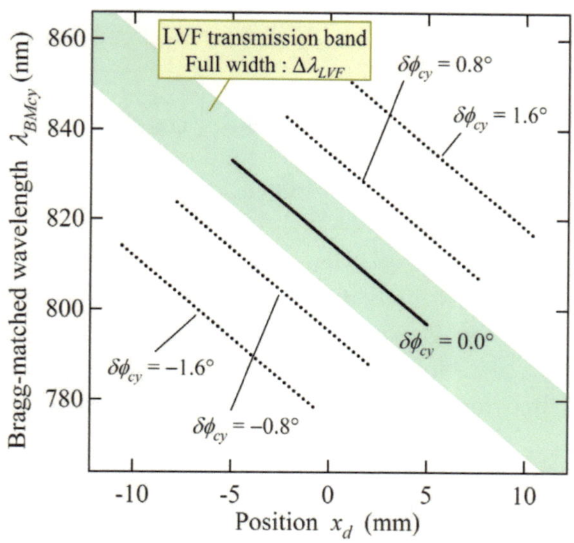

Fig. 17. The concept of the selective detection method. The dotted part of the Bragg-matched line corresponds to the portion of the image that will not be detected by the imager.

Note that the variable range of $\delta\lambda_{off}$ is now independent of the targeting position (x_{si}, y_{si}) within the input image because of the transmitting property of the LVF. In this case, the restriction of $\delta\beta_{soff}$ in Eq. (71) should be changed to

$$\left|\delta\beta_{soff}\right| \le \tan\beta_{s0}\frac{\Delta\lambda_{LVF}}{\lambda_{p0}} + \frac{\lambda_w}{n_w\cos\beta_{s0}L_x}. \tag{112}$$

From Eqs. (73) and (112), we find that the off-Bragg diffraction will occur in the range

$$\left|\delta\phi_{cy}\right| \le \frac{\tan\beta_{s0}}{2\cos\phi_{cz}}\frac{\Delta\lambda_{LVF}}{\lambda_{p0}} + \frac{\lambda_w}{2n_w\cos\phi_{cz}}\left(\frac{1}{\cos\beta_{s0}L_x} + \frac{1}{\sin\beta_{s0}L_z}\right). \tag{113}$$

Therefore, in order to avoid inter-page crosstalk noise, the crystal rotation angle should be set larger than $\delta\phi_{cyMin(SelDet)}$; that is,

$$\begin{aligned}\delta\phi_{cyMin(SelDet)} &= \frac{\tan\beta_{s0}}{2\cos\phi_{cz}}\frac{\Delta\lambda_{LVF}}{\lambda_{p0}} + \frac{\lambda_w}{2n_w\cos\phi_{cz}}\left(\frac{1}{\cos\beta_{s0}L_x} + \frac{1}{\sin\beta_{s0}L_z}\right)\\ &= \frac{\Delta\lambda_{LVF}}{2\Delta\lambda_{BM}}\delta\phi_{cyMin(PCR)} + \delta\phi_{cyMin(mono)},\end{aligned} \tag{114}$$

where $\delta\phi_{cyMin(mono)}$ and $\delta\phi_{cyMin(PCR)}$ are the minimum crystal rotation angle in the monochromatic case and the PCR method, respectively; and we use Eq. (21) to derive the second equality. If we ignore the $\delta\phi_{cyMin(mono)}$, we see that $\delta\phi_{cyMin(SelDet)}$ is improved by a factor of $2\Delta\lambda_{BM}/\Delta\lambda_{LVF}$ than $\delta\phi_{cyMin(PCR)}$. Note that the probe spectral width $\Delta\lambda_p$ need not be equal to the spectral width required for the full image reconstruction $\Delta\lambda_{BM}$ because Eq. (114) no longer includes $\Delta\lambda_p$.

Furthermore, the intra-page crosstalk noise will be also suppressed if $\Delta\lambda_{LVF}$ is sufficiently small. Since the off-Bragg diffraction will occur only when $\delta\beta_{soff}$ satisfies both the inequalities in Eqs. (60) and (62) under the range of Eq. (111), the restriction of $\delta\beta_{soff}$ should be written as

$$\left|\delta\beta_{soff}\right| \le \min\left[\tan\beta_{s0}\frac{\Delta\lambda_{LVF}}{\lambda_{p0}} + \frac{\lambda_w}{n_w\cos\beta_{s0}L_x}, \frac{\lambda_w}{n_w\sin\beta_{s0}L_z}\right]. \tag{115}$$

Therefore, the maximum off-Bragg deviation angles $\delta\alpha_{soffMax(SelDet)}$ and $\delta\beta_{soffMax(SelDet)}$ are obtained as follows:

$$\delta\alpha_{soffMax(SelDet)} = \frac{\lambda_w}{n_w L_y} \tag{116}$$

$$\delta\beta_{soffMax(SelDet)} = \min\left[\tan\beta_{s0}\frac{\Delta\lambda_{LVF}}{\lambda_{p0}} + \frac{\lambda_w}{n_w\cos\beta_{s0}L_x}, \frac{\lambda_w}{n_w\sin\beta_{s0}L_z}\right]. \tag{117}$$

If we adopt $\delta\alpha_{soffMax(SelDet)}$ in Eq.(116), $\delta\beta_{soffMax(SelDet)}$ in Eq. (117), and $\delta\phi_{cyMin(SelDet)}$ in Eq. (114) as $\delta\alpha_{s1bit}$, $\delta\beta_{s1bit}$, and $\delta\phi_{cy1page}$ in Eq. (96), respectively, then V_{K1bit} for the selective detection method is written as

$$V_{K1bit(SelDet)} = \frac{1}{F[\Delta\lambda_{LVF}]}\frac{8\pi^3}{L_xL_yL_z}\left\{1+\min\left[\frac{\xi}{F[\Delta\lambda_{LVF}]},\frac{F[\Delta\lambda_{LVF}]}{\xi}\right]\right\}, \tag{118}$$

where

$$F[\Delta\lambda_{LVF}] \equiv \left(\frac{\Delta\lambda_{LVF}}{2\delta\lambda_{offMax}}+1\right)^{-1} \tag{119}$$

is the improvement factor of the storage capacity. When the condition $\Delta\lambda_{LVF} >> 2\delta\lambda_{offMax}$ holds, Eq. (118) can be simplified to

$$V_{K1bit(SelDet)} = \frac{8\pi^3}{L_xL_yL_z}\frac{\Delta\lambda_{LVF}}{2\delta\lambda_{offMax}}. \tag{120}$$

Comparing this with $V_{K1bit(PCR)}$ in Eq. (106), we find that the storage capacity will be improved by a factor of $2\Delta\lambda_{BM}/\Delta\lambda_{LVF}$. On the other hand, when $\Delta\lambda_{LVF} << 2\delta\lambda_{offMax}$ is satisfied, $V_{K1bit(SelDet)}$ becomes identical to $V_{K1bit(mono)}$. Therefore, the storage capacity in the PCR method will be completely recovered if we use an LVF having a sufficiently small $\Delta\lambda_{LVF}$. However, it should be noted that, in this case, the diffraction efficiency will also decrease with decreasing $\Delta\lambda_{LVF}$. This is because the assumption used to derive Eq. (57) is no longer valid, and the integration range should be modified to $-\Delta\lambda_{LVF}/2 \le \delta\lambda_{off} \le \Delta\lambda_{LVF}/2$:

$$I_{di} \propto \int_{-\frac{\Delta\lambda_{LVF}}{2}}^{\frac{\Delta\lambda_{LVF}}{2}} \mathrm{sinc}^2\left(\frac{\pi\delta\lambda_{off}}{\delta\lambda_{offMax}}\right) I_p\left(\lambda_{pi}+\delta\lambda_{off}\right) d\left(\delta\lambda_{off}\right) \approx \frac{\Delta\lambda_{LVF}}{\Delta\lambda_p} I_{p0}. \tag{121}$$

In order to avoid such reduction in the diffraction efficiency, by considering the spectral width $\delta\lambda_{offMax}$ of the signal diffracted wave, we should set $\Delta\lambda_{LVF} = 2\delta\lambda_{offMax}$. In this case, $V_{K1bit(SelDet)}$ becomes

$$V_{K1bit(SelDet)} = \frac{16\pi^3}{L_xL_yL_z}\left\{1+\min\left[2\xi,\frac{1}{2\xi}\right]\right\}. \tag{122}$$

Since the relation $0 < \min[\xi, 1/\xi] \le 1$ holds, we see that the storage capacity at $\Delta\lambda_{LVF} = 2\delta\lambda_{offMax}$ will reach about half that of the monochromatic case.

From Eqs. (95) and (118), the resultant storage capacity in the selective detection method is

$$N_{bit(selDet)} = \frac{8\pi n_w{}^3 L_xL_yL_z}{\lambda_w{}^3}\frac{F[\Delta\lambda_{LVF}]\sin\beta_{s0}\sin 2\beta_{s0}}{1+\min\left[\frac{\xi}{F[\Delta\lambda_{LVF}]},\frac{F[\Delta\lambda_{LVF}]}{\xi}\right]}\delta\beta_{s\max}. \tag{123}$$

In most cases, $\delta\beta_{sMax}$ is limited by the spectral width $\Delta\lambda_p$ of a given probe light source. Thus, if Eq. (26) is substituted into Eq. (123), we obtain

$$N_{bit(selDet)} = \frac{16\pi n_w^{\ 3} L_x L_y L_z}{\lambda_w^{\ 3}} \frac{F[\Delta\lambda_{LVF}]}{1+\min\left[\dfrac{\xi}{F[\Delta\lambda_{LVF}]}, \dfrac{F[\Delta\lambda_{LVF}]}{\xi}\right]} \frac{\sin^3\beta_{s0}}{\left(1+\dfrac{\tan\beta_{s0}}{\tan\beta_p}\right)} \frac{\Delta\lambda_p}{\lambda_{p0}}. \tag{124}$$

For example, if $\lambda_w = 532$ nm, $\lambda_{p0} = 815$ nm, $\Delta\lambda_p = 50$ nm, $\beta_{s0} = 30°$, $n_w = 1$, $L_x = L_y = L_z = 10$ mm, and $\Delta\lambda_{LVF} = 2\delta\lambda_{offMax} = 0.043$ nm, then the resultant storage capacity $N_{bit(selDet)}$ is 1.3 Tbit. Therefore, in this case, the storage capacity in the selective detection method reaches 40 percent of the theoretical limit for the monochromatic case.

Finally, the storage capacity in Eq. (124) is plotted as a function of $\Delta\lambda_{LVF}$ in Fig. 18. The improvement starts at $\Delta\lambda_{LVF} = 100$ nm, then the storage capacity increases linearly with increasing $\Delta\lambda_{LVF}$, and finally, it asymptotically approaches the theoretical limit of the monochromatic case. Note that the kink observed at $\Delta\lambda_{LVF} = 0.06$ nm is derived from the minimum function; that is to say, from that point on, the storage capacity further increases due to the reduction of the intra-page crosstalk noise $\delta\beta_{soffMax}$ in Eq. (117).

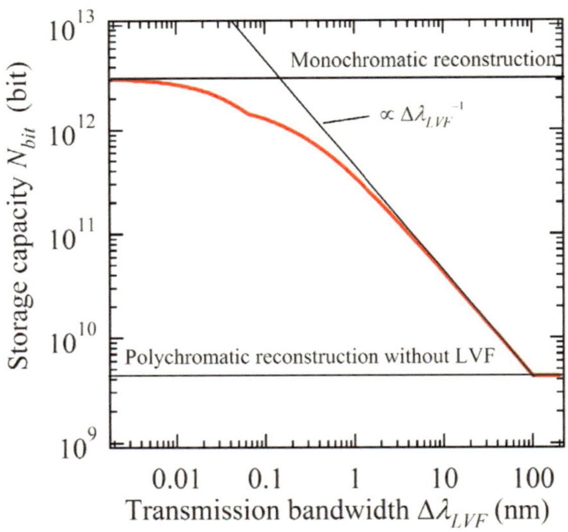

Fig. 18. Storage capacity as a function of transmission bandwidth $\Delta\lambda_{LVF}$. The calculation parameters are as follows: $\lambda_w = 532$ nm, $\lambda_{p0} = 815$ nm, $\Delta\lambda_p = 50$ nm, $\beta_{s0} = 30°$, $n_w = 1$, and $L_x = L_y = L_z = 10$ mm.

5. Summary

We have developed the theory of holographic reconstruction with polychromatic light. In particular, focusing on its application to holographic memory, the required spectral width, distortion of the reconstructed image, diffraction efficiency, intra- and inter-page crosstalk, and storage capacity were investigated. The obtained results are summarized below.

The required spectral width and image distortion

A larger input image (x_{sMax}, y_{sMax}) or smaller focal length f_s require a larger spectral width $\Delta\lambda_{BM}$ to reconstruct the whole image. Generally the required spectral width depends on both directions x_{sMax} and y_{sMax}, but if the signal, reference, and probe beams lie in the same plane ($\alpha_p = 0°$), it becomes independent of y_{sMax}. Furthermore, the condition $\alpha_p = 0°$ minimizes the image distortion, where the image is magnified by the wavelength ratio μ_0 in the y_d-direction, but is unchanged in the x_d-direction.

Multiplexing method

Crystal angle multiplexing used in combination with peristrophic multiplexing is the most suitable multiplexing method in polychromatic reconstruction (PCR). Other multiplexing methods are also possible in principle, but most of them require a complicated system to read a target page because the diffracted wave $\mathbf{e}_{d0}$ will be pointed in a different direction at each multiplexed page.

Diffraction efficiency

Since the grating component can diffract only a limited spectral component, the diffraction efficiency, which is defined here as the ratio of the diffracted power by one grating component to the total power of the input probe beam, is much smaller than in the conventional monochromatic readout. This may be an unavoidable drawback of the PCR method.

Intra- and Inter-page crosstalk noise

Due to the wide spectral width of the probe beam, the hologram dimension L_x does not influence the intra-page crosstalk noise, and thus the input image area suffering from the intra-page crosstalk noise is slightly enlarged. On the other hand, the inter-page crosstalk noise is a crucial problem in the PCR method. The angular Bragg selectivity is greatly degraded, and thus, the storage capacity decreases by more than two orders of magnitude. However, this problem can be solved by using the selective detection method mentioned below.

Selective detection method and the achievable storage capacity

Taking advantage of the wavelength difference after the crystal rotation, we can detect the signal image alone, even if the noise images overlap with the target signal image. The storage capacity when employing the selective detection method reached 40 percent of the theoretical limit for the monochromatic case.

PCR is a unique and useful method for holographic memory systems. While other nondestructive readout methods have failed to achieve a high storage capacity, we theoretically proved that PCR, when used in combination with the selective detection method, enables us to achieve both nondestructive readout and a high storage capacity, simultaneously.

6. References

Barbastathis, G. & Psaltis, D., (2000). Volume holographic multiplexing methods, In: *Holographic data storage*, Coufal, H.J., Psaltis, D., & Sincerbox, G.T., pp. 21-62, Springer, 978-3540666912, New York.

Champagne, E.B. (1967). Nonparaxial imaging magnification and aberration properties in holography. *Journal of the Optical Society of America*, Vol. 57, No. 1, pp.51-55

Curtis, K., Pu, A., & Psaltis, D. (1994). Method for holographic storage using peristrophic multiplexing. *Optics Letters*, Vol. 19, No. 13, pp.993-994

Fujimura, R., Shimura, T., & Kuroda, K. (2007). Polychromatic reconstruction for volume holographic memory. *Optics Letters*, Vol. 32, No. 13, pp.1860-1862

Fujimura, R., Shimura, T., & Kuroda, K. (2010). Multiplexing capability in polychromatic reconstruction with selective detection method. *Optics Express*, Vol. 18, No. 2, pp.1091-1098

Gulanyan, E.K., Dorosh, I.R., Iskin, V.D., Mikaelyan, A.L., & Maiorchuk, M.A. (1979). Nondestructive readout of holograms in iron-doped lithium niobate crystals. *Soviet Journal of Quantum Electronics*, Vol. 9, No. 5, pp.647-649

Külich, H.C. (1987). A new approach to read volume holograms at different wavelengths. *Optics Communications*, Vol. 64, No. 5, pp.407-411

Petrov, M.P., Stepanov, S.I., & Kamshilin, A.A. (1979). Holographic storage of information and peculiarities of light diffraction in birefringent electro-optic crystals. *Optics and Laser Technology*, Vol. 11, No. 3, pp.149-151

van Heerden, P.J. (1963). Theory of optical information storage in solids. *Applied Optics*, Vol. 2, No. 4, pp.393-400

Part 3

Holographic Devices

11

Application of Holograms in WDM Components for Optical Fiber Systems

Alfredo Martín Mínguez and Paloma R. Horche
ETSIT-Universidad Politécnica de Madrid
Spain

1. Introduction

Coarse Wavelength Division Multiplexing (CWDM) technologies are being widely deployed internationally in metropolitan and access networks due to the increased demand for delivering more bandwidth to the subscriber, created by the need of enhanced services, (Koonen, 2006). For metro, and mainly for access networks applications, an increment in capacity may be achieved with a cost-effective multiplexing technology without the need for the high channel counts and closely spaced wavelengths typically used in long haul networks. A channel space of 20 nm, as proposed in the G. 694.2 ITU Rec., can be used relaxing the processing tolerances and potentially lowering the cost of components. CWDM technology reaches those requirements and it has been proposed for these applications. It is in this context that holographic optical devices have a potential use.
This chapter describes the theory, design, and experimental results of a generic multipurpose device that can operate as a tunable wavelength filter, wavelength multiplexer and wavelength router. This device could be especially useful in optical network applications based on both Coarse and Dense Wavelength Division Multiplexing technology (CWDM/DWDM). The enabling component is a Ferro-electric Liquid Crystal (FLC) Spatial Light Modulator (SLM) in which dynamic holograms are implemented in real time. As a consequence, the device will be able to carry out different functions according to the hologram recorded on the SLM. The great advantage of this device is polarization insensitivity in the region of operation, allowing low cross-talk and simple handling. As hologram management is the basis for this device, some topics in the Computer Generated Hologram (CGH) design process are commented on and general guidelines are also considered.
Laboratory experiments have demonstrated the capability of a phase FLC-SLM, with the great advantage of polarization insensitivity operation, to diffract the incident light according its wavelength and hologram patterns, for the use in the former applications.
Two typical applications of this technology are described: the first one is a design of an equalized holographic Reconfigurable Optical Add-Drop Multiplexer (ROADM), where this device can address several wavelengths at the input to different output fibers, according to the holograms stored in a SLM (Spatial Light Modulator), all the outputs being equalized in power; the second one is dealing with the design of an holographic router with loss compensation and wavelength conversion whose main application is in Metro networks in the interconnection nodes. This device uses a SOA (Semiconductor Optical Amplifier), in the non linear region, to do the wavelength conversion and, in addition, to supply the gain in order to compensate for the intrinsic losses of the holographic device.

2. Operating principle

The working principle of a holographic device design is based on the wavelength dispersion produced in a diffraction grating element (Agrawal, 2002). When a polychromatic light reaches a diffraction grating, there is an angular dispersion (diffraction) according to the incident light wavelength.
Equation (1) expresses the relationship between the diffraction angle and the wavelength of the incident light λ:

$$\sin \Phi = \frac{m\lambda}{d} \tag{1}$$

by considering the incident light perpendicular to the grating, Φ is the diffracted light angle, *m* is the diffraction order and *d* the grating spatial period.
The light diffracted, in a far field approximation, follows the Fourier transform distribution and the intensity for the different diffraction orders, *m*, is proportional to $sinc^2(\Phi d/\lambda)$; the separation between diffraction orders is given by $\lambda R/d$, where *R* is the distance between the binary transmissive diffraction grating and the Fourier plane (Kashnow, 1973).
Most diffraction grating elements are not practically useful for changing the spatial period or the wavelength. A way to allow these variations is, by using a Spatial Light Modulator (SML), to implement on it a Computer Generated Hologram (CGH). The pixelated structure of the SLM produces the effect of a two-dimensional diffraction grating when the device is illuminated with a coherent light. In the SLM every ferro-electric liquid crystal (FLC) pixel can be electro-optically configured to provide a phase modulation to the incident light. Therefore, by managing the hologram on the SLM and its spatial period a programmable diffraction grating is obtained.
In optical fiber communications, wavelengths around 0.8 - 1.6 µm are used. Thus, an SLM pixel pitch close to these wavelength values is required. Unfortunately, current commercial SLMs do not have enough resolution. Therefore, to solve this limitation, a fixed diffraction grating with a low spatial period, together with the SLM giving a high resolution filter, is used (Parker et al., 1998).

	2-Phases		4-Phases	
	% darkness	PHASE (rad)	% darkness	PHASE (rad)
black	100	0	100	π/4
grey1	-	-	66	3π/4
grey2	-	-	33	-3π/4
white	0	π	0	-π/4

Table 1. Relationship between phases and contrast

2.1 2 and 4-phases holograms

Different types of holograms can be used (Horche & Alarcón, 2004) in the SLM. In order to optimize losses, phase holograms are preferred instead of amplitude holograms due to its intrinsic 3 dB of loss and 4-phase holograms are used instead of 2-phase (binary) holograms because of its greater efficiency (40.5% → 81%), which is proportional to $\text{sinc}^2(\pi/M)$, where M is the number of phases. Table 1 summarizes the relationships between phase and contrast for 2 and 4 phase holograms.

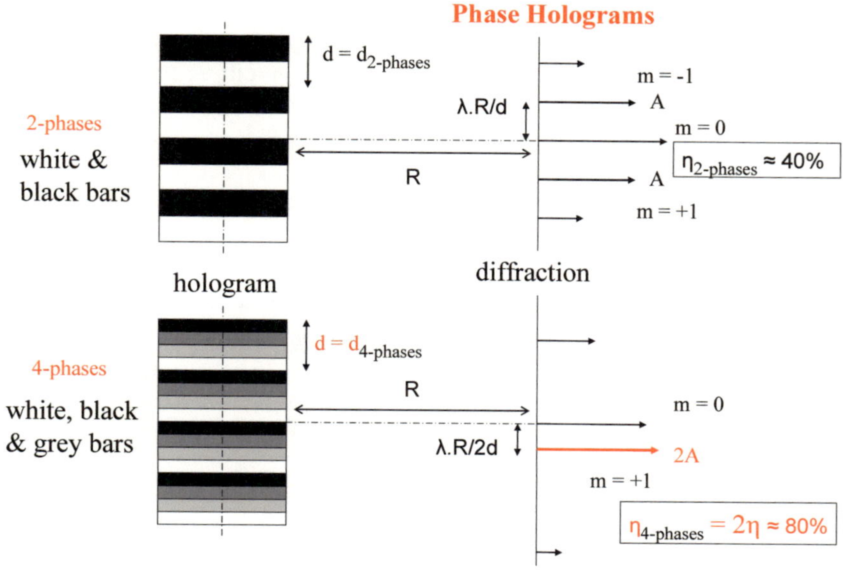

Fig. 1. two/four-phases bars holograms

Fig 1 shows a bars hologram for 2 and 4-phases and their diffraction target in a far field approach. As we can see, the main difference in the holograms is the grey bars in the 4-phases holograms; in this case there is a white bar, a black bar and two different grey bars for addressing the 4-phases (π/4, 3π/4, -3π/4, -π/4); with regard to the diffraction target. Another characteristic is the loss of the symmetry for the diffraction orders.

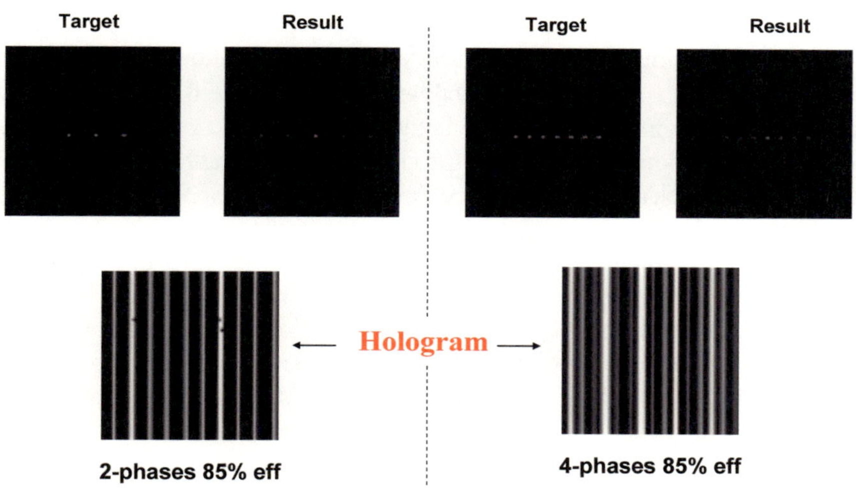

Fig. 2. Examples of 2/4-phases holograms and diffraction targets

In Fig. 2 examples of calculated holograms are shown. The program calculates the inverse Fourier transform (F.T.)$^{-1}$ of the diffraction target (result) by an annealing optimization algorithm. In this case both holograms have a calculated efficiency of 85% and the grey bars are clearly visible in the figure. In the following Section some guidelines about design of holograms by computer are given.

2.2 Computer generated hologram design

Taking into account the former considerations and by implementing a hologram on the SLM where its spatial period can be modified in real time, we obtain a programmable diffraction grating.
The relationship between the hologram and its Fourier Transform function are:

Hologram → F.T. → Diffraction target

Diffraction target →F.T⁻¹ → Hologram

In order to implement the CGH, holograms are calculated by using a program based on a variation of the widely adopted simulated annealing optimization algorithm (Dames, Dowling et al., 1991), (Broomfield, Neil et al., 1992) whose cost function to minimize the calculation error is:

$$C = \sum_t \frac{(I_i^2 - A^2)^2}{A_i^2} \tag{2}$$

where I_i^2 is the calculated spot intensity for the diffraction order i; A_i^2 is its defined intensity and A^2 is the average intensity for the diffraction target spots; t is the number of process calculations.
There are three steps in a CGH design process:

1. Target definition: the target is the diffraction pattern that is to be obtained from the SLM. Depending on the use: filter, switch or others, this target is usually an array or a matrix of spots. This is the input for the program.
2. Fourier transform calculation: the program calculates the inverse Fourier transform $(F.T.)^{-1}$ of the target. The optimization algorithm compares the FT of the hologram with the defined target improving the efficiency at each calculation time. Hologram pixels are flipped between the amplitude values 0, 1 (or phase 0, π) to reduce an error function, (2), specifying the difference between the desired target in the Fourier plane and the reconstruction obtained from the current state of the hologram, improving the efficiency at each calculation (Efficiency defined as: $\eta = \Sigma$ *m orders diffracted light /total incident light*).
3. Finally, CGH implementation in an optical substrate, using a photographic film or SLM.

The CGH designed for this work is a black & white bars pattern implemented onto a Spatial Light Modulator, where there are only two possible states: "1" for white (total transparency or π phase shift) and "0" for black (total darkness or 0 phase shift). Fig. 3 shows the original diffraction target (a), an array of spots with different light intensities (non uniform, as in Fig. 3a), and three consecutive holograms (b, c, d), calculated by the program carrying out the inverse FT according to the algorithm efficiency. A 45% efficiency is an initial calculation value and close to 90% efficiency is practically the best result in the optimization process.
During the calculation of the hologram, the program can find out different holograms which match the diffraction target. It is possible to change, dynamically, the initial conditions (original diffraction target and efficiency, optimization process parameters), to change the direction for the optimization process allowing the algorithm to escape from local minima and reach the correct hologram.

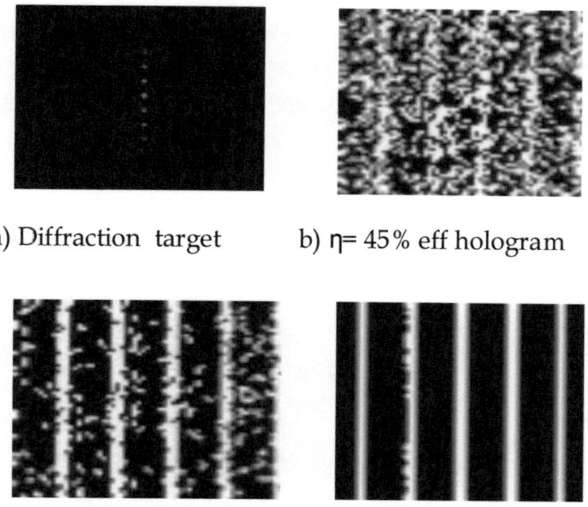

a) Diffraction target b) η= 45% eff hologram

c) η =7 0% eff hologram d) η=90% eff hologram

Fig. 3. Hologram calculation process according to the algorithm efficiency η. a) diffraction target; b), c) and d) are calculated holograms with η = 45%, 70% and 90%, respectively

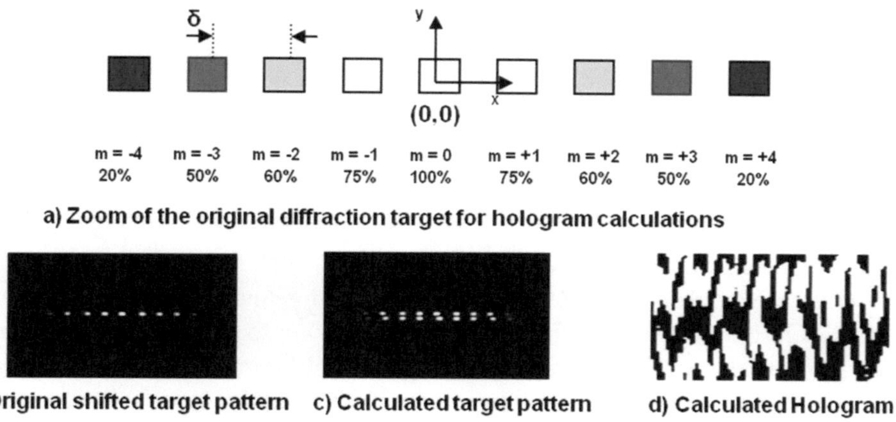

Fig. 4. a) "Zoom" of the original diffraction target, b) original shifted diffraction pattern along the y axis, c) calculated diffraction target and d) corresponding hologram when the original pattern is shifted

Computer calculations are very sensitive to the geometrical distribution of the original diffraction target. A very slight misalignment on it (centre: $x = 0$, $y = 0$) can produce a hologram completely different from the correct one. This effect is shown in Fig. 4 when the original array of spots (Fig. 4a) is shifted by 30% of spot separation δ (Fig. 4b), along the vertical axis y; the calculated target (Fig. 4c) is an array of spots "duplicated" and "shifted" instead of a singular one.

To avoid small misalignments, along the x axis, of the output fibers array positions , with impact on the efficiency, we can optimize the hologram pattern, introducing an offset in the bar positions to correct them (Crossland et al., 2000) An offset of 5% of the hologram period would impact the efficiency up to a 40%.

For the operation of holographic devices after the generation of holograms, it is necessary to configure with them the SLM. To perform the switching operation a closed control between the holographic component (SLM) and the computer is needed to assign the correspondent hologram from a local database. This procedure is represented in Fig. 5, where a switching control acts over the PC-SLMs interface.

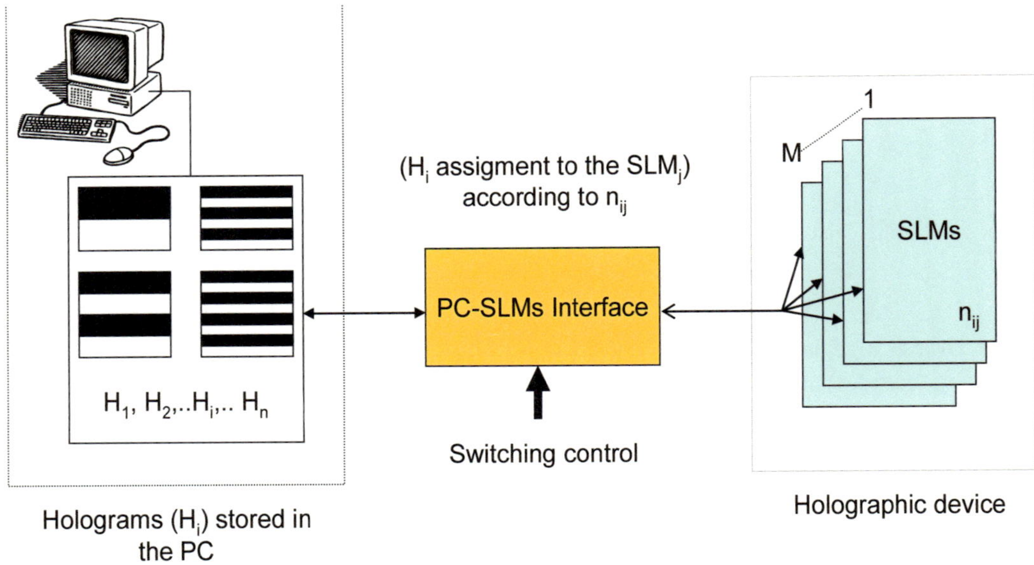

Fig. 5. Tunable holographic device: switching operation

3. Dynamic holographic device design

In order to design a holographic optical device a "4f" structure is chosen using a transmissive SLM and fixed grating. Fig. 6 illustrates the device used in the present work. The previously calculated CGH (black and white bars) is loaded onto the SLM via a PC-based interface. The SLM-FLC and fixed grating are illuminated by light coming from a singlemode optical fiber collimated by means of a lens. A second lens produces the replicated array of spots explained above on the back focal plane of the lens.

In our experiments, we are interested only in the array of spots corresponding to the first order of diffraction. Therefore, the output optical fibers array is placed in the back focal plane of the lens at a certain angle in order to optimize the coupling. Because of the small size of the singlemode fiber radius, it acts as a spatial light filter.

Output fibers $F_1,..,F_{10}$, must be located at the Fourier lens plane in order to receive the maximum light intensity of the diffracted beams. The relationship between the system diffraction angles (Parker et al., 1998) is in agreement with the expression:

$$\arctan\left(\frac{x}{f}\right) = \arcsin\left(\frac{\lambda}{d}\right) + \arcsin\left(\frac{\lambda}{H}\right) \quad (3)$$

where x is the distance of the output optical fiber from the optical axis, f is the focal length of the lens, d is the spatial period of the fixed grating and H is the hologram spatial period, which relationship with D, the size of the pixel, and N, the number of pixels in one dimension of the SLM is given in (4):

$$H = \frac{ND}{n} \qquad 0 < n < \frac{N}{2} \tag{4}$$

where n is the integer number of black & white bar pairs and depends on the type of hologram (pattern). For small angles, equation (3) can be simplified as follows:

$$\lambda \approx \frac{x}{f} \cdot \frac{1}{\left(\frac{n}{ND} + \frac{1}{d}\right)} \tag{5}$$

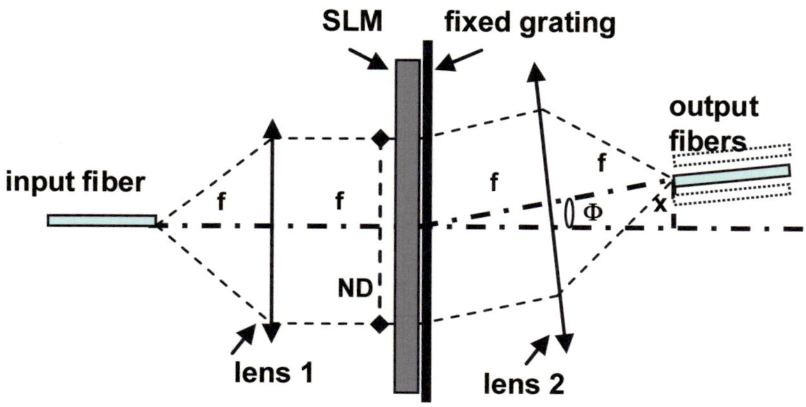

Fig. 6. "4f "dynamic holographic device with a transmissive SLM and fixed grating.

When the other holographic device parameters are fixed, λ only depends on n, as can be shown from (5). According to fixed or variable values for n or/and x, different applications for our device can be considered giving an idea of the device's versatility (see Table 2).
In the following sections, we design a generic multipurpose device based on the experimental scheme explained above that can operate as a tunable wavelength filter, wavelength multiplexer and wavelength router, by simply modifying in real time the CGH loaded on the SLM. The PC-based interface used to load the CGH on to the SLM also serves to calculate the different patterns needed. The electronic interface allows an automatic program to be developed for loading different patterns when they are needed.

n	x	Application
fix	fix	Holographic band-pass filter
variable	fix	Tunable holographic band-pass filter
fix	variable	Demultiplexer 1x M
variable	variable	λ router 1x M

Table 2. Different device applications

In all cases, the central wavelength channel, λ_0, is obtained for $n = N/4$ in (5), and the operating wavelength range $\Delta\lambda_f$ is given by:

$$\Delta\lambda_f = \lambda(n=0) - \lambda\left(n = \frac{N}{2}\right) \tag{6}$$

The -3dB passband width, *BW*, for each wavelength channel tuned, is limited by the output fiber characteristics and the wavelengths coupled inside the core diameter φ_{core}. Taking this into account, and from (3), the following expression relates the bandwidth *BW* for every wavelength channel tuned in the filter and the focal distance *f* of the lens according the optical power coupled into the output optical fiber (Parker et al., 1998):

$$f > \phi_{core}\frac{d}{BW}\times\left(1 - \frac{\lambda_0^2}{d^2}\right)^{3/2} \tag{7}$$

In order to obtain minimum losses, the collimated light through the SLM has to illuminate the maximum quantity of pixels. As its intensity distribution has a Gaussian profile, it is sufficient that $1/e^2$ beam bandwidth illuminates the SLM aperture. According to optical Gaussian laws, the following condition is reached:

$$ND = 4\lambda_0\frac{f}{\pi\phi_{core}} \tag{8}$$

For commercial FLC-SLMs, available pixel size *D* is > 5 μm and the number of pixels, *N*, usually is from 250 to 1000. From expressions (5) and (7), it is possible to calculate the *x* value and λ_{max} and λ_{min} for the operating range of tuning.

4. Tunable holographic filter application

In order to design a tunable holographic filter with a -3dB passband width, *BW*, of 1 nm (125 GHz), for each wavelength channel tuned, we take *d* = 3.5 μm for the spatial period of the fixed grating. To use the same device for CWDM/DWDM, a SLM with a value of *N* = 720 and *D* = 7 μm for the spatial period, is chosen. The output singlemode fibers used in our device have a core diameter, φ_{core}, of 9 μm. Then, from (7), *f* must be greater than 23.9 mm. As a practical value we assume *f* = 25 mm.

Table 3 summarizes the filter figures for CWDM systems applications where channels are allocated between λ_{min}= 1290 nm and λ_{max}= 1590 nm, with central wavelength λ_0 = 1431 nm, and for DWDM systems (λ_{min}= 1530 nm and λ_{max}= 1590 nm, λ_0= 1551 nm).

CWDM		DWDM
1270 -1590	$\Delta\lambda$ (nm)	1510 -1590
1	BW (nm)	1
1431	λ_0 (nm)	1551
25.00	*f* (mm)	25.00
11.499	*x* (mm)	12.463
24.71	Φ (°)	26.51
1591	λ_{max} (nm)	1591
1311	λ_{min} (nm)	1531

Table 3. Device parameters for CWDM(DWDM) systems

The operation as a tunable CWDM/DWDM filter is obtained by changing the hologram period, n. From the output fibers (F_1 *to* F_{10}), a CWDM tunable filter uses F_4 and a DWDM tunable filter uses F_8 (see Fig. 6).
In Table 4 tuning between 1311 < λ < 1591 nm (CWDM, output fiber F_4) is obtained for 17 < n < 328 and between 1531 < λ < 1591 nm (DWDM, output fiber F_8) is reached for 138 < n <201.

a)

n	**328**	**276**	**227**	**180**	**136**	**94**	**55**	**17**
λ (nm)	1311	1351	1391	**1431**	1471	1511	1551	1591

b)

n	**201**	**180**	**159**	**138**
λ (nm)	1531	**1551**	1571	1591

Table 4. a) Filter operation for CWDM (F4) and b) for DWDM (F8)

4.1 Wavelength response

For CWDM applications the holographic filter has a tuning range of $\Delta\lambda_f$ =1591 -1311 = 280 nm with a -3dB passband of 1 nm. In Fig. 7 the transmission response is shown, according to (Parker et al., 1998). For wavelengths very close to the centre, the shape is Gaussian ($\lambda < \lambda_0$ +/- 1,5 nm); from these wavelengths the shape is like a Bessel function and the zero convergence is slower (> 20 dB for $\lambda > \lambda_0$ +/- 1,5 nm ; > 40 dB for $\lambda > \lambda_0$ +/- 5 nm).
Table 5 shows, in case of CWDM systems, the different values of n and corresponding central wavelengths separated by 40 nm, from 1311 to 1591 nm. In this case, an adjacent channel isolation > 50 dB is achieved and the complete filter tuning range is covered according to the type of hologram.

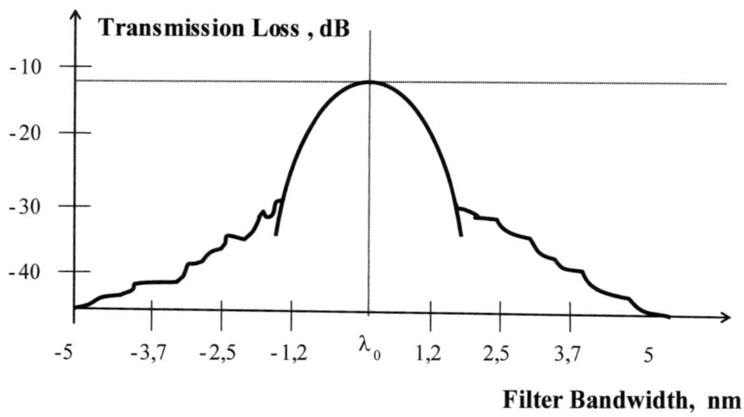

Fig. 7. Holographic filter shape. For wavelengths close to the central value, $\lambda < \lambda_0$ +/-1.5 nm, the shape is Gaussian and like a Bessel function for wavelengths $\lambda > \lambda_0$+/-1.5 nm

n	180	136	94	55
λ_0 (nm)	1431	1471	1511	1551
BW (nm)	1	1	1	1
Δλ (-3 dB)	1430.5-1431.5	1470.5 -1471.5	1510.5-1511.5	1550.5-1551.5
Δλ (>-50 dB)	1421-1441	1461-1481	1501-1521	1541-1561

Table 5. Tunable pass band filter (CWDM)

This feature allows the possibility of a multiple pass band filter in the same optical fiber, but with an increment of losses penalty according to the expression 10 log C (dB), where C is the number of simultaneously tuned channels (Parker et al., 1998). In this case, $C = 4$ and therefore the increment of losses in the device is: Δ losses = 10 log 4 = 6 dB.

4.2 Loss estimation

There are three different sources of loss in this holographic device:

a. SLM losses

An FLC-SLM works as a dynamic full π-binary phase fixed grating hologram with a diffraction efficiency $\eta = 36.5\%$ (4.38 dB) for the first diffractive order ($m = 1$) and a FLC switching angle of 45°.

Another cause of losses is the insertion of the hologram, for a phase SLM, as a result of the light polarization plane and FLC switching angle different to 45° (theoretical optimal angle); at least another 2 dB are lost, assuming a good alignment of the collimated input light and the FLC pixels.

b. Fixed grating losses

The diffraction efficiency for a fixed grating, binary π-phase, is $\eta = 36.5\%$ for the first diffractive order (m=1). That means a loss of 4.38 dB.

c. Fiber/lens coupling efficiency

A fiber/lens coupling efficiency of 50% is a good approximation; therefore another 3 dB of losses have to be added (2 dB, with very good alignment).

Losses can be improved using multiple-phase or blazed gratings; in this case the efficiency can reach $\eta \approx 80$ - 90% and the losses decrease to 1.5 dB (Ahderom, Raisi et al., 2002).

The estimated total losses of the device are: SLM losses + fixed grating losses + fiber/lens coupling losses = (4.38+2) + 4.38 + 3 =13.76 dB and with loss optimization: (1.5+2)+1.5+2= 7 dB can be obtained.

5. WDM (wavelength division multiplexing) application

We can use this device as a 1x M demultiplexer, where M is the number of output fibers. For this, a fixed value of n is used and the output fibers are located in certain x positions. Output fibers (9/125 μm) must be placed in agreement with the diffracted angles Φ, according to input wavelengths and they have to be separated at least $\Delta x = 125$ μm.

From (5), we can calculate the Δx taking the value of center to center wavelength channel separation, $\Delta\lambda$, into account:

$$x \approx \lambda f\left(\frac{n}{ND}+\frac{1}{d}\right) \Rightarrow \Delta x \approx \Delta\lambda f\left(\frac{n}{ND}+\frac{1}{d}\right) \tag{9}$$

In order to design a compatible device with the frequency grid provided in ITU-T G.694.1/G.694.2 Rec. for CWDM/DWDM systems, a 1x4 demultiplexer ($M = 4$) for DWDM [1] with $\Delta x = 161$ μm and a 1x8 demultiplexer ($M = 8$) for CWDM with $\Delta x = 321$ μm, can be implemented.

[1] In this case, $\Delta\lambda$'s < 20 nm are not feasible due to the physical dimension of the device (i.e. $\Delta\lambda = 2$ nm and $\Delta x = 161$ μm → f,focal distance of the lens, = 25 cm).

Table 6 summarizes the fiber positions in order to demultiplex the wavelengths used in the CWDM/DWDM systems. A CWDM system uses F_1, F_2, F_3, F_4, F_5, F_6, F_8 and F_{10} and a DWDM uses F_7, F_8, F_9, and F_{10} output fibers (see Fig. 6). It is necessary to emphasize that a better performance as demultiplexer could be implemented if only this function is required. For example, we could design a demultiplexer device with channel separation smaller than 50 GHz (Parker, Cohen et al., 1997). However, the novel idea is to design a compatible CWDM/DWDM device able to carry out different functions.

Output fiber	**λ (nm) CWDM**	**λ (nm) DWDM**	**x(μm)**
F_1	1311		10 535
F_2	1351		10 856
F_3	1391		11 178
F_4	1431		11 499
F_5	1471		11 821
F_6	1511		12 142
F_7		1531	12 303
F_8	1551	1551	12 463
F_9		1571	12 624
F_{10}	1591	1591	12 785

Table 6. CWDM/DWDM demultiplexers (n = 180)

6. Wavelength routing application

Maintaining output fibers in the same place as shown in Table 6, if n value (type of hologram) is properly varied, a certain wavelength coming from the input fiber can be routed to any one of the output fibers. As an example, Table 7 highlights the n values for routing λ_0 = 1431nm (CWDM) and λ_0 = 1551 (DWDM) towards an output fiber; these values have been calculated from (10), considering the variation of n according to Δx:

$$n \approx \left(\frac{x}{\lambda f} - \frac{1}{d}\right) ND \Rightarrow \Delta n \approx \left(\frac{\Delta x}{\lambda f}\right) ND \tag{10}$$

For Δx = 161 μm, Δn was calculated by using (10) resulting in Δn = 21 and for Δx = 321 μm, Δn is 45. Therefore, the device is a 1x8 λ router in case of CWDM and a 1x4 λ router for DWDM systems. It is necessary to highlight that the positions of the fibers are compatible with all applications and that the crosstalk resulting from high-order diffraction beams ($m = \pm 2, \pm 3,$.) are outside of the locations of the output array fibers ($\Delta\Phi$ = 4°), (Horche, 2004).

Output fiber	CWDM n value $\lambda_0 = 1431$ nm	DWDM n value $\lambda_0 = 1551$ nm
F_1	44	
F_2	89	
F_3	135	
F_4	180	
F_5	225	
F_6	270	
F_7		159
F_8	315	180
F_9		201
F_{10}	360	222

Table 7. CWDM/DWDM routers

7. Basic experimental results

In this section two complementary experiments have been made. The first one is related to diffraction patterns measurements for different bars holograms and the second one to a SLM characterization for holographic filters, demultiplexers and routers use with reference to the devices whose design and characteristics have been described in the previous sections. Due to the unavailability of components in the laboratory with the characteristics previously described, the experimental optical bench is somewhat different from the appropriate one, but, the measurements obtained are in agreement with the calculations.

In order to carry out the measurements, the experimental lab bench showed in Fig. 8 was used; it is in agreement with the structure of Fig. 6, but, in this case, we used a reflective SLM instead of a transmissive one; therefore, it is necessary to include a polarizing beam splitter in order to direct the reflected beam to the lens. Due to the "spatial invariability" of the Fourier transform, it is not necessary to illuminate the entire SLM active surface to reproduce the diffraction pattern; taking this into account we can select, by a diaphragm aperture, the SLM zone where the incident light is focused. The characteristics of a commercial binary phase SLM are shown in Fig. 8(c).

As optical sources, a green He-Ne laser and a tunable Argon laser with $\lambda_g = 528.7$ nm (green) and $\lambda_b = 462.6$ nm (blue) have been used. These wavelengths have been selected because they belong to the visible spectrum and the correct alignment of the system is easier, a critical factor in the experiment. In this case, a detector-array (6.3 × 4.7 mm) of a CCD camera is placed at the "*focal plane*", as an image sensor, to analyze the results.

A single personal computer, PC, is used to generate the CGHs for the design process described previously, and they are loaded onto the SLM by changing its pixels state; the diffracted patterns were stored in the same PC, where they could be observed and processed. To recalculate the new output fibers position, the distance for the diffraction order (x) is derived from (5) without the fixed grating:

$$x = \lambda \cdot \frac{f}{H} \tag{11}$$

where H, defined in (4), is the hologram spatial period, but now, the maximum value for $n = N/2 = 128$, the size of the pixel, $D = 15\ \mu m$ and the number of pixels in one dimension of the

SLM, $N = 256$, have been taken into account. Fig. 9 summarizes the new fiber positions (F_1 *to* F_8) with a separation of $\Delta x = 176.23$ µm between them.

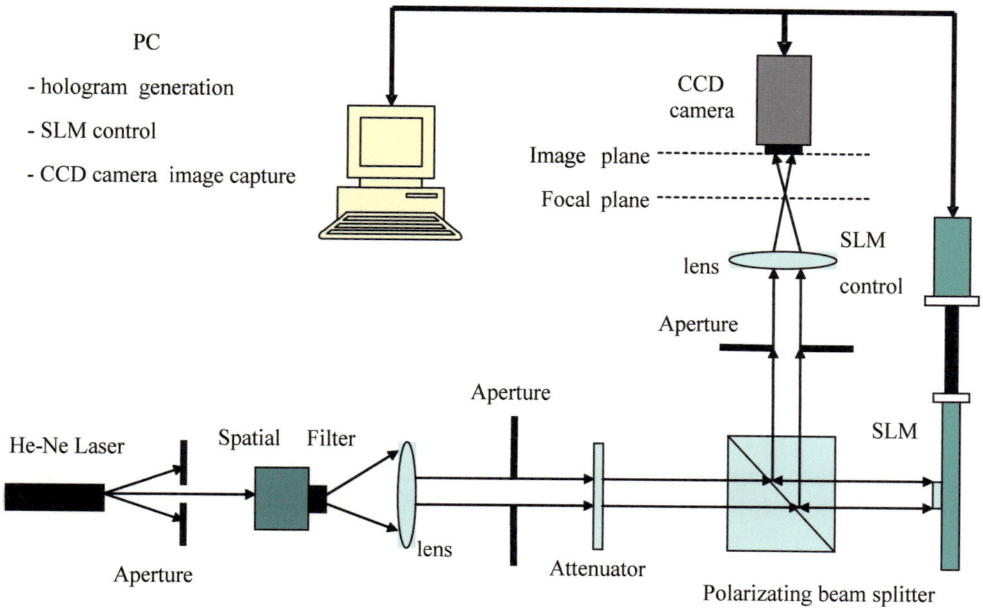

a)

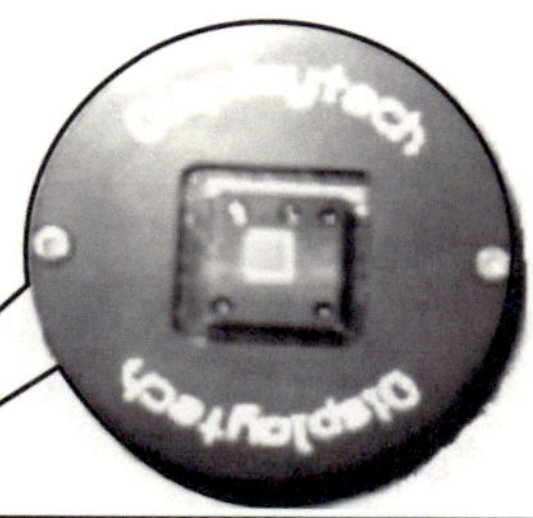

N	256
D	15 µm
Fill factor	87 %
Active area	3,84 x 3,84 mm

b) c)

Fig. 8. (a) Optical bench diagram; (b) experimental optical bench and c) binary phase SLM characteristics

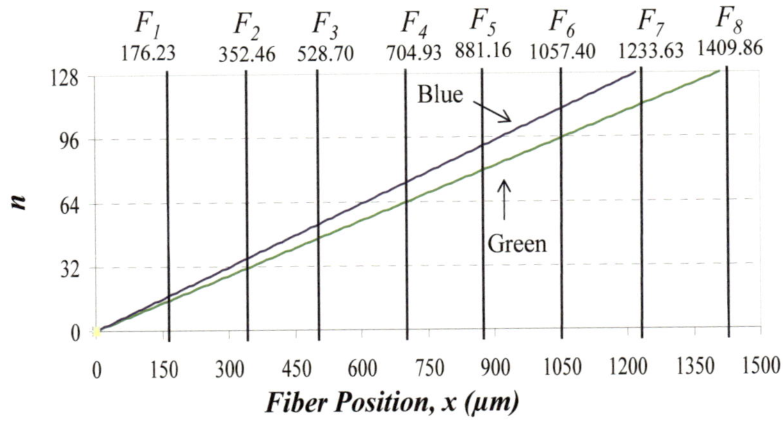

Fig. 9. New output fiber positions for the experimental measurements with two different λ's: λ_g = 5287 nm (green) and λ_b = 4626 nm (blue)

7.1 SLM characterization for wavelength routers

When we implement different holograms, according to n values, in the SLM and an incident light wavelength is illuminating the pixels, different diffracted angles are obtained; by placing an array of fibers at the output, in the focal plane, a λ-router is implemented.

To test the capability of a commercial SLM as a part of a dynamic holographic router, the holographic setup shown in Fig. 8 (a) was implemented; a photo of the experimental optical bench is shown in Fig. 8(b). For this test the optical source was an He-Ne laser at λ = 528.7 nm (green wavelength) and the lens focal length, 8 cm.

In order to route the green wavelength to the F_4 output fiber, according to Fig. 9, it is necessary to load a CGH-(A) with a spatial period H corresponding to n = 64 in the SLM and for routing the same wavelength to the F_8 output fiber, a CGH-(B) with n = 128 was calculated and implemented onto the SLM.

7.2 SLM characterization for filters and demultiplexers

Other measurements, to test the capability of a commercial SLM as a part of a dynamic holographic device, have been done with the holographic setup shown in Fig. 8 (Alarcón, 2004). According to Fig. 9, if a CGH black & white bars type hologram with $n = 128$ is loaded onto the *programmable* SLM, a blue wavelength channel will reach the F_7 output fiber and a green wavelength channel will reach the F_8 output fiber.

The diffracted light spots distance, calculated from (9) without a fixed grating, is:

$$\Delta x \approx \Delta\lambda f\left(\frac{n}{ND}\right) \tag{12}$$

where: $\Delta\lambda$ = 66.1 nm; f = 8 cm; n = 128; N = 256 and D = 15 μm.

In Fig.10 light spots captured by CCD camera, from the CGH with n = 128 (black and white bars) are shown. In this case the tunable Argon laser with the blue and green colors has been used. The experimental diffracted light spot distances were x_{blue} = 1233.6 μm (F_7) and x_{green} = 1409.8 (F_8) μm, separated by Δx = 176.25 μm according to Fig. 9, in good agreement with (12). Therefore, in this way, we can build an optical 1x2 demultiplexer.

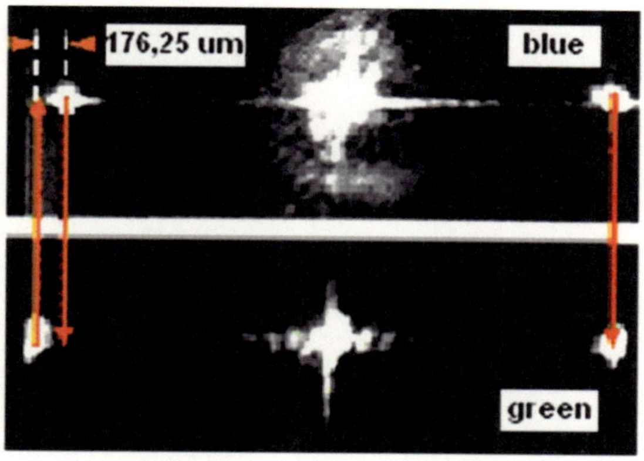

Fig. 10. Diffracted wavelengths with a phase FC-SLM. A tunable Argon laser with λg = 528.7 nm (green) and λb = 462,6 nm (blue) is used. An Δx = 176.25 μm (diffracted wavelengths separation) is obtained

The central light spot is due to the zero diffraction order $m = 0$, with the maximum light intensity diffracted ($x = 0$); it can be reduced with a SLM with better performance by impacting on the total insertion losses reduction.
The temporal response of the system was also measured. The SLM optical switching time was estimated to be roughly 250 μs, as the sum of the electric storage and FLC material response times (Alarcón, 2004). We also noticed a damped response when low-frequency switching is carried out; this is probably due to relaxation of the FLC molecules.

8. Design of equalized holographic ROADMs for application in CWDM metro networks

These type of ROADMs are designed for application in CWDM (Coarse Wavelength Division Multiplexing) networks, where the distance between the different wavelengths allow the use of DML (Direct Modulation Lasers) without cooling, reducing the cost and the tolerances of the network components. Application in METRO networks and its interconnection with some PON (Passive Optical Network), as a part of the access to the subscriber, is reviewed.
Different technologies have been proposed for the implementation of ROADMs (Ma & Kuo, 2003), (Homa & Bala, 2008). Each of them has its own advantages and drawbacks. The main characteristic of holographic ROADMs is the easy way of changing the tuning and power level of the signal at the output fibers by the dynamic implementation of different holograms on the SLM according to the requirements of the network management.

8.1 Holographic ROADM structure

The working principle of the dynamic holographic device is based on the wavelength dispersion produced in a diffraction component (grating, spatial light modulator) as explained in Section 3.
We use for this application a phase reflective spatial light modulator (SLM) and a fixed transmissive diffraction grating to select the corresponding output wavelength from an set of channels in the input, as shown in Fig 11. The active element of the SLM is a Ferroelectric Liquid Crystal (FLC) with a low switching time (less than 50 μs) that allows a real time

operation. The role of the fixed diffraction grating is to provide more wavelength tuning range and greater total diffraction angle.

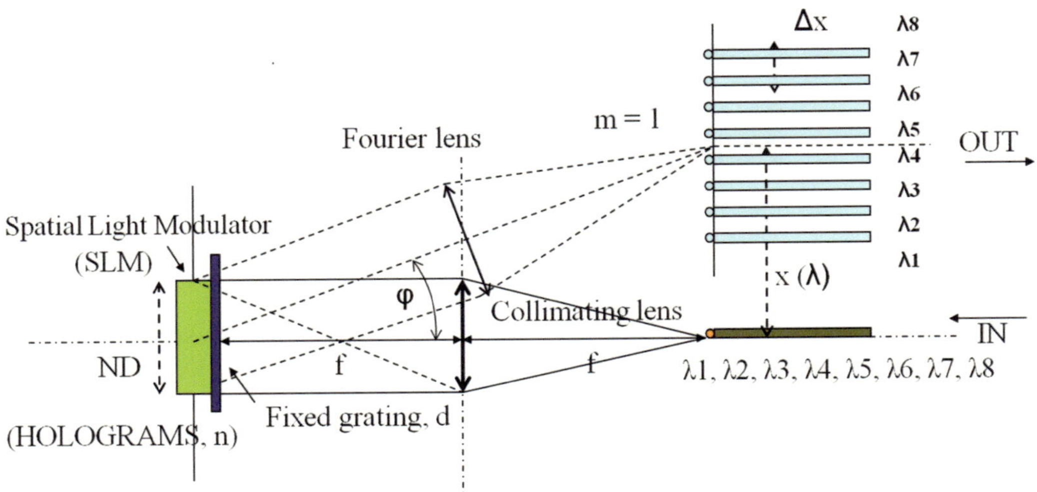

Fig. 11. Reflective holographic router

One of the reasons because we have chosen this type of "2f-folded"implementation, , is the reduced size of the device in comparison with the other possible structure, "linear-4f", where the length of the optical axis is four times the focal distance of the lens used. Its working operation has been described in Section 3.

8.2 Holographic ROADM design

8.2.1 Dynamic wavelength tuning

At the input of the router there are different wavelengths λ_1, λ_2,.... λ_n according to some ITU Rec. For the design of this holographic router, these wavelengths are in agreement with the G.694 Rec. for use in CWDM systems. The range of wavelengths is from 1271 nm to 1611 nm with 20 nm as separation between channels; 4, 4+4, 8, 12 and 16 groups of channels are specified distributed along the complete range.

In a holographic router the tuning of this wavelength range is achieved by changing the spatial period of the hologram ND/n, where n is the number of pairs of bars (2-phases) or number of four bars (4-phases), N is the number of pixels and D the size of the SLM pixels. The expression which allows the selection of the output wavelength λ, according to the physical parameters and structure of the device, is (Martin Minguez & Horche, 2007):

$$\lambda \approx \frac{x}{f} \cdot \frac{1}{\left(\frac{n}{ND} + \frac{2}{(M/2).d} \right)} \tag{13}$$

where x is the distance from the optical axis to the output fiber, f is the focal length of the lens, d is the spatial period of the fixed diffraction grating and M is the number of phases. Fig. 12 shows some tuned wavelengths according to different values of n, for a typical holographic device.

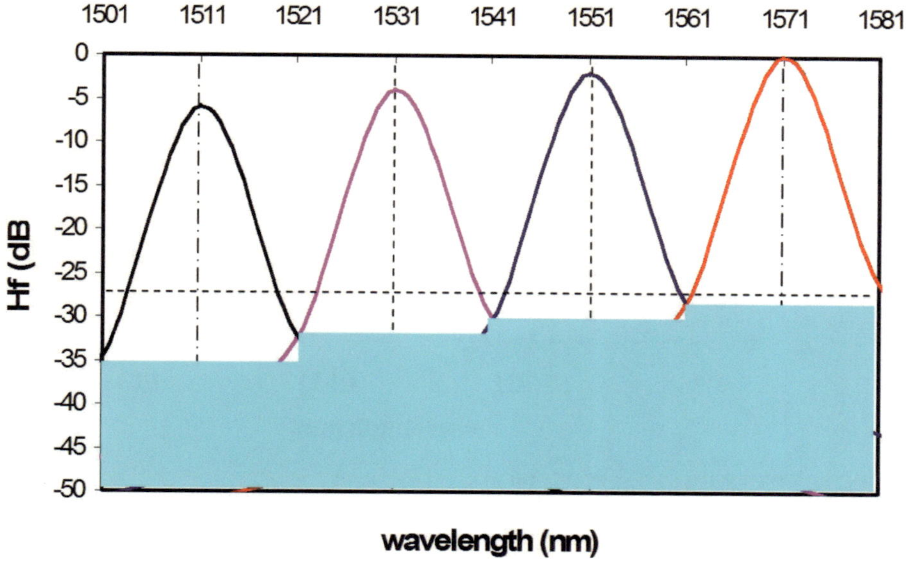

Fig. 12. Four different tuned wavelengths at the output of the holographic router

As we have commented in Section 4.1, for wavelengths close to the central value, the filter response is very similar to the Gauss filter; for wavelengths far from the central value, the filter response is similar to a 3rd order Bessel filter with less out band attenuation. Both of them have a linear phase characteristic, which means a constant group delay. These simulations are in agreement with experimental measurements shown in (Parker et al.., 1998).

8.2.2 Holographic device losses

The losses produced in this holographic router, as we have commented before, are due to the following causes:

- *diffraction loss*: the total light diffraction is composed of the transmissive diffraction in the grating (twice) and the reflexive diffraction in the SLM. Using a 4-phases SLM and a grating with 1st order intensity efficiency of about 80%, the total losses are 10 x log (3x.0.8) ≈ 3 dB.
- *intrinsic SLM loss*: it is due to the liquid crystal (LC) switching angle different from optimal and the coverage of SLM aperture (*1/e² of ND x ND*). A typical value is 2 dB.
- *fibre/lens coupling*: by considering 90% efficiency, 1 dB is added.

In total, with an optimized holographic device, a loss about 6 dB has to be taken into account.

8.2.3 Channel power equalization

Power equalization at the all output channels is necessary to compensate the different response of the network components and distances for the used channel wavelengths. To reach it and to compensate for the holographic device losses, a gain component, such as a Semiconductor Optical Amplifier (SOA), has to be employed. The total equalization takes into account the gain-wavelength variation of this amplifier, ΔG_A, whose typical response is drawn in Fig. 13, (the maximum gain, G_A is about 20~25 dB).

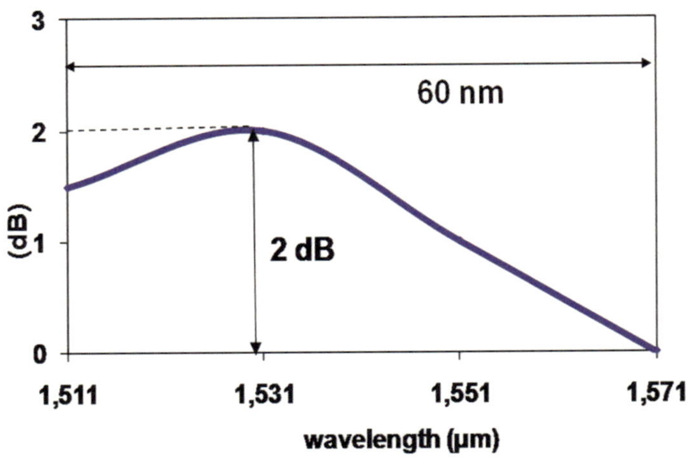

Fig. 13. Incremental gain (ΔG_A) of a typical CWDM SOA

The target is to have at the output fibers a net loss of 0 dB (GT), according to the equation:

$$G_T = G_A - \Delta G_A - 10\log(number.of.channels) - L_{HR} - \Delta At = 0 \tag{14}$$

where ΔAt is the total attenuation range for channels to be equalized at the input of the device; L_{HR} is the intrinsic holographic router losses (≈ 6 dB) and the term ΔL_{HR} =10 x log (number of channels) has taken into account the additional loss due to the mixed holograms utilized for equalizing all the input channels. This point will be explained in detail in the following paragraphs. Fig. 14 shows the structure of an Equalized Holographic ROADM (EH-ROADM) for 4 input channels with full routing of them to the 4 output fibers. A way to obtain at the output fibers tuned wavelengths with different relative attenuation between them is to control the losses due to the SLM aperture, as pointed out in Fig. 15.

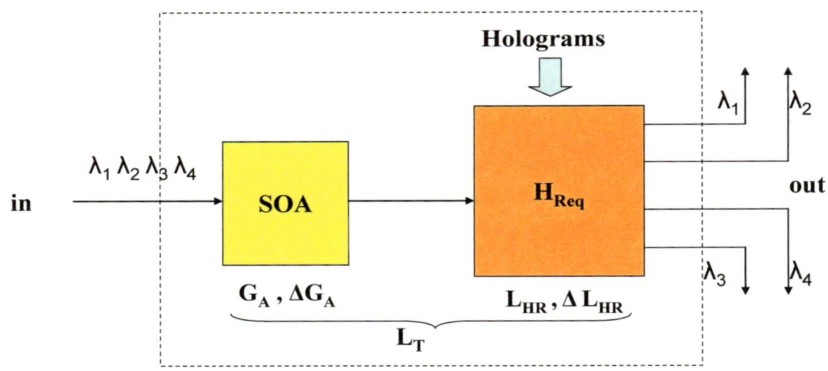

Fig. 14. Equalized Holographic ROADM 1x4

The minimum losses due to the SLM aperture are obtained when the incident light, with a Gaussian distribution, fills the complete surface *ND x ND* of the SLM. Therefore, the losses are proportional to the quantity of SLM aperture illuminated by the collimated light coming from the lens, as in Fig. 15. A practical way to reach the former condition is by changing the size of the hologram according to the number of active pixels.

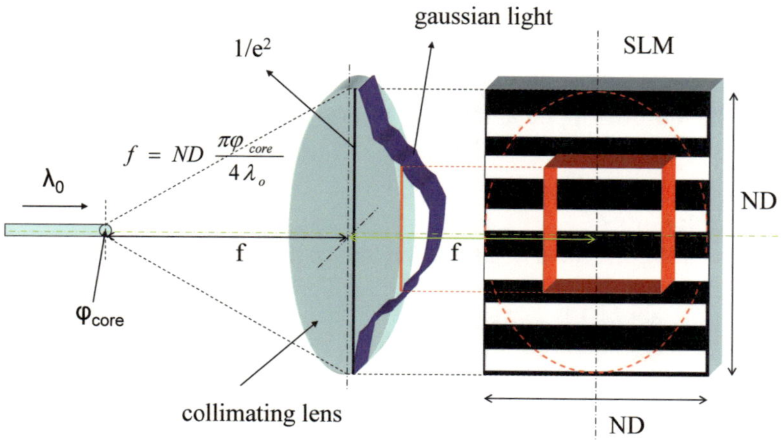

Fig. 15. Losses in the incident light due to different ND x ND hologram apertures

8.2.4 Mixed hologram operation

The EH-ROADM is able to select at the output fibers any combination of wavelengths at the input fiber, from all input wavelengths in just one output fiber to each input wavelength at the corresponding output fiber, including all intermediate cases. This operation mode is done by the selection in the SLM of a mixed hologram composed of all individual holograms corresponding to each input wavelength.

Fig. 16 shows an example for three input wavelengths and its holograms, formed, in this case, by black and white bars (2-phases). For every input wavelength (channel) a hologram is assigned, where n_i (spatial period) produces the pass-band filter for the channel and N_i

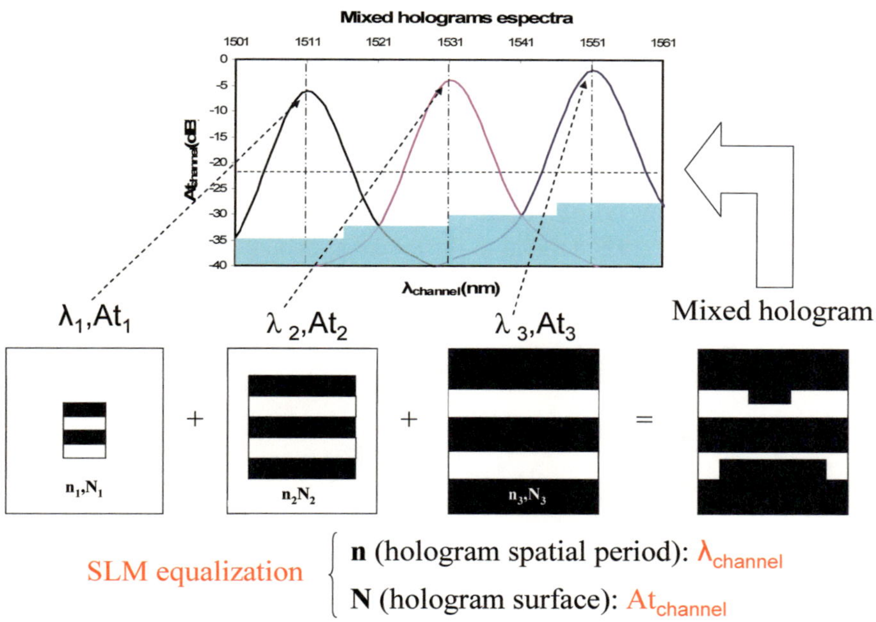

Fig. 16. Mixed holograms operation

takes into account the number of active pixels to reach the correct attenuation, $At_{channel}$, to equalize the output signals. This mixed hologram produces the additional loss in the holographic router, *10 x log (number of channels)*. A more fitted equalization can be obtained by monitoring the outputs with a feed-back loop to adjust the size of the holograms according to the wanted output signal level.

8.2.5 Design calculations

Having chosen the SLM, the focal length of the lens, *f*, to illuminate with collimated light the complete active surface *ND x ND* of the SLM (see Fig.15), is related to the number of pixels *N* and their size *D* according to expression (8):

$$f = \pi \phi_{core} \frac{ND}{4\lambda_0} \tag{15}$$

where Φ_{core} is the input fiber core diameter and λ_0 the central wavelength in the operation region. The 3 dB pass-band filter bandwidth of the device, *BW*, is (Parker et al., 1998)

$$BW \geq \phi_{core} \frac{(M/2).d}{f} \cdot \left(1 - \frac{\lambda_0^2}{(M/2.d)^2}\right)^{3/2} \tag{16}$$

if the condition $8D >> d$ is reached, where *d* is the fixed grating spatial period. For our calculations, we have a reflective 4-phases SLM with N = 1024 and D = 8 μm (ND = 8.192 mm). Then, the focal distance for the lens is 37.655 mm and the $BW \geq$ 1.52 nm (190 GHz), *d* being = 6.5 μm, the spatial period of a 4-phases transmissive diffraction grating and Φ_{core} = 9 μm the core diameter of a singlemode fiber. By changing *d* we can adjust the *BW* of the holographic filter.

In the expression (13) the selected wavelength of operation is calculated. The value of *n* is varied from $n = 0$ (for maximum wavelength) and $n = N/4$ (for minimum wavelength); the central wavelength λ_0 is obtained when $n = N/8$. For the design of a 1x4 router working in the upper band of the CWDM grid, 1471-1611 nm, we take λ_0 = 1541 nm. In this case λ_{min} = 1407 nm and λ_{max} = 1693 nm; these values cover the entire CWDM upper band. The distance from the optical axis to the output fiber, (see Fig. 11), where the 1st order of the total diffraction is produced, is x = 9.808 mm and the total diffraction angle $\varphi \approx 14.6^{o}$.

From ITU G.694 Rec., all CWDM channels are spread $\Delta\lambda$ = 20 nm to allow Direct Modulated Lasers (DML) wavelength variation with temperature and filter tolerance; therefore, an $\Delta\Lambda = (8\text{-}1).\Delta\lambda$ = 1611-1471 = 140 nm range assumes ΔX = 1260 μm, according to the relation:

$$\Delta x \approx \Delta\lambda . f \left(\frac{n}{ND} + \frac{2}{(M/2).d} \right) \tag{17}$$

This is the maximum interval, at the output plane, where all the output fibers have to be placed. That mean, a separation between fibers of $\Delta x = \Delta X / (8\text{-}1)$ = 1260 / 7 = 180 μm. Single mode fibers have a cladding diameter of 125 +/- 1 μm, so, we can reduce this separation to aprox. 130 μm, if it is needed.

	n_{ij} (holograms)				
	λ_3 = 1511 nm	λ_4 = 1531 nm	λ_5 = 1551 nm	λ_6 = 1571 nm	x_j (μm)
Fibre 3	113	95	78	61	**9538**
Fibre 4	139	121	103	85	**9718**
Fibre 5	165	146	128	110	**9898**
Fibre 6	191	172	153	135	**10078**
N_i (pixels)	**256÷1024**	**256÷1024**	**256÷1024**	**256÷1024**	

Table 8. Holograms and active pixels for an EH_ROADM 1x4

In an equalized holographic router, the directing of the input wavelengths to the output fibers is done by the choice of three parameters: n_{ij} for wavelength tuning, N_i for power equalization and Δx_j for placing the output optical fibers. Subscript *i* is related to the number of input wavelengths and subscript *j* related with the number of output fibers. Haqving fixed the separation between fibers, in our case Δx = 180 μm, we obtain the corresponding value of n_{ij} from (13), according to the input wavelength(s) and output fiber(s) considered. As we are managing different sets of n_{ij} values, all of them have to be different in order to avoid cross-talk between wavelengths on different output fibers.
Table 8 shows the holograms (n_{ij}) and number of active pixels (N_i) for a 4-channels grid according to the ITU G.695 Rec. For instance, in Fig. 11, a mixed hologram 113+95+78+61 addresses the 4 input wavelengths (λ_3+ λ_4+ λ_5+ λ_6) to the output fibre 3; a mixed hologram 113+121+128+135 addresses λ_3 to fibre 3, λ_4 to fibre 4, λ_5 to fibre 5 and λ_6 to fibre 6. In each case, every λ_i has the corresponding N_i range to assure the power equalization at the output.
Table 9 is a summary of the losses in the device (SOA+EH-ROADM) according to the different input channels, whose variation in wavelength is in agreement with Fig. 15. In this case, there is a net gain of 10 dB to compensate for the power variation due to different paths of the input channels along the network. The N_i range, 256÷1024, in Table 8 is to compensate a total of 12 dB of attenuation; with a step of ΔN_i = 16 the ripple at the output is < 0.5 dB.

Intrinsic router loss (dB)	**6**		L_{HR}
Additional mixed hologram loss (dB)	**6**		ΔL_{HR} **[10 x log (4)]**
Max. gain of SOA, G_A (dB)	**24**		**@ 1531 nm**
Min. gain of SOA, G_A (dB)	**22**		**@ 1571 nm**
λ (nm)	ΔG_A (dB)	Total loss (dB)	
1511	**1.5**	**13.5**	
1531	**2**	**14**	
1551	**1**	**13**	
1571	**0**	**12**	
Total min. net gain, G_T (dB)		**10**	

Table 9. SOA gain, EH_ROADM losses and total net gain

8.2.6 CWDM METRO networks application

The use of tunable holographic devices in Access and Metro networks, like demultiplexers or routers has been studied in different papers (Koonen, 2006), (Martin Minguez & Horche, 2010). In Fig. 17 an application for the equalized holographic ROADM is represented.

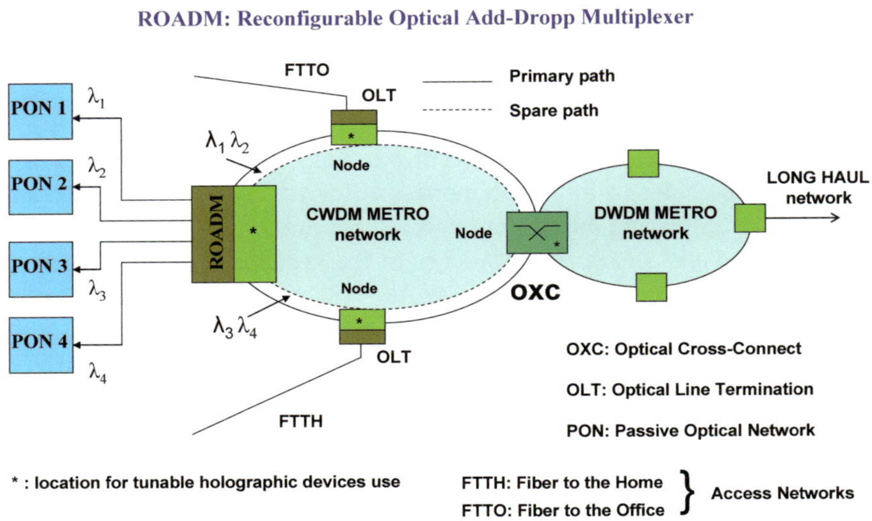

Fig. 17. Application of an EH_ROADM in a CWDM METRO network

A double ring CWDM METRO topology is used to connect this primary access network, through an Optical Line Termination (OLT), with some Fiber to the Office (FTTO) or Fiber to the Home (FTTH) networks with Passive Optical Network (PON) structure; on the other side, a connection to a DWDM METRO network, by an OXC (Optical Cross Connect) with λ conversion, is provided. The target is to address the wavelengths of the double ring network, λ_1, λ_2, λ_3 and λ_4 to four different PONs with the possibility of wavelength reallocation.

9. Design of an holographic router with λ conversion and losses compensation

Fig. 18 shows a device composed of a Semiconductor Optical Amplifier (SOA) and a holographic wavelength router. The SOA performs the wavelength conversion by a non linear operation using the Cross Gain Modulation (XGM) method. An incident wavelength, λi, modulated by a digital signal is combined with the wavelength λ_{CWj} generated by a tunable laser (CW) into the SOA. At the amplifier output, according to different CWj wavelengths, λ_{CWj} signals are obtained modulated with the digital signal from the incident λi wavelength. These λ_{CWj} signals are also amplified and inverted.
The holographic wavelength router, depending on the input signal, λ_{CWj},, and the generated hologram *(nij)* stored in the SLM, addresses this signal to the assigned output. As has been stated, this technology has the drawback of high insertion losses (less than 10 dB, using an optimized device). In order to solve this problem, by combining a SOA with the holographic router, this insertion loss is compensated with the amplifier gain in the saturation zone of operation. A parameter to control in the SOA operation, is related to the amplified spontaneous emission (ASE) because of the impact on the signal distortion.

Fig. 19 shows the simulation of this device, composed of three different blocks: a CW tunable laser, a wavelength conversion semiconductor optical amplifier and a wavelength holographic router. In Fig. 20, the response of the Wavelength Conversion and Routing Holographic Device (WCR-HD) is represented for a 2.5 Gb/s input signal, λi = 1540 nm, which is converted to an output signal, λo = 1520 nm, where the losses of the holographic router are compensated by the gain of the SOA.

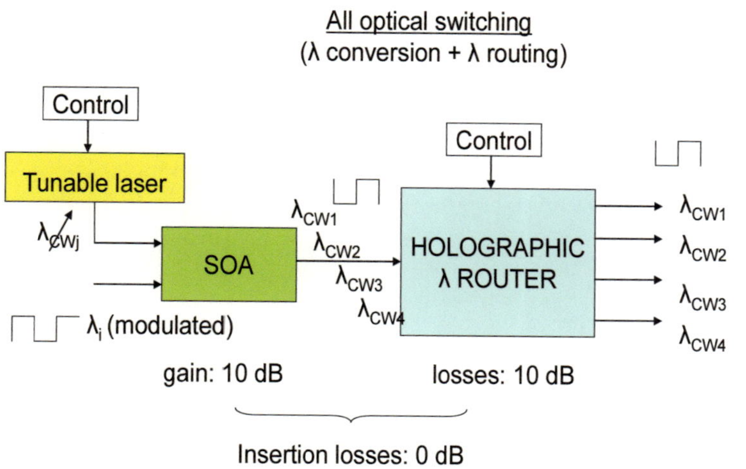

Fig. 18. Device composed of an optical λ converter and a holographic λ router

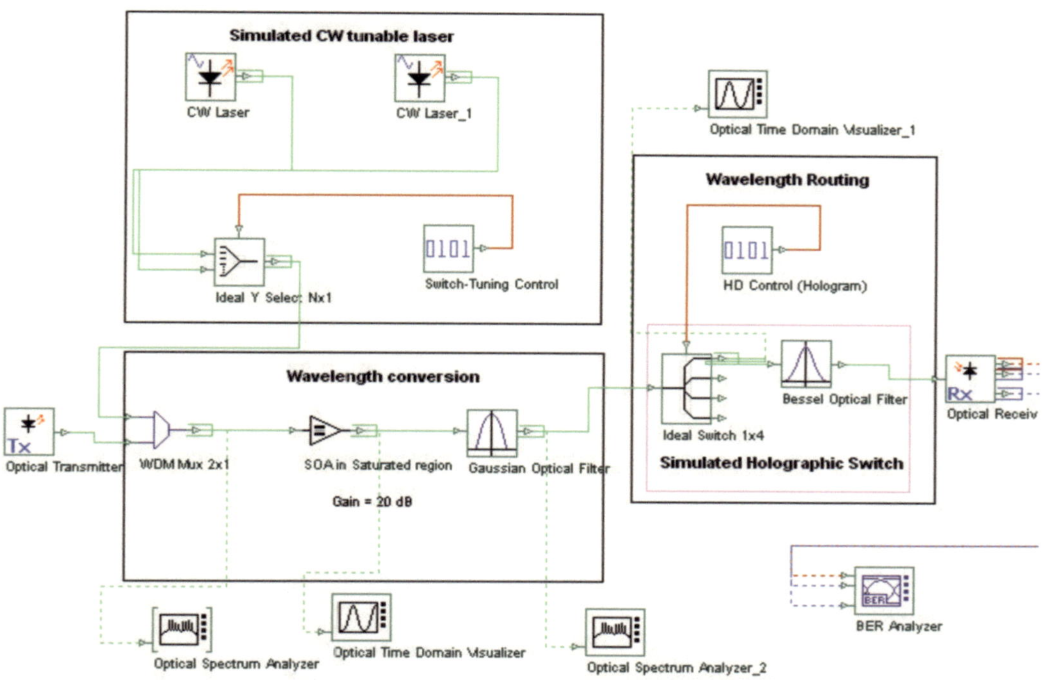

Fig. 19. Wavelength Conversion and Routing Holographic Device (WCR-HD) simulation

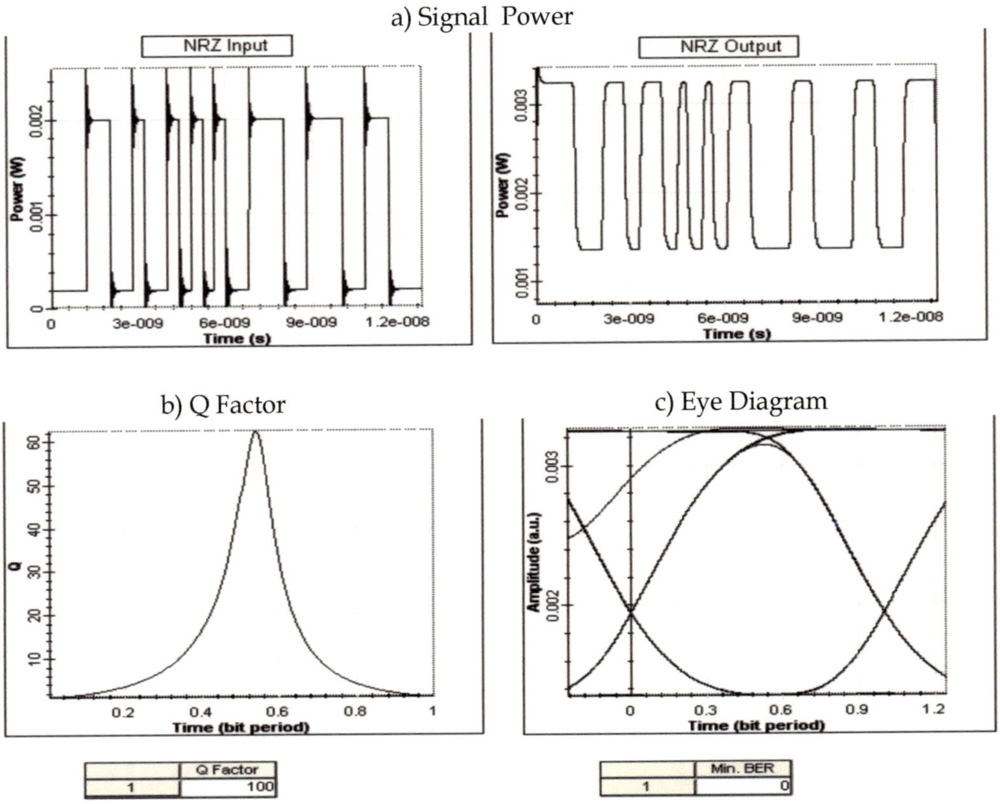

Fig. 20. WCR-HD response for a 2.5 Gbit/s input signal: a) λi = 1540 nm, with wavelength conversion λo = 1520 nm, and losses compensation, b) Q factor ≈ 100 and c) BER ≈ 0.

10. Conclusion

In this chapter the design of a singular device for use both in CWDM/DWDM systems has been studied. Applications such as, tunable optical filters, demultiplexers and wavelength routers, using holographic SLM technology, have been reviewed taking into account ITU-T G.694.1 and G.694.2 Recs. for central wavelengths allocation.

Application of Computed Generated Hologram design (CGH) to CWDM/DWDM systems has been studied and some comments about this hologram generation technique and its results have been made in order to highlight the phases of the process implementation and the issues related to the diffraction target misalignment and the use of 2-4 phase holograms, etc.

The novel idea in this work is the design of a compatible CWDM/DWDM device able to carry out different multiplexing functions. As we commented before, a better device performance as a tunable filter, demultiplexer or router could be implemented if only one of these functions is required.

The design of equalized holographic ROADM devices for applications in CWDM optical networks has been developed. By using a mixed hologram, corresponding to the combination of several input wavelengths, the tuning of a broad range of wavelengths has

been obtained allowing the full routing of several channels from the input fiber to the outputs. As it is possible to change the active pixels in the SLM for each hologram, in order to maintain a fixed output power level, channel equalization has been reached. Intrinsic losses of the device have been optimized using 4-phases holograms whose diffraction efficiency, for the 1st order, is twice that of binary holograms.
Also, the ROADM size has been minimizing by using a "2f-folded" instead of a "linear-4f" for the optical structure. To reduce the total insertion losses of the holographic device a SOA has been added increasing the input power range for equalization. An example of use of these ROADM devices in CWDM Metro and Access Networks (PONs) has been reviewed.
Another example of application is dealing with the design of a holographic router with losses compensation and wavelength conversion, whose main application is in the interconnection nodes of Metro networks. This device uses a SOA (Semiconductor Optical Amplifier), in the non-linear region, to do the wavelength conversion and, in addition, to supply the gain in order to compensate for the intrinsic losses of the holographic router. Other applications in Metro networks like path protection between nodes or switch matrix for ring networks interconnection could be implemented showing the versatility of these devices (Tibuleac & Filer, 2010).
Laboratory experiments testing the capability of a phase FLC-SLM to be used in these devices have been carried out and results show that, for different types of holograms, the possibility of distributing several wavelengths depends on the diffracted angle and, therefore, enabling the building of filters, demultiplexers or wavelength routers.

11. Acknowledgment

The authors gratefully acknowledge the support of the MICINN (Spain) through project TEC2010-18540 (ROADtoPON).

12. References

Ahderom S.; Raisi M. et al.. (Jul 2002) "Applications of Liquid Crystal Spatial Light Modulators in Optical Communications", 5th International Conference on High Speed Networks and Multimedia Comm., 3-5, pp 239-242

Agrawal, G.P. (2002) "Fiber-Optic Communication Systems" (Third Edition), Wiley Interscience

Alarcón A., (2004) "Dynamic holography applications in SLMs based systems", Master Thesis, ETSITM, Universidad Politécnica de Madrid

Broomfield S.; Neil M. et al. (1992)"Programmable binary phase-only optical device based on ferroelectric liquid crystal SLM", Electronics Letters, vol. 28 (1), pp. 26-28

Crossland W. A. et al.. (Dec 2000)"Holographic Optical Switching: The ROSES Demostrator" IEEE/OSA, J. of Lightwave Tech., vol 18 , no 12, pp 1845-1853

Dames M.; R. Dowling et al. (1991) "Efficient optical elements to generate intensity weighted spot arrays: design and fabrication", Applied Optics vol. 30, no (19), pp. 2685-2691

Homa J. & Bala K., (Jul 2008) "ROADM Architectures and Their Enabling WSS Technology" IEEE Communications Magazine, pp 150-153

Horche P.R.; Alarcón A. et al.. (2004), "Spatial Light Modulator holographic filter for WDM systems", International Union of Radio Science URSI'04

Kashnow R.A. & Bigelow J.E. (Oct 1973) "Difracttion from a Liquid Crytal Phase Grating", Applied Optics, vol. 12, no. 10, pp. 2303-2304

Koonen T.,(May 2006) "Fiber to the Home/Fiber to the Premises: What, Where, and When?" Proceedings of the IEEE, Vol. 94, No. 5, pp 911-934

Ma X, & Kuo G.S., (Nov 2003) "Optical Switching Technology comparison: optical MEMs vs other technologies", IEEE Optical Communications, S16-S23

Martin Minguez A. & Horche P.R, (Jun 2007) "A multipurpose WDM holographic device for filtering, wavelength multiplexing and routing applications", Photonic Network Communications Journal, Vol 13, nº 3, pp 313-322

Martin Minguez A. & Horche P.R., (2010) "Tunable Holographic Components in WDM Optical Networks", Optical and Quantum Electronics J., Volume 42, Issue 1, pp 45-67

Parker M. C.,A.D. Cohen et al., (1997) "Programmable Holographic Elements for WDM", IEE Colloquium on Optoelectronic Integ. and Switching- Ref Nº 1997/372, 10/1-6

Parker M. C. et al. (Jul 1998) "Dynamic Digital Holographic Wavelength Filtering", IEEE/ /OSA, J. of Lightwave Tech., vol. 16, no 7, pp. 1259-1270

Tibuleac S. & Filer M., (Feb 2010) "Transmission Impairments in DWDM Networks with Reconfigurable Optical Add-Drop Multiplexers", J. of Lightwave Technology, vol 28, nº 4, pp 557-568

12

Polarization-Selective Substrate-Mode Volume Holograms and Its Application to Optical Circulators

Jing-Heng Chen[1], Kun-Huang Chen[1] and Der-Chin Su[2]
[1]*Feng Chia University*
[2]*National Chiao Tung University*
Taiwan

1. Introduction

Optical circulators [Ramaswami et al., 2009; Hecht, 2005; Mynbaev & Scheiner, 2000] are important nonreciprocal devices that can direct a light from one port to another in only one direction. They are essential components in the construction of fundamental network modules, such as optical add-drop multiplexers, dispersion-compensation, optical amplifiers, and time-domain reflectometry. Different kinds of design of optical circulator have been proposed [Iwamura et al., 1979; Shirasaki et al., 1981; Yokohama et al., 1986; Koga, 1994; Wang, 1998]. According to the operation principles, optical circulators can be divided into three types, traditional, waveguide, and holographic. The traditional optical circulators mainly apply spatial walk-off polarizers (SWPs) [Nicholls, 2001], Faraday rotators (FRs), and half-wave plates (Hs) to implement its function. The waveguide optical circulators utilize a waveguide Mach-Zehnder interferometer to implement the function of SWPs. The holographic optical circulators apply holographic optical elements to replace traditional SWPs. Accordingly, the spatial walk-off polarizer is a key component in the design of optical circulator that significantly influences the performances and cost of a device.

Traditional spatial walk-off polarizers are essentially birefringent crystals that can split an optical beam into two orthogonally polarized beams. However, birefringent crystals suffer from challenges of highly optical qualities, crystal manufacturing, and hard optical fabrications. The highly optical qualities mean high transparency and optical uniformity for a wide spectrum range, high birefringence, and enough hardness. The main crystal growth thechnologies are Czochralski method and Verneuil process. The hard fabrications include x-ray orientation, slicing, polishing, coating, cleaning, testing, packaging, and related processes. Therefore, the cost is hard to down. In addition, limited by the finite birefringence, the beam splitting distance is small. Therefore, the device length is hard to shorted.

Compare to the crystal-type SWPs, polarization-selective substrate-mode volume holograms (PSVHs) [Huang, 1994] have a large splitting angle and several superior advantages. A PSVH are phase volume holograms stacked on a glass or plastic substrate and signals transmit in the substrate by total internal reflection. With this planar structure, PSVHs have advantages of easy fabrication, low cost, high efficiency, compactness, easy coupling, and easily to combine with other elements. Due to these merits, PSVHs had been widely applied in several optical systems, such as optical sensing, optical data storage, imaging system, and switching network. In 2003, the PSVHs was firstly proposed to replace crystal-type SWPs in a four-port optical ciculator [Chen et al., 2003]. In the application, these PSVHs are consequently termed as holographic spatial walk-off polarizer (HSWP). Due to the introduction of HSWPs, the fabricated four-port optical circulator has advantages of polarization-independence, compactness, high isolation, low polarization mode dispersion, low cost, and easy fabrication.

However, the feasibility of conventional PSVHs is usually limited by the finite refractive index modulation strength of a recording material. The common solution is to increase the thickness of the recording material, in order to compensate the shortage of the refractive index modulation strength in the phase modulation term. However, under the thickness condition of thick material, the distortion effect of interference fringe is worsened. An ideal holographic recording condition hinges on the thin thickness of a recording material with a high refractive index modulation strength. Actually these cannot be completed in both respects. To overcome the problems, based on the coupled-wave theory and the structure of substrate-mode hologram, a special design of PSVHs was proposed with a relatively large splitting angle near 90° [Chen et al., 2008]. With this design, a low refractive index modulation strength is required, which can be easily achieved with common recording materials. In addition, this design should bear all merits of conventional PSVHs.

As the design of optical communication systems becomes more and more complex, an optical circulator with many input and output ports has become highly desirable. However, the port numbers for presently most commercial optical circulators are limited. In 2004, based on holographic spatial- and polarization-modules (HSPMs), two kinds design of holographic-type multi-port optical circulator were also proposed [Chen et al., 2004; Chen et al., 2004]. The HSPM is consisted of two HSWPs, an half-wave plate (H), and a Faraday rotator (FR). The merits of these designs include polarization-independence, compactness, high isolation, low polarization mode dispersion, and easy fabrication. Furthermore, the number of port can be scaled up easily.

Accordingly, this chapter devotes to introduce the polarization-selective substrate-mode volume hologram in several respects and its novel applications in design of optical circulator. The second section, according to coupled-wave theory, will clearly describe the principle and characteristic of conventional PSVHs; a modified design method of PSVHs will also be described to overcome the shortage in refractive index modulation strength. The third section will introduce the applications of PSVH to optical circulators. The principle and operation characteristic of a four-port optical circulator will be introduced. The following context will introduce the principles and operation characteristic of holographic spatial- and polarization-modules (HSPMs) and their applications to multi-port optical circulators. All the design details will be described and their characteristic will be discussed. Finally, the fourth section is conclusion.

2. Polarization-selective substrate-mode volume holograms

2.1 Conventional polarization-selective substrate-mode volume holograms

Figure 1 shows the structure of the conventional polarization-selective substrate-mode hologram [Huang, 1994] which is composed of four volume holograms, input grating coupler (H_I), polarization beam splitter hologram (H_{PBS}), output grating couplers (H_{OS} and H_{OP}), and two substrates. An unpolarized light is incident on H_I normally, and is diffracted into H_{PBS} at a special angle. The output diffraction lights of H_{PBS} are split into *s*- and *p*-components which are perpendicular to each other. These two components are then total internal reflected (TIR) at the base of the substrate and are diffracted and coupled out normally by H_{OS} and H_{OP}, respectively. Therefore, the *s*- and *p*-polarized lights are successfully separated.

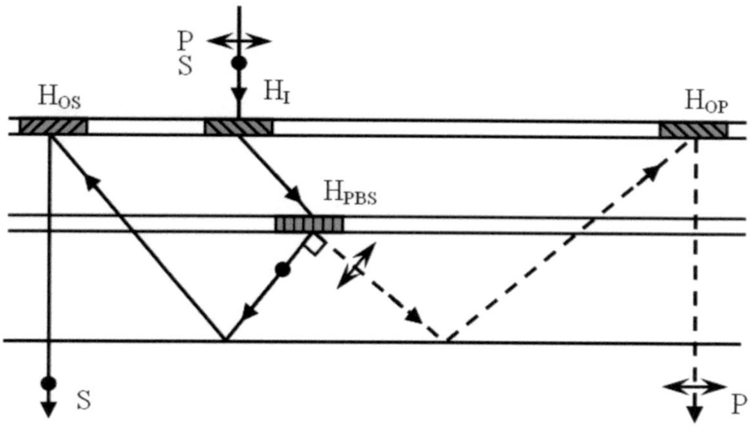

Fig. 1. Schematic representation of the conventional polarization-selective substrate-mode volume hologram.

In this structure, H_I, H_{PBS}, H_{OS}, and H_{OP} are actually transmission-type phase volume holograms and can be designed according to the coupled-wave theory [Kogelnik, 1969]. For a transmission-type phase volume hologram, as shown in Fig. 2, the relation between the diffraction efficiencies of *s*- and *p*-components can be written as

$$\eta_{s,p} = \sin^2 \upsilon_{s,p}, \tag{1}$$

where the modulation parameters for *s*- and *p*-components, υ_s and υ_p, are given as

$$\upsilon_s = \frac{\pi N_1}{(\cos\theta_{r1}\cos\theta_{r2})^{1/2}}, \tag{2}$$

$$\upsilon_p = \upsilon_s \cos(\theta_{r2} - \theta_{r1}), \tag{3}$$

and

$$N_1 = \frac{n_1 d}{\lambda}. \tag{4}$$

N_1 is the effective index modulation in which λ is the reconstruction wavelength, d is the thickness of the recording material, and n_1 is the refractive index modulation. θ_{r1} and θ_{r2} are corresponding angles of the reconstruction and the diffraction beams in the recording material, respectively.

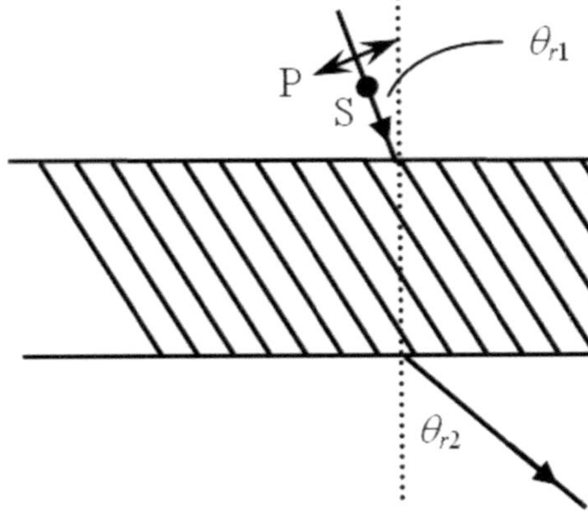

Fig. 2. Reconstruction geometry of the phase volume hologram: S, *s*-polarization field; P, *p*-polarization field.

In the case of normal incident, *i.e.* $\theta_{r1}=0°$, eqs. (2) and (3) can be reduced as

$$\upsilon_s = \frac{\pi N_1}{(\cos\theta_{r2})^{1/2}}, \tag{5}$$

$$\upsilon_p = \upsilon_s \cos(\theta_{r2}). \tag{6}$$

Accordingly, the design of H_I requires $\eta_s=\eta_p \geq 90\%$ that can be solved by eqs. (1)-(6). The function of H_{PBS} requires $\eta_s=100\%$ and $\eta_p=0$ or $\eta_s=0$ and $\eta_p=100\%$. The functions of H_{OP} and H_{OS} require $\eta_p=100\%$ ($\eta_s=0$) and $\eta_s=100\%$ ($\eta_p=0$), respectively.
However, in order to satisfy the requirements of H_{OP} and H_{OS}, the parameters υ_s and υ_p stand on the following conditions: (1) $\upsilon_s=[m+(1/2)]\pi$ and $\upsilon_p=m\pi$ (for $\eta_s=100\%$ and $\eta_p=0$); (2) $\upsilon_s=m\pi$ and $\upsilon_p=[m-(1/2)]\pi$ (for $\eta_s=0$ and $\eta_p=100\%$), where m is a positive integer. Under these conditions, the values of related parameters m, θ_{r2}, and N_1 are listed in Table 1. In order to fulfill the required TIR inside the substrates, only conditions at m=1 are valid. However, the feasibility of fabricating these elements is usually limited by the finite refractive index modulation strength n_1 of a recording material. Therefore, an alternative design method is described below to overcome this drawback.

		m				
		1	2	3	4	5
θ_{r2}	Condition (1)	48.2°	36.9°	31.0°	27.3°	24.6°
	Condition (2)	60.0°	41.4°	33.5°	29.0°	25.8°
N_1	Condition (1)	1.22	2.24	3.24	4.24	5.24
	Condition (2)	0.707	1.73	2.74	3.74	4.74

Table 1. Related parameters for condition (1) ($\eta_s=100\%$ and $\eta_p=0$) and condition (2) ($\eta_s=0$ and $\eta_p=100\%$).

2.2 Alternative design of polarization-selective substrate-mode volume holograms

Shown in Fig. 3 is a schematic representation of the proposed polarization-selective substrate-mode volume hologram [Chen et al., 2008] which is consisted of a transmission-type phase volume holographic grating and a substrate. Its grating structure is designed in such a way that either of the *s*- or *p*-polarized component of a normal incident beam at A is transmitted straight through the grating and the substrate (channel 1) while the other orthogonally polarized component is completely diffracted into the substrate with a large diffraction angle θ_{r2} which is larger than the critical angle θ_c at the interface of recording material and substrate. In this way, the diffracted beam is totally reflected at point B and hits the grating again at point C. This beam is totally reflected at point C, and the reflected beam from point C satisfies the Bragg condition of the grating. The propagation direction of the reflected beam is in parallel to that of the beam diffracted by the grating at point A. Because the structure of the grating at point C is the same as that at point A, the diffracted beam at point C will be in parallel to the input beam at point A; that is, the output beam passes normally through the substrate (channel 2). The detail of the beam propagation at point C is shown in the upper right circle of Fig. 3. Consequently, two orthogonally polarized parallel beams with the separation of length AC=2d(tanθ_{r2}) can be obtained in which d is the thickness of the recording material.

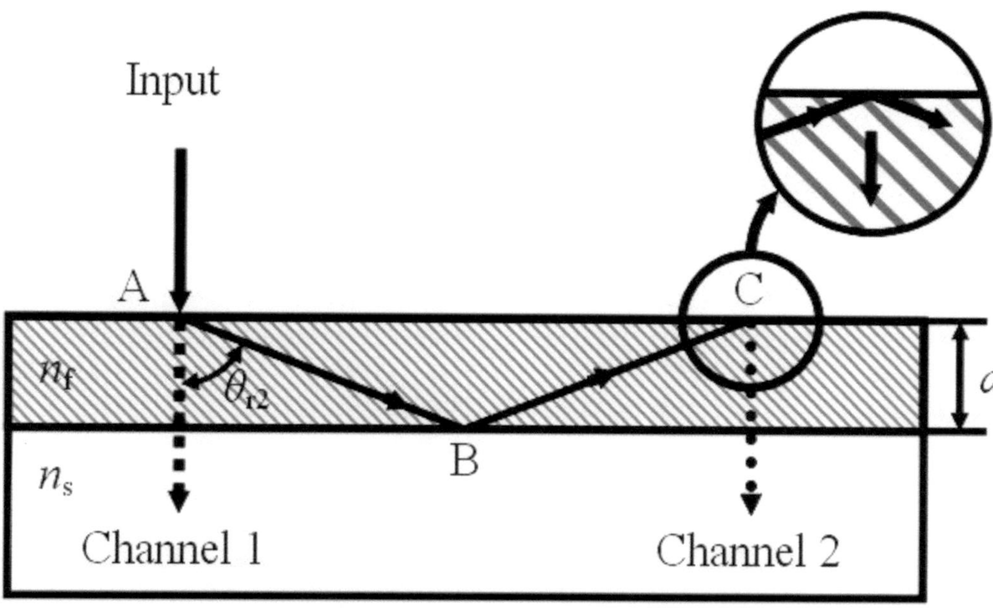

Fig. 3. Schematic representation of the proposed polarization-selective substrate-mode volume hologram (I).

In addition, in the same principle, we can properly choose the substrate with its refractive index equally that of the recording material (n_s=n_f). Under this condition, the light propagation details in Fig. 3 change as shown in Fig. 4. The light separated distance becomes AC=2t(tanθ_{r2}) in which t is the thickness of the substrate. In generally, the structure in Fig. 3 is suitable for integrated optical systems, and that in Fig. 4 can be applied in common optical systems for the purpose of more compactness.

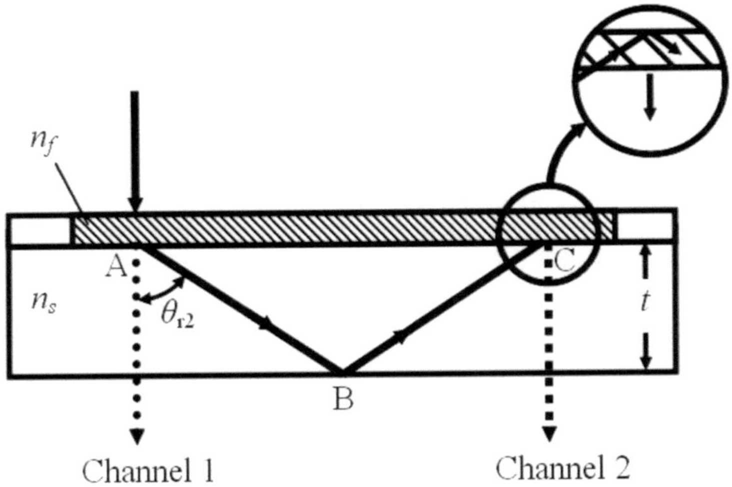

Fig. 4. Schematic representation of the proposed polarization-selective substrate-mode volume hologram (II).

According to eqs. (1), (5), and (6), the relation between the diffraction efficiencies of *s*- and *p*-components can be rewritten as

$$\eta_s = \sin^2(\frac{x}{a}), \tag{7a}$$

and

$$\eta_p = \sin^2(ax), \tag{7b}$$

where

$$x = \frac{\pi n_1 d}{\lambda}, \tag{8a}$$

and

$$a = (\cos\theta_{r2})^{1/2}. \tag{8b}$$

It is obvious from eqs. (7a) and (7b) that the diffraction efficiencies of *s*- and *p*- components oscillate in the form of a sine square function asynchronously of which the primitive periods are $T_s=a\pi$ and $T_p=\pi/a$, respectively. Therefore, when θ_{r2} has a large diffraction angle near 90°, the parameter a has a relative small value. This condition results a smaller value of T_s and a larger value of T_p, and the peak values of *s*- and *p*- diffraction efficiencies leave far away each other. The smaller value of T_s means a smaller required phase modulation. Therefore, in the condition of a small phase modulation value n_1d, we can obtain a desired result of η_s=100% and η_p~0 and complete the purpose of polarization beam splitting effectively. Shown in Fig. 5 is the relation of diffraction efficiencies v.s. x considering θ_{r2}=85°. It is obviously that when the value of x equals 0.46, corresponding to an effective index modulation N_1=0.15, we can obtain η_s=100% and $\eta_p\cong$1.89%.

Shown in Fig. 6 is a preliminary measurement result of a fabricated element for *p*-polarized input signal. The thickness *t* of the substrate is 1.50mm, and the light separating distance is about 26mm. The technique of shorter wavelength construction for longer wavelength reconstruction is applied for the fabrication of the holographic polarization selective element. A 532nm solid-state laser was applied as the exposure light source. Silver-halide recording material (VRP-M, Slavich) is used for the fabrication of this element designed with θ_{r2}=83.5° for 632.8nm. The related recording material parameters of *n* and *d*, before and after post-processing are measured by an optical thin film analysis system (Model: nkd-6000™, aquila Instruments Ltd.) The measured parameters are n_{f1}=1.60 (@ 532nm), n_{f2}=1.66 (@ 632.8nm), d_1=5.70μm, and d_2=5.35μm. Therefore, the ideal value of phase modulation n_1d is 0.11μm. In addition, in order to easy the operation, a right-angle prism with specification of 150×150×50mm is introduce for the exposure light guiding. Some castor oil (n=1.48, @20°C) is used as index-matching oil. Due to the large diffraction angle, a BaSF2 glass substrate (n_s=1.66, Producer: Schott Glaswerke and Schott Glass Technologies) with the same refractive index of recording material is used avoiding the reflection at the interface of recording material and substrate.

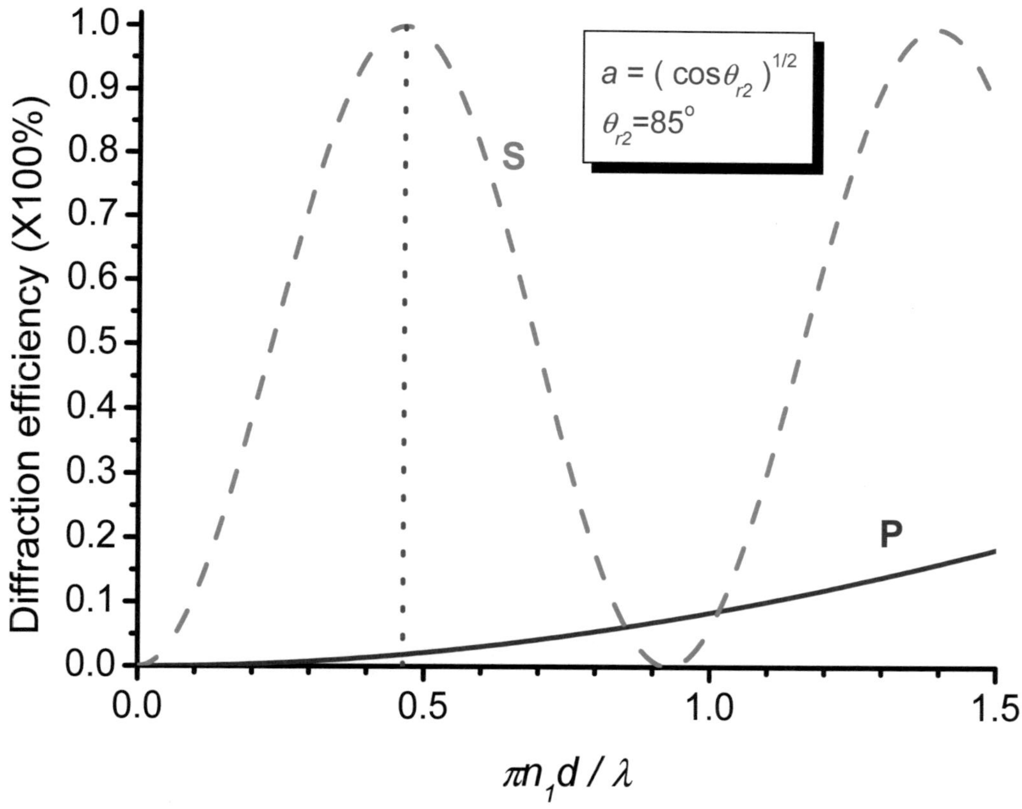

Fig. 5. The relation of *x* v.s. diffraction efficiencies, η_s and η_p, considering θ_{r2}=85°.

Fig. 6. Transmission image for *p*-polarized input signal.

In fabrications, the value of refractive index modulation n_1 relates the exposure time. Therefore, according to eqs. 8(a) and 8(b), knowing the reconstruction wavelength, the recording material thickness, and the diffraction angle, the refractive index modulation n_1 can be obtained

$$n_1 = \frac{T_s}{2} = \frac{\lambda(\cos\theta_{r2})^{1/2}}{2d}. \tag{9}$$

In the previously mentioned case, the estimated refractive index modulation n_1 is 0.02 that can be obtained by controlling the exposure time experimentally.
In addition, according to eqs. (7a) and (7b), the extinction ratio (ER) of channel 1 and channel 2 can be defined as

$$ER_1 = \frac{1-\eta_p}{1-\eta_s}, \tag{10a}$$

and

$$ER_2 = \frac{\eta_s^2}{\eta_p^2} = \left[\sin\left(\frac{\pi\cos\theta_{r2}}{2}\right)\right]^{-4}. \tag{10b}$$

According to eq. (10a), ER_1 >>1 can be obtained easily with this design. From eq. (10b), it is obvious that ER_2 is related with the diffraction angle θ_{r2}. Therefore, ER_2 has a lager value as θ_{r2} is larger. Shown in Fig. 7 is the relation of ER_2 v.s. θ_{r2}. It can be seen that the value of ER_2 is larger than 1000, when θ_{r2} is larger than 83.5°. In the same mentioned case, the diffraction efficiencies of the s- and *p*-components are about 83% and 5%, and the calculated extinction ratio of channel 1 and channel 2 are 5.58 and 275, respectively. The preliminary experimental results show the validity of the proposed method. The experimental errors mainly come from the optical setup, the process of optical exposure, and the post-processing.
In addition, the conventional polarization-selective substrate-mode volume holograms are designed with θ_{r2}=48.19° and 60°. The comparisons of the improved method and conventional method are listed in Table 2. Considering the commercial holographic recording materials, the maximum values of refractive index modulation of dichromated

gelatin and silver-halide material are not excess 0.08 and 0.03, respectively. Therefore, the condition of finite phase modulation $n_1 d$ will cause these elements hard to be realized by conventional method. This situation is especially serious in the near infrared for optical communications. The improved method not only can solve the problem but also has all merits of conventional substrate-mode volume holograms such as compactness, plane structure, easily light collimation, easily fabrication, and low cost.

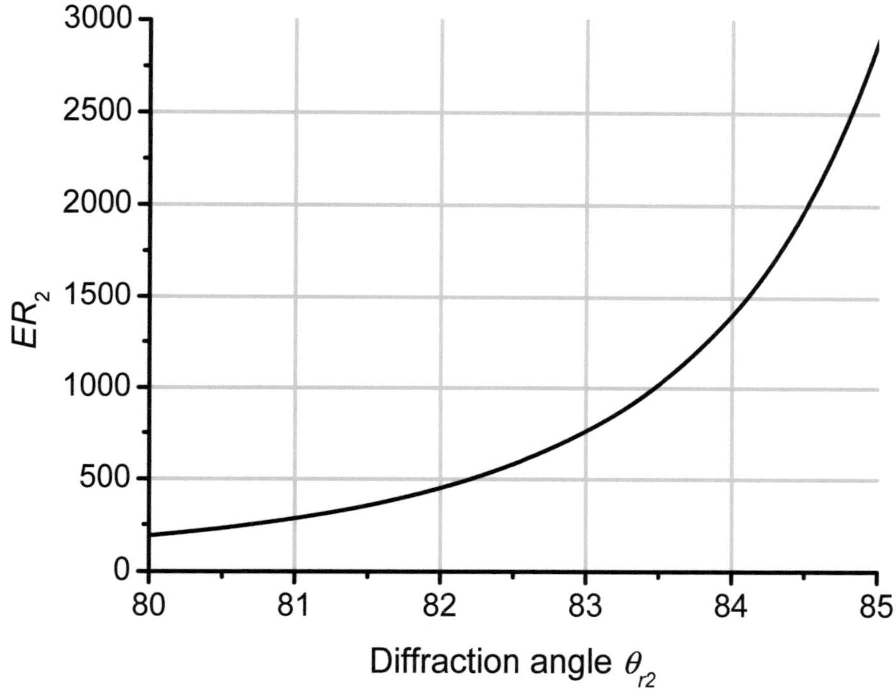

Fig. 7. The theoretical relation of ER_2 v.s. θ_{r2}.

	Conventional method				Improved method	
θ_{r2}	48.19°		60°		θ_{r2}	
(η_s, η_p)	(100%, 0)		(0, 100%)		(η_s, η_p)	
$n_1 d/\lambda$	1.22		0.71		$n_1 d/\lambda$	
Maximum index modulation strength	n_{DCG}<0.08	n_{SH}<0.03	n_{DCG}<0.08	n_{SH}<0.03	n_{DCG}<0.08	n_{SH}<0.03
Satisfied thickness d/λ	>15.31	>40.82	>8.84	>23.57	>1.85	>4.92

Table 2. Comparisons for the improved method and conventional method; n_{DCG} : maximum index modulation strength of DCG; n_{SH} : maximum index modulation strength of silver-halide material.

3. Holographic-type optical circulators

3.1 Spatial- and polarization-modules

As shown in figure 8, the spatial- and polarization- module (SPM) is composed of two spatial walk-off polarizers (SWPs), a 45° Faraday rotator (FR), and a 45° half wave-plate (H). For easy understanding, an orthogonal *x*-*y*-*z* coordinate is introduced, symbol ⊕ represent an unpolarized light, and symbols ⊖ and ⦶ represent *h*-polarized (*y*-polarized) and *v*-polarized (*x*-polarized) lights, respectively.

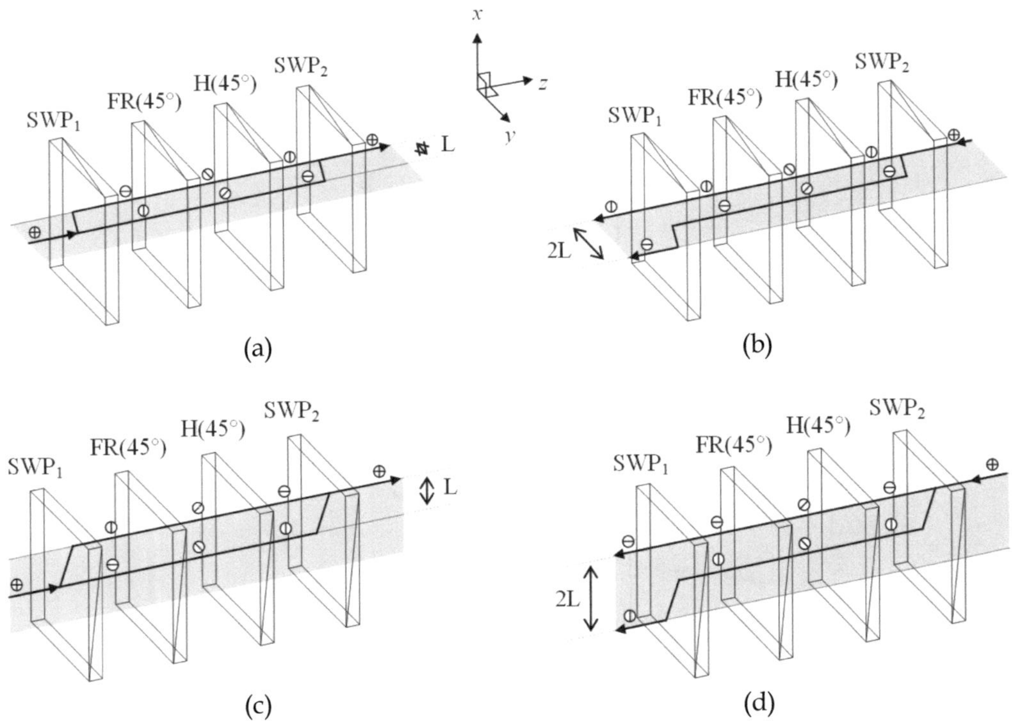

Fig. 8. Structure and operation characteristic for (a), (b) SPM_y and (c), (d) SPM_x.

From figure 8(a), when an unpolarized light is incident into the SWP_1 in +*z* direction, the transmitted light is divided into two orthogonally polarized components, *h*- and *v*-polarized lights, respectively. These two lights then pass through the 45° FR and 45° H. Therefore, their statuses of polarization (SOPs) are rotated 90° in total. Continuing their journey, they enter the SWP_2 and are recombined together with a lateral shift *L* in −*y* direction. On the other hand, Fig. 8(b) shows that when an unpolarized light is incident into the SWP_2 in −*z* direction, the transmitted light is similarly divided into two orthogonally polarized components, *v*- and *h*-polarized lights. These two lights then sequentially pass through the same H and FR. Their SOPs are rotated −45° by the H and +45° by the FR. Because Faraday rotator is a nonreciprocal element, their SOPs are rotated 0° in total. Therefore, the *v*-polarized light transmits the SWP_1 directly and the *h*-polarized light transmits the SWP_1 with a lateral shift 2*L* in +*y* direction. In Figs. 8(a) and 8(b), because the shifts of transmitted light of the SPM are in *y*-direction, this operation type SPM is defined as SPM_y. Based on the same principle, when the SPM_y is clockwise rotated

90° with respect to $+z$ axis (viewing from SWP_1 to SWP_2), the shifts of transmitted light of the SPM are in x-direction, as shown in Figs. 8(c) and 8(d). Accordingly, this operation type SPM is defined as SPM_x.

3.1.1 Parallel connection of two SPMs

As shown in figures 9(a) and 9(b), two SPM_xs are connected, *i.e.* parallel connection of two SPMs.

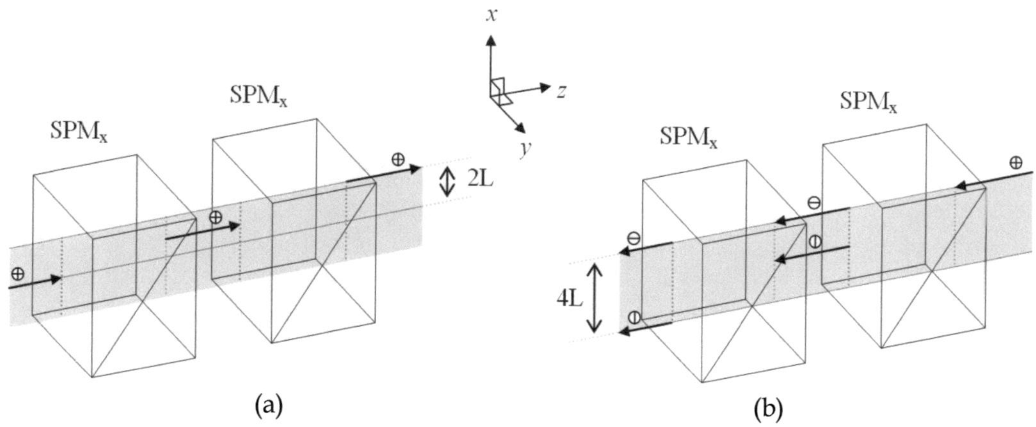

Fig. 9. Structure and operation characteristic of two connected SPM_xs for (a) forward and (b) backward transmissions.

In Fig. 9(a), when an unpolarized light is incident into the module in $+z$ direction, the transmitted unpolarized light is spatially shifted $2L$ in $+x$ direction. On the other hand, Fig. 9(b) shows that when an unpolarized light is incident into the module in $-z$ direction, the h-polarized light transmits the module directly and the v-polarized light transmits the module with a lateral shift $4L$ in $-x$ direction. Consequently, Fig. 10 shows that when an unpolarized light is shuttled between the two sides of the module, the h- and v-polarized components are separated in two opposite directions gradually in x-z plane at y=0. The corresponding x coordinates of the h- and v-polarized components at two sides of the module can be expressed as

$$\begin{bmatrix} x_{h(2n-1)} \\ x_{v(2n-1)} \end{bmatrix} = \begin{bmatrix} 2(n-1)L \\ 2(1-n)L \end{bmatrix}, \text{ (for an odd port)} \tag{11}$$

and

$$\begin{bmatrix} x_{h(2n)} \\ x_{v(2n)} \end{bmatrix} = \begin{bmatrix} 2nL \\ 2(2-n)L \end{bmatrix}, \text{ (for an even port)} \tag{12}$$

where subscripts h and v denote the h- and v-polarized components, ($2n$-1) and ($2n$) indicate the port numbers, and n is a positive integer. Accordingly, the module can sequentially guide and separate the forward and backward transmitted lights in a z-shape.

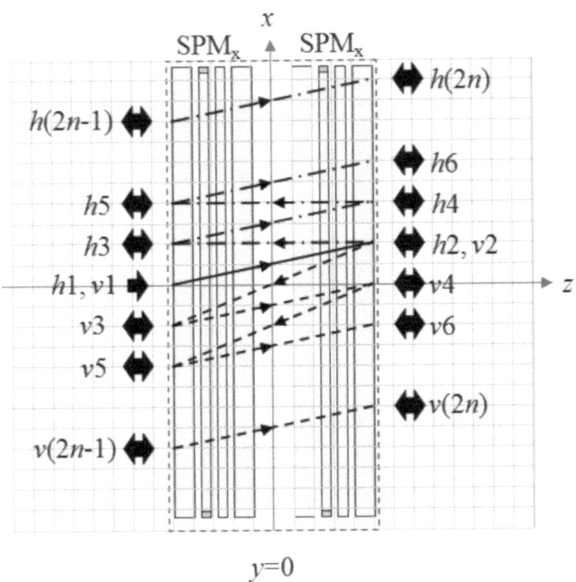

Fig. 10. Operation characteristic for an unpolarized light shuttled between the two sides of two connected SPM_xs.

3.1.2 Orthogonal connection of two SPMs

Similarly, as shown in figures 11(a) and 11(b), a SPM_y and a SPM_x are sequentially connected, *i.e.* orthogonal connection of two SPMs. In Fig. 11(a), when an unpolarized light is incident into the module in $+z$ direction, the transmitted unpolarized light is spatially shifted L in $+x$ and $-y$ directions, respectively. On the other hand, Fig. 11(b) shows that when an unpolarized light is incident into the module in $-z$ direction, the h-polarized light transmits the module with a lateral shift $2L$ in $+y$ direction and the v-polarized light transmits the module with a lateral shift $2L$ in $-x$ direction. Consequently, Fig. 12 shows that when an unpolarized light is shuttled between the two sides of the module, the h- and v-polarized components are separated in two opposite directions gradually along two slanted lines $y=x$ and $y=x\text{-}2$, respectively. The corresponding (x, y) coordinates of the h- and v-polarized components at two sides of the module can be expressed as

$$\begin{bmatrix} x_{h(2n-1)} & y_{h(2n-1)} \\ x_{v(2n-1)} & y_{v(2n-1)} \end{bmatrix} = \begin{bmatrix} (n-1)L & (n-1)L \\ (1-n)L & (1-n)L \end{bmatrix}, \text{ (for an odd port)} \tag{13}$$

and

$$\begin{bmatrix} x_{h(2n)} & y_{h(2n)} \\ x_{v(2n)} & y_{v(2n)} \end{bmatrix} = \begin{bmatrix} nL & (n-2)L \\ (2-n)L & -nL \end{bmatrix}. \text{ (for an even port)} \tag{14}$$

Accordingly, the module can sequentially guide and separate the forward and backward transmitted lights in another z-shape.

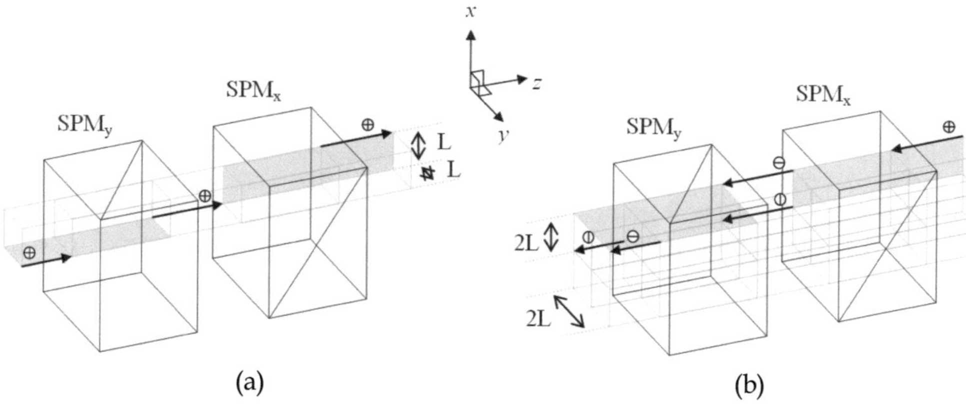

Fig. 11. Structure and operation characteristic of the connected SPM$_y$ and SPM$_x$ for (a) forward and (b) backward transmissions.

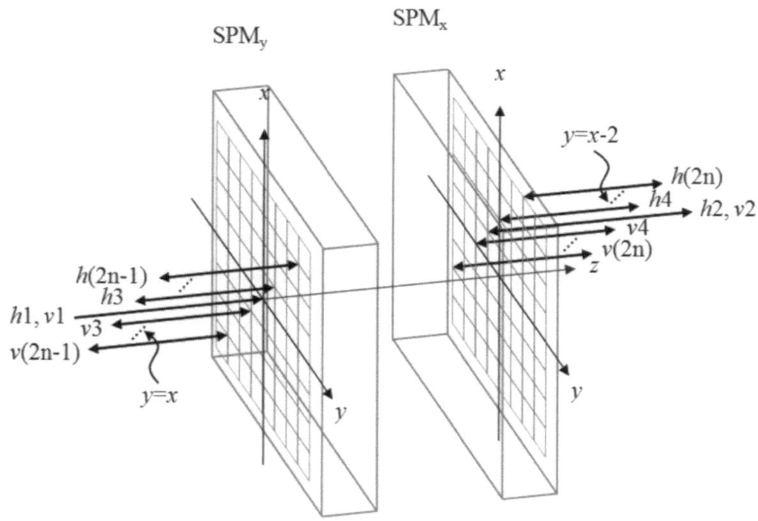

Fig. 12. Operation characteristic for an unpolarized light shuttled between the connected SPM$_y$ and SPM$_x$.

3.2 Four-port polarization-independent optical circulator

Shown in Fig. 13 is the design of proposed holographic-type 4-port polarization-independent optical circulator. In the design, based on a previously described SPM, two polarization-selective substrate-mode volume holograms are applied to implement the function of spatial walk-off polarizers. They are consequently termed as holographic spatial walk-off polarizers (HSWPs). The two identical HSWPs face the opposite directions as shown in the figure. Besides the SPM, this optical circulator also consists of four reflection prisms (RPs) and six polarization-beam splitters (PBSs). If an input beam is normally incident on HSWP$_1$ from Port 1, as shown in Fig. 13(a), then the *s*- polarized component passes through HSWP$_1$ directly and the *p*- polarized component also passes through HSWP$_1$

after two diffractions and two total internal reflections. Next, these two orthogonally polarized components pass through FR and H. Their state of polarization (SOP) are rotated a total of 90°, +45° by FR and +45° by H. For easy understanding, a circle with a bisecting line is used to represent the associated SOP of the light after propagating through each component. Symbols ⦶ and ⊖ represent the electric field lies in the planes perpendicular (*s*-polarization) and parallel (*p*-polarization) to the paper plane respectively, and the symbol ⊕ represents the light beam has both *s*- and *p*-polarized components. The beams finally enter $HSWP_2$ and then recombine together with the similar diffraction and total internal reflection effects in $HSWP_1$ and reach Port 2.

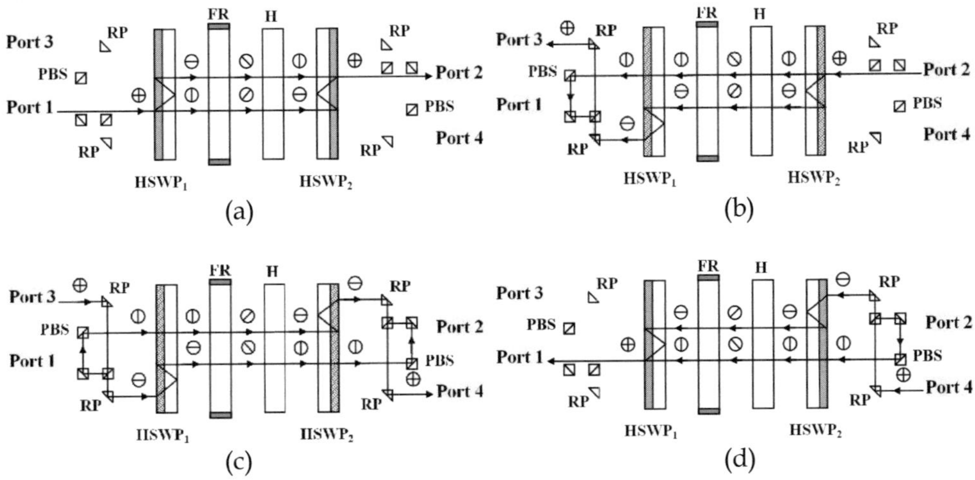

Fig. 13. Structure and operation principles of the 4-port polarization independent optical circulator.

On the other hand, if an input beam is incident normally on $HSWP_2$ from Port 2 as shown in Fig. 13(b), then the *s*-polarized component passes through $HSWP_2$ directly, and the *p*-polarized component also passes through $HSWP_2$ after two diffractions and two total internal reflections. These two orthogonally polarized components pass through H and FR. Their SOPs are rotated -45° by H and +45° by FR, a total of 0°. The *s*-polarized component passes through $HSWP_1$ and is reflected by three PBSs and one PR, and enters Port 3. The *p*-polarized component is diffracted and total internal reflected similarly in $HSWP_1$ and propagates through one RP, one PBS, and another RP. Finally, it arrives at Port 3 and recombines with the *s*-polarized component. Two other similar operations for the routes of Port 3→Port 4 and Port 4→Port 1 can be done with the introduction of additional RPs and PBSs, as shown in Fig. 13(c) and 13(d) respectively.

If the PBSs are located accurately in the configurations of Fig. 13(b) and 13(d), there will be no optical path difference between *s*- and *p*-polarizations for any route. Hence, this optical circulator can function as a polarization-independent 4-port optical circulator without polarization mode dispersion (PMD).

Listed in Table 3 are parameters for a prototype of 4-port optical circulator which were estimated from the diffraction efficiencies of fabricated HSWPs and the transmittances of FR and H. The diffraction efficiencies of η_s and η_p are 3% and 90%, respectively; the transmittances of FR and H are 0.95 and 0.97, respectively. Because η_s and η_p are slightly different from theoretical values, the transmittances of two orthogonally polarized components are slightly different in the routes of Port 2→Port 3 and Port 4→Port 1.

	Out Port			
In Port	1	2	3	4
1	14.26[a]	2.09[b]	47.91[c]	14.18[c]
2	11.91[c]	14.26[a]	2.02[b]	47.91[c]
3	28.24[c]	14.15[c]	14.26[a]	2.09[b]
4	2.02[b]	28.24[c]	44.92[c]	14.26[a]

Table 3. Associated parameters (in Decibels) for the prototype 4-port optical circulator with wavelength 1300 nm; [a]Return losses; [b]Insertion losses; [c]Isolations.

Since the fabricated HSWPs have no anti-reflection coatings, there is about 4% reflection loss at each boundary. If they are anti-reflection coated, then the reflection losses should be decreased to 0.1%. In addition, if the holographic exposure and the post-processing procedure are controlled more accurately, the HSWPs may have the theoretical diffraction efficiencies, i.e., $\eta_s \cong 0\%$ and $\eta_p \cong 100\%$. Under these two possible improved conditions, the performance of this 4-port optical circulator can be enhanced greatly for demands of a commercial device. Moreover, if K and $\Delta\lambda$ are the magnitude of grating vector $\vec{K}$ and the wavelength shift with respect to the central wavelength λ_r, the diffraction efficiencies of a transmission-type phase volume hologram for the *s*- and *p*-polarization states near the Bragg condition are given as [Kogelnik, 1969]

$$\eta_i = \frac{\sin^2(\sqrt{\nu_i^2 + \xi^2})}{(1+\xi^2/\nu_i^2)} \quad (i = s, p), \tag{15}$$

with

$$\xi = \frac{-\Delta\lambda K^2 d}{8\pi n_{f2}\cos\theta_d}, \tag{16a}$$

$$K = (\frac{4\pi n_{f2}}{\lambda_r})\cdot \sin\frac{\theta_d}{2}. \tag{16b}$$

Substituting the experimental conditions n_1=0.054, d=17μm, λ_r=1300nm, θ_{r2}=60°, and n_{f2}=1.48 (at λ_r=1300nm) into eq. (15), the theoretical curves of diffraction efficiencies versus wavelengths for the HSWP is shown in Fig. 14. It is obvious that the bandwidth with η_p>90% and $\eta_s \cong 0\%$ at 1300nm central wavelength is as large as 20nm. It is also possible to design the central wavelength at other ranges for optical communications.

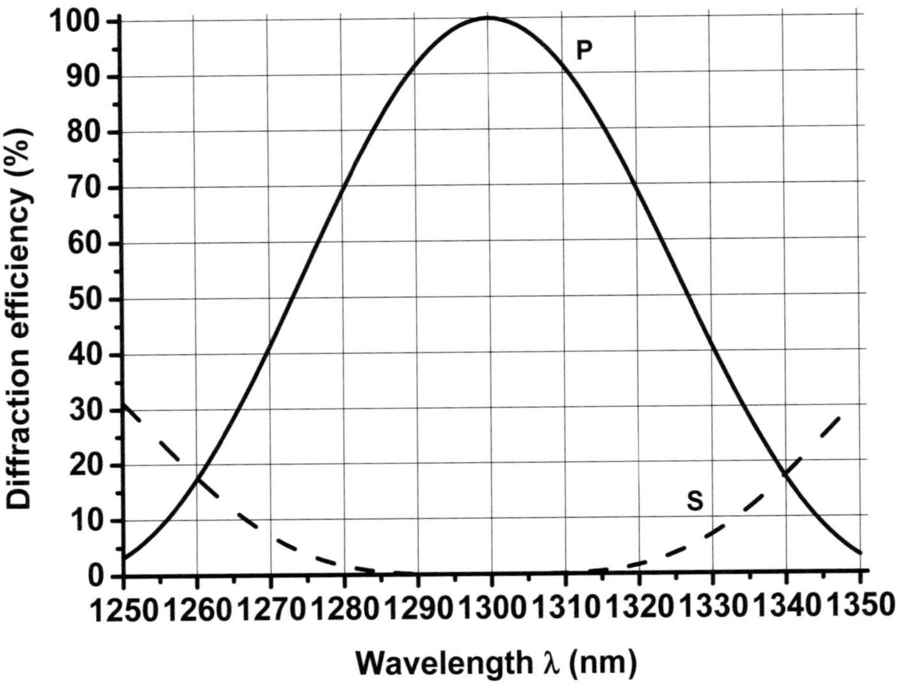

Fig. 14. Calculated diffraction efficiencies of the HSWP versus wavelength at 1300 nm central wavelength.

3.3 Multi-port polarization-independent optical quasi-circulator

Based on two connected SPM_xs, it is obvious that if some polarization-beam splitters (PBSs), and reflection prisms (RPs) are introduced appropriately at the corresponding positions of the *h*- and *v*-components, a multi-port optical quasi-circulator can be obtained. Shown in Fig. 15 is an optical quasi-circulator with 2n-ports consisting of a pair of $HSPM_x$s, PBSs, and RPs. According to equations (11) and (12), the introduced PBSs and RPs at the *j*-th port are located at (x_{PBSj}, z_{PBSj}) and (x_{RPj}, z_{RPj}), which can be expressed as

$$\begin{bmatrix} x_{PBS(2n-1)} & z_{PBS(2n-1)} \\ x_{RP(2n-1)} & z_{RP(2n-1)} \end{bmatrix} = \begin{bmatrix} 2(n-1)L & (-2n-4)L \\ 2(1-n)L & (-2n-4)L \end{bmatrix}, \text{ (for an odd port)} \quad (17)$$

$$\begin{bmatrix} x_{PBS(2n)} & z_{PBS(2n)} \\ x_{RP(2n)} & z_{RP(2n)} \end{bmatrix} = \begin{bmatrix} 2nL & (2n+4)L \\ 2(2-n)L & (2n+4)L \end{bmatrix}, \text{ (for an even port)} \quad (18)$$

where *n* is a positive integer. Figure 15(a), (b), (c), and (d) show the routes of port 1→port 2, port 2→port 3, port 3→port 4, and port (2n-1)→port 2n, respectively. In these figures, symbols ◪ and ◢ represent a PBS and a RP. Other propagation routes can be obtained based on the similar principle.

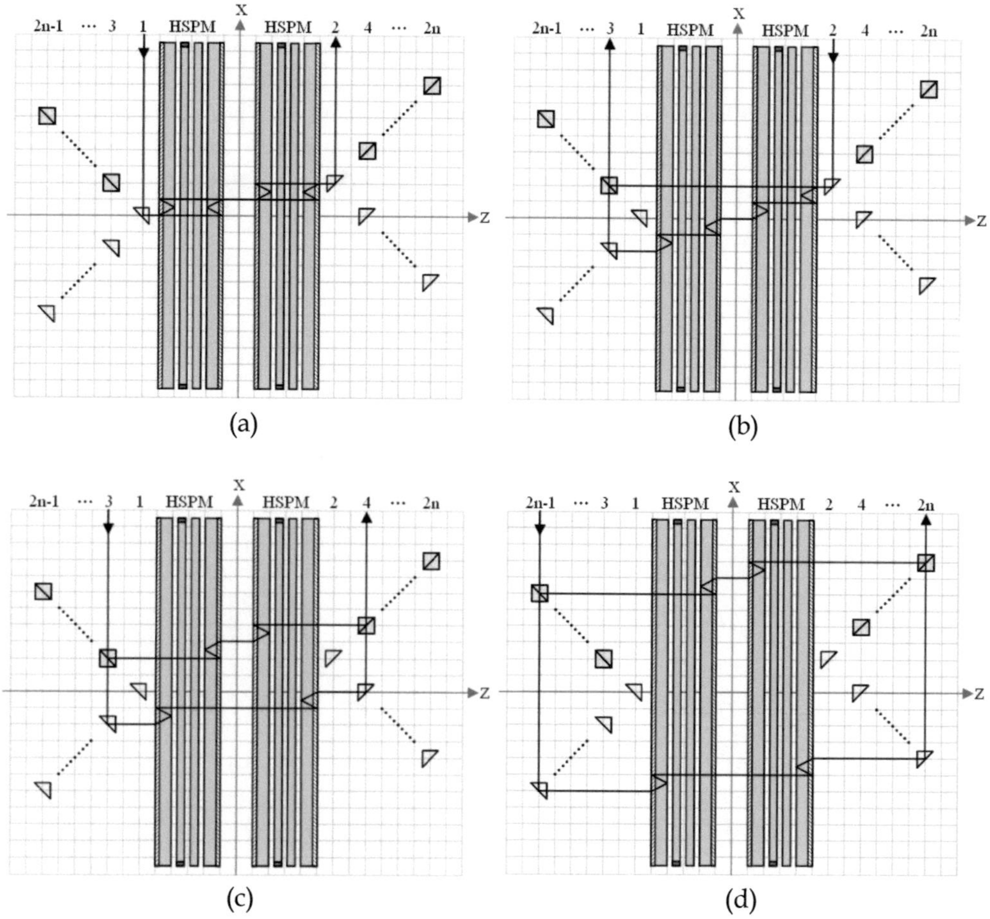

Fig. 15. Structure and operation principles of the proposed multi-port optical quasi-circulator.

However, in the design of Fig. 15, the optical path of the *p*- component is larger than that of the *s*-component. This optical path difference might cause polarization mode dispersion (PMD) to blur the transmission signal. Therefore, in order to solve the PMD problem, the original optical guiding paths in Fig. 15 must be changed. As shown in Fig. 16, two different guiding modules can be introduced for the odd and even ports, respectively, which are composed of PBSs and RPs. The designs of these two guiding modules with specifications (*Length*×*Width*) of $(4n\text{-}3)L\times0.31(n\text{-}1)L$ and $(4n\text{-}4)L\times0.31(n\text{-}1)L$ for an odd and an even port are shown in Fig. 17(a) and (b), respectively. These guiding modules are located at (x_{Mj}, z_{Mj}) which can be expressed as

$$[x_{M(2n-1)} \quad z_{M(2n-1)}] = [2(1-n)L \quad (-2n-4)L], \text{ (for an odd port)} \qquad (19)$$

$$[x_{M(2n)} \quad z_{M(2n)}] = [2nL \quad (2n+4)L], \text{ (for an even port)} \qquad (20)$$

where n is a positive integer larger than 1. The coordinate in equation (19) corresponds to the center of the RP (in red color) in the odd-port guiding module; the coordinate in equation (20) corresponds to the center of the PBS (in green color) in the even-port guiding module. When the guiding modules are appropriately introduced, the optical path differences between the h- and v-components can be reduced to zero. Therefore, the PMD problem can be solved. Fig. 16(a), (b), (c), and (d) show the routes of port 1→port 2, port 2→port 3, port 3→port 4, and port (2n-1)→port 2n, respectively. Other propagation routes can be obtained based on the similar principle.

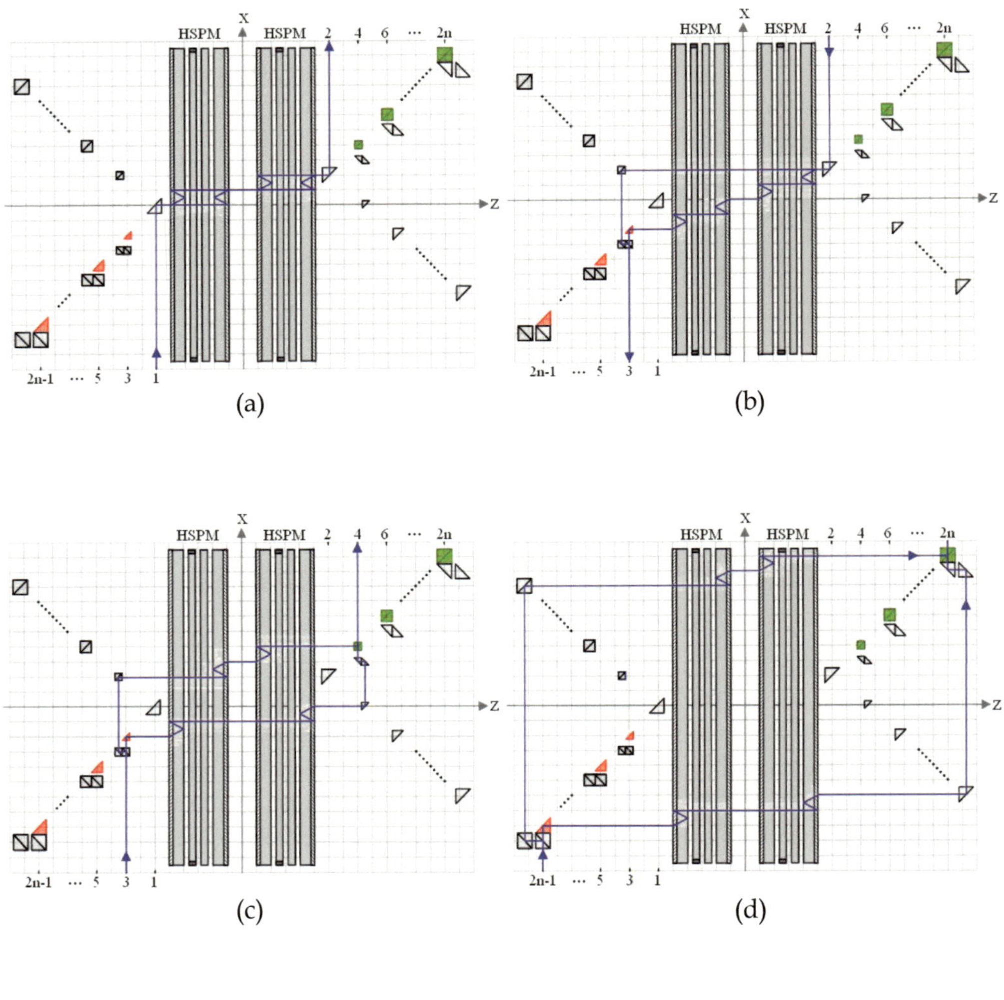

Fig. 16. Structure and operation principles of the proposed multi-port optical quasi-circulator without polarization mode dispersion.

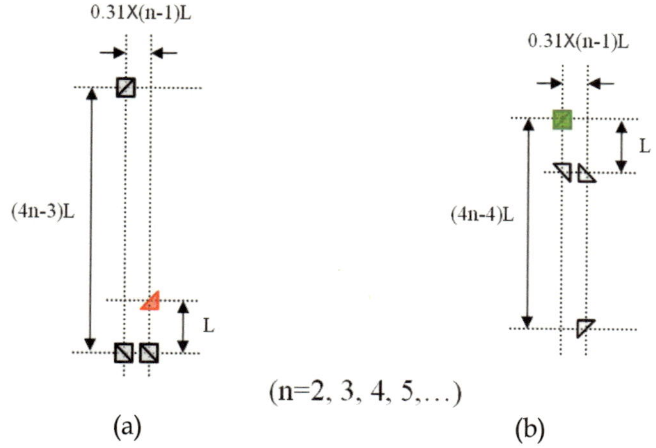

Fig. 17. PBSs and RPs guiding modules for (a) the odd ports; (b) the even ports.

Use the fabricated HSWPs mentioned in Section 3.2, a prototype of 6-port polarization-independent optical quasi-circulator for 1300nm can be assembled. In addition to a pair of HSPMs, it needs another eight polarization-beam splitters and ten reflection prisms to complete the function of this 6-port optical quasi-circulator. The characteristic parameters of this prototype device are estimated and listed in Table 4.

In Port	Out Port					
	1	2	3	4	5	6
1	14.26[a]	4.18[b]	>25.36[c]	>25.36[c]	>25.36[c]	>25.36[c]
2	>26.92[c]	14.26[a]	3.90[b]	>26.92[c]	>26.92[c]	>26.92[c]
3	>25.36[c]	>25.36[c]	14.26[a]	4.18[b]	>25.36[c]	>25.36[c]
4	>26.92[c]	>26.92[c]	>26.92[c]	14.26[a]	3.90[b]	>26.92[c]
5	>25.36[c]	>25.36[c]	>25.36[c]	>25.36[c]	14.26[a]	4.18[b]

Table 4. Associated parameters (in Decibels) for the prototype 6-port optical quasi-circulator with wavelength 1300 nm; [a]Return losses; [b]Insertion losses; [c]Isolations.

In order to solve the PMD problem, two different guiding modules composed of PBSs and RPs can be appropriately introduced for the odd and even ports, respectively. However, if more compact modules are desired, these guiding devices should become smaller simultaneously. The result will increase the difficulty of device assembling. Expediently, we can increase the beam splitting distance L ($L=2t\tan\theta_d$) by increasing the thickness of the substrate to reduce the assembling difficulty. Another reliable method is to operate this device beginning with a high number port.

3.4 Improved multi-port polarization-independent optical quasi-circulator

Based on two connected SPM_y and SPM_x, it is obvious that if reflection prisms (RPs) and polarization beam-splitters (PBSs) are introduced appropriately to guide the light beams in and out of the module, an alternative multi-port optical quasi-circulator can also be obtained. Only one RP should be added at port 1 and port 2, separately. For other ports, each port needs two RPs and one PBS. According to eqs. (13) and (14), the introduced RPs and PBS at the j-th port are located at (x_{RP1j}, y_{RP1j}), (x_{RP2j}, y_{RP2j}) and (x_{PBSj}, y_{PBSj}), which can be expressed as

$$\begin{bmatrix} x_{RP1(2n-1)} & y_{RP1(2n-1)} \\ x_{RP2(2n-1)} & y_{RP2(2n-1)} \\ x_{PBS(2n-1)} & y_{PBS(2n-1)} \end{bmatrix}\Bigg|_{z=-L} = \begin{bmatrix} (n-1)L & (1-n)L \\ (1-n)L & (n-1)L \\ (n-1)L & (n-1)L \end{bmatrix}\Bigg|_{z=-L}, \text{ (for an odd port)} \tag{21}$$

$$\begin{bmatrix} x_{RP1(2n)} & y_{RP1(2n)} \\ x_{RP2(2n)} & y_{RP2(2n)} \\ x_{PBS(2n)} & y_{PBS(2n)} \end{bmatrix}\Bigg|_{z=L} = \begin{bmatrix} nL & (2-n)L \\ (2-n)L & nL \\ nL & nL \end{bmatrix}\Bigg|_{z=L}, \text{ (for an even port)} \tag{22}$$

These equations are still valid for port 1 and port 2 to determine the position of its associated RP. Shown in Fig. 18 is a 5-port polarization-independent optical quasi-circulator consisting of a pair of $HSPM_x$ and $HSPM_y$, 3 PBSs, and 8 RPs. Figures 18(a) and 18(b) show the routes of port 1→port 2 and port 4→port 5, respectively. Based on the same principle, other propagation and expanded routes can also be obtained.

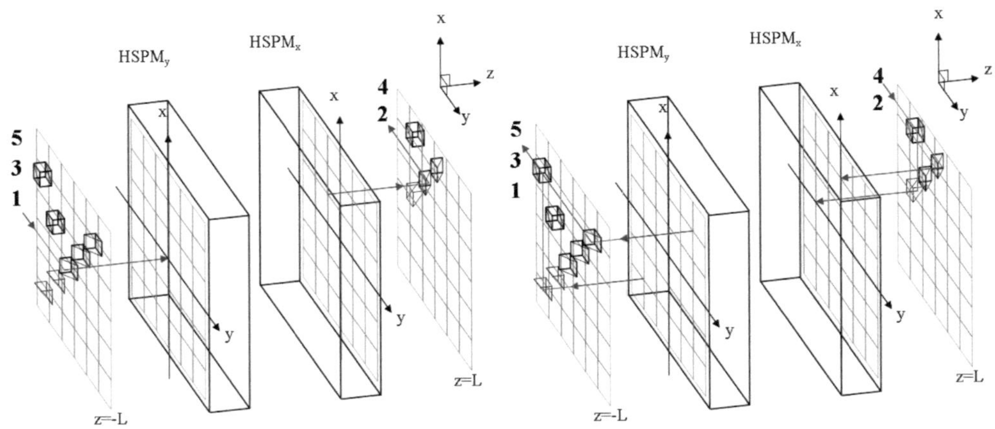

Fig. 18. Structure and operation principles of the proposed 5-port polarization-independent optical quasi-circulator.

A 5-port polarization-independent optical quasi-circulator for 1300nm can be assembled with the fabricated HSWPs. In addition to a pair of orthogonal HSPMs, it needs another three PBSs and eight RPs to complete the function of this device. Its associated losses and isolation values are estimated and listed in Table 5. The isolation values are in the range from 20 to 54dB. The return losses and the insertion losses are about 14dB and about 3dB, respectively. Return losses mainly come from the interface reflections that influence the isolation values directly. If the applied HSWPs are anti-reflection coated and are fabricated under accurate fabrication processes, the return losses could be over 50 dB and the diffraction efficiencies may reach theoretical values, i.e., η_h=0% and η_v=100%. Under these two improved conditions, the performance of this 5-port optical quasi-circulator can be enhanced greatly with isolation values larger than 51dB and insertion losses smaller than 0.9dB.

In Port	Out Port 1	2	3	4	5
1	14.26[a]	3.26[b]	>20.46[c]	>37.65[c]	>54.85[c]
2	>54.85[c]	14.26[a]	3.26[b]	>20.46[c]	>37.65[c]
3	>37.65[c]	>54.85[c]	14.26[a]	3.26[b]	>20.46[c]
4	>54.85[c]	>54.85[c]	>54.85[c]	14.26[a]	3.26[b]

Table 5. Associated parameters (in Decibels) for the prototype 5-port optical quasi-circulator with wavelength 1300 nm; [a]Return losses; [b]Insertion losses; [c]Isolations.

Compared with the design in Section 3.3, only the second HSPM is rotated 90° clockwise in this improved device. So this design still has all the advantages of the previous one. In addition, because two orthogonally polarized components have the same numbers of diffractions and total internal reflections in this design, their optical path lengths are all the same. Consequently, only fewer PBSs and RPs are required to guide the light beams in and out of the module. Hence the optical configuration is simpler and it is easier to be assembled. Moreover, the optical paths and the optical elements are not restricted in the same plane as the previous design in Section 3.3. So the light leakages producing by the unideal diffraction efficiencies of the HSWPs can not enter any port. The crosstalk between any ports can be avoided. Hence this design has higher isolations. In this device, polarization-selective substrate-mode volume holograms are used to replace conventional crystal spatial walk-off polarizers. Accordingly, compared with conventional quasi-circulators, this design has merits of compactness, easy fabrication, and low-cost. So it has high potential in optical communications.

4. Conclusion

In this chapter, polarization-selective substrate-mode volume holograms were introduced which are applied in several novel designs of optical circulator. Alternative design method of polarization-selective substrate-mode volume holograms was also introduced for overcoming the finite refractive index limit in practical holographic recording materials. The described optical circulators have advantages of polarization-independent, compactness, high isolation, low polarization mode dispersion, easy fabrication, and low cost. In addition, the port number of the proposed multi-port device can be expanded easily. High application potential of these devices in optical communications is expected. Their commercialization finally relies on available high-performance holographic recording materials.

5. Acknowledgment

These researches were supported partially by grants from the National Science Council, Taiwan, ROC, under contracts NSC 94-2215-E-035-003, NSC 94-2215-E-035-009, NSC 96-2221-E-035-052, NSC 97-2221-E-035-016-MY3, and the Lee & MTI Center for Networking at National Chiao Tung University, Taiwan, ROC.

6. References

Ramaswami, R.; Sivarajan, K.; & Sasaki, G. (2009). Optical Networks: A Practical Perspective. 3rd ed. Morgan Kaufmann, ISBN 9780123740922, San Francisco, USA

Hecht, J. (2005). Understanding fiber optics. 5th ed. Prentice Hall, ISBN 9780131174290, New Jersey, USA

Mynbaev, D. & Scheiner, L. (2000). Fiber-Optic Communications Technology. Prentice Hall, ISBN 9780139620690, New Jersey, USA

Iwamura, H. et al. (1997). Simple polarisation-independent optical circulator for optical transmission systems. *Electronics Letters*, Vol. 15, pp. 830-831

Shirasaki, M. et al. (1981). Compact polarization-independent optical circulator. *Applied Optics*, Vol. 20, pp. 2683-2687

Yokohama, I. et al. (1986). Polarisation-independent optical circulator consisting of two fibre-optic polarising beam splitters and two YIG spherical lenses. *Electronics Letters*, Vol. 22, pp. 370-372

Koga, M. (1994). Compact quartzless optical quasi-circulator. *Electronics Letters*, Vol. 30, pp. 1438-1440

Wang, L. (1998). High-isolation polarization-independent optical quasi-circulator with a simple structure. *Optics Letters*, Vol. 23, pp. 549-551

Nicholls, J. (2001). Birefringent crystals find new niche in WDM networks. *WDM Solutions*, R&D Review 2001, pp. 33-36

Huang, Y. (1994). Polarizing-selective volume holograms: general design. *Applied Optics*, Vol. 33, pp. 2115-2120

Chen, J. et al. (2003) Holographic spatial walk-off polarizer and its application to a 4-port polarization-independent optical circulator. *Optics Express*, Vol. 11, pp. 2001-2006

Chen, J. et al. (2008) An alternative design of holographic polarization-selective elements. *Proceedings of SPIE*, Vol. 7072 , pp. 707210-1, San Diego, California, USA

Chen, J. et al. (2004). Multi-port polarization-independent optical circulators by using a pair of holographic spatial- and polarization- modules. *Optics Express*, Vol. 12, pp. 601-608

Chen, J. et al. (2004). Improved N-port optical quasi-circulator by using a pair of orthogonal holographic spatial- and polarization- modules. *Optics Express*, Vol. 12, pp. 6553-6558

Kogelnik, H. (1969). Coupled wave theory for thick hologram gratings. *Bell System Technical Journal*, Vol. 48, pp. 2909-2947

13

Holographic Synthesis of Diffraction Free Beams and Dark Hollow Beams

G. Martínez Niconoff[1], P. Martínez Vara[2],
J. Muñoz Lopez[1] and A. Carbajal Dominguez[3]
[1]*Instituto Nacional de Astrofísica, Óptica y Electrónica, INAOE, Departamento de Óptica, Grupo de Óptica Estadística, Puebla*
[2]*Benemérita Universidad Autónoma de Puebla, BUAP, Facultad de Ingeniería, Puebla*
[3]*Universidad Juárez Autónoma de Tabasco, UJAT, Cunduacan, Tabasco, México*

1. Introduction

Since its discovery, holographic techniques had been proved to be one of the most powerful tools for controlling the optical field in almost all the optical areas [1,2,3]. Just to mention some ones, these techniques had been used in the design of hybrid lenses to decrease or avoid aberrations, in optical metrology it had been used to analyze the vibration modes in order to detect potential risks. The recording of the holographic interaction on photo resins it had been possible to design surface roughness, where the statistical features of the surface can be controlled with the parameters of the recording interaction such as time of exposition, angle of recording, wavelength, etc. The influence of these parameters is manifested in the power spectrum associated to the scattered field. A very interesting application of this technique was implemented by West-O'Donnell [4] in order to describe the relation of the scattered field with surface plasmon modes. The experiment was obtained by depositing a metal thin film over the hologram surface. This technique paves the way to implement a two dimensional surface plasmon optics [5,6]. This configuration allows the study of interesting physical properties such as enhanced backscattering also the apparition of satellite peaks in the scattered field [7,8]. This process can be extended by depositing a stack of layers of dielectric thin film-metal thin film, generating in this way a holographic roughness wave guides with application to communication systems.
In the holographic context, a very interesting subject consists in the research of novel material of holographic register. In this same context, recently has been reported the recording of the interaction between optical waves on rubidium vapor, which allows the control of light with light [9,10]. Other important materials of holographic register are the photorefractive crystals. In this kind of material, non-linear features can be induced and controlled. For example shift frequency, control of fluorescence time [11,12].
The purpose of this work is to present, in a plain way, the use of a photorefractive crystal as holographic recording material. The hologram consists in the interference between diffraction free beams. The interest in this topic comes from the fact that in the physical optics context, a contemporary topic research consists in the trapping and manipulation of particle conductors or/and dielectrics by means of optical fields also as atom/ion trapping

[13,14]. This problem implies the generation of diffraction free beams [15,16] and dark hollow beams [17] with geometries and phase easily tunable [18]; more details concerning to the applications of these beams can be founded in [19].

In the present chapter, we describe a holographic technique to generate these kinds of optical fields by incorporating spatial filtering techniques. The resulting optical fields can be diffraction free beams and/or dark hollow beams. The spatial filtering is implemented by controlling the kind of illumination which allows us to add or avoid, in the diffraction field some spatial frequencies.

The transmittance function for generate the diffraction field is obtained by interfering two zero order Bessel beams with non-common axis. The irradiance distribution is recorded on a holographic plate which consists in a photorefractive crystal. In this configuration it is indistinct which is the object beam and which is the reference beam. This configuration is analogous to one point hologram as was described by Hariharan [20]. By illuminating the hologram with the reference beam, the resulting optical field consists in a set of diffraction free beams propagating in different directions. We show below that, for certain values of the recording angle a beam is generated that propagate quasi-parallel to the surface. We will show that the holographic reconstruction with other kinds of illumination generates a variety of non-diffracting beams. The simplest case is for illumination with a plane wave illuminating in the same direction as the recording Bessel beams. In this configuration, we are adding the zero order spatial frequency. In consequence, reverse contrast in the optical field respect to the zero order Bessel beam is expected. This simple technique allows us to generate dark hollow beams. Changing the kind of illumination we can generate arrays of diffraction free fields and/or dark hollow beams whose profile and phase features are easily tunable. Also, the kind of illumination allows us to incorporate partially coherent features.

The experimental results showed are for a holographic transmittance generated with the interference between two zero order Bessel beams and reconstructing with the same reference beam. The emerging optical field consists in a set of beams where one of them emerges quasi-parallel to the surface. The other experimental result are by illuminating the hologram with a plane wave. The beam that emerges in the same direction that the plane wave is a dark hollow beam.

2. Description of diffraction free beams

In this section we show an overview of the theoretical synthesis for diffraction free beams propagating in free space. More details of this were described by Patorsky [21] and reference there in.

The starting point is to solve the Helmholtz equation using the operator's formalism. The Helmholtz equation is

$$\nabla^2\phi + K^2\phi = 0\,; \tag{1}$$

where $K = 2\pi / \lambda$ is the wave number. The analysis presented is for a reference system separable in the coordinate of propagation, without loss of generality, it is considered as the z-axis. In this system, the representation for the Helmholtz equation takes the form

$$\frac{\partial^2\phi}{\partial z^2} + \nabla^2{}_{\perp}\phi + K^2\phi = 0 \tag{2}$$

Defining the operator $\hat{A}=\left(\nabla_{\perp}^{2}+K^{2}\right)^{1/2}$, the equation (2) is similar to the differential equation for the harmonic oscillator

$$\frac{\partial^{2}\phi}{\partial z^{2}}+A^{2}\phi=0\,. \tag{3}$$

Solving explicitly for the z coordinate, we have that the amplitude function is given by

$$\phi(x,y,z)=\exp\{iAz\}f(x,y)\,. \tag{4}$$

In this expression *f(x,y)* is an arbitrary function. Letting now that *f(x,y)* be an eigenfunction of the operator $\hat{A}$ with eigenvalue β, i.e.

$$\begin{aligned}\hat{A}f(x,y)&=\beta f(x,y)\\ \hat{A}^{2}f(x,y)&=\beta^{2}f(x,y)\\ \hat{A}^{n}f(x,y)&=\beta^{n}f(x,y)\end{aligned} \tag{5}$$

the amplitude solution takes the simple form

$$\varphi(x,y,z)=f(x,y)\exp\{i\beta z\}, \tag{6}$$

The eigenfunction *f(x,y)* describe the profile of the beam and it satisfies the eigenvalue equation

$$\nabla_{\perp}^{2}f(x,y)+\left(K^{2}-\beta^{2}\right)f(x,y)=0\,. \tag{7}$$

The solution obtained following these treatment generates a kind of waves whose profile does not changes when it is propagating along the coordinate of propagation, as can be deduced from equation (6). This is the reason that these mode solutions (6) are known as diffraction free beams. The equation (7) can be analyzed in frequency space taking its Fourier transform. Is easy to show that its frequency representation stay on a circle, given by

$$-4\pi^{2}F(u,v)\left(u^{2}+v^{2}\right)+\left(K^{2}-\beta^{2}\right)F(u,v)=0 \tag{8}$$

In last representation we have that $F(u,v)=\int_{-\infty}^{\infty}\int_{-\infty}^{\infty}f(x,y)\exp\{i2\pi(xu+yv)\}dxdy$ being the space frequencies $u=\cos\alpha/\lambda, v\cos\beta/\lambda$. Then we have that the profile of the diffraction free beam is given by

$$f(x,y)=\int_{-\infty}^{\infty}\int_{-\infty}^{\infty}\delta\left(u^{2}+v^{2}-d^{2}\right)F(u,v)\exp\{-i2\pi(xu+yv)\}dudv\,; \tag{9}$$

where δ is the Dirac delta function and $d=\frac{\left(K^2-\beta^2\right)^{1/2}}{2\pi}$. It must be noted that the general expression for the diffraction free beams is of the form $\varphi(x,y,z)=f(x,y)\exp\{i\beta z\}$ where *f(x,y)* satisfies equation (9).
Two simple cases can be identified from last equation. The first case is trivial and it occurs when the radius of the circle is zero, *d=0*, and *F(u,v)* is a constant. The diffraction free beam corresponds with a plane wave. The second case occurs when the radius is different of zero and *F(u,v)* is a constant. For this case we have that the profile of the beam is

$$f(x,y)=J_0\left(2\pi d\sqrt{x^2+y^2}\right). \tag{10}$$

This means that the profile is a zero order Bessel function. This profile is shown in figure (1).

a)

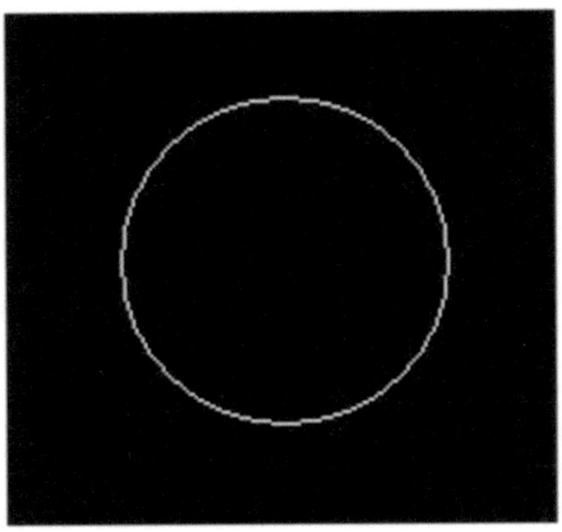

b)

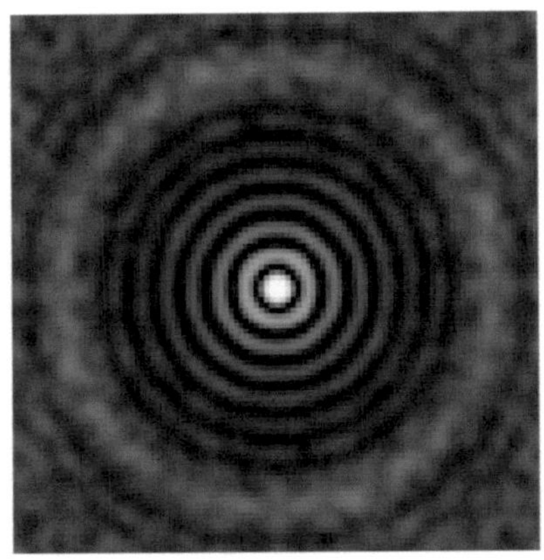

Fig. 1. a) Annular transmittance in frequency space. b) Profile in z=0 for the zero order Bessel beam.

3. Holographic recording with two zero order bessel beams

In general, holography consists in two stages. One of them is the recording process between the interference of two optical fields. The irradiance associated is recorded on a media and the irradiance distribution generates the hologram. The other stage consists in the reconstruction process. In general it is obtained by illuminating the hologram using a reference beam. In this section, we describe these two stages as follows. The hologram is generated by the interference between two zero order Bessel beams. The irradiance is recorded on a photorefractive crystal. The reason to use this media is to avoid the development process required in the traditional holographic techniques. During the reconstruction process, we use two kinds of reference beam. One of them is illuminating with a zero order Bessel beam and another is illuminating with a plane wave. The mathematical analysis is as follows.
From the previous section we have that the spatial frequency representation for diffraction free fields must be on some regions of a circle of radius d. This same result was obtained by

Durnin [22]. With this condition, the synthesis of diffraction free beams using a Fourier transform system is very simple.
The mathematical representation for the zero order Bessel beam propagating along the z-coordinate is given by

$$\phi(x,y,z) = AJ_0\left(2\pi d\sqrt{x^2+y^2}\right)\exp\{i2\pi\gamma z\}, \tag{11}$$

where $\gamma = \sqrt{\frac{1}{\lambda^2} - d^2}$, d is constant. By means of a rotation respect the *y-axis*, we have that the amplitude function for the coherent superposition of two zero order Bessel beams with non common axis propagating on *x-z* plane is given by

$$\begin{aligned}\phi(x,y,z) = AJ_0\left(2\pi d\sqrt{(ax+bz)^2+y^2}\right)\exp\{i\xi(-ax+bz)\} + \\ AJ_0\left(2\pi d\sqrt{(ax-bz)^2+y^2}\right)\exp\{i(\alpha x+\beta z)\}\end{aligned}, \tag{12}$$

where $a = \cos\theta$ and $b = \sin\theta$ $\alpha = \gamma\cos\theta; \beta = \gamma\sin\theta$. The irradiance is obtained taking the square modulus of the amplitude function, it takes the form

$$\begin{aligned}I(x,y,z) = J^2{}_0\left(\mu\sqrt{(ax+bz)^2+y^2}\right) + J^2{}_0\left(\mu\sqrt{(ax-bz)^2+y^2}\right) + \\ J_0\left(\mu\sqrt{(ax+bz)^2+y^2}\right)J_0\left(\mu\sqrt{(ax-bz)^2+y^2}\right)\exp\{-i2\xi ax\} + \\ J_0\left(\mu\sqrt{(ax+bz)^2+y^2}\right)J_0\left(\mu\sqrt{(ax-bz)^2+y^2}\right)\exp\{i2\xi ax\}\end{aligned} \tag{13}$$

where $\mu = 2\pi d$. The irradiance distribution is recorded on a photorefractive crystal, generating the hologram. We assume that the transmittance function of the hologram is proportional to the irradiance distribution $t(x,y)=hI(x,y)$, where h is a constant. For further analysis this constant is not relevant and it will be omitted. The holographic plate is placed on $z=0$ plane, consequently the hologram has a transmittance function given by

$$\begin{aligned}t(x,y) = J^2{}_0\left(\mu\sqrt{(ax)^2+y^2}\right) + J^2{}_0\left(\mu\sqrt{(ax)^2+y^2}\right) + \\ J_0\left(\mu\sqrt{(ax)^2+y^2}\right)J_0\left(\mu\sqrt{(ax)^2+y^2}\right)\exp\{-i2\xi ax\} + . \\ J_0\left(\mu\sqrt{(ax)^2+y^2}\right)J_0\left(\mu\sqrt{(ax)^2+y^2}\right)\exp\{i2\xi ax\}\end{aligned} \tag{14}$$

The experimental set up for the generation of the hologram also as the parameters involved are shown in figure (2.a).

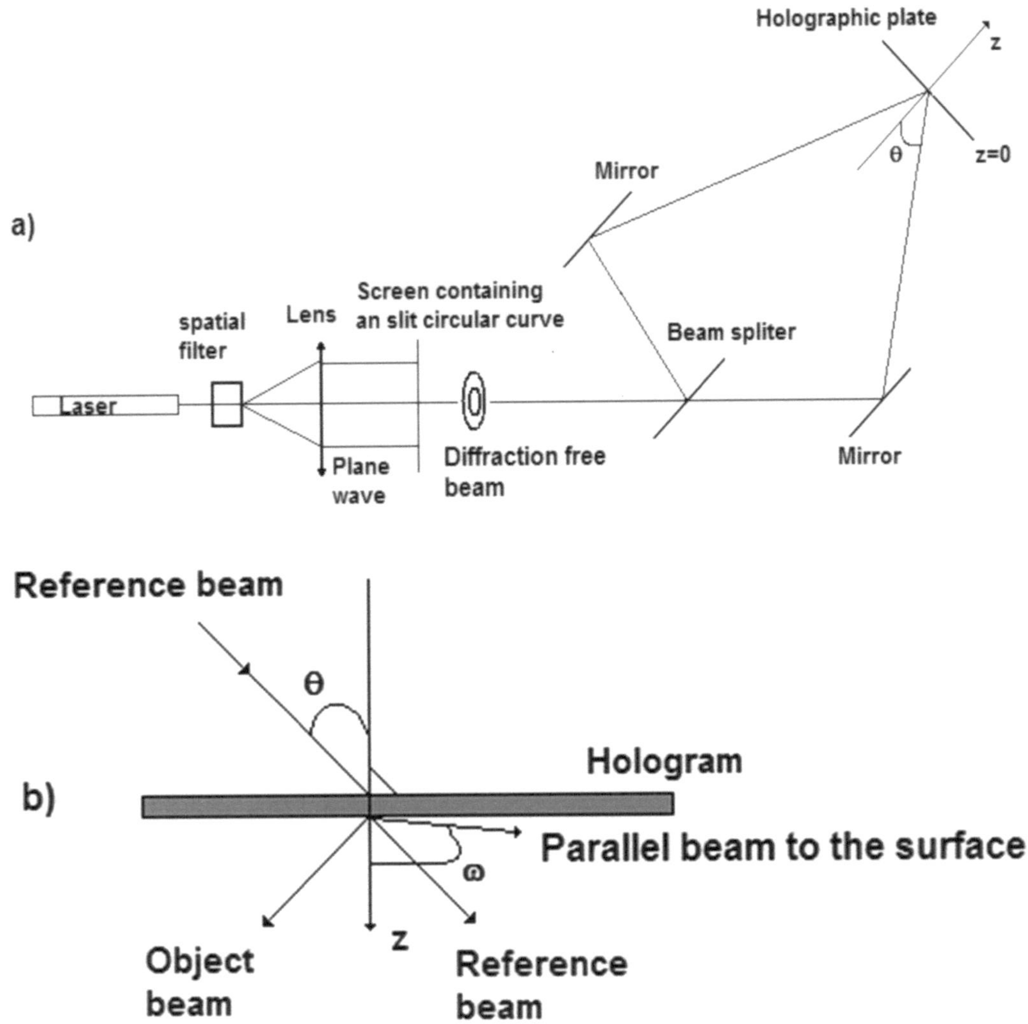

Fig. 2. In a) Experimental Set up for generate the hologram, the radius of the circular slit is 2 mm. approximately and the recording angle θ is in the range [15°-20°]. In b), The arrows represent the direction of propagation for the set of diffraction free beams generated during the reconstruction process.

4. Holographic reconstruction

During the reconstruction process, we illuminate with a zero order Bessel beam, the optical field can be obtained from the diffraction integral representation, where a set of optical beams can be identified. The amplitude function for the diffraction optical field is

$$\phi(x,y,z)=\iint 2J^2{}_0\left(\mu\sqrt{(ax)^2+y^2}\right)J_0\left(\mu\sqrt{(ax)^2+y^2}\right)\exp\{-i\xi ax\}\frac{\exp\{ikr\}}{r}dxdy+$$
$$\iint J^2{}_0\left(\mu\sqrt{(ax)^2+y^2}\right)J_0\left(\mu\sqrt{(ax)^2+y^2}\right)\exp\{i\xi ax\}\frac{\exp\{ikr\}}{r}dxdy+ \quad (15)$$
$$\iint J^2{}_0\left(\mu\sqrt{(ax)^2+y^2}\right)J_0\left(\mu\sqrt{(ax)^2+y^2}\right)\exp\{-i\xi ax\}\frac{\exp\{ikr\}}{r}dxdy$$

The physical meaning of each integral is as follows. The first integral corresponds to the reference beam, whose profile is an optical field similar to the zero order Bessel beam. The second integral represents the "object beam" and the third term corresponds to a beam forming an angle ω=*arcsin(3sin*θ*)*. These beams are sketched in figure (1.b). One interesting feature is that we can generate a beam, quasi-parallel to the surface transmittance. This is obtained if the recording angle is in the range [15°-20°] for this case $\omega \cong 90°$. In figure (3) we show the experimental result for the quasi-parallel beam. In figure (3.a) the beam has the structure of a speckle pattern, it was obtained for a reconstruction angle of *20°*. The origin of the speckle pattern is because the interaction of the beam with the roughness of the surface. The speckle pattern can be controlled by a tilt in the reconstruction beam. In Fig. (3.b) we show the experimental result for an angle of *15°*. In this case, the beam propagates no-parallel to the surface and its profile resembles a zero order Bessel beam. An interesting topic that remains to be solved consists in describing the non-diffracting features of such speckle-beams.

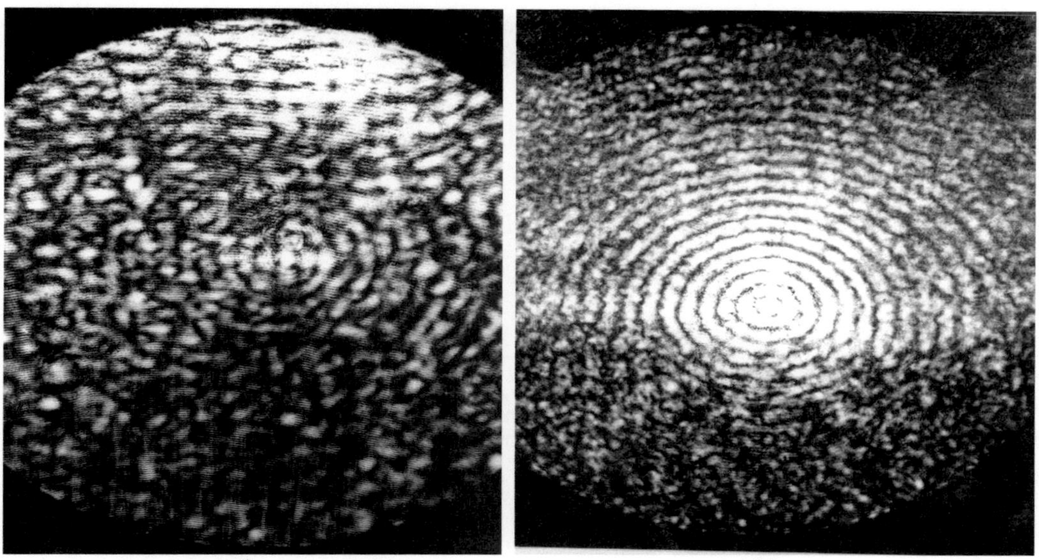

a) b)

Fig. 3. Experimental results for the beams propagating quasi-parallel to the hologram surface. In a) the angle of the illumination beam is $\theta+\Delta\theta$, and a diffraction free speckle pattern is obtained, in b) the illumination beam is $\theta-\Delta\theta$ and a zero order Bessel beam is obtained.

5. Diffraction synthesis of the dark hollow beams

An interesting feature is that, spatial filtering techniques can be implemented if we change the kind of illumination during the reconstruction process. From equation (14) it is clear that the optical field does not have the diffraction zero order. If we illuminate the holographic transmittance with a plane wave, we are adding the diffraction zero order to the optical field, consequently the new optical field must change its profile, having in particular reverse contrast. In figure (4) we show the experimental results obtained when a plane wave illuminates the hologram, in the same direction as the reference beam. For this case, the reverse contrast in the zero order Bessel beam is expected, generating a hollow dark beam.

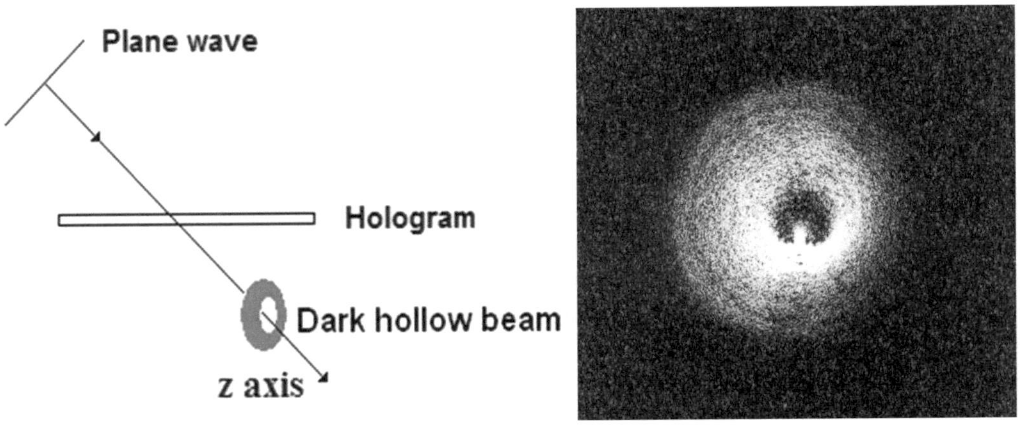

Fig. 4. Optical field generated by illuminating the hologram with a plane wave. The experimental result corresponds with a dark hollow beam.

The amplitude representation for a dark hollow beam on a arbitrary point *X(x,y,z)* can be obtained from the angular spectrum model, this representation is

$$\begin{aligned}\phi(X) &= \int\int_{-\infty}^{\infty}\left(\delta(u,v)-\delta\left(u^2+v^2-d^2\right)\right)\exp\{i2\pi(xu+yv+zp)\}dudv \\ &= \exp\{ikz\}-\left(J_0\left(2\pi d\sqrt{x^2+y^2}\right)\right)\exp\{i\gamma z\}\end{aligned} \tag{16}$$

The negative sign between the δ-functions corresponds with a phase change between diffraction beams emerging from the transmittance function and the illumination beam. This is a consequence of the Fresnel equations. The irradiance associated is given by

$$I(X)=1+J^2_0\left(2\pi d\sqrt{x^2+y^2}\right)-2J_0\left(2\pi d\sqrt{x^2+y^2}\right)\cos(k-2\pi\gamma)z. \tag{17}$$

The period of the cosine term is very large and it is given

$$\begin{aligned}P &= \frac{2\pi}{k}-2\pi\gamma \\ &\approx \frac{2}{\lambda d^2}\end{aligned}$$

The consequence of this large period is that in $z \geq 0$ region, the irradiance can be approximated as

$$I(X) = 1 - J^2{}_0\left(2\pi d\sqrt{x^2 + y^2}\right),$$

that explains the contrast reverse. It must be noted for this approximation that dark hollow beams are not depending on the z coordinate. For this reason, dark hollow beams may be interpreted as quasi-diffraction free beams. The experimental results shown in figure (4) are in very good agreement with the theoretical model presented.

6. Final remarks

In this chapter we have described a simple method using holographic techniques in order to generate a variety of diffraction free beams and dark hollow beams. A fundamental part of the study consists in the generation of the boundary condition for the optical field characterized by a transmittance function. This is obtained by interfering two zero order Bessel beams. However, our model can be easily extended to other kind of diffracting free beams.

During the reconstruction process, the transmittance function can be illuminated with different kinds of optical fields incorporating in this way spatial filtering techniques. The analysis showed allowed us to generate simultaneously a set of diffraction free beams and dark hollow beams with profiles and phases easily tunable. An important fact is that, we can generate diffracting free beams quasi-parallel to the hologram surface. This offers the possibility to study the self regeneration property of the electromagnetic field associated to the diffraction free beams, also, this beam offers the possibility to be implemented as surface optical twisters to transfer angular momentum perpendicular to the surface for the development of micro-motors and it can be implemented for the synthesis of surface plasmon waves.

7. References

[1] Hariharan, P, "Colour Holography", Progress in optics 20, Elsevier Science, 1983, Chapter 4, p.p. 265-321 North Holland.

[2] Christian Frere and Olof Bryngdahl, " Computer-Generated Holograms: Reconstruction of curves in 3-D", Opt. Comm., Vol 60, 6, pp. 369-372

[3] Feng Gao, Jianhua Zhu, Qizhong Huang, Yixiao Zhang, Yangsu Zeng, Fuhua Gao, Yongkang Guo, Zheng Cui, "Electron-beam lithography to improve quality of computer- generated hologram" Microelectronic Engineering, Vol. 61-62, 2002 pp. 363-369

[4] C. S. West and K. A. O'Donnell, "Observations of backscattering enhancement from polaritons on a rough metal surface," J. Opt. Soc. Am. A ,12, 390-397 (1995)

[5] G. Martinez-Niconoff, J. A. Sanchez-Gil, H. H. Sanchez Hernandez and A. Perez Leija, "Self Imaging and caustic in two dimensional plasmon optics" Opt. Comm., 281, Vol. 8 (2008), p.p. 2316-2320

[6] G. Martinez Niconoff, P. Martinez-Vara, J. Muñoz-Lopez, J.C. Juarez M and A. Carbajal-Dominguez "Partially coherent surface plasmon modes" Journal of the European of the Optical Society, 6,1109(2011)

[7] G. Bernine, "Small-amplitude perturbation theory for two dimensional surfaces, in light scattering and nanoscale surface roughness", Ed. A. A. Maradudin (Springer-Verlag, Ney York, 2007).

[8] T.A. Leskova, M. Leyva-Lucero, E. R. Mendez, A.A. Maradudin, I.V. Novikov, "The surface enhanced second harmonic generation of light from a randomly rough metal surface in the Kretschmann geometry" Opt. Comm., 183, 529, 2000

[9] E. Korolev, "Dynamic holography in alkali metal vapour", Optics and Laser Technology, Vol. 28, 4, (1996), 277-284

[10] 10.- Ryan M. Camacho, Preveen K. Vudyasetu A and John C. Howell, "Four-wave-mixing stopped light in hot atomic rubidium vapour", Nature photonics 3,103 (2009)

[11] D.L. Mill, "Nonlinear Optics: Basic concepts" 2nd ed. (Springer-Verlag, New York, 1998).

[12] R.W. Boyd, "Nonlinear Optics", 3rd ed. (Academic press, 2008)

[13] J. Arlt, V. Garces-Chavez, W. Sibbett and K. Dholakia, "Optical micromanipulation using a Bessel light beam", Opt. Comm, 197, 239-245 (2001)

[14] D. P. Rhodes, G.P.T. Lancaster, J. Livesey, D. Mcgloin, J. Arlt and K. Dholakia, "Guiding a cold atomic beam along a co-propagating and oblique hollow light guide", Opt. Comm., 214, 247-254 (2002)

[15] J. Durnin, "Exact solutions for diffraction free beams I: the scalar theory" J. Opt. Soc. Am. A, Vol 4, 651-654 (1987)

[16] Andrey S. Ostrovsky, Gabriel Martinez Niconoff and Julio C. Ramirez San-Juan, "Generation of light string and light capillary beams", Opt. Comm., 207, (2002), 131-138

[17] Gabriel Martinez Niconoff, Julio C. Ramirez San-Juan, Patricia Martinez Vara, Adrian Carbajal Dominguez and Andrey S. Ostrovsky., "Generation of partially coherent diffraction free fields with tunable geometry", J. Opt. Soc. Am. A, Vol. 21, No. 4, (2004)

[18] Jianping Yin, Weijian Gao and YifunZhu, "Generation of dark hollow beams and their applications", Progress in Optics 45, Elsevier Science 2003, Chapter 3,pp 119. North Holland.

[19] Andrey S. Ostrovsky, Gabriel Martinez Niconoff, Julio C. Ramirez San Juan, "Coherent Mode representation of propagation-invariant fields", Opt. Comm., 195 (2001), 27-34

[20] Hariharan, P., "Optical Holography: Principles, Techniques, and Applications". Cambridge University Press., New York, USA, 1984, pp 23-25

[21] K. Patorski, "The self Imaging phenomenon and its applications" Progress in Optics Vol. 27 Chapter 1, pp 3-108 Elsevier Science 1990 North Holland.

Part 4

Holograms in Security Applications

14

Optimization of Hologram for Security Applications

Junji Ohtsubo
Shizuoka University
Japan

1. Introduction

Optical pattern recognition for validation and security verifications has been one of the important issues of optical image processing for the past decade. As one technique, a method of joint transform correlation (JTC) has frequently been used for optical security systems for identification of biometric images. In that method, an image such as a fingerprint pattern is encrypted by a random key mask, and the joint Fourier transform of the image to be encrypted and the key pattern conforms an encrypted hologram. The hologram is printed on a card or a document for authenticity, such as a credit card or a passport. The hologram is read when and where necessary and decoded by using the same key that is used for the encryption. In practical optical security systems, a digital technique is used for the image encryption, since the time required to calculate an encryption pattern is not a critical factor and the encryption of an image and printing of the encrypted hologram on a card may be done offline. On the other hand, fast processing is required to decode an encrypted image and verify it. Accordingly, the optical technique is very suitable for such processing.

In this chapter, we discuss optical security systems suitable for the use of holograms. For the congeniality of the method with real optical systems, binary holograms are frequently used in those systems. By the binarization of hologram, the reconstructed hologram is greatly degraded and, therefore, the optimization of hologram is required for the identification of a reference image. We study the method of the optimization of binary holograms based on a simulated annealing technique. The method of the simulated annealing usually takes a long time to reach a correct estimate, so that the fast optimization for binary hologram is applied. We also demonstrate an image decryption by the optimization of a binary hologram when both the hologram and the key for decryption are embedded in real electronic displays with periodic lattice structures. Finally, we discuss a technique to obtain an exact decryption image based on a phase-encoding technique, which enables easier realization for the practical applications in optical security systems. Even in this technique, the optimization of phase-encoded hologram plays an important role to obtain a good image-reconstruction.

2. Optical security systems

In this section, we discuss a fundamental optical security system treated throughout this chapter. The optical security system under consideration is shown in Fig. 1 (Yamazaki &

Ohtsubo, 2001). The system consists of three parts. Fig. 1(a) is an encryption system of a target image. An image, for example a finger print image, is encrypted with an encoding key with holographic technique and the encoded image is binarized according to a certain rule to make it easier for reading the image and to match the post processing system. The encrypted binary image printed on a security card is optically decrypted with the decoding key in Fig. 1(b). The decoded image is compared with a test image for identification with optical joint transform correlation as shown in Fig. 1(c). The advantage of the optical method in a security system is the fast processing for decoding an encrypted image and identifying it. In practical optical security systems, a digital technique may be used for the image encryption, since the time required to calculate an encryption pattern is not essential and the encryption of an image and printing the encrypted hologram on a card may be done by offline. On the other hand, the fast processing is required for decoding an encrypted image and verifying it. Accordingly, optical technique is very suited for such processing.

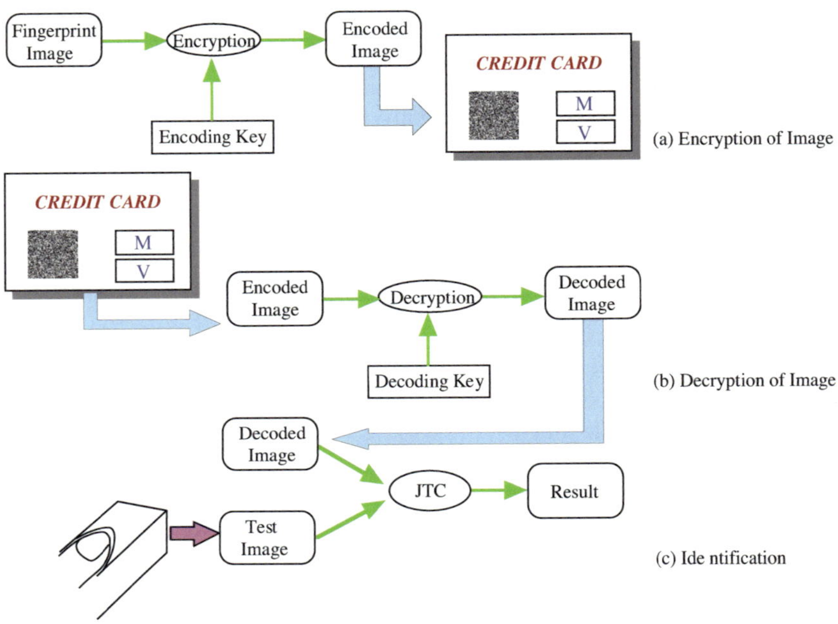

Fig. 1. Schematic example of optical security system. (a) Image encryption, (b) Image decryption, and (c) identification systems.

The encrypted image is binarized to print it, for example on a credit card, and it is read when and where necessary. Binarization of hologram may be performed according to the sign of each pixel value of the encrypted hologram. By the binarization of hologram, we can avoid the load for pre-processing of an encrypted hologram for the decryption and to make a robust optical security system. Due to the binarization of hologram, the encrypted image may be degraded, so that the optimization of the encrypted binary hologram is essential. In the decryption process, the reconstruction of the encrypted image should be quickly processed. Therefore, we assume an optical image processing in the decryption system. Also, the image identification with a test image should be quickly done. We also assume an optical system for image identification. Image identification based on optical methods is an attractive issue, however the methods, especially the methods of optical correlation, are well established (Kobayashi & Toyoda,1999). So we do not treat them in detail in this chapter. Therefore, in the following, we will discuss how to make an encryption hologram suitable for optical decryption and how to optimize it.

3. Encryption and decryption of hologram

3.1 Theory of image encryption and decryption of hologram

The method of image encryption for optical security system, which is considered here, is a common one already proposed (Javidi & Horner, 1994, Refregier & Javidi, 1995, and Javidi, 1997). In the image encryption process, an image with a random phase mask is jointly Fourier transformed with another random pattern, which is a key for encryption and decryption, as shown in Fig. 2 and a hologram is formed at the Fourier plane (Yang & Kim, 1996, Javidi et al., 1996, and Unnikrishnan et al., 1998). The phase random mask put in front of the image to be embedded plays a role for scrambling the image, however it little affects the reconstruction of hologram, since the reconstruction is only the intensity of them. The holographic fringe terms are given by

$$H(u,v) = F(u,v)G^*(u,v)\exp(-i4\pi dv) + F^*(u,v)G(u,v)\exp(i4\pi dv) \tag{1}$$

where $F(u,v)$ and $G(u,v)$ are the Fourier transformed functions of the image with a random phase, $f(x,y)$, and the random pattern $g(x,y)$ used as an encryption key, respectively, $2d$ is the separation between the centers of two functions, * denotes the complex conjugate, and $H(u,v)$ is the resultant hologram. We, here, only consider the AC components of the hologram. The image with a random phase is written by

$$f(x,y) = f_0(x,y)\exp\{ib(x,y)\} \tag{2}$$

where $f_0(x,y)$ is the original image function and $b(x,y)$ is also a random function but different from $g(x,y)$. To make an encrypted image, we assume a digital synthesis of the hologram. Fig. 3 is an example of sets of patterns used in the numerical simulations. Fig. 3(a) shows a fingerprint image with 64×64 pixels and Fig. 3(b) is a random pattern used for the encryption key. We here assume 8-bit gray scale for the fingerprint image and (1,-1) binary random pattern (i.e., equivalently $(0,\pi)$ phase) as the encryption key. As a random phase mask multiplied to the image in the input plane is also assumed to be $(0,\pi)$ random phase different from the random key pattern.

To reconstruct the image from the encrypted pattern, the hologram is illuminated by the same random phase pattern used for the encryption as shown in Fig. 4. For the decryption corresponding to Eq.(1), we obtain

$$\begin{aligned} p(x,y) = {} & f(x,y-d) \otimes g(-x,-y) \otimes g(x,y) \\ & + f(-x,-y) \otimes g(x,y) \otimes g(x,y) \otimes \delta(x,y+3d) \end{aligned} \tag{3}$$

where $\otimes$ denotes the convolution operation. The first term in Eq.(3) is the reconstructed image since the convolution between $g(-x,-y)$ and $g(x,y)$ reduces to a delta function due to the random nature of the function. On the other hand, the second term is the convolution between the image and the random function and it is a noise term in the reconstruction. Two terms can be spatially separated with each other. Thus, the encrypted image is successfully decrypted in the image plane without noise terms.

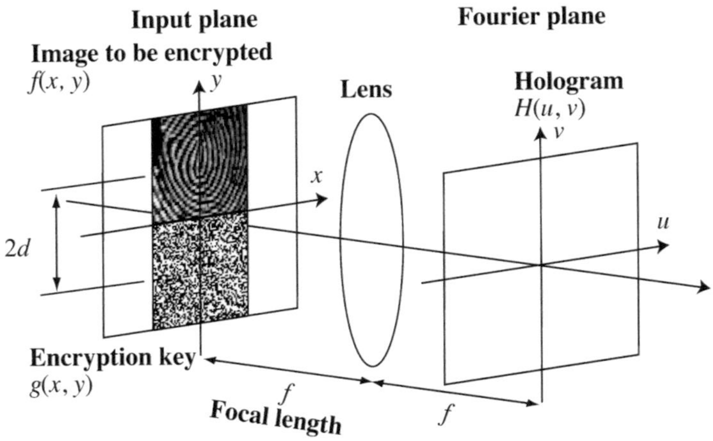

Fig. 2. Optical encryption system using joint Fourier transform

Fig. 3. (a) Original input image and (b) random key pattern for encryption.

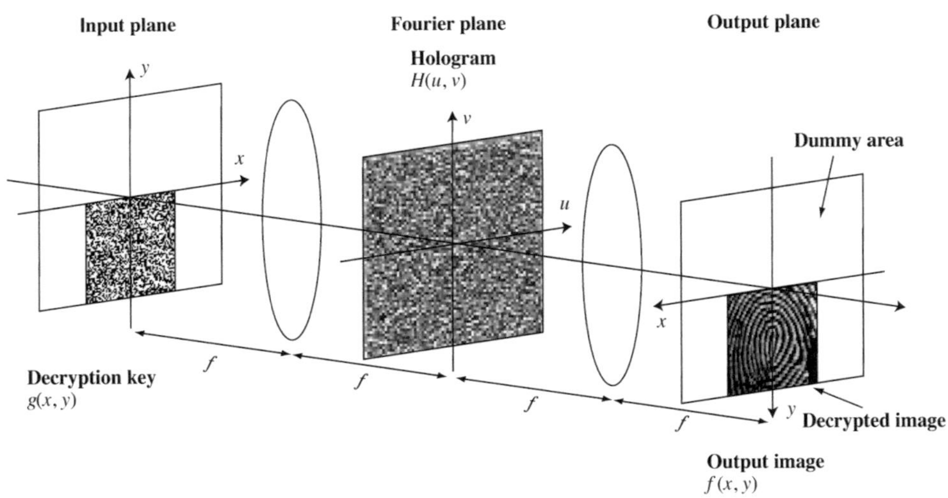

Fig. 4. Optical decryption system

3.2 Reconstruction of binary hologram

In actual applications, such as in credit card identification, a binary hologram is suited for acquiring electronic image and digital-electronic pre-processing. Thus, the use of the binarization of hologram is one of excellent methods to make a robust security system, so that we employed a binary hologram as an encrypted pattern in the following. The calculated hologram is binarized according to the sign of each element of the composite pattern. Fig. 5 is the result of the encrypted binary hologram. In the figure, the hologram that has a (0,1) binary distribution is printed on a card as a black and white pattern. However, for the reconstruction of the binary hologram, the value of each pixel is assigned to +1 ($\exp(i0)$) when the pixel hologram has a value of 1, while it is -1 ($\exp(i\pi)$) for 0. A hologram that has $(0,\pi)$ phase distribution can be easily realized by using a phase modulation spatial light modulator such as a parallel aligned liquid crystal display. The hologram that has 0 and π phase distribution has the advantage in the reconstruction, since the zero-th order diffraction is eliminated in the reconstruction pattern. Fig. 5(a) shows the calculated binary hologram corresponding to the original image with the random key pattern in Fig. 3. Fig. 5(b) is the decrypted pattern. The hologram and the reconstructed image have the size of 256×256 pixels. The lower noisy part in the figure is the second term in Eq.(3). The binarization of hologram is suited for printing it on a credit card in practical use. However the image is not completely reconstructed because of the binarization for the original hologram as shown in Fig. 5(b). As a result, the ability for the identification between the reconstructed and reference images is deteriorated. Therefore, the optimization of the binary hologram is expected to obtain a good reconstructed image. The method is discussed in the following section.

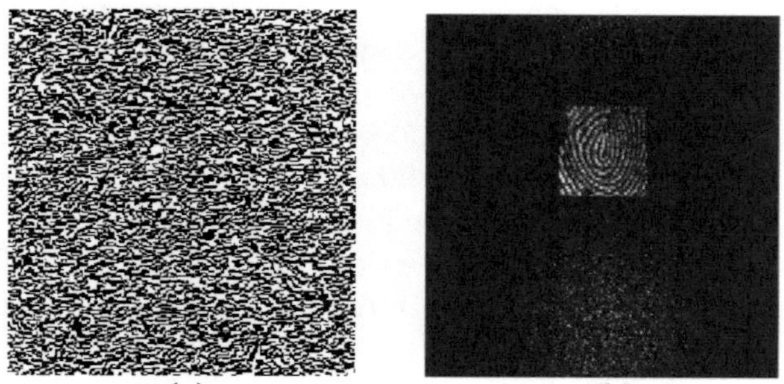

(a) (b)

Fig. 5. (a) Binary phase hologram and (b) reconstruction of hologram with random key pattern.

4. Optimization of binary hologram

4.1 Procedure for optimization of binary hologram

For an optical security system considered here such as in a credit card identification system, the image encryption may be performed in off-line. In that case, the hologram to be printed such as on a credit card may not necessarily be made by the optical method. The encryption of an image and the optimization of the hologram to reconstruct a good image can be performed on digital computer. We here discuss the optimization of the encrypted hologram to obtain a good reconstruction of it based on the numerical method. The method employed for the optimization is a simulated annealing like technique (Kirkpatrick et al., 1983, Ohtsubo & Nakajima, 1991, and Bättig et al., 1992). In ordinary sense, a small perturbation is applied to each pixel of a analogue-valued hologram in a simulated annealing method and the reconstructed image is gradually improved through the iterations (Metropolis et al., 1953, and Aarts and Korst, 1990). However, the method proposed here is somewhat different from the strict simulated annealing technique, since a large perturbation for each pixel value of 1 (zero-phase) or -1 (pi-phase) flipping is applied to each pixel.

In the simulated annealing in the present method, $(0,\pi)$ phase of each pixel in the hologram elements is flipped 0 to π or vice versa as a perturbation. Then the cost function is calculated and the perturbation is accepted or not according to the simulated annealing. The cost function defined here is the mean-square error between the original image intensity to be reconstructed and the estimated one and is given by

$$E = \iint \left| |f_0(x,y)|^2 - \alpha |f_n(x,y)|^2 \right|^2 dxdy \tag{4}$$

where $f(x,y)$ is the amplitude of the original image to be reconstructed (defined by Eq.(2)), $f_n(x,y)$ is the n-th estimate, and the scaling factor α is defined by

$$\alpha = \frac{\iint |f_0(x,y)|^2 \, dxdy}{\iint |f_n(x,y)|^2 \, dxdy} \tag{5}$$

The cost function is evidently zero when the estimate converges to the original image.

The basic flow of the simulated annealing employed here is shown in Fig. 6. According to the diagram, each step in the simulated annealing is described as follows;

Step 1. As an initial input for the iteration, a binary phase hologram calculated from Eq.(1) is used. The hologram is reconstructed and the initial cost function E (E_{old}) is calculated. The reference for the reconstruction is the random key pattern used for the encryption. The temperature for the annealing is set with a relatively high value.

Step 2. The perturbation is applied to one of the pixels of the hologram and the other pixels are remained unchanged. The phase is flipped 0 to π or π to 0. Then the estimated hologram is reconstructed and the new cost function E_{new} is calculated.

Step 3. The difference between the cost functions before and after the perturbation $\Delta E = E_{new} - E_{old}$ is calculated. If $\Delta E < 0$, the new phase is accepted and the cost function is retained as an old cost function for the next perturbation. Otherwise ($\Delta E \geq 0$), the acceptance or rejection is stochastically determined according to the Boltzmann distribution

$$P = \exp(-\frac{\Delta E}{T}) \tag{6}$$

where T is the temperature of the annealing. If $P < r$ (r is a random number between 0 and 1), the perturbation is accepted. On the other hand, it is rejected when $P \geq r$

Step 4. Step 4: Steps 2 and 3 are repeated for every pixel.

Step 5. If the cost function of each iteration has still a large value, the temperature for the annealing is lowered and the next iteration is performed again. If the cost function is lowered enough, the iteration is stopped.

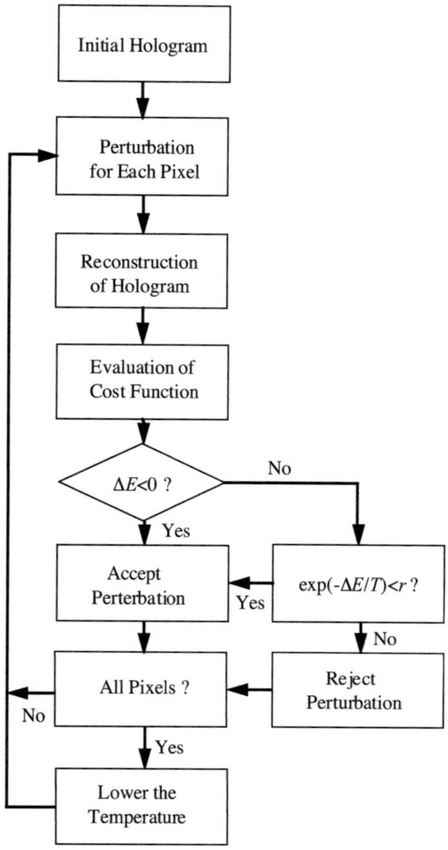

Fig. 6. Algorithm for optimizing hologram by simulated annealing method.

he process is almost the same as a usual simulated annealing method except for the random flipping of the (0,π) phase pattern at the second step. When the cost function becomes small and temperature is sufficiently cool down, the obtained pattern should be a good estimate for the hologram that well reproduces the original image to be reconstructed and thus the optimization of the hologram is realized.

4.2 Results of optimization

In the simulations, the optimization of the binary phase hologram for the fingerprint image as shown in Fig. 3(a) is performed. The reference for the reconstruction of the hologram is the random (1,-1) pattern in Fig. 3(b). Three cooling schedules of the temperature are used in the simulations, i.e., T=0 (this corresponds to no annealing), T=1/exp(n), and T=0.5/(1+n) (n being the iteration number). Fig. 7(a) shows the variations of the cost functions for the iteration number during the simulated annealing. When the annealing process exists ($T \neq 0$), the value of the cost function once increases with increase of the iteration number and reaches its maximum point. Then it decreases for further increase of the iteration number. In usual simulated annealing, the cost function monotonically decreases with increase of the iteration. However, in our case, the annealing process once trapped to a local minimum, since the perturbation to be added to the image is different from that of ordinary simulated annealing. From the comparison between the two cooling schedules (T=1/exp(n) and T=0.5/(1+n)), a rapid cooling rate is rather effective for the optimization of the hologram. The cost function monotonically and rapidly decreases without trapping any local minimum when T=0, that is, no annealing process. The point of the simulated annealing method is the moderate perturbation with random fluctuations to escape local minima in the energy function. The reference to construct the hologram in the method is a binary random pattern and the hologram itself is also a random like pattern, so that random fluctuations are automatically given without introducing the stochastic process (Step 3) in the iteration. This may come from a random nature of the reconstruction process in the present method. For either case, the optimized hologram well reproduces a fingerprint image very close to the original one. Fig. 7(b) is the result of the optimized hologram at T=0.

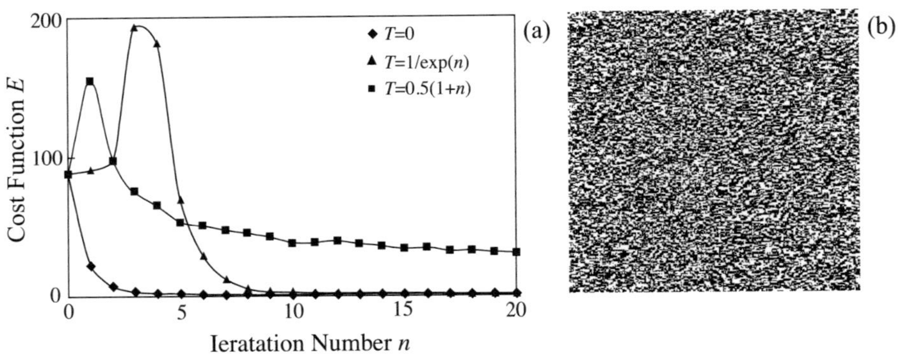

Fig. 7. (a) Variations of cost functions for each cooling schedule and (b) optimized binary hologram at T=0.

The degree of the reconstruction for the optimized hologram is tested by a joint transform correlation method. Fig. 8 shows the results. Fig. 8(a) is the original fingerprint image (left) and the joint transform correlation between the same patterns (right). The result of the correlation is a one-dimensional scan along the correlation peaks. For the calculation of the correlation function, the power spectrum is filtered by a band-pass filter in the Fourier plane to eliminate the unwanted noise floor. The zero-th order correlation peak is normalized to 255 level and the value of the correlation peak is 63. Fig. 8(b) is the decrypted image from the original binary hologram (not optimized one) and its correlation with the original fingerprint image. The fingerprint image is vague compared with the original one due to the binarization of the hologram and its correlation peak value is only 39. Starting from the hologram corresponding to Fig. 8(b), the simulated annealing is performed along the procedure discussed in the previous section. Fig. 8(c) shows the reconstructed fingerprint image by the optimized hologram for T=0 and its correlation with the original image. The image is perfectly recovered (compare the pattern in Fig. 8(a)) and its correlation peak is 63, which is completely the same value with the correlation between the original images. For the iteration cycle of $n \geq 5$ the hologram is almost optimized. The optimization is also successful for the cooling schedule of T=1/exp(n) and the fingerprint image having the correlation value of 63 is obtained for $n \geq 10$. The reconstructed image is much improved for the cooling schedule of T=0.5/(1+n), however, the optimization speed is slower and the value of the correlation is 58 at n=20.

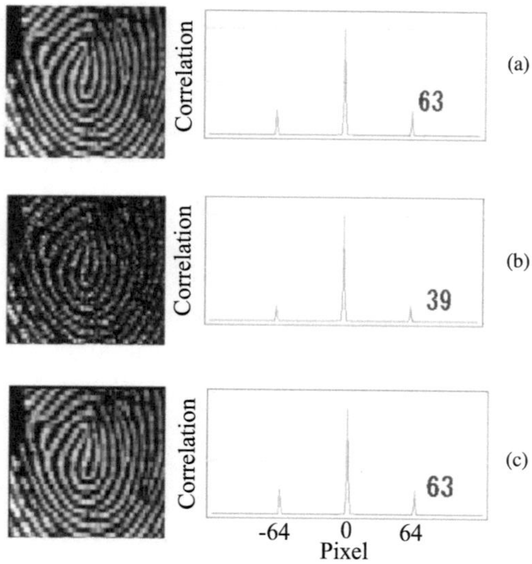

Fig. 8. Reconstructed images (left) and correlations with original image (right). (a) Original fingerprint image, (b) reconstructed image without optimization, and (c) reconstructed image with optimization.

5. Fast optimization method

5.1 Algorithm of fast optimization

Though the method based on the simulated annealing algorithm discussed in the previous section is very effective for the optimization of encrypted holograms, it is time consuming. Here, we propose an error correction method by which we can perform very fast optimization of encrypted binary holograms (Nakayama & Ohtsubo, 2007). The simulated annealing technique introduced in the previous section is a like Boltzmann machine system, while the proposed error correction method here is similar to a back propagation technique in neural net works. Before discussing the method, we first define two terms; 'decrypted image' is the part of the decrypted area where the pattern corresponding to the encrypted image is reconstructed as shown in Fig. 4. The 'dummy area' is the rest of the pattern in the output plane. The complex amplitude of the output image reconstructed from the binary encrypted hologram after n iterations is given by $f_n(x,y) = f_{0n}(x,y)\exp\{ib_n(x,y)\}$. Then, we introduce the error function $e(x,y)$ for the decrypted image;

$$|\beta f_n(x,y) + e(x,y)|^2 = |f_0(x,y)|^2 \tag{7}$$

where the coefficient β is defined by

$$\beta = \sqrt{\frac{\iint |f_0(x,y)|^2 \, dxdy}{\iint_{\text{decrypted area}} |f_n(x,y)|^2 \, dxdy}} \tag{8}$$

The function $e(x,y)$ defines the difference between the decrypted and original images and can be given by

$$\begin{aligned} &e(x,y) = (|f_0(x,y)| - \beta|f_n(x,y)|)\exp\{ib_n(x,y)\} && \text{within decrypted image} \\ &e(x,y) = 0 && \text{in dummy area} \end{aligned} \tag{9}$$

In this method, the error information is used not only for evaluating the intensity of the decrypted image but also for selecting the flipping pixels in the hologram plane. For the concern of intensity, the phase of the decrypted image may be ignored as far as the optimization is performed only in the decryption plane. However, the phase $b_n(x,y)$ is included in the error function. The error projected back into the hologram plane, namely generated from $e(x,y)$ by the inverse Fourier transform operation, is affected by the phase as shown in the following and it plays a crucial role in this algorithm.
Also, the error function is accounted in the hologram plane. The expression of the error function $\Delta H(u,v)$ in the hologram plane is deduced from the following relation;

$$\begin{aligned} \beta f_n(x,y) + e(x,y) &= \beta \mathrm{IFT}[H_n(u,v)G(u,v)] + \mathrm{IFT}[\mathrm{FT}[e(x,y)]] \\ &= \mathrm{IFT}[\beta H_n(u,v)G(u,v) + \mathrm{FT}[e(x,y)]] \\ &= \beta \mathrm{IFT}\left[\left\{H_n(u,v) + \frac{\mathrm{FT}[e(x,y)]}{\beta G(u,v)}\right\}G(u,v)\right] \\ &= \beta \mathrm{IFT}[\{H_n(u,v) + \Delta H(u,v)\}G(u,v)] \end{aligned} \tag{10}$$

where $H_n(u,v)$ is the hologram after n-th iterations and FT and IFT are the forward and inverse Fourier transform operations. Then, the error function in the hologram plane reads

$$\Delta H(u,v) = \frac{\mathrm{FT}[e(x,y)]}{\beta G(u,v)} \tag{11}$$

Equation (11) indicates that $\Delta H(u,v)$ is like as a gradient vector to the locally or globally optimal solution with respect to $H_n(u,v)$. However, we cannot directly use the error information ΔH for the correction for $H_n(u,v)$, since the hologram is a binary nature and $\Delta H(u,v)$ is a continuous complex valued function. Therefore, we require some modifications for the application of the error correction method for the binary encrypted hologram.
In accordance with the above discussion, we adopt the following processes for the optimization of a binary hologram;

Step 1. At first, $\Delta H(u,v)$ is calculated by Eq. (11). Then, each pixel of the hologram is ranked by the magnitude of the value $|\mathrm{Re}[\Delta H(u,v)]|$ for the optimization, namely, the pixel with the largest value of $|\mathrm{Re}[\Delta H(u,v)]|$ has the highest priority for the flip.

Step 2. On the descending order of the priority for the error correction, each pixel value is flipped according to the following rules;

$$\begin{aligned} &\text{if } H_n(u,v) = -1 \text{ and } \mathrm{Re}[\Delta H(u,v)] > 0, \text{ then } H_{n,new}(u,v) = 1 \\ &\text{if } H_n(u,v) = 1 \text{ and } \mathrm{Re}[\Delta H(u,v)] < 0, \text{ then } H_{n,new}(u,v) = -1 \\ &\text{else } no \text{ flip} \end{aligned} \tag{12}$$

where $H_{n,new}(u,v)$ is a temporal hologram to be tested for the new reconstruction, which is generally called *neighborhood* in the field of the combinatorial optimization theory. This flipping trial is continued until it reaches a certain number.

Step 3. When the image decrypted from $H_{n,new}(u,v)$ is closer to the original image than that from $H_n(u,v)$, $H_{n,new}(u,v)$ is adopted as a new $H_n(u,v)$. Then, the above process is repeated. In this stage, we introduce the measure for the optimization of binary hologram and define the cost function in the decryption plane as

$$E = \iint |e(x,y)|^2 \mathrm{d}x\mathrm{d}y \,. \tag{13}$$

The optimization is performed so as to lower the above cost function.
The steps 1~3 are the essentials of the proposed algorithm. The method has many advantages over the existing optimization methods. Firstly, the most effective pixel to be corrected is selected for the optimization. Secondly, not only one pixel but also multiple pixels can be flipped at the same time on the basis of the priority order for the error correction. Therefore, we can expect faster calculation for the optimization of encrypted holograms.

5.2 Multiple-flip algorithm

In ordinary single flipping for pixel elements in the simulated annealing like method, it usually takes a long time to obtain a good reconstruction. On the other hand, the error correction method has a merit of simultaneous flipping of multi-pixels for the correction of the binary hologram. We could perform a faster optimization for encrypted hologram by the method of the error correction. In the multiple-flip algorithm, many pixels of the encrypted hologram are selected according to the priority order calculated from the error function and the respective pixels are simultaneously flipped. Then, the hologram is reconstructed and the iteration is either accepted or rejected for the evaluation of the cost function. The iteration is stopped either when the cost function is sufficiently lowered or when the process is trapped to a local minimum.
Fig. 9 shows the results of the cost function for a fixed multiple-flip algorithm. Two multiple-flip schemes are plotted; one is a 7-flip and the other is a 655-flip. The optimization stopped at 1818 iterations for the 7-flip, while it stopped only at 19 iterations for the 655-flip. The number of iterations to reach an optimized hologram is drastically reduced for a large number of simultaneous flipping, however the iteration stopped at a higher value of the cost function and the quality of image is rather poor compared with that for a lower number of the multiple-flip algorithm. The final iteration number for the 655-flip algorithm is 1/96 of that for the 7-flip algorithm. The actual calculation time on the computer is 1/86. Therefore, as a rough estimate, the calculation time is inversely proportional to the number of pixels to be flipped at the same time. The final costs for the 655-flip and the 7-flip are 117.67 and 67.78, respectively, while it is only 9.47 for the single-flip algorithm. Accordingly, the quality of the decrypted image for a higher number of the multiple-flip is worse than that of the lower case. There exists a trade-off between the final quality of the decrypted image and the calculation speed for the optimization. To show the trade-off, the dependence of the final iteration number and the cost function were investigated for various numbers of the multiple-flip algorithm. Fig. 10 is the results for two different fingerprint images. Similar trends have been observed for respective fingerprint images.

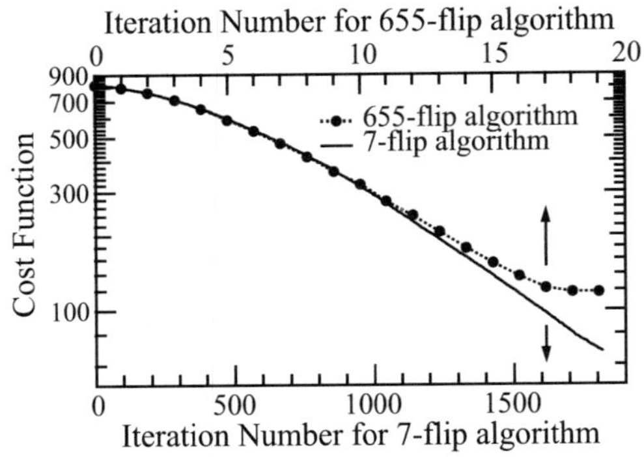

Fig. 9. Cost function for flip number in a fixed multiple-flip algorithm

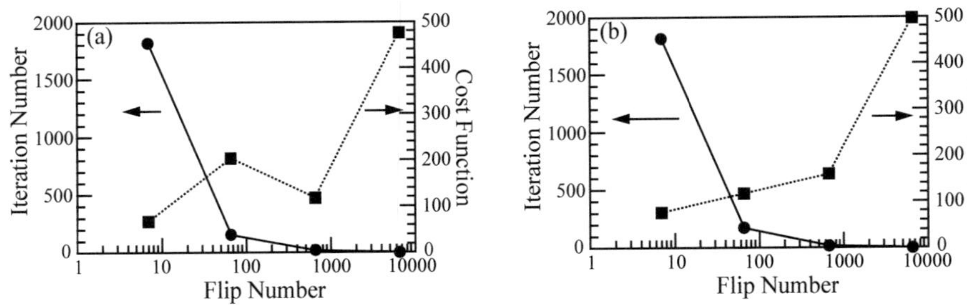

Fig. 10. Dependence of final iteration number and its cost on the number of simultaneous multiple flipping pixels. (a) corresponds to the fingerprint image in Fig. 3(a), and (b) is also for the fingerprint image in Fig. 12(a).

5.3 Variable multiple-flip algorithm

To avoid the trap of a local minimum and obtain a good decryption, we employ a scheme of a variable multiple-flip algorithm. In this algorithm, the flip number is dynamically changed. We first set the flip number at a certain value and started the iteration. When the iteration stops due to the trapping of a local minimum, the flip number is lowered and the next iteration is repeated. This process is repeated until the cost function is much lowered.

Fig. 11 is the results of the algorithm for the variable multiple-flip number. Fig. 11(a) is the decrypted image for the final optimized hologram with the encryption key and Fig. 11(b) is that with a wrong key. Fig. 11(c) is the change of the cost function for the iteration number. The flip number was at first chosen to be 655 and it was lowered as 66 after the trapping a local minimum. Then, the iteration was continued as lowering the flip number as 7, 4, and finally 2. The boundaries of the changing points are indicated as broken lines. We calculated the joint transform correlation between the obtained pattern and the original fingerprint image. The value of the correlation peaks calculated in Fig. 11(a) is 0.243 (normalized by the initial cost to unity), while that of the simulated annealing method of single flipping is 0.247. Almost the same quality of the reconstruction as the simulated annealing method is achieved by the current technique. We also compared the calculation time both for the proposed and the previous methods. For the optimization of the simulated annealing method, the calculation time at which the value of cost function became 0.243 was evaluated. The calculation times for the proposed and previous methods are 1149 and 39364 s, respectively, i.e., the calculation time of the proposed method is only 2.9 % of that of the simulated annealing method. Thus, we can attain a very fast optimization of encrypted holograms by using the error correction method with the variable multiple-flip algorithm without loosing the quality of the reconstruction. Finally, the variable multi-flip algorithm was applied to other types of images. Fig. 12 shows the results. It is clear that the proposed algorithm is applicable not only to fingerprint image but also to other images (binary logo of Shizuoka University and stamp of Chinese characters).

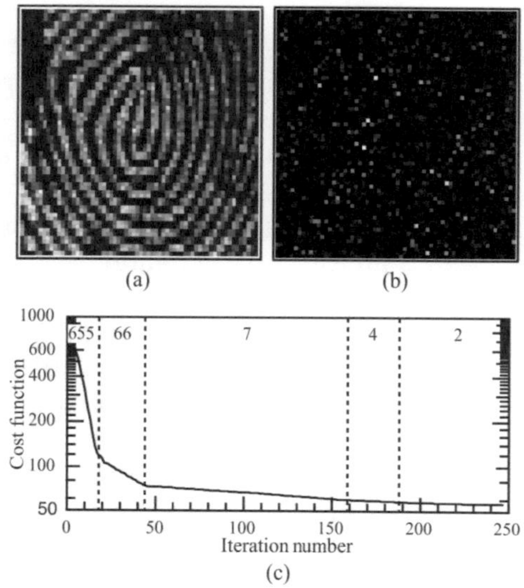

Fig. 11. Results for variable multiple-flip algorithm. Decrypted images from optimized hologram (a) with the encryption key and (b) with a wrong key. (c) Cost function for the iteration number.

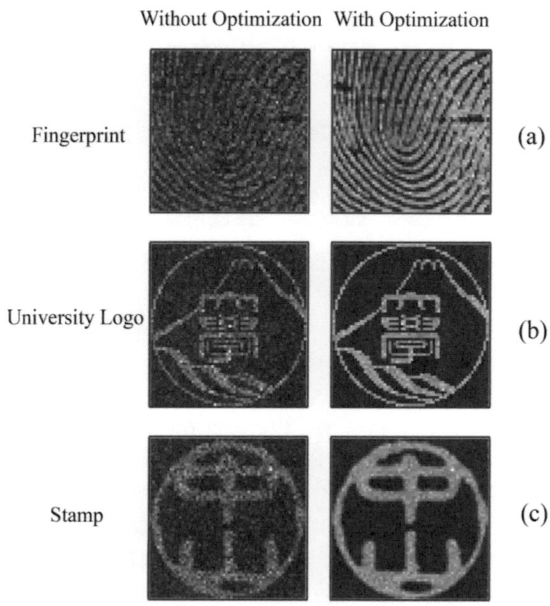

Fig. 12. Decrypted images from optimized holograms for various image structures in variable multiple-flip algorithm.

6. Optimization of hologram in real systems

6.1 Optical devices in real optical security systems

In the previous section, a binary encrypted hologram is used for easiness of optical reading and the degraded decryption image due to the binarization is successfully recovered by the optimization of hologram. In that system, we assume that not only a hologram but also a decryption key are displayed through electronic imaging devices. Even when we use optically addressed spatial light modulators (SLMs) with smart-pixel in optical security systems, we must inevitably use some electronic devices for the projection of images onto the SLMs. For example, a liquid crystal television display panel is frequently used for such purpose either of amplitude or phase modulation device. Such an imaging device has a lattice structure, in which the clear aperture is less than 100 % and only a limited light by the lattice structure passes through each pixel. Fig. 13 shows an example of a microscopic image of a liquid crystal display used as an electronically addressed SLM. The aperture ratio is about 0.55 in this case. In the joint transform system using real electronic devices, the decrypted image is greatly degraded due to the presence of the lattice structure, if we use an encrypted hologram generated from pure numerical calculation without considering the lattice structure. Or at worst case, we could not reconstruct any original image due to the degradation (Ohtsubo & Fujimoto, 2002). In this section, we apply the method of the simulated-annealing like optimization for binary hologram in real optical security systems and demonstrate successful decryptions of original images in the presence of lattice structures. The optical security system we treat here is the joint Fourier transform system. Both for the displays or projections of the hologram and the decryption key, we consider the use of electronically addressed SLMs such as liquid crystal television panels.

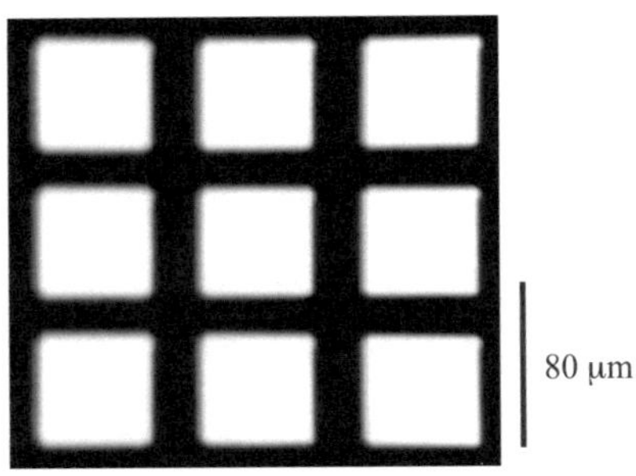

Fig. 13. Example of lattice structure of a liquid crystal television display used for electronically addressed spatial light modulator.

6.2 Holographic reconstruction in the presence of lattice structure

Here, we suppose the use of electronic display devices for the decryption of the encrypted hologram in a real optical security system. We assume that each pixel of an image in the input plane has a clear aperture ratio of 25 %. Actual display devices, for example LCTV panel, may have a rather larger value of the clear aperture, however, we take the value for the easiness of the numerical simulations. Fig. 14 shows the result of the numerical simulation for such a case. Fig. 14(a) is the same hologram as in Fig. 7(b), but it is embedded into the periodic lattice structure. The total size of the pattern in the input plane is expanded to 256x256 due to the presence of the periodic lattice structure. Fig. 14(b) is the same random key pattern as that in Fig. 3(b), but it is also embedded into the periodic lattice structure.

Using the hologram and the random key pattern in Figs. 14(a) and (b), the decryption was performed. The result is shown in Fig. 14(c). We cannot see any information of the original pattern of the fingerprint image. In the optical security system discussed here, the decryption is not a simple reconstruction of hologram, such as illumination by a plane wave. The hologram is illuminated by the Fourier transform of the random key pattern. Therefore the illumination of the Fourier transform of the periodic lattice structure greatly affects the performance of the reconstruction of hologram. It has less redundancy compared with a simple holographic reconstruction done by a plain wave illumination. Without considering the lattice structure, at worst case, we cannot extract any information from the reconstructed pattern as shown in Fig. 14(c).

(a) (b) (c)

Fig. 14. Image decryption in the presence of opaque lattice structures on the hologram and the decryption key pattern. (a) Hologram, (b) decryption key pattern, and (c) decryption of image.

6.3 Optimization of hologram in the presence of lattice structure

The procedure for optimization of hologram in the presence of a lattice structure is almost the same as discussed in section 3. Starting from a hologram as shown in Fig. 14(a) together with a decryption key in Fig. 14(b) all including the lattice structures, the optimization of hologram to obtain good reconstruction is performed following the step 1~5 as discussed in section 3. Throughout the following optimization, the decryption key pattern, which also has a periodic lattice structure, is not changed and is assumed to be the same pattern as shown in Fig. 14(b). On the other hand, starting from the encryption hologram shown in Fig. 14(a), the value of each pixel of the hologram is modified by flipping from +1 to -1 or vice versa. In each flipping, we test the newly decrypted image as to whether it gives rise to a good reconstruction or not. The flipping is successively repeated for every pixel. If the cost function for the optimization still has a large value, the next iteration is performed. When the value of the cost function is sufficiently lowered, the iteration stops. Then, the image is optimized to reach a good estimation.

In the numerical simulation for the optimization, the area of the image to be compared with the decrypted pattern is expanded to 64x64 pixels due to the presence of the periodic opaque lattice structure. Therefore, we used the fingerprint image with 64x64 pixels as the ideal target image as shown in Fig. 15(a). Fig. 15(b) is the optimized hologram calculated by the proposed method when it contains the lattice structure. Using the optimized hologram together with the random key pattern in Fig. 14(b), we obtain the decrypted pattern as shown in Fig. 15(c). We can successfully decrypt an image close to the original one, though the periodic multiple images are reconstructed. The multiple images are originated due to the presence of the periodicity of the lattice structure in the display panels. The value of the cost function rapidly decreases in a few iterations and it reaches almost less than 10 % of the initial cost. At the final iteration number, the cost function is less than 5 % of the initial cost. The cost function of 5 % is considered as an enough criterion for the reconstruction of the original images. Indeed, the correlation between the original and decrypted images over 95 % is obtained at this criterion and the reconstructed image can be used for the identification in the security system (Nakayama & Ohtsubo,2007).

Flipping each pixel value of the hologram on the SLM through a computer control, we can optically and electronically perform the same optimization as done by the numerical simulation discussed here. In actual situation, a lattice structure to display an image is not only the issue to be overcome. However, based on the same principle proposed here, structures of image acquisition devices, a misalignment between optical acquisition and display devices, and even aberrations though optical elements can be compensated by the optimization of binary hologram. Then we can obtain optically an optimized binary hologram for image decryption for a particular optical system.

(a) (b) (c)

Fig. 15. Optimization of hologram to obtain the exact image. (a) Original image to compare the estimate of the decryption, (b) optimized hologram, and (c) result of the decryption from the optimized hologram.

7. Phase-coding method and optimization of hologram

7.1 Method of phase-coding

A hologram with a reference of a random pattern has been used for security encoding of biometric patterns. The methods have been proved to be useful by numerical simulations. Phase-coding techniques have widely been used due to the suitability for optical encoding and decoding with high degree of security (Neto & Sheng, 1996, Javidi & Ahouzi, 1998, Towghi et al., 1999, Tan et al., 2000, and Mogensen &Gluckstad, 2000 & 2001). However, they have some difficulties due to the lack of efficient optical devices used in the systems. In those methods, an image for authenticity is encrypted as a hologram and the hologram is illuminated and reconstructed by a random decode key that is the same as the encryption key and, thus, the methods have some difficulties for actual optical implementation (Yan & Kim, 1996). For example, the size of each pixel of a decryption key must be exactly matched to the corresponding area of a hologram in the Fourier space. To avoid the difficulty, Park *et al.* (2001) proposed a technique to obtain a decrypted image by a simple joint Fourier transform of an encrypted pattern and a decryption key. Another difficulty is the availability of optical devices with sufficient resolution having a large dynamic range to implement a compact optical security system. Optical security systems to verify the

authenticity such as credit cards and passport identifications are usually used with electronic imaging systems. Therefore, diffraction effects induced by array structures of electronic addressed spatial light modulators as input optical devices may also degrade the image quality of the reconstruction of holograms.
We here discuss a simpler method of image encryption and decryption in this section, which is different from one in the previous sections. In this method, an encryption of an image for the identification is done by a digital technique, and the decryption and the identification are assumed to be performed by optical systems. We here focus on the method of encryption and decryption of images. The degree of security for an encryption may not be so high, since the information of an original image is partially obtained from the reconstructed phase from the complex encrypted pattern. Therefore, we further consider the use of real-valued encrypted pattern instead of complex-valued encrypted pattern to enhance the degree of the security. Binarization of a complex-valued encrypted pattern that is obtained from the real-valued data is an another common technique to match optical read-out of images in practical optical security systems. Then, we also perform a binarization of an encrypted pattern and propose the method of the optimization for it to obtain a good quality image of the decryption.

7.2 Phase-coding: Theory

We here discuss the theoretical background of the phase-coding method. At first we assume that an original image $f_0(x,y)$ to be encrypted has binary values. The image is transformed to a phase pattern and the phase-coded pattern is multiplied by a binary random phase mask. The distribution of the total coded image is written by

$$f(x,y) = \exp[i\pi f_0(x,y)] \cdot \exp[i\pi r_0(x,y)] \tag{14}$$

where $r_0(x,y)$ is a random function that plays a role for the encryption key and both the functions $f_0(x,y)$ and $r_0(x,y)$ take values of "0" or "1." The Fourier transform of the coded function $f(x,y)$ is an encrypted pattern. Since the original image is transformed to a phase pattern and it is also scrambled by a random phase, one cannot exactly extract the information of the original image from the encrypted pattern without knowing the random key. It is noted here that the encrypted pattern has a complex-valued function and we need a complex representation of the encrypted pattern for the implementation in optical security systems. However, as discussed later, one can extract the original image with good quality from the real-valued binarized data of the encrypted pattern.
In the Fourier space, the amplitudes of the encrypted pattern and the Fourier-transform of the random phase key are simply added together for the decryption. Writing the encrypted pattern defined by the Fourier transform of Eq. (14) as $F(u,v)$, the amplitude distribution of the addition is given by

$$H(u,v) = F(u,v) + R(u,v) \tag{15}$$

where $R(u,v)$ is the Fourier transform of the function $r(x,y)=\exp\{i\pi r_0(x,y)\}$. The function $R(u,v)$ plays the role for the decryption key. Also it is noted that the decryption key is a complex-valued function. Then, the decryption is simply performed by an inverse-Fourier transform for the added pattern. The intensity of the inverse-Fourier transform $h(x,y)$ of Eq. (15) is easily calculated and written by

$$|h(x,y)|^2 = |f(x,y)|^2 + |r(x,y)|^2 + f(x,y)r^*(x,y) + f^*(x,y)r(x,y) \\ = 2 + 2\cos[\pi f_0(x,y)] \quad (16)$$

Since we assumed that the original image has a binary value either "0" or "1," we obtain a negative binary image as a decryption for the input that has a value "4" or "0." The merit of the method is that the inverse-Fourier transform of Eq. (15) is exactly equal to the original image itself and the decrypted image does not contain a zero-th order diffraction or other components.

The method is straightforwardly applicable to a gray scale image. For a gray scale image, we need no change of the procedure for encryption and decryption. If analogue values of the function $f_0(x,y)$ are normalized by its maximum value and they are distributed between 0 and 1, we can also obtain a negative image of the input as easily understand from Eq. (16), though the decrypted image is nonlinearly transformed with a cosine function. But the nonlinearlity can be easily compensated by a digital method due to one to one correspondence between the intensity distributions of the two images. Alternatively, the original image $f_0(x,y)$ may be nonlinearly transformed in advance by an arccosine function to give rise to the exact distribution of the image in the encryption process.

7.3 Decryption of image by phase-coding method

The method is applied to a gray scale image and the real and imaginary parts of the hologram (not a binary hologram) are used in this subsection. Fig. 16 shows the simulation result. A fingerprint image in Fig. 16(a) has a gray scale distribution between 0 and 1. The image is encrypted with a random phase pattern and, then, decrypted following the theory. Finally, we obtain a negative image of the original one as shown in Fig. 16(b). The nonlinearlity of the image distribution due to a cosine function is not compensated in the figure. The size of the original image is also 64x64 pixels and the original intensity level of 8-bit gray scale is normalized to be unity.

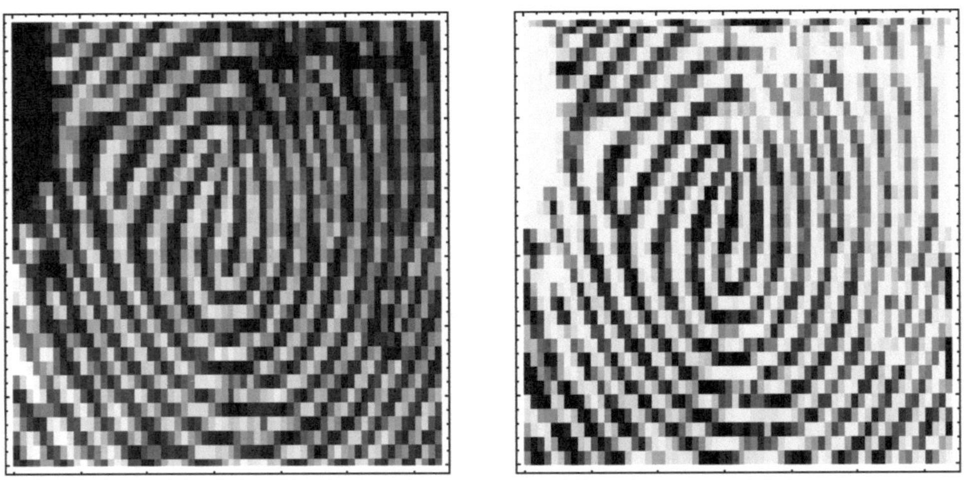

(a) (b)

Fig. 16. Encryption (a) and decryption (b) for gray scale fingerprint image based on phase-coding method.

The degree of the security to hide an image behind a random mask is usually not so high when a phase function is reconstructed from an encrypted pattern. Furthermore, the representation of complex values of an encrypted pattern may not be suited for the implementation of practical optical security systems, since optical images are usually read out through an electronic interface. Therefore, we consider the use of a real-valued data from a complex encrypted pattern. For that purpose, we consider only real parts of an encrypted pattern and a decryption key. Using the real-valued data, the original image is exactly decrypted as shown in Fig. 17. Fig. 17(a) is an input fingerprint image to be decrypted. Here, we used an inverted gray scale and nonlinearly processed (arccosine transformed) fingerprint image to obtain the exactly expected image (compare it with Fig. 16(a)). The converted image is placed only in the area quarter of the input plane, since a mirror image is reconstructed due to the use of real-valued patterns in the following operation. Therefore, the total pixel size of the patterns used throughout the simulations is 128x128. Fig. 17(b) is a random key corresponding to the function $r_0(x,y)$. The encrypted pattern and the decryption key are calculated according to Eq. (14) and the Fourier transform of the random phase function, respectively. Next, we take the real parts of the Fourier transformed patterns. They are shown in Figs. 17(c) and (d). Then, the real-valued functions in Figs. 17(c) and (d) are added together and the result is inversely Fourier-transformed. The decrypted image is shown in Fig. 17(e). We obtain the exact image as a decryption, however a mirror image is also reconstructed due to the lack of the information of the imaginary parts of the encrypted image and the decryption key. From the standpoint of the degree of security, the use of a real-valued data from a complex encrypted image is not secure, since the imaginary part of the complex pattern can be easily reconstructed from its real part. Therefore, the binarization of encrypted image is essential for enhancing the degree of the security.

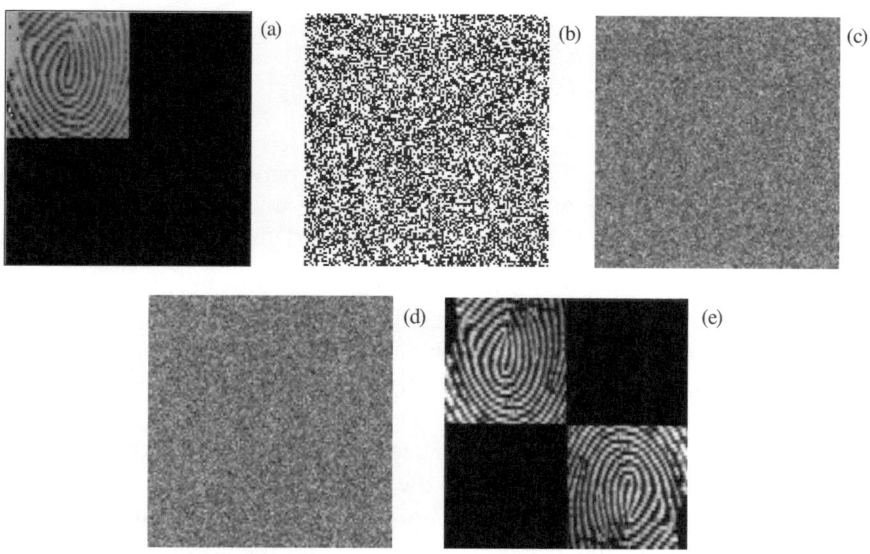

Fig. 17. Decryption using real-valued patterns. (a) Original image. The fingerprint image is inverted and the gray scale of each pixel is nonlinearly transformed by an arccosine function (compare it with Fig. 2(a)). (b) Encryption key, (c) real part of the encrypted pattern, (d) real part of the decryption key, and (e) decrypted image.

7.4 Binarization of encrypted image and its optimization

In actual applications, the format of an encrypted pattern must be congenial to an electronic interface in decryption process. Fast processing is only required for image decryption and identification. The binarization of a real-valued data also greatly enhances the degree of security for encryption and decryption in optical security systems as have already been discussed. We again employ the method of the binarization for an encrypted pattern. Fig. 18 shows the result of the binarizations both for the encrypted pattern and the decryption key. The binary encrypted pattern is obtained from the signs of the real part values of the original encryption pattern in Fig. 17(c). Namely, when a real part of each pixel of the encrypted pattern is positive, we set the pixel value to be +1 (optical phase is 0), while it is –1 (optical phase is π) for a negative value. The binary decryption key is made from the pattern in Fig. 17(d) as the same manner. Here we assume the use of optically addressed spatial light modulators (SLMs) with phase modulation in actual optical systems. The binarized patterns still have the original information. Then, adding the two binary patterns of the encrypted pattern and the decryption key shown in Figs. 18(a) and (b), the decryption of the image is performed. The result is shown in Fig. 18(c). Though we can recognize a dim structure of the fingerprint image, the decrypted image is greatly degraded due to the binarizations of the encrypted pattern and the decryption key.

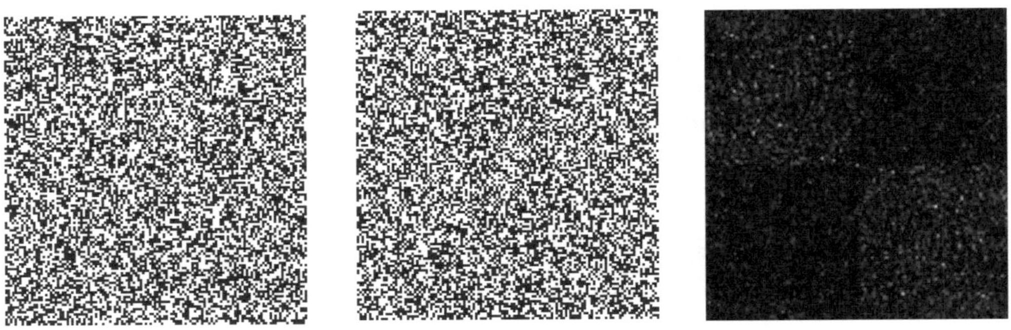

(a) (b) (c)

Fig. 18. Decryption using binarized patterns of encryption pattern and decryption key. (a) Binarized encryption pattern, (b) binarized decryption key, and (c) decrypted image. The patterns are binarized according to the signs of the real part values.

Therefore, we consider the optimization of the binary encrypted pattern. In the optimization, the decryption key is not changed and remains the same binary pattern as shown in Fig. 18(b) throughout the iterations. Starting from the binary encrypted pattern shown in Fig. 18(a), the value of each pixel of the pattern is flipped form +1 to –1 or vice versa. In each flipping, we test the newly decrypted image whether it gives rise to a good reconstruction or not. Then, the image is optimized to reach a good estimate. The method is one like a simulated annealing technique and the detail of the method is the same as that in the pervious section. The area to be compared for the optimization in the decryption image plane is only a quarter of the original pattern where the ideal fingerprint image is reconstructed. Figs. 19(a) and (b) show the optimized binary encrypted pattern and the result of the decrypted image, respectively. From the comparison between Figs. 19(b) and 18(c), the optimization goes well and almost the same image as the original fingerprint image is recovered as easily recognized from the comparison with the pattern in Fig. 17(e).

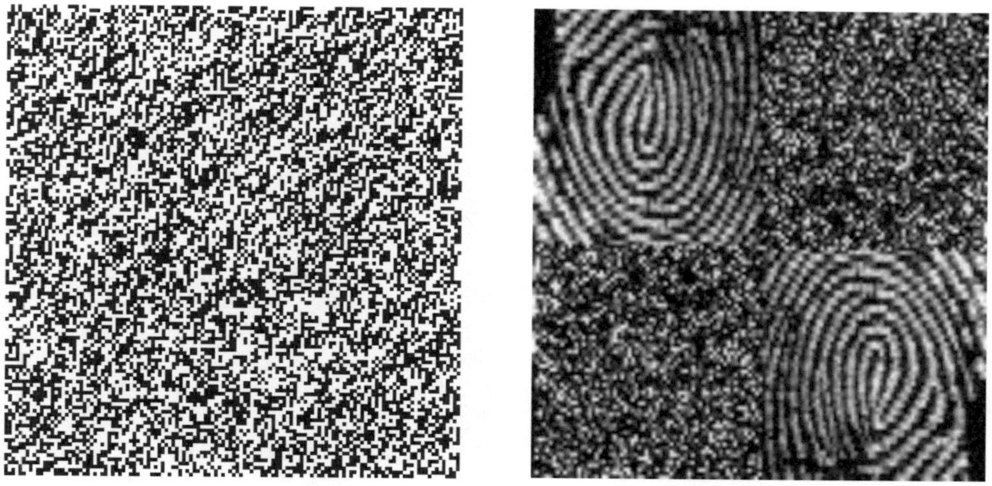

(a) (b)

Fig. 19. Optimization of encrypted pattern. (a) Optimized binary hologram for Fig. 4(a) and (b) decryption of it.

Beside of the security problems in the proposed method, we must consider the congeniality of the method with optical systems. One of the merits of optics is a complex representation of information. Therefore, encryption and decryption of a complex image is easily performed and a high quality image of the decryption can be obtained in optical security systems. However, real systems include electronic interfaces and a complex image must be replaced by real-valued patterns. Furthermore, analogue write-in and read-out of optical data sometimes cause difficulties, for example a gray scale of a pattern must be faithfully recorded in the write-in process and the analogue gray level must be correctly scaled in read-out process. Reducing such problems, binarization of a complex-valued pattern is used as a common technique. However, the quality of a reconstructed pattern usually results in degraded one. Therefore, the optimization for a binary encrypted pattern is essential in optical security systems. As already discussed, we can obtain an excellent decrypted image based on a statistical optimization technique. In the system, we employed an optical phase encoding method, however we can alternatively consider an amplitude encoding technique. For example, instead of 0 (+1) or π (-1) phase encoding of the pattern, a binary amplitude, i.e. amplitude of 0 or 1, can be used. However, the amplitude encoding in the proposed method leads to the same result as that of the phase encoding except for a zero-th order diffraction spot in the reconstructed plane, since the addition of an encrypted pattern and a decryption key takes only three values, 0, 1, and 2. In the meantime, that for the phase coding has also three values of -2, 0, and 2 and the decrypted image reduces to the same pattern as that of the amplitude encoding. Therefore, we here employed the phase encoding technique. An optical phase modulation device is commercially available. The device can be used as a phase modulation spatial light modulator over 2π modulation depth. Generation and addition of phase images can be optically performed by using phase-controlled SLMs together with electronically addressed liquid crystal display panels as input imaging devices. Thus, optical decryption from encrypted pattern proposed here is easily implemented by using such devices. Fig. 20 is an example of the experimental results using

LCTV SLM. Fig. 20(a) is the experimentally reconstructed image and the nurerical reconstruction image is shown in Fig. 20(b) for comparison.

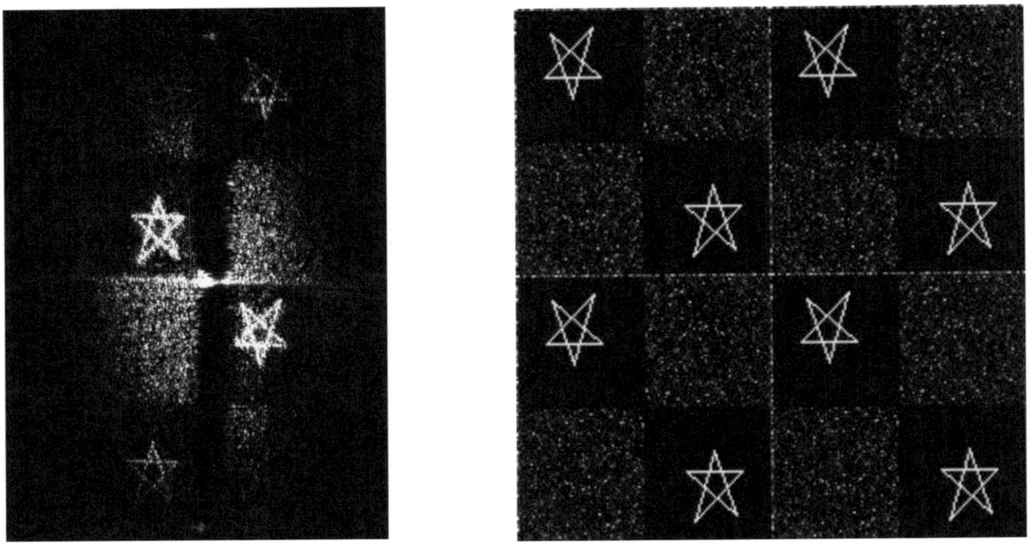

(a) (b)

Fig. 20. Experimental result of optical phase-coding method. (a) Optical and (b) simulation results.

8. Conclusion

We have studied optical security systems for practical applications in verification of the authenticity, such as in a card system. The advantages of the optical method in a security are the fast decoding of an encrypted image and the identification of it. Firstly, we study a common method of joint transform correlation for optical security systems and the optimization of binary holograms and proved that the optimization of a hologram will be a powerful tool in the systems. As an alternative method, a phase-coding technique is introduced to congenital with the uses of real optical devices in optical security systems. Originally, the merit of optical systems is fast processing of decryption of hologram and identification of it. However, the methods discussed here, i.e., not only image encryption, but also decryption of hologram and identification of it, can be applicable in all-digital techniques using a recent fast computer.

9. References

Aarts E.; & Korst J. (1990). *Simulated Annealing and Boltzmann Machines*, John Wiley & Sons, Chichester

Bättig R. K.; Guest C. C.; Schaefer S. R.; & Toms D. J. (1992). Simulated Annealing of Binary Holograms for the Interconnection of Single-Mode Sstructures, *Appl. Opt.*, Vol.31, pp.1059-1066

Javidi B.; & Horner J. L. (1994). Optical Pattern Recognition for Validation and Security Verification, *Opt. Eng.*, Vol.33, pp.1752-1756

Javidi B.; Zhang G. S.; & Li J. (1996). Experimental Demonstration of the Random Phase Encoding Technique for Image Encryption and Security Verification, *Opt. Eng.*, Vol.35, pp.2506-2512

Javidi B. (1997). Securing Information with Optical Technologies, *Phys. Today*, Vol.50, pp.27-32

Javidi B.; & Ahouzi E. (1998). Optical Security System with Fourier Plane Encoding, *Appl. Opt.*, Vol.37, pp.6247-6255

Kipatrick S.; Gelatt Jr. C. D.; & Vecchi M. P. (1983). Optimization by Simulated Annealing, *Science*, Vol.220, pp.671-679

Kobayashi Y.; & Toyoda H. (1999). Development of an Optical Joint Transform Correlation System for Fingerprint Recognition, *Opt. Eng.*, Vol.38, pp.1250-1210

McCallum B. C. (1990). Blind Deconvolution by Simulated Annealing, *Opt. Commun.*, Vol.75, pp.101-105

Metropolis N.; Rosenbluth A.; Rosenbluth M.; Teller A.; & Teller E. (1953). Equation of State Calculations by Fast Computing Machines, *J. Chem. Phys.*, Vol.21, pp.1087-1092

Mogensen P. C.; & Gluckstad J. (2000). Phase-Only Optical Encryption, *Opt. Lett.*, Vol.25, pp.566-568

Mogensen P. C.; & Gluckstad J. (2001). Phase-Only Optical Encryption of a Fixed Mask, *Appl. Opt.* Vol.40, pp.1226-1235

Nakayama K.; & Ohtsubo J. (2007). Fast Optimization of Binary Encrypted Hologram Based on Error Correction Method in Optical Security Systems, *Opt. Rev.*, Vol.14, pp.290-296

Neto L. G.; & Sheng Y. (1996). Optical Implementation of Image Encryption using Random Phase Encoding, *Opt. Eng.*, Vol.35, pp.2459-2463

Ohtsubo J.; & Nakajima K. (1991). Image Recovery by Simulated Annealing with Known Fouier Modulus, *Opt. Commun.*, Vol.86, pp.265-270

Ohtsubo J.; & Fujimoto A. (2002). Practical Image Encryption and Decryption by Phase-Coding Technique for Ooptical Security Systems, *Appl. Opt.*, Vol.41, pp.4848-4855

Ohtsubo J.; & Fujimoto A. (2007). Optimization of Binary Hologram Degraded by Periodic Lattice Structure of LCTV Panel in Real Optical Security Systems, *Opt. Rev.*, Vol.14, pp.266-270

Park S. J.; Kim J. Y.; Bae J. K.; & Kim S. J. (2001). Fourier-Plane Encryption Technique Based on Removing the Effect of Phase Terms in a Joint Transform Correlator, *Opt. Rev.*, Vol.8, pp.413-415

Refregier P.; & Javidi B. (1995). Optical Image Encryption Based on Input Plane and Fourier Plane Random Encoding, *Opt. Lett.*, Vol.90, pp.767-769

Tan X.; Matoba O.; Shimura T.; Kuroda K.; & Javidi B. (2000). Secure Optical Storage that uses Fully Phase Encryption, *Appl. Opt.*, Vol.39, pp.6689-6694

Towghi N.; Javidi B.; & Luo Z. (1999). Fully Phase Encrypted Image Processor, *J. Opt. Soc. Am. A*, Vol.16, pp.1915-1927

Unnikrishnan G.; Joseph J.; & Singh K. (1998). Optical Encryption System that Uses Phase Conjugation in a Photorefractive Crystal, *App. Opt.*, Vol.37, pp.8181-8186

Yamazaki M.; & Ohtsubo J. (2001). Optimization of Encrypted Holograms in Optical Security Systems, *Opt. Eng.*, Vol.40, pp.132-137

Yang H-G.; & Kim E-S. (1996). Practical Image Encryption Sscheme by Real-Valued Data," *Opt. Eng.*, Vol.35, pp.2473-2478

15

Nanophotonic Hierarchical Holograms: Demonstration of Hierarchical Applications Based on Nanophotonics

Naoya Tate et al.*
The University of Tokyo, Japan

1. Introduction

Recently, Italy's National Committee for Cultural Heritage found some microscopic codes in Mona Lisa's pupils by using a magnifying glass (Lorenzi, 2010). Experts have pointed out that the codes may represent several messages, including the initials of Leonardo Da Vinci, "*LV*". On the other hand, how and why such microscopic messages were embedded in her pupils has not been revealed yet. The most interesting part of this *other Da Vinci code* is that, although techniques for microscopic fabrication and retrieval had not been generally established in the early 16th century, the concept of embedding secret messages in a macro-scale view already existed—or as we now say, "The best place to hide a leaf is in a forest".

Present-day techniques for realizing this concept involve the ideas of *covertness* and *overtness*. The former means not showing something openly, and the latter means the opposite. As Da Vinci showed 500 years ago in Mona Lisa' s pupils, microscopic optical techniques are suitable for embedding secret messages in a macro-scale optical observation, because the hierarchical structure inherent between different levels of the optical scale can be implemented simply, and the levels are functionally independent of each other. For instance, confidential information can be hidden in any of the physical attributes of light, such as phase, wavelength, spatial frequency, or polarization, so that one kind of anti-counterfeiting is represented (Javidi et al., 1994; Refregier et al., 1995; Rakuljic et al., 1992).

Holography, which generates natural three-dimensional images consisting of a number of diffracted light beams, is one of the most common anti-counterfeiting optical techniques (Renesse et al., 1998). In the case of a volume hologram, the surface of the hologram is ingeniously designed into a complicated structure that diffracts incident light in specific directions. A number of diffracted light beams can form an arbitrary three-dimensional image. Because these structures are generally recognized as being difficult to duplicate,

* Makoto Naruse[1,2], Takashi Yatsui[1], Tadashi Kawazoe[1], Morihisa Hoga[3], Yasuyuki Ohyagi[3], Yoko Sekine[3], Tokuhiro Fukuyama[3], Mitsuru Kitamura[3] and Motoichi Ohtsu[1]
[1]*The University of Tokyo, Japan*
[2]*National Institute of Information and Communications Technology, Japan*
[3]*Dai Nippon Printing Co. Ltd., Japan*

holograms have been widely used in the anti-counterfeiting of bills, credit cards, etc. However, conventional anti-counterfeiting methods based on the physical appearance of holograms are less than 100% secure (McGrew et al., 1990). Although they provide ease of authentication, adding another security feature without causing any deterioration in the appearance is quite difficult.

Many existing optical devices and systems, not just holography, operate based on the phenomena of *propagating* light. Therefore, their performance is generally limited by the diffraction of light (Zhdanov et al., 1998). The critical difficulty in improving the function of conventional holograms is that they are also bounded by the diffraction limit. However, with recent advances in nanophotonics, especially in systems utilizing optical near-field interactions, several optical devices and systems can be designed at densities beyond those conventionally constrained by the diffraction limit (Ohtsu et al., 2008). Because several physical parameters of propagating light are not affected by nanometric structures, the conventional optical responses in the optical far-field are not affected by these structures either. Essentially, this means that another functional layer in the optical near-field regime can be added to conventional optical devices and systems without any effect on their primary quality, such as reflectance, absorptance, refractive index, or diffraction efficiency.

Here, we propose a *nanophotonic hierarchical hologram* as a typical demonstration of this concept. The nanophotonic hierarchical hologram is a functionally improved version of a conventional hologram that works in both the optical far- and near-fields (Tate et al., 2008). Moreover, a *nanophotonic code*, which is physically a subwavelength-scale shape-engineered metal nanostructure, is embedded in the hierarchical hologram to implement a near-mode function (Tate et al., 2010). In this chapter, the basic concept of the nanophotonic hierarchical hologram with embedded nanophotonic codes and the fabrication of a sample device are described. In particular, since the proposed approach involves embedding a nanophotonic code *within* the patterns of the hologram, which is basically composed of one-dimensional grating structures, clear polarization dependence is found compared with the case where it is not embedded within a hologram or an arrayed structure. There are also other benefits with the proposed approach: a major benefit is that the existing industrial facilities and fabrication technologies that have been developed for conventional holograms can be fully utilized, yet allowing novel functionalities to be added to the hologram.

2. Hierarchical nanophotonic system

2.1 Nanophotonics

Nanophotonics is a novel technology that utilizes the optical near-field, the electromagnetic field that mediates the interactions between closely spaced nanometric matter (Ohtsu et al., 2008). As shown in Fig. 1(a), optical near-fields are the elementary surface excitations on nanometric particles, which are induced by incident propagating light. By exploiting optical near-field interactions, nanophotonics has broken through the integration density restrictions imposed on conventional optical devices by the diffraction limit of light (Fig. 1(b)). This higher integration density has enabled realization of *quantitative* innovations in photonic devices and optical fabrication technologies (Nishida et al., 2007; Ozbay et al., 2006). Moreover, *qualitative* innovations have been accomplished by utilizing novel functions and phenomena made possible by optical near-field interactions that are otherwise unachievable with conventional propagating light (Ohtsu et al., 2008).

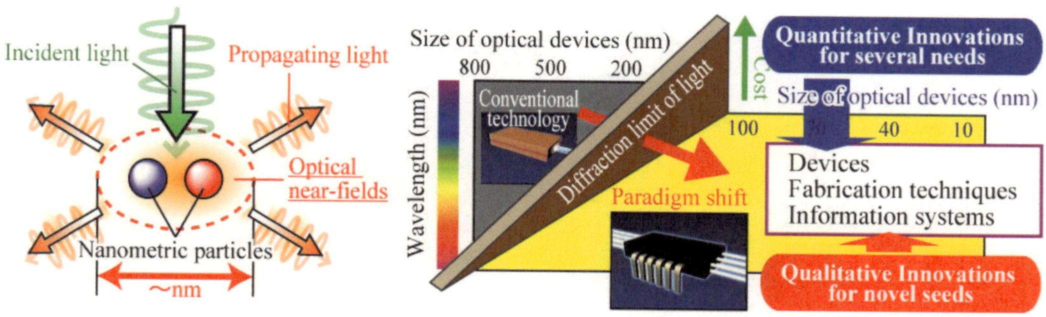

(a) (b)

Fig. 1. (a) Generation of optical near-fields, and (b) development of nanophotonics. Because optical near-fields do not involve the optical diffraction limit and they exhibit characteristic features that depend on direct interaction with materials, both *quantitative* and *qualitative* innovations can be achieved.

2.2 Hierarchy based on nanophotonics

Hierarchy in optical near-fields is one of the most appealing attributes for making innovative devices and systems based on nanophotonics. Naruse et al. investigated the hierarchy within the scale of optical near-fields, whose distribution is represented by a Yukawa function (Ohtsu et al., 2008), by investigating the size of materials and their associated optical near-fields (Naruse et al., 2005). The optical near-field response at a given scale is the result of interactions between the retrieval probe and the nanometric materials, and it is correlated with the materials involved at that scale. This feature has been exploited in various applications, for example, hierarchical optical memories where shape-engineered nanostructures provide two-layer responses in optical near-fields (Naruse et al., 2008). Moreover, besides the sizes of the materials, the shape, alignment, and composition are also important physical properties for engineering hierarchical systems. By suitable arrangement of such properties, several characteristic distributions of optical near-fields can be revealed (Naruse et al., 2008; Tate et al., APB2009; Tate et al., OptExp2009). These characteristics exert a large influence on the retrieval, and it means that various retrieval layers can be independently implemented in the same device in the form of a *nanophotonic hierarchical system* (Fig. 2).

From the point of view of the optical security, each layer is defined as an independent information layer. This enables a security layer structure in which nano-scale layers implement covertness and macro-scale layers implement overtness. The former is technically difficult to access and is non-duplicatable, whereas the latter is easy to access and is mass-producable.

3. Nanophotonic hierarchical hologram

3.1 Concept

We can see the hierarchy of optical near-fields and far-fields because optical near-field interactions are distinguishable with propagating light. This characteristic feature has led to hierarchical optical system designs, such as *nanophotonic hierarchical holograms* (Tate et al., 2008), where independent functions are associated with both optical near- and far-fields in the same device. Figure 3 shows the basic concept of the hierarchical hologram.

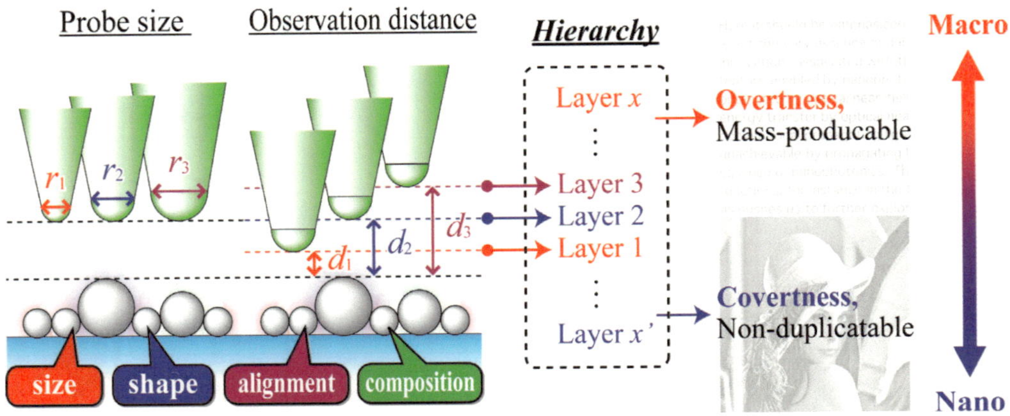

Fig. 2. Conceptual image of hierarchy based on nanophotonics, and innovative optical security system utilizing the hierarchical structure. The important point is that, because induced optical near-fields are the result of interactions between nanometric structures, the hierarchical property can be designed by adjusting the size, shape, alignment, and composition of the nanometric structures.

In a nanophotonic hierarchical hologram, the physical scale of the nanometric structural changes is less than 100 nm, whereas the physical scale of the elemental structures of the hologram is larger than 100 nm. In principle, a structural change occurring at the subwavelength scale does not affect the optical response function, which is dominated by propagating light. Therefore, the visual aspect of the hologram is not affected by such a small structural change on the surface. Additional data can thus be written by engineering structural changes in the subwavelength regime so that they are only accessible via optical near-field interactions (we call this *near-mode* retrieval) without having any influence on the optical response obtained via the conventional far-field light (what we call *far-mode* retrieval). By applying this hierarchy, new functions can be added to conventional holograms.

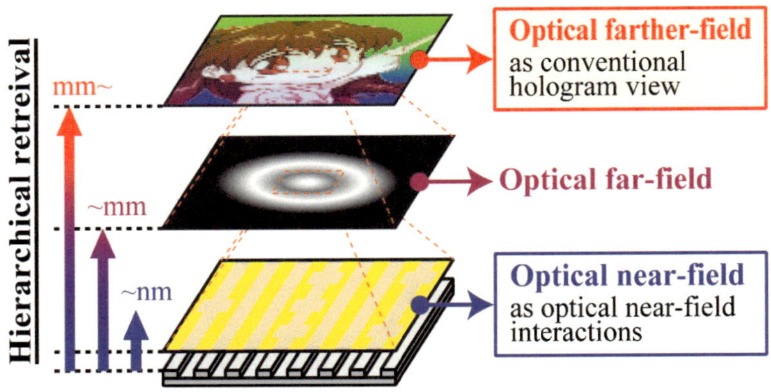

Fig. 3. Basic concept of functional hierarchy of nanophotonic hierarchical hologram. In principle, no interference occurs between each layer, because optical near-field interactions are distinguishable from the conventional hologram view, which consists of diffracted propagating light.

3.2 Near-field – Far-field hierarchy

First of all, in actual use of the hierarchical hologram, it is necessary to demonstrate that a nanometric structural change does not affect the optical response in the far-mode retrieval. To verify this, 500 nm-pitch Si diffraction gratings, in which nanometric depressions were embedded in the grid structures as near-mode data, were fabricated by using electron-beam (EB) lithography. A single isolated depression and multiple periodic depressions were embedded in each grating, as shown in Fig. 4(a) and (b), respectively. The size of each depression was less than 50 nm. The fabricated diffraction gratings were illuminated by the light from a He-Ne laser (λ = 633 nm), and the intensity of each diffracted beam was measured. Figure 4(c) shows the diffracted light pattern from the grating with periodic depressions. Large regular diffraction spots due to the grid structure and a number of tiny spots due to the periodic depressions were observed in the diffraction pattern. The diffraction efficiencies of each order of diffraction spots for each grating, calculated from the measured optical intensity of each spot, are shown in Fig. 4(d). The first-order diffraction efficiencies of a grating with no embedded depressions, the grating with the isolated depression, and the grating with the periodic depressions were 9.8 %, 9.7 %, and 10.3 %, respectively, showing a relative difference of less than 10 % between gratings with and without embedded depressions. No large differences were evident in the other diffraction orders either. This result shows that embedding the nanometric fabricated structures did not have a large effect on the function of conventional optical devices, that is, on their far-mode retrieval.

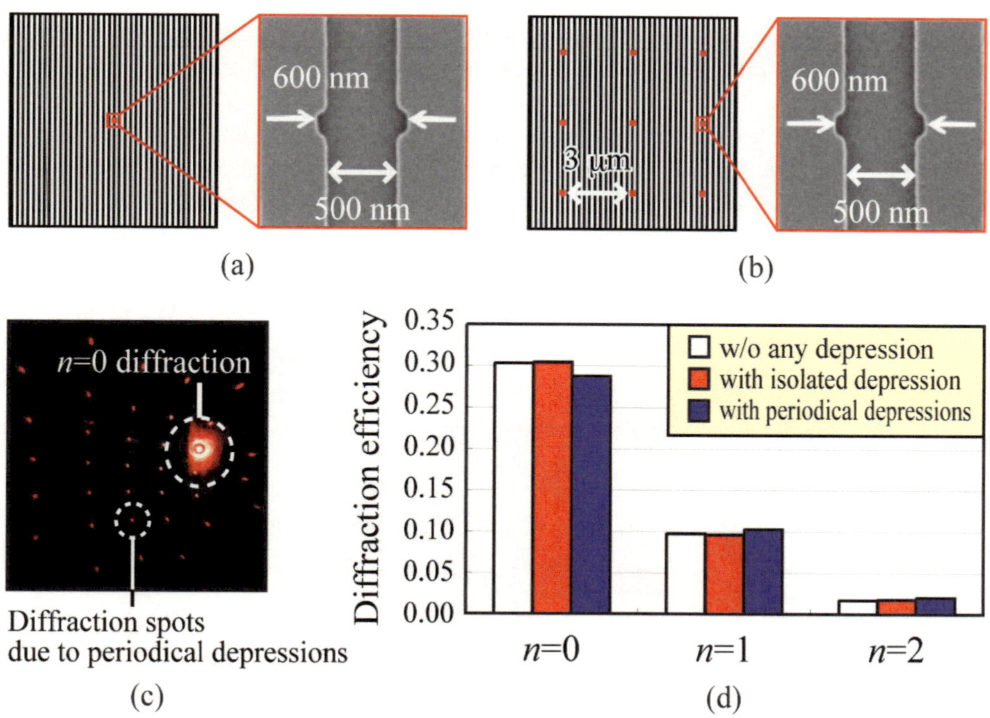

Fig. 4. SEM images of fabricated gratings with (a) single isolated depression and (b) multiple periodic depressions. (c) Diffraction pattern from the grating with periodic depressions. (d) Compared diffraction efficiencies of each grating, where n represents the diffraction order.

3.3 Retrieval of nanophotonic code

For a practical demonstration of near-mode retrieval, the pattern of induced optical near-fields generated by irradiating an embedded nanometric structure with light is defined as a *nanophotonic code*. An optical near-field is a non-propagating light field generated in a space extremely close to the surface of a nanometric structure. Because the light distribution depends on several parameters of the structure and the retrieval setup, various types of coding can be considered. Moreover, several novel features of nanophotonics, such as energy transfer (Ohtsu et al., 2008) and hierarchy (Naruse et al., 2005), may be exploited to achieve further functional improvements of nanophotonic codes.

As shown in Fig. 5, we created a sample device to experimentally demonstrate the retrieval of a nanophotonic code within a hologram. The entire device structure, whose size was 15 mm × 20 mm, was fabricated by EB lithography on a Si substrate, followed by sputtering a 50 nm-thick Au layer, as schematically shown in the cross-sectional profile in Fig. 5.

Our prototype device was essentially based on the design of Virtuagram®, developed by Dai Nippon Printing Co., Ltd., Japan, which is a high-definition computer-generated hologram composed of binary-level one-dimensional modulated gratings. As indicated in the left-hand side of Fig. 5, we could observe a three-dimensional image of the earth from the device. Within the device, we formed slightly modified square or rectangular nanometric structures embedded in the original hologram structure, so that near-mode information carried by the nanometric structures was accessible only via optical near-field interactions. The unit size of each embedded structure ranged from 40 nm to 160 nm. For comparison, they were embedded both outside and inside the grid structures, as shown in Fig. 6(a) and (b), respectively.

Note that the original hologram was composed of arrays of one-dimensional grid structures, extending along the vertical direction in Fig. 6 (a). To embed the nanometric structures, the grid structures were partially modified in order to implement the nanophotonic codes. Nevertheless, the grid structures remained topologically continuously connected along the vertical direction. On the other hand, the nanophotonic codes were always isolated from the continuous original grid structures. These geometrical characteristics produce interesting polarization dependence, which is discussed in detail in Sec. 3.3.1.

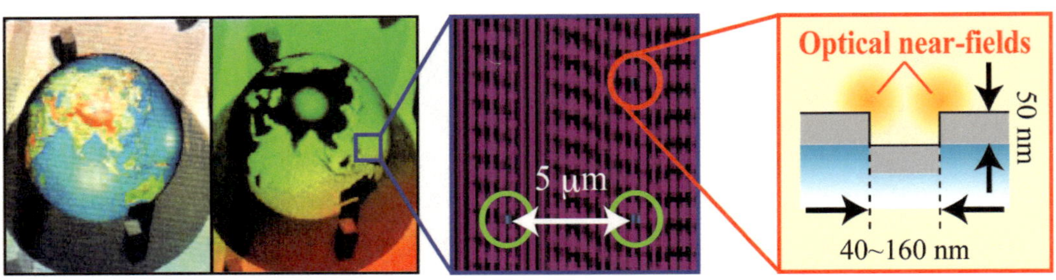

Fig. 5. Schematic diagram of the sample device for demonstration of nanophotonic hierarchical hologram with a nanophotonic code embedded within the embossed structure of Virtuagram®.

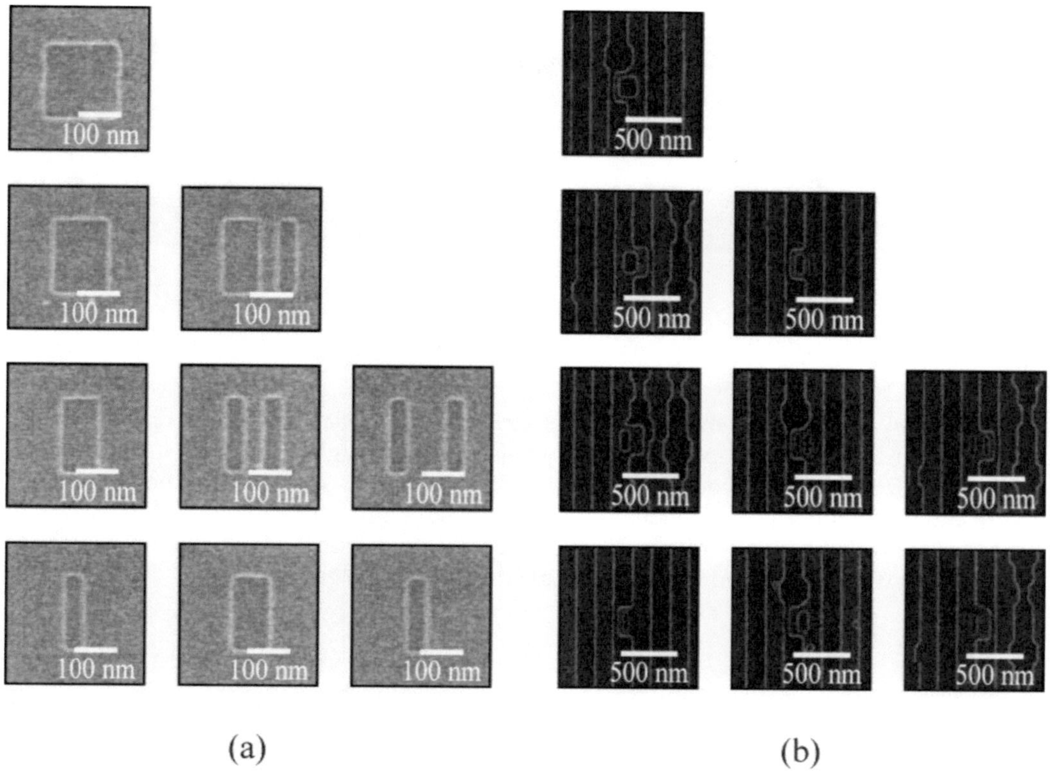

Fig. 6. SEM images of embedded nanometric structures (a) outside and (b) inside the grid structures of the original hologram.

3.3.1 Numerical evaluations

Before conducting the experimental demonstration with fabricated samples, the electric fields at the surfaces of nanometric structures were numerically calculated by the finite-difference time-domain (FDTD) method. As shown in Figs. 7, two types of calculation models were created in order to examine polarization dependency in retrieving the nanophotonic code. The embedded nanometric structure was represented by a square aperture whose side length was 150 nm, shown near the centre of the model. On the other hand, in the model shown in Fig. 7(b), the pitch of the periodic one-dimensional wire-grid structure was 150 nm, and the depth was 100 nm, which models the typical structure of an embossed hologram, and an aperture of the same size as that in Fig. 7(a) was embedded. The material of structures was assumed to be Au, and the structures were assumed to be irradiated with polarized plane waves coming from far above the structures. The wavelength was set to 785 nm. Periodic-conditioned computational boundaries were located 1.5 μm away from the center of the square-shaped aperture. By comparing those two cases, we can predict the effect of the existence of grid structures serving as the environmental structures on nanophotonic code retrieval. Also, we chose the square-shaped structure, which is isotropic in both the x and y directions in order to clearly evaluate the effects of the environmental structures and ignore the polarization dependency originating in the structure of the nanometric aperture itself.

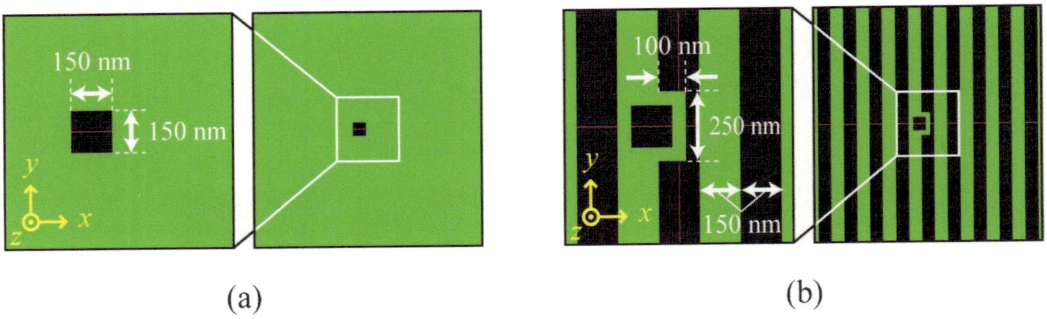

Fig. 7. Calculation model of embedded nanometric structure (a) without and (b) with environmental grid structure.

Figure 8(a), (c) and (b), (d) show calculated electric field intensity distributions on the surfaces of the structures assuming x-polarized and y-polarized input light irradiation to each model, respectively. As shown, although the model *without* the grid structure did not reveal any polarization dependency, the model *with* the grid structure revealed evident polarization dependency for the x- and y-polarized irradiation. Moreover, enhanced electric field intensity was obtained with x-polarized irradiation to the model with the grid structure.

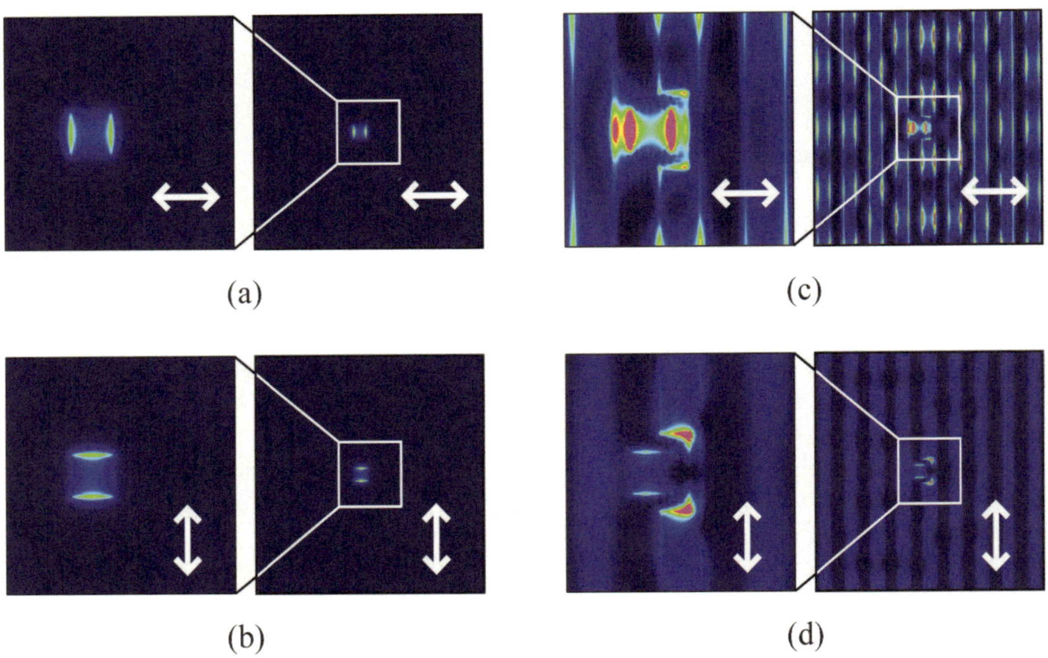

Fig. 8. Calculated intensity distribution of electric field produced by (a), (c) x-polarized light and (b), (d) y-polarized light input to the models without ((a), (b)) and with ((c), (d)) the grid structures.

In order to quantitatively investigate how the environmental grid structure affected the electric field in the vicinity of the nanometric structure and the influence of input light polarization, the average electric field intensity was evaluated in the area of the nanometric structure, denoted by $\langle I \rangle_{\text{code}}$. The average electric field intensity in the area including the surrounding areas is denoted by $\langle I \rangle_{\text{env}}$. More specifically, $\langle I \rangle_{\text{code}}$ represents the average electric field intensity in the 0.6 μm × 0.6 μm area covering the nanophotonic code, as shown by the dotted square in Fig. 9(a), whereas $\langle I \rangle_{\text{env}}$ indicates that in the 2.5 μm × 2.5 μm area indicated by the dashed square in Fig. 9(a). Figure 9(b) summarizes the calculated values of $\langle I \rangle_{\text{code}}$ and $\langle I \rangle_{\text{env}}$ with each model, respectively shown by the left and right bars.

First, the polarization dependencies are investigated. As shown in Fig. 8, in the case of the nanometric structure embedded in the environmental grid structure, evident polarization dependency was observed for both $\langle I \rangle_{\text{code}}$ and $\langle I \rangle_{\text{env}}$ in Fig. 9(b), too. Figure 9(c) compares the ratio of $\langle I \rangle_{\text{code}}$ with x-polarized input light to that with y-polarized input light for the embedded and isolated structures. As shown, $\langle I \rangle_{\text{code}}$ with x-polarized input light was about two times larger than $\langle I \rangle_{\text{code}}$ with y-polarized input light. On the other hand, the isolated nanometric structure did not show any polarization dependency.

Second, from the viewpoint of facilitating recognition of the nanophotonic code embedded in the hologram, it is important to obtain a kind of higher *recognizability* for the signals associated with the nanometric structures. In order to evaluate this recognizability, here we define a numerical figure-of-merit R_{num} as

$$R_{\text{num}} = \frac{\langle I \rangle_{\text{code}}}{\langle I \rangle_{\text{env}}} \times \langle I \rangle_{\text{code}}, \tag{1}$$

which yields a higher value with higher contrast with respect to $\langle I \rangle_{\text{code}}$ and $\langle I \rangle_{\text{env}}$ (indicated by the term $\langle I \rangle_{\text{code}} / \langle I \rangle_{\text{env}}$) and with higher signal intensity (indicated by $\langle I \rangle_{\text{code}}$). Figure 9(d) shows the calculated R_{num} in the case of y-polarized light input to thetwo types of models. The result indicates that the nanometric structure embedded in the environmental grid structure is superior to that of the isolated structure in terms of the recognizability defined by eq. (1), as shown in Figure 9.

We consider that such polarization dependency and enhanced recognizability are due to the environmental grid structure that extends along the vertical direction. The input light induces oscillating surface charge distributions due to coupling between the light and electrons in the metal. The y-polarized input light induces surface charges along the vertical grid; since the grid structure continuously exists along the y-direction, there is no chance for the charges to be concentrated. However, in the area of the embedded nanometric structure, there is a structural discontinuity in the grid; this results in higher charge concentration only at the edges of the embedded nanometric structure. On the other hand, the x-polarized input light sees structural discontinuity along the horizontal direction due to the vertical grid structure, as well as in the areas of the embedded structures. It turns out that charge concentration occurs not only in the edges of the embedded structures but also at other horizontal edges of the environmental grid structure. Therefore, by comparing this with the y-polarized model, enhancement of the electric field intensity and its polarization dependency is evident. In contrast to these nanometric structures embedded in holograms, for the isolated square apertures, both the x-polarized input light and the y-polarized input light have equal effects on the nanostructures.

These mechanisms indicate that such nanophotonic codes embedded in holograms could also exploit these polarization and structural dependences, not only for retrieving near-mode information via optical near-field interactions. For instance, we could facilitate near-mode information retrieval using suitable input light polarization and environmental structures.

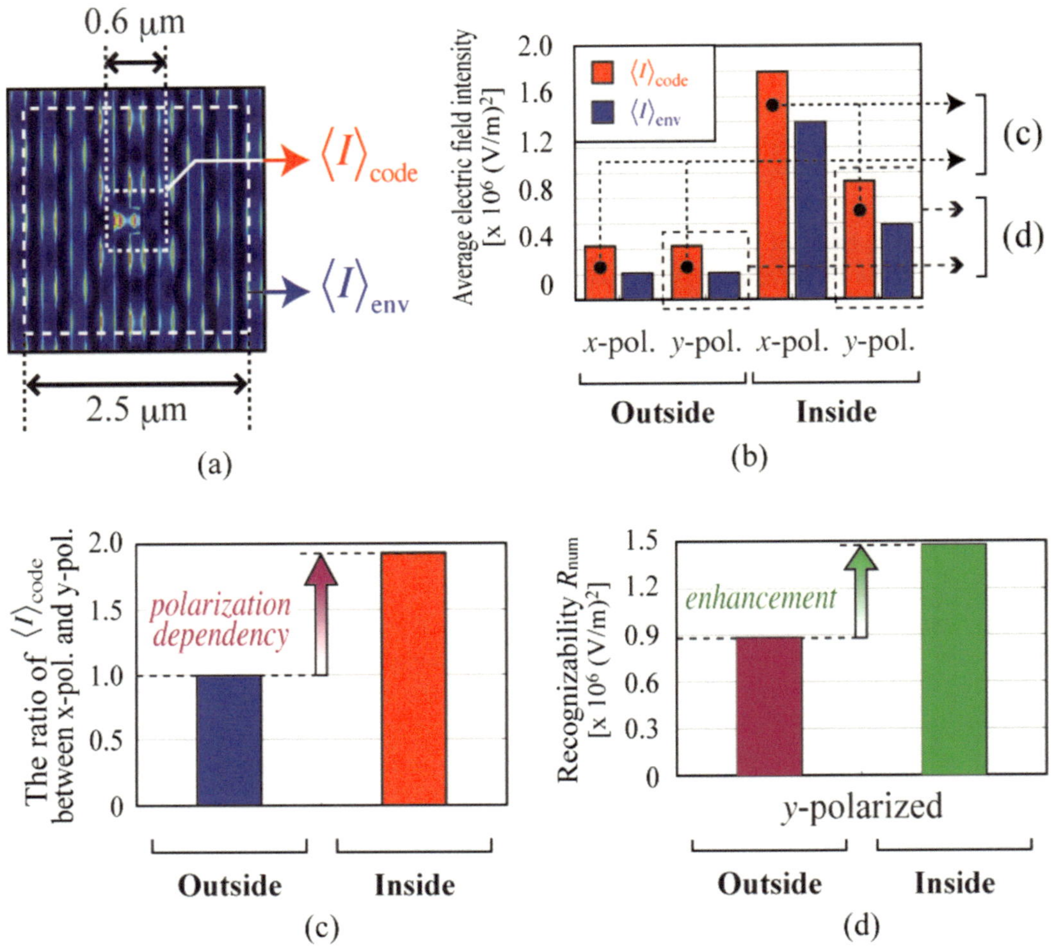

Fig. 9. (a) Schematic diagram explaining definition of average electric field intensities $\langle I \rangle_{code}$ and $\langle I \rangle_{env}$, and (b) their graphical representations in each calculation model. Evident polarization dependency was exhibited in the case of the nanometric code embedded in environmental structures. (c) The ratio of $\langle I \rangle_{code}$ with x-polarized input light to that with y-polarized input light for the embedded and isolated structures. (d) Numerical visibility R_{num} in two types of models with y-polarized input light. The result indicates that the visibility of the nanophotonic code was greatly enhanced by embedding it in the environmental structure.

3.3.2 Experimental demonstration

In the experimental demonstration, the optical responses of sample devices, shown in Figs. 6, during near-mode observation were detected using a near-field optical microscope (NOM). A schematic diagram of the detecting setup is shown in Fig. 10. The NOM was operated in an illumination-collection mode with a near-field probe having a tip with a radius of curvature of 5 nm. The fiber probe was connected to a tuning fork. Its position was finely regulated by sensing a shear force with the tuning fork, which was fed back to a piezoelectric actuator of the probe stage. The observation distance between the tip of the probe and the sample device was set at less than 50 nm. The light source used was an LD with an operating wavelength of 785 nm, and scattered light was detected with a photomultiplier tube (PMT). A Glan-Thompson polarizer (extinction ratio 10^{-6}) selected only linearly polarized light as the radiation source, and a half-wave plate (HWP) rotated the polarization.

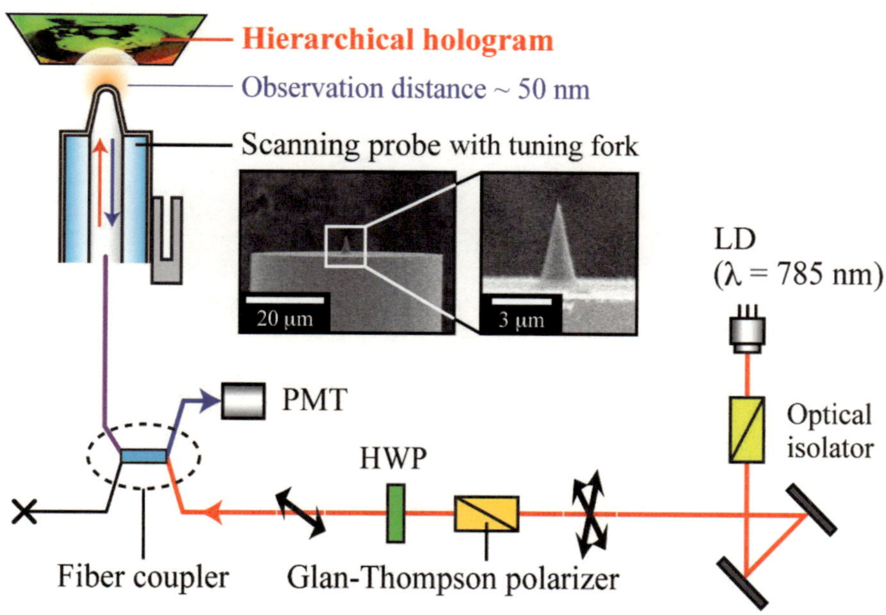

Fig. 10. Schematic diagram of the experimental setup for retrieving a nanophotonic code.

Figures 11(a) and (b) show retrieved results of nanophotonic codes that were outside and inside the environmental grid structures of the hologram, respectively, using a linearly polarized radiation source rotated by 0 degree to 180 degree at 20-degree intervals. As is evident in Fig. 11(a), although small and noisy intensity distributions were obtained, clear polarization dependence was observed in Fig. 11(b); for example, from the area of the nanophotonic code located in the center, a high-contrast signal intensity distribution was obtained with polarizations around 80 degree.

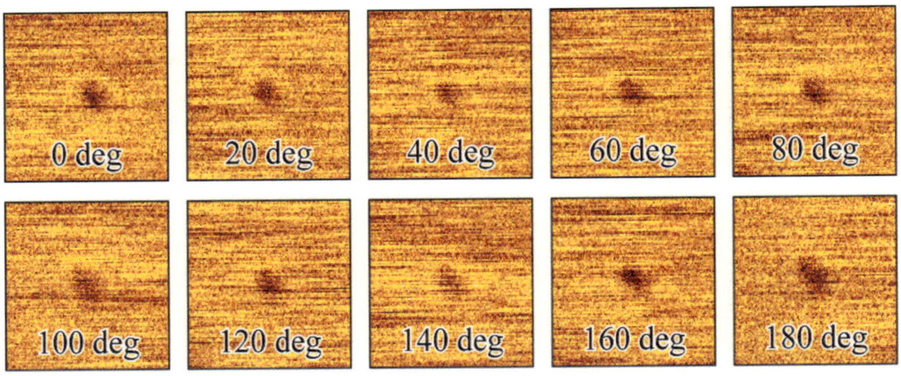

(a)

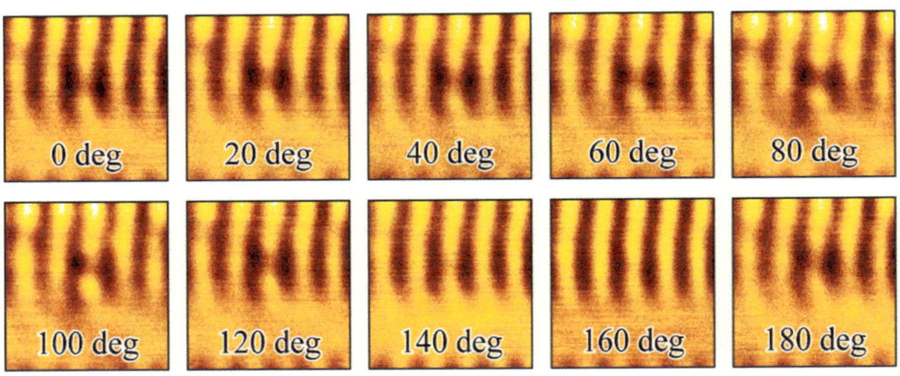

(b)

Fig. 11. Observed NOM images of optical intensity distributions of retrieved nanophotonic codes (a) outside and (b) inside the environmental grid structure.

To quantitatively evaluate the characteristics of the embedded nanophotonic code, we investigated two kinds of intensity distribution profiles from the observed NOM images. One is a horizontal intensity profile along the dashed line in Fig. 12(a), which crosses the area of the nanophotonic code, denoted by $I(x)$, where x represents the horizontal position. The other was also an intensity distribution as a function of horizontal position x; however, at every position x, we evaluated the average intensity along the vertical direction within a range of 2.5 μm, denoted by $\langle I(x) \rangle_{\text{env}}$, which indicates the environmental signal distribution. When a higher intensity is obtained selectively from the area of the nanophotonic code, the difference between $I(x)$ and $\langle I(x)\rangle_{\text{env}}$ can be large. On the other hand, if the intensity distribution is uniform along the vertical direction, the difference between $I(x)$ and $\langle I(x)\rangle_{\text{env}}$ should be small. Thus, the difference between $I(x)$ and $\langle I(x)\rangle_{\text{env}}$ indicates the visibility of the nanophotonic code. We define an experimental recognizability R_{exp} as

$$R_{\text{exp}} = \sum_{x} \left| I(x) - \langle I(x)\rangle_{\text{env}} \right| . \tag{2}$$

Figures 12(b) shows R_{exp} as a function of input light polarization based on the NOM results shown in Figure 11.

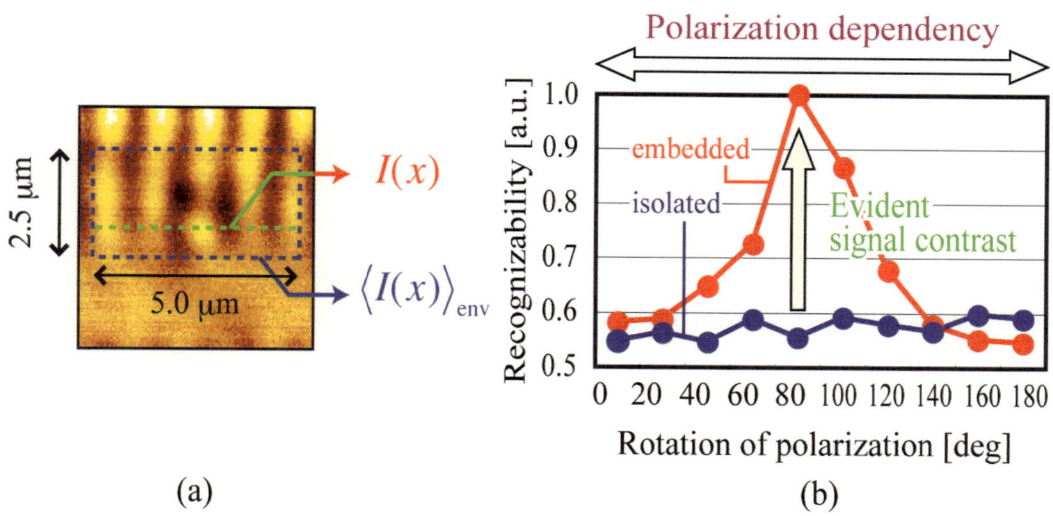

Fig. 12. (a) Definition of $I(x)$ and $\langle I(x)\rangle_{\text{env}}$ for numerical evaluation, and (b) calculated experimental recognizability, R_{exp}, of embedded nanometric structure (red curve) and that of isolated nanometric structure (blue curve). Evident recognizability and polarization dependency were exhibited.

The nanophotonic code embedded in the hologram exhibited much greater polarization dependency, as indicated by *embedded* in Fig. 12(b), where the maximum R_{exp} was obtained at 80-degree input polarization. On the other hand, only slight polarization dependency was observed with the isolated nanophotonic code, as indicated by *isolated*. These characteristics of retrieving the nanophotonic code in the environmental grid structure agree well with the numerical results in Fig. 9.

4. Summary

In this chapter, we described the basic concept of a nanophotonic hierarchical hologram and a nanophotonic code embedded in a hologram as an implementation of a hierarchical hologram. One of the most notable characteristics of our proposed approach is embedding a nanophotonic code within the patterns of a hologram composed of one-dimensional grid structures; it yields clear polarization dependence compared with an isolated nanophotonic code that is not embedded within a grid structure. These features were successfully demonstrated both numerically and experimentally.

In future research, the relation between the retrieved optical intensity distributions and the design of the nanometric structures may come to be understood, including their environmental conditions. Such insights should allow us to propose, for instance, an optimized strategy for implementing nanophotonic codes, or a strategy that is robust against errors that possibly occur in the fabrication and/or retrieval processes (Naruse et al., 2009).

From the point of view of practical use of these concepts in novel security devices, because embedding and retrieval of a nanophotonic code require highly advanced technical know-how, this approach can also improve the strength of anti-counterfeiting measures. The number of layers can be increased in the near-mode observation to further extend the hierarchical function. The optical near-field interaction between multiple nanometric structures produces a characteristic spatial distribution depending on the size, alignment, etc. Therefore, various optical signal patterns can be observed depending on the size of the fiber probe, and another layer can be added in near-mode observation (Naruse et al., 2007; Naruse et al., 2005). A simpler method for achieving such a hierarchical function is required, without using optical fiber probe tips. The scale for such retrieval might be set at the submillimeter range, as represented by *optical far-field* in Fig. 3. These aspects are currently being investigated by the authors.

Finally, the concept of hierarchy based on nanophotonics may produce innovations not only in optical security but also for other optical applications, such as lenses and jewellery. Adding extra functions in this way creates value-added media with only a slight impact on the primary functions. However, a trade-off occurs between the nanometric fabrication conditions (e.g., size and pitch) and the impact on the primary functions. Adequate discussion is needed to investigate these trade-offs in each application for their practical use.

5. Acknowledgments

This work was supported in part by a comprehensive program for personnel training and industry–academia collaboration based on projects funded by the New Energy and Industrial Technology Development Organization (NEDO), Japan, the Global Center of

Excellence (G-COE) Secure-Life Electronics project, and Special Coordination Funds for Promoting Science and Technology sponsored by the Ministry of Education, Culture, Sports, Science and Technology (MEXT), Japan.

6. References

Lorenzi, R. (2010). Do Mona Lisa' s eyes hide a secret code?, *Discovery News*, 13 December

Javidi, B. & Horner, J. L. (1994). Optical pattern recognition for validation and security verification, *Optical Engineering*, Vol. 33, pp. 1752-1756, 1994

Refregier, P. & Javidi, B. (1995). Optical image encryption based on input plane and Fourier plane random encoding, *Optics Letters*, Vol. 20, pp. 767-769, 1995

Rakuljic, G. A., Leyva, V., & Yariv, A. (1992). Optical data storage by using orthogonal wavelength-multiplexed volume holograms, *Optics Letters*, Vol. 17, pp. 1471-1473, 1992

Van Renesse, R. L. (1998). *Optical document scanning*, Altech House Optoelectronics Library, 1998

McGrew, S. P. (1990). Hologram counterfeiting: problems and solutions, *Proceedings of SPIE, Optical Security and Anticounterfeiting Systems*, Vol. 1210, pp. 66-76, 1990

Zhdanov, G. S., Libenson, M. N., & Martsinovskii, G. A. (1998). Optics in the diffraction limit: principles, results, and problems, *Physics-Uspekhi*, Vol. 41, pp. 719-722, 1998

Ohtsu, M., Kobayashi, K., Kawazoe, T., Yatsui, T., & Naruse, M. (2008). *Principles of Nanophotonics*, Taylor and Francis, Boca Raton, 2008

Tate, N., Nomura, W., Yatsui, T., Naruse, M., & Ohtsu, M. (2008). Hierarchical hologram based on optical near- and far-field responses, *Optics Express*, Vol. 16, pp. 607-612, 2008

Tate, N., Naruse, M., Yatsui, T., Kawazoe, T., Hoga, M., Ohyagi, Y., Fukuyama, T., Kitamura, M., & Ohtsu, M. (2010). Nanophotonic code embedded in embossed hologram for hierarchical information retrieval, *Optics Express*, Vol. 18, pp. 7497-7505, 2010

Ozbay, E. (2006). Plasmonics: Merging Photonics and Electronics at Nanoscale Dimensions, *Science*, Vol. 311, pp. 189-193, 2006

Nishida, T., Matsumoto, T., Akagi, F., Hieda, H., Kikitsu, A., & Naito, K. (2007). Hybrid recording on bit-patterned media using a near-field optical head, *Journal of Nanophotonics*, Vol. 1, 011597, 2007

Naruse, M., Yatsui, T., Nomura, W., Hirose, N., & Ohtsu, M. (2005). Hierarchy in optical near-fields and its application to memory retrieval, *Optics Express*, Vol. 13, pp. 9265-9271, 2005

Naruse, M., Yatsui, T., Kawazoe, T., Tate, N., Sugiyama, H., & Ohtsu, M. (2008). Nanophotonic Matching by Optical Near-Fields between Shape-Engineered Nanostructures, *Applyed Physics Express*, Vol. 1, 112101, 2008

Tate, N., Sugiyama, H., Naruse, M., Nomura, W., Yatsui, T., Kawazoe, T., & Ohtsu, M. (2009). Quadrupole-Dipole Transform based on Optical Near-Field Interactions in Engineered Nanostructures, *Optics Express*, Vol. 17, 11113, 2009

Tate, N., Nomura, W., Yatsui, T., Naruse, M., & Ohtsu, M. (2009). Hierarchy in optical near-fields based on compositions of nanomaterials, *Applied Physics B*, Vol. 96, pp. 1-4, 2009

Naruse, M., Inoue, T., & Hori, H. (2007). Analysis and Synthesis of Hierarchy in Optical Near-Field Interactions at the Nanoscale Based on Angular Spectrum, *Japanese Journal of Applied Physics*, Vol. 46, pp. 6095-6103, 2007

Naruse, M., Hori, H., Kobayashi, K., Ishikawa, M., Leibnitz, K., Murata, M., Tate, N., & Ohtsu, M. (2009). Information theoretical analysis of hierarchical nano-optical systems in the subwavelength regime, *Journal of the Optical Society of America B*, Vol. 26, pp. 1772-1779, 2009

Part 5

Signal Processing

16

Photonic Microwave Signal Processing Based on Opto-VLSI Technology

Feng Xiao[1], Kamal Alameh[1,2] and Yong Tak Lee[2,3]
[1]*Electron Science Research Institute, Edith Cowan University,*
[2]*Department of Nanobio Materials and Electronics,*
Gwangju Institute of Science and Technology,
[3]*Department of Information and Communications,*
Gwangju Institute of Science and Technology,
[1]*Australia*
[2,3]*Korea*

1. Introduction

The processing of radio frequency (RF) and microwave signals in the optical domain is an attractive approach to overcome the bottlenecks of bandwidth, power loss, and electromagnetic interference (EMI) encountered in conventional electronic signal processing systems. A wide range of emerging RF signal processing applications require specifically high resolution, ultra-wide bandwidth, wide-range tunability, and fast reconfigurability. While these requirements are difficult to achieve using conventional all-electronic processing, they are feasible with photonics-based signal processing.

Holography is an historic technology that allows the light scattered from an object to be recorded and reconstructed so that the object can reappear when a reference optical beam illuminates a hologram used to record that object. Holography has a wide range of applications such as optical signal storage and retrieval and information processing. The application of reconfigurable phase holograms to realize optical beam steering is an attractive field for either optical engineering or fundamental research. The reconfigurable phase holograms are calculated from the targeted beam steering pattern and implemented using spatial light modulator. Opto-VLSI processors are one of these devices that can dynamically generate phase holograms and perform optical beam steering.

Although full-electronic RF signal processing is very flexible and controllable, it is experiencing the bottlenecks of bandwidth and EMI. Processing microwave signals in the photonic domain can overcome the bottlenecks in the electronic signal processing. However the current technologies of microwave photonics have limited flexibility and reconfigurability. The Opto-VLSI technology is a novel and potential discipline that combine benefits of photonic devices and the intelligence plus processing capabilities of Very-Large-Scale Integrated (VLSI) circuits. It integrates intelligence into photonic systems providing a new foundation for a future information processing and communication systems and networks. This book chapter will discuss a new methodology of expanding the use of Opto-VLSI from the conventional paradigm of optical beam processing to a new frontier of photonics-based RF signal processing, leading to a wide applications such as future broadband RF signal filters and beamformers for wireless systems for many wireless- and mobile-communication-related areas.

The application of holography in processing RF and microwave signals, which is implemented through an Opto-VLSI processor, offers many advantages over traditional microwave photonic methods in terms of flexibility and reconfigurability. The phase hologram is generated according to the applications of RF signal processing and uploaded onto the Opto-VLSI processor.

In this chapter, we discuss the use of holography for the applications of RF and microwave signal processing. This book chapter is organised as follows. In Section 2 a brief background on Opto-VLSI processors is provided. Section 3 discusses the use of holography to realize tunable microwave filters. In Section 4 the adaptive generation of true time delay (TTD) based on uploading appropriate phase hologram onto an Opto-VLSI processor is discussed. In Section 5 architecture of a tunable Opto-VLSI-based beamformer for phased array antennas is discussed. We conclude this chapter in Section 6.

2. Opto-VLSI processor

An Opto-VLSI processor is an array of liquid crystal (LC) cells driven by a Very-Large-Scale-Integrated (VLSI) circuit. Figure 1 shows a typical layout and a cell design of an 8-phase Opto-VLSI processor. It is driven by digital holographic diffraction gratings capable of steering/shaping incident optical beams as illustrated in Fig. 2. The voltage level of each pixel can individually be controlled by using a few memory elements that select a discrete voltage level and apply it, through the aluminium mirror electrode, across the LC cell. A transparent Indium-Tin Oxide (ITO) layer is used as the second electrode, and a quarter-wave-plate (QWP) layer is deposited between the LC and the aluminum mirror to accomplish polarization-insensitive operation. The ITO layer is generally grounded and a voltage is applied at the reflective electrode by the VLSI circuit below the LC layer. Opto-VLSI processors are electronically controlled, software-configured, polarization independent and are cost effective because of the high-volume manufacturing capability of VLSI chips as well as their capability of controlling multiple fiber ports in one compact Opto-VLSI module; they are also very reliable since beam steering is achieved with no mechanically moving parts.

By driving the Opto-VLSI with blazed gratings of different pitches, as shown in Fig. 3, optical beam steering can be achieved. The diffraction (or steering) angle for an Opto-VLSI processor, αm, is given by (Xiao *et al.*, 2008):

$$\alpha_m = \arcsin(\frac{m\lambda}{d}) \tag{1}$$

where m is the diffraction order (usually only the first order is considered), λ is the light wavelength in vacuum, and d is the grating period. By addressing each pixel independently a phase hologram can be synthesized leading to optical beam steering, beam shaping or multicasting. For example, a 4-phase Opto-VLSI processor having a pixel size of 5 microns can steer a 1550 nm laser beam by a maximum angle of around ±4°. The maximum diffraction efficiency of an Opto-VLSI processor depends on the number of discrete phase levels that the VLSI can accommodate. The theoretical maximum diffraction efficiency is given by (Dammann, 1979):

$$\eta = \mathrm{sinc}^2\left(\frac{\pi n}{M}\right) \tag{2}$$

where $n = gM + 1$ is the diffraction order ($n = 1$ is the desired order), and g is an integer. Thus an Opto-VLSI processor with binary phase levels can have a maximum diffraction efficiency of 40.5%, while a four phase level allow for efficiency up to 81%. The higher diffraction orders (which correspond to the cases $g \neq 0$) are usually unwanted crosstalk signals, which must be attenuated or properly routed outside the output ports to maintain high signal-to-crosstalk performance.

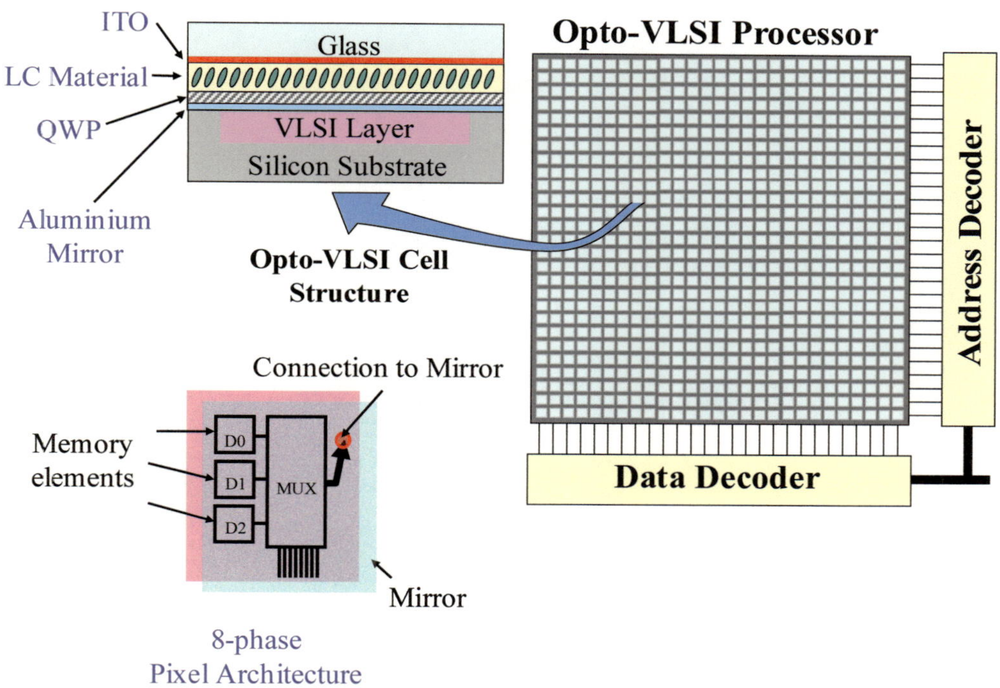

Fig. 1. Typical 8-phase Opto-VLSI processor and LC cell structure design.

Adaptive optical beam steering can be achieved by reconfiguring the phase hologram uploaded onto the Opto-VLSI processor. Recent advances in low-switching-voltage nematic LC materials and layer thickness control have allowed the incorporation of a thin quarter-wave-plate (QWP) layer between the LC and the aluminium mirror to accomplish polarization-insensitive multi-phase-level Opto-VLSI processors (Manolis *et al.*, 2002), as shown in Fig. 1. In addition, with current 130nm VLSI fabrication processes, VLSI chips featuring 24mm×24mm active area, maximum switching voltage of 3.0 volts, and pixel size of 5 microns, can be realised. Depositing low-switching-voltage electro-optic materials and QWP over such VLSI chips, can realize a polarization-insensitive Opto-VLSI processor that has a diffraction efficiency of 87% (0.6 dB loss) and a maximum steering angle of more than ±4.0°.

There have been different algorithms reported for the optimization of Opto-VLSI phase holograms to achieve effective beam steering, including simulated annealing and projection methods. In our study, a modified simulated annealing method that can achieve accurate beam steering with low crosstalk is adopted (Yen-Wei Chen *et al.*, 2000).

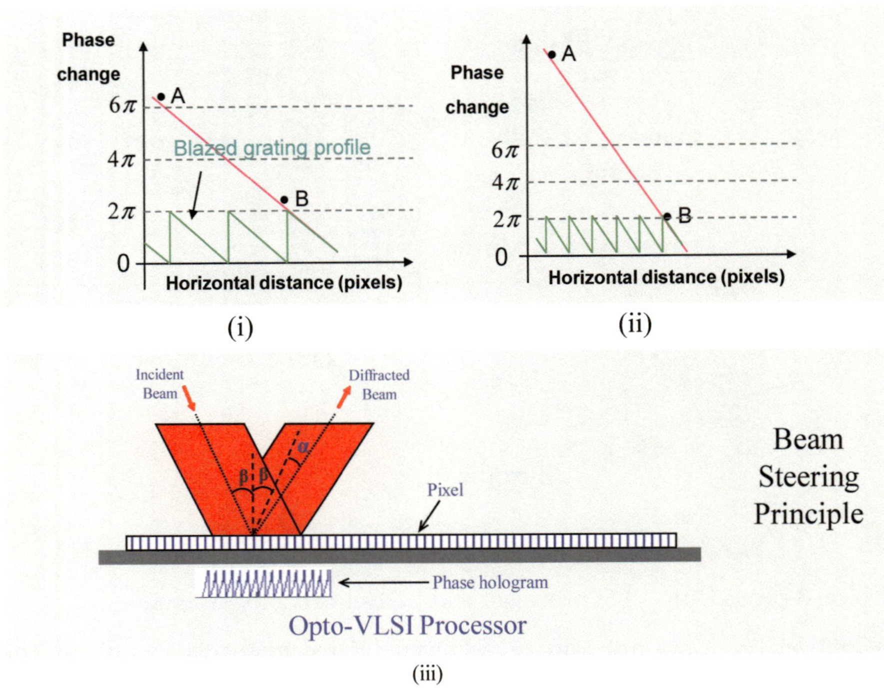

Fig. 2. Principle of beam steering through variable-pitch blazed grating generation.

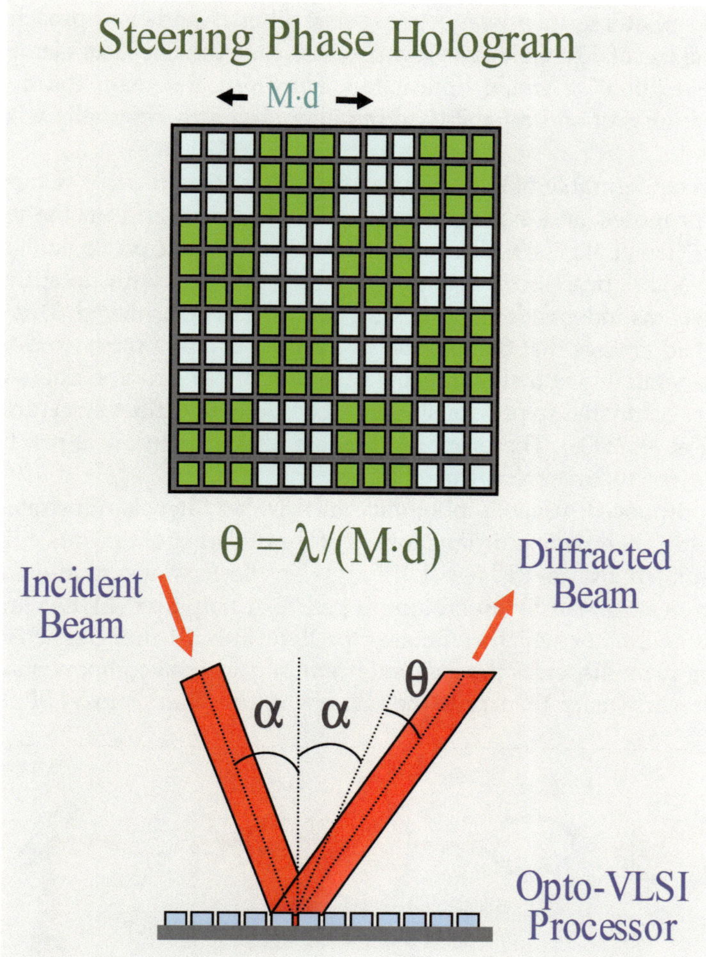

Fig. 3. Steering and multicasting capabilities of an Opto-VLSI processor.

3. Tunable photonic microwave filters

Photonics-based microwave and millimeter-wave filters offer advantages such as ultra-wide bandwidth, immunity to electromagnetic interference, and lightweight (Seeds&Williams, 2006; Capmany et al., 2006; Minasian, 2006). These advantages open new opportunities in a wide range of potential applications especially when high selectivity, resolution, wide tunability, and fast reconfigurability characteristics are required.

In recent years, numerous reconfigurable coherent-free photonic microwave transversal filter structures have been proposed and demonstrated, where multi-wavelength source is employed to suppress the optical interference in conjunction with modifying the optical tap weights or the time-delay increment between taps (Capmany et al., 1999; Polo et al., 2003; Ortigosa-Blanch et al., 2006; Hunter&Nguyen, 2006; Ning et al., 2007; Blals&Yao, 2008). Spectral slicing of an RF-modulated broadband optical source has been employed to generate different wavebands. However, the use of Bragg gratings or arrayed waveguide gratings for realizing spectral slicing results in fixed time-delay increments, which limit the tunability of the photonic microwave transversal filter. Another approach to generating optical taps is the use of a tunable laser array, where each tunable laser element is dedicated to control the weight of a single optical tap. However, the main disadvantage of this approach is the high cost and reliability of the filter structure, especially when the number of taps increases.

As a powerful reconfigurable holography technology, a novel tunable photonic microwave filter has been proposed and experimentally demonstrated based on the use of an Opto-VLSI processor (Xiao et al., 2009). Through computer-generated phase holograms uploaded onto the Opto-VLSI processor, arbitrary spectral slicing with adaptive wavelength separations as well as independent tap weight control can be achieved. This demonstration has a significant advantage that the time-delay increment, tap numbers and tap weights can be adjusted independently and simultaneously, simply by electronics. This structure has the highest flexibility compared to previously reported microwave filter structures (Capmany et al., 2006; Zheng et al., 2006). The proposed tunable filter structure is a practical solution to realizing flexible and tunable microwave filters.

In Fig. 4, the proposed tunable photonic microwave filter is illustrated through an experimental setup. A broadband light source of amplified spontaneous emission (ASE) is externally modulated by an RF signal through an electro-optic modulator (EOM). The modulated light is amplified by an erbium doped fiber amplifier (EDFA) and routed via a circulator into a collimator which collimates the light into a 1-mm diameter beam. A 1200 line/mm grating plate disperses the incident beam into spectral components along different directions and linearly maps them onto the active window of an Opto-VLSI processor.

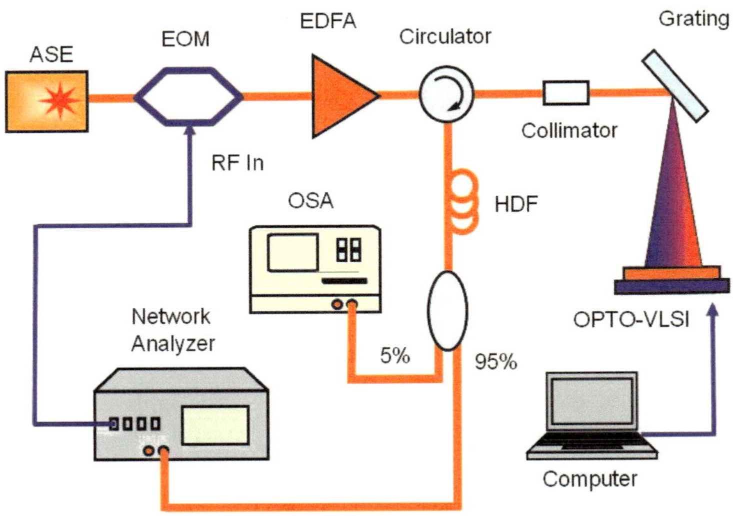

Fig. 4. Experimental setup for the tunable photonic microwave filter structure.

The wavelength of the optical field incident onto the Opto-VLSI processor varies along the pixels, which can be logically partitioned into pixel blocks by programming the Opto-VLSI processor. The spectral component, falling within the specific pixel block of the Opto-VLSI processor, can be either steered back along the incidence path thus coupling it back into the fiber collimator with minimum attenuation, or deliberately steered "off-track" so that its power is partially coupled back into the fiber collimator leading to an appropriate optical attenuation for that spectral component. Therefore, by manipulating the phase hologram of individual pixel block, the power of each waveband component can be independently adjusted according to the required tap weights; and by partitioning the pixels into appropriate blocks, the optical taps with specific wavelength spacing can be picked out from the wideband light source while all the other wavebands are steered off-track and attenuated dramatically (Xiao *et al.*, 2008). For example, by applying an appropriate phase holograph consisting of 5 different blazed gratings, 5 different wavebands could be steered back and coupled to the collimator, as shown in Fig. 5. In other words, by configuring the phase hologram employed to the Opto-VLSI processor, tap weights and tap separations can be adjusted simultaneously, independently and continuously.

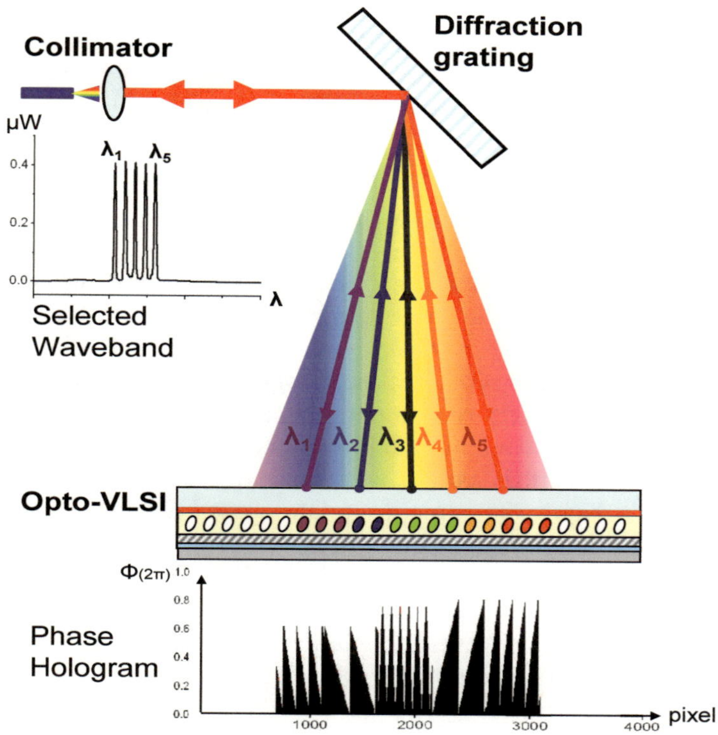

Fig. 5. The principle of optical waveband selection.

The optical taps that are coupled back into the fiber collimator are fed, through the circulator, to a high dispersion optical fiber, where the different RF-modulated wavebands experience different delay times. The delayed RF-modulated wavebands are finally photo-detected by a photo-receiver built into a Network Analyzer, which displays the microwave filter response. An optical spectrum analyzer (OSA) is also used as a monitor driven by a small fraction of the light detected by the photodiode of the Network Analyzer.

In the experiments, the RF signal generated by the Network Analyzer was used to intensity modulate the broadband ASE source using a JDS Uniphase electro-optical modulator of 4-GHz bandwidth. A 256-phase-level one-dimensional Opto-VLSI processor was used, which has 1×4096 pixels, with 1 μm pixel size and 0.8 μm dead spacing between adjacent pixels. A Labview software was specifically developed to appropriately partition the pixel blocks so

that the optimum wavebands are selected. The pixel blocks are then driven by optimized phase holograms (blazed gratings) that steer the selected wavebands so that they are coupled back into the collimator. By optimizing the size and the phase profile of each pixel block, any desirable weight and time-delay increment is synthesized after the wavebands are launched into the high dispersion fiber (HDF), which has dispersion coefficient 382.5ps/nm and insertion loss 4.6 dB.
The response of the filter structure shown in Fig. 4 can be expressed as:

$$H(f) = \sum_{r=0}^{M} a_r \exp[-j2\pi r f \tau] \tag{3}$$

where f is the RF frequency, M is the number of the detected RF-modulated wavebands, a_r is the rth tap weight, which is proportional to the optical power of the rth waveband, and τ is the time delay between adjacent wavebands introduced by the high dispersion fiber. The time delay, τ, can also be expressed in terms of the dispersion of the HDF as

$$\tau = \alpha \cdot \Delta\lambda \tag{4}$$

where α denotes the dispersion coefficient of the HDF, and $\Delta\lambda$ is the adjacent waveband separation.
The principle of the photonic microwave filter was demonstrated using 5 optical taps. Figures 6(a), 6(d), and 6(g) show the phase holograms applied to the Opto-VLSI processor to generate a constant time-delay increment with variable filter weights. The Opto-VLSI processor was partitioned into 5 pixel blocks corresponding to 5 wavebands, and the size of each pixel block was 400 pixels, resulting in a center-to-center waveband separation of 3.60 nm. For each pixel block, an optimum phase hologram was appropriately uploaded, so that the power level of the specific waveband was attenuated to an appropriate intensity. As shown in Figure 6(a), an appropriate phase hologram for each pixel block was employed to generate a normalized weight profile of [1, 1, 1, 1, 1]. Fig. 6(b) shows the selected wavebands measured by the OSA, and Fig. 6(c) shows the corresponding filter response. Note that the waveband coupled back into the collimator depends on the steering angle associated to this waveband and the numerical aperture of the collimator, and that the separation between adjacent wavebands depends on their pixel block sizes. The measured linewidth for each waveband was about 0.5 nm. Note that the spectral range illuminating each pixel block was around 3.6 nm. However, because this spectral range was diverging, only 0.5nm of this range was actually steered back and coupled into the collimator. By reconfiguring the phase holograms uploaded onto the various pixel blocks, the filter weights were varied to [0.4, 0.8, 1, 0.8, 0.4], and then to [0.2, 0.8, 1, 0.8, 0.2] as shown in Figs. 6(e) and 6(h). The corresponding filter responses are shown in Figures 6(f), and 6(i), respectively, where changes in rejection band and bandwidth are demonstrated, as a result of waveband attenuation.
Note that in Figures 6(c), 6(f) and 6(i) the solid lines denote the experimental results, which agree well with the simulation results calculated from Eq. (3,4), shown in dashed lines. All the filter responses shown in Fig. 6 exhibit a free-spectral range (FSR) of about 722 MHz and this is in good agreement with the specified dispersion coefficient of the HDF used in the experiments. The time delay increment, which is the product of the waveband separation and dispersion coefficient of the HDF, was 1.38 ns.

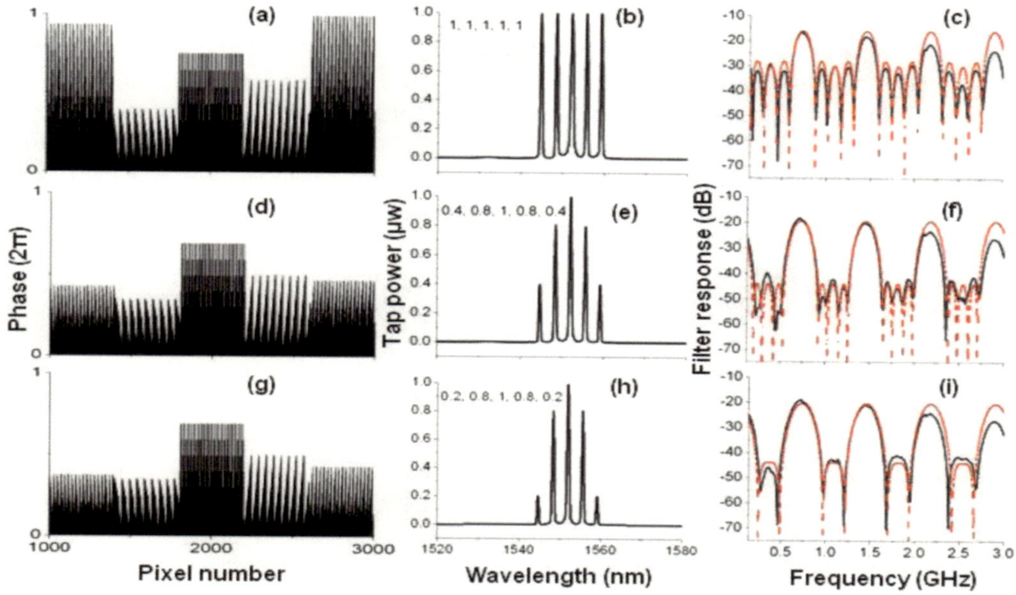

Fig. 6. Microwave filter tuning through tap weight control. (a), (d), and (g) phase hologram applied to the Opto-VLSI processor; (b), (e), and (h) selected RF-modulated wavebands; (c), (f), and (i) measured (solid line) and simulated (dashed line) filter responses for each case.

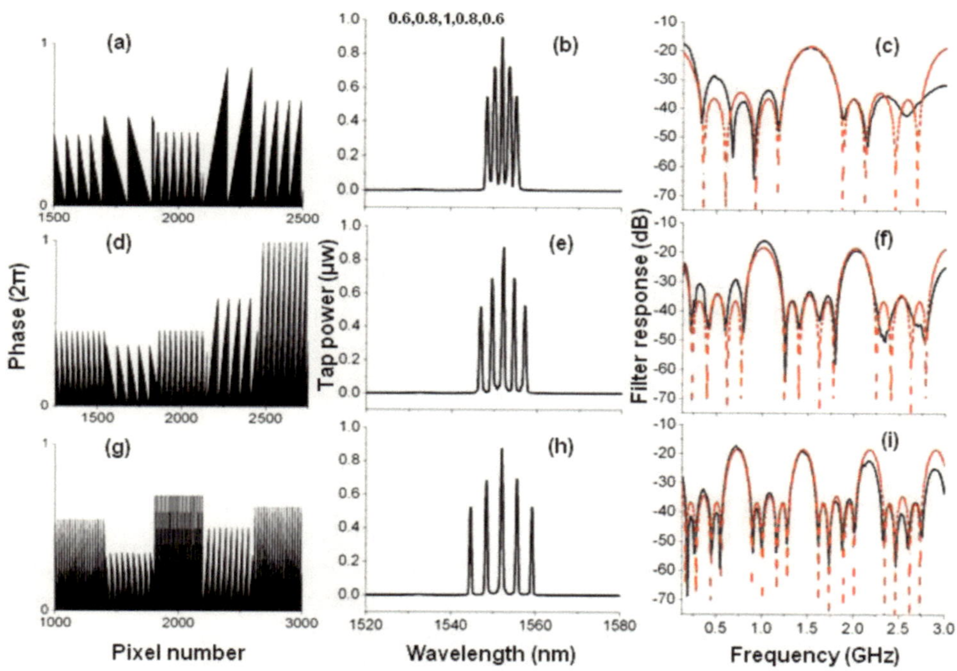

Fig. 7. Microwave filter tuning through time-delay increment control. (a,d,g) phase hologram applied to the Opto-VLSI processor; (b,e,h) selected RF-modulated wavebands; (c,f,i) measured (solid line) and simulated (dashed line) filter responses for each case.

The unique capability of the proposed tunable filter structure is its ability to continuously tune the time delay increment. This is accomplished by changing the spacing between adjacent pixel blocks (i.e., the pixel block size and center). Figures 7(a), 7(d), and 7(g) show the phase holograms applied to the Opto-VLSI processor to synthesize a normalized weight profile of [0.6, 0.8, 1.0, 0.8, 0.6] but different time delay increments. The corresponding wavebands are shown in Figures 7(b), 7(e), and 7(h). For these three cases, the sizes of the pixel blocks were 200, 300, and 400 pixels, respectively, corresponding to waveband separations of 1.72 nm, 2.60 nm, and 3.60 nm, respectively. Note that by optimizing the phase hologram for each pixel block, a normalized weight profile [0.6, 0.8, 1, 0.8, 0.6] was maintained in all cases. Figures 7(c), 7(f), and 7(i) show the corresponding measured (solid) and simulated (dashed) filter responses. It is evident from Fig. 7 that the tuning of the waveband separation (ie. the time delay increment) controls the free-spectral-range (FSR) as well as the bandwidth of the filter. For example, in Fig. 7, the FSR was reduced from 1.52 GHz, to 1.01 GHz, and then to 722MHz, and the filter bandwidth dropped from 216 MHz to 142 MHz and then to 86 MHz, when the time-delay increment was increased from 0.66 ns, to 0.99 ns, and then to 1.38 ns, respectively. Note that a good agreement between the simulated and measured filter responses is displayed in Fig. 7.

By investigating the filter responses shown in Figures 6(c, f, i) and Figures 7(c, f, i), one can see that the spectral response fades remarkably as the RF frequency increases. This is due to the interaction between the dispersion of the HDF and the nonzero optical bandwidth of each waveband (Taylor *et al.*, 2007; Pastor *et al.*, 2003a). This limitation is inherent to all filters that use spectral slicing. The bandwidth of the sliced wavebands in Fig. 6 and Fig. 7 was about 0.5 nm. Preliminary experimental results have shown that a narrower waveband bandwidth can be achieved using a higher dispersion grating plate and/or a larger distance between the grating and the Opto-VLSI processor.

Even though the optical tap generation involves fiber to free-space and free-space to fiber coupling, standard passive micro-assembly can be used to realise low-cost, robust and stable alignment without the need of automated high- precision stages (Baxter, 2006). Note that the maximum number of optical taps that can be generated depends on the bandwidth of the ASE source, the size of the active window of the Opto-VLSI processor, and the waveband separation. For our current experimental system, which based on the 1-D Opto-VLSI processor, up to 12 optical taps with the waveband separation of 2.60 nm can be generated. Note that, by using a 2-D Opto-VLSI processor, one can increase the number of optical taps significantly.

4. Photonic true-time delay

Many current applications, such as RF imaging systems and wireless communications, are required to exhibit improved resolution, wider angular scans, and wide bandwidths. To obtain an optimal radiation pattern for broadband transmission, the signals received by or transmitted from the antenna array must be accurately time-compensated via true-time RF delay generation (Meijerink *et al.*, 2010; Jiang *et al.*, 2005). Currently electronic phase shift for phased-array antenna features the advantages of flexibility and reconfigurability. However, they are limited in their processing speed and bandwidth. In addition, the electronic phase shift control is accomplished through metallic coaxial cables or waveguide feeds which are heavy, bulky, lossy, and susceptible to electromagnetic interference (EMI) and crosstalk. The technology of digital signal processor (DSP) is currently thwarted either by the limited resolution and the narrow bandwidth of analogue-to-digital converter, or by high power consumption of broadband analogue-to-digital converter.

The processing of radio frequency (RF) and microwave signals in the optical domain is an attractive approach to overcome the bottlenecks encountered in conventional electronic signal processing systems (Capmany *et al.*, 2005). A wide range of emerging RF signal processing applications require specifically high resolution, wide-range tunability, and fast reconfigurability. These requirements are difficult to achieve using conventional all-electronic processing, but feasible with photonics-based signal processing.

The use of photonics-based true time delay units has extensively been investigated in the last decade for applications ranging from modern microwave radar to wireless communication systems. In particular, broadband microwave phased-array antennas require the generation of variable true-time delays at each antenna element to realize beam or null steering, and optical fibers have been the best candidates for true-time delay synthesis. Compared with all-electrical techniques, optical true-time delay generation offers the advantages of broader bandwidth, lower insertion loss, higher phase stability, smaller size, lighter weight, and excellent immunity to both electromagnetic interference and crosstalk (Frigyes&Seeds, 1995; Italia *et al.*, 2005b; Y. Chen&Chen, 2002; Rideout *et al.*, 2007). Several approaches have been adopted to realise tunable true-time delay units, including the use of in-fiber chirped Bragg gratings (FBGs) (Italia *et al.*, 2005b), white cells or fiber delay lines in conjunction with MEMS (Mital *et al.*, 2006b; Anderson *et al.*, 2006; Vidal *et al.*, 2006), integrated optical waveguides (C. M. Chen *et al.*, 2010), optically-switched fiber delay structures (Tong&Wu, 1998), dispersion-enhanced photonic-crystal fibers (Jiang *et al.*, 2005), and higher-order mode dispersive multi-mode fibers (Raz *et al.*, 2004). However, most of these reported true-time delay architectures have mainly been used for realising beam steering in phased array antennas, and therefore, they do not have the flexibility to simultaneously generate multiple arbitrary true-time delays. In addition, such architectures can only generate discrete true-time delays, making them impractical for broadband null steering (Zmuda *et al.*, 2000). Zmuda et al. have reported a few adaptive true-time delay architectures based on the use of multiple tunable lasers in conjunction with high-dispersion fibres for the implementation of broadband nulling in microwave phased arrays (Zmuda *et al.*, 2000; Zmuda *et al.*, 1998). However, the use of multiple tunable lasers requiring continuous calibration makes the system implementation very expensive and impractical.

Recently, a novel true-time delay unit has been demonstrated through uploading appropriate holograms onto an Opto-VLSI processor to synthesize multiple arbitrary time delays (Juswardy et al., 2009). This true-time-delay unit, which consists of a broadband optical source using Amplified Spontaneous Emission (ASE) and high dispersion fibers, has the capability to generate multiple true-time delays for several antenna elements simultaneously, making it attractive for broadband null-steering in phased array antennas.

The principle of the phased-array antenna architecture shown in Fig. 8 is demonstrated using the experimental setup illustrated in Fig. 4. In this setup, a broadband ASE source was modulated, via a JDS Uniphase electro-optical modulator (EOM) with a half-wave voltage of 6 V, by an RF signal which was generated using a 20 GHz network analyzer. The RF-modulated optical signal was amplified by an EDFA and collimated at 1-mm diameter and then launched onto a diffractive grating plate. The latter demultiplexed the collimated ASE beam into multiple RF-modulated wavebands, which were then mapped onto the active window of a 256-phase-level 1×4096-pixel Opto-VLSI processor of 1-μm pixel size and 0.8-μm dead spacing between adjacent pixels.

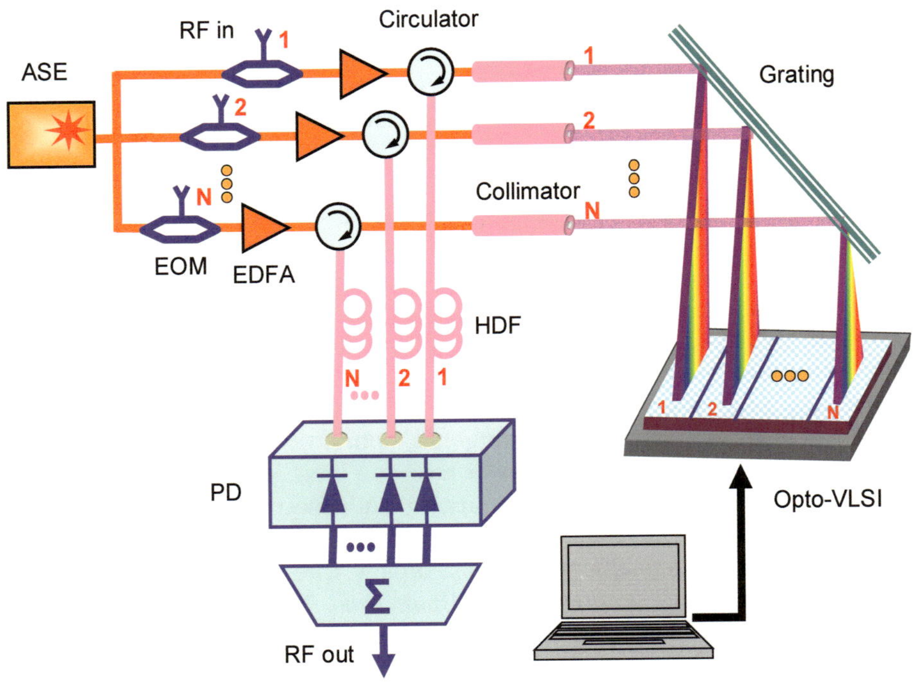

Fig. 8. Experimental setup used to demonstrate tunable time delay generation.

Labview software was specially developed to generate optimised phase holograms that couple the specific RF-modulated wavebands back into the fiber collimator, and at the same time equalize their intensities by changing the maximum phase levels applied to the different pixel blocks. The selected wavebands were then routed via a circulator to a 22-km high dispersion fiber (HDF) of dispersion coefficient 382.5 ps/nm and insertion loss 4.6 dB. An optical spectrum analyzer (OSA) was used to monitor the spectrum detected by a photodiode built in the Network Analyzer, as illustrated in Fig. 4.

One of the attractive features of the Opto-VLSI-based tunable time delay architecture is its ability to generate multiple RF delays without the need for RF splitters. Furthermore, the amplitude weight of each generated RF delay sample can simultaneously be controlled. This architecture offers excellent flexibility in applications such as phased-array null steering because multiple true-time RF delays for each antenna element can simultaneously be synthesized using computer generated holograms.

In order to measure the true-time-delay between the wavebands, the network analyzer was set to measure the RF response produced after the photodetection of the delayed RF-modulated wavebands. The transfer function that results from detecting M wavebands can be described as (Capmany et al., 2005):

$$H(f)=\sum_{r=0}^{M} a_r \exp[-j2\pi rf\tau] \tag{5}$$

Where f is the RF frequency, M is the number of the detected RF-modulated wavebands, a_r is the rth tap weight, which is proportional to the optical power of the rth waveband, and τ is the time delay between adjacent wavebands introduced by the high dispersion fiber. The free spectral range of the transfer function is given by the following equation (Capmany et al., 2005):

$$f_{FSR} = \frac{1}{\tau} \tag{6}$$

The time delay, τ, can also be expressed in terms of the dispersion of the HDF as

$$\tau = \alpha \cdot \Delta\lambda \tag{7}$$

where α denotes the dispersion coefficient of the HDF, and $\Delta\lambda$ is the wavelength separation between the centers of two adjacent wavebands.

Note that, from Eq. 7, the time delay, τ, depends on the dispersion coefficient of the dispersion medium, therefore, a higher dispersion medium results in a longer maximum attainable time delay.

The RF insertion losses of the whole tunable true-time delay system, defined as the RF power ratio between input and output of the RF signal, can be approximately expressed as (Capmany et al., 2006)

$$T_{RF} = \frac{P_{RFout}}{P_{RFin}} = \left(\frac{\pi P_{opt} T_{opt} Z_0}{2V_\pi} R \right)^2 \tag{8}$$

Where Z_0 is the effective EOM RF input impedance or resistance of the EOM electrode, V_π is the voltage for a π-radian optical phase shift, R (A/W) is the photodetector responsivity, Popt is the input continuous wave (CW) optical power to the EOM, and Topt is the optical power transmission parameter that embraces all the optical losses and/or gain in the optical processing including the EOM insertion losses. In this experiment, the RF insertion loss is mainly due to the free-space optical system including the fibre collimator, the diffraction grating and the Opto-VLSI processor, which contributes around 12.5dB loss. Furthermore, the high dispersion fibre (HDF) and the EOM have insertion loss 4.6dB and 3.8dB, respectively. The total optical insertion loss of the entire system is around 21dB, whereas the EDFA provides a low gain of about 10dB due to saturation. The overall RF insertion loss in the experiment was about 25dB. However, this RF insertion loss can be compensated for by the use of an optical amplifier of 12.5dB gain placed after the HDF (before photodetection).

Different phase holograms, which are depicted in Fig. 9(a), were applied to the Opto-VLSI processor to generate five equally-separated RF-modulated optical wavebands with different wavelength separations as shown in Fig. 9(b). Fig. 9(c) shows the measured RF responses for seven true-time delay generation scenarios, corresponding to wavelength separations of 1.74nm, 2.64nm, 3.66nm, 4.32nm, 5.16nm, 5.88nm and 6.84nm, respectively.

The measured free spectral ranges of the various RF responses shown in Fig. 9(c) were used to calculate the true time delays, using Eq. 6. In addition, the measured waveband spacings (Fig. 9(b)) were also used to calculate the time delays for each scenario using Eq. 7. Table 1 summarizes the free spectral ranges of the various measured RF responses, the measured waveband separations, and their corresponding time delays calculated using Eq. 6 and Eq. 7, for the different scenarios. Excellent agreement between the true-time delays calculated using Eq. 6 and Eq. 7 is displayed in Table 1.

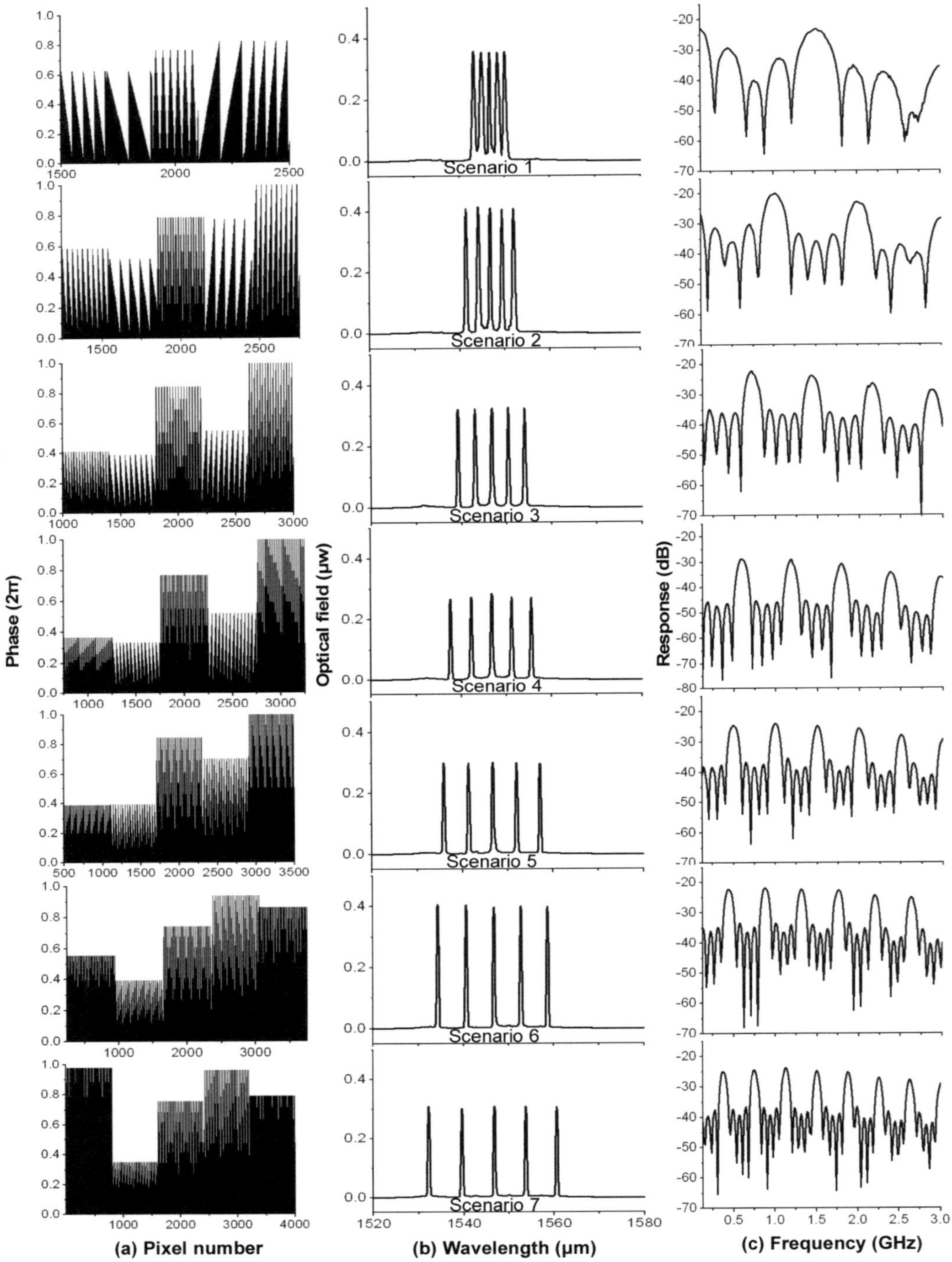

Fig. 9. (a) Opto-VLSI hologram, (b) Optical spectrum of 5 RF-modulated wavebands, (c) Measured RF responses due to the photodetection of the RF-modulated wavebands displayed in (b).

Note that the spectral response fades as the RF frequency increases, as shown in Fig. 9(c). This is due to the interaction between the dispersion of the HDF and the nonzero optical bandwidth of each waveband (Taylor *et al.*, 2007; Pastor *et al.*, 2003b). The bandwidth of the sliced wavebands in Fig. 9(b) was about 0.5 nm, and this caused about 4dB spectral walk-off at 3GHz. This limits the practical bandwidth of the proposed true-time delay system. However if the sliced wavebands are narrow, the walk off effect becomes insignificant as reported in (Pastor *et al.*, 2003b). In addition, our preliminary experimental results have shown that a narrower waveband can be achieved using a higher dispersion grating plate and/or a larger distance between the grating and the Opto-VLSI processor.

Note that the number of multiple time delays that can be generated simultaneously depend on (i) the spectral width of the ASE source, (ii) the maximum delay time (i.e., the maximum wavebands separation), and (iii) the size of the active window of the Opto-VLSI processor. The larger the size of the active window of the Opto-VLSI processor, the larger the number of time delays that can be generated. On the other hand, the more wavebands are required (more nulls), the smaller the maximum waveband separation that could be achieved, thus limiting the null angle which can be synthesized. Note, however, that for a certain number of nulls, the required number of wavebands for each antenna element is fixed. In this case the ASE source should have sufficient spectral width to ensure the synthesis of arbitrary null angles.

Finally, the experimental results shown in Fig. 9 and Table 1 confirm the ability of the Opto-VLSI-based true time delay unit to adaptively generate arbitrary RF delays for broadband null steering of phased array antennas. In order to generate multiple RF true-time delays for each antenna element, the Opto-VLSI processor was driven by optimised phase holograms that select and couple appropriate RF-modulated wavebands into a HDF that simultaneously delays the selected wavebands. To measure the generated time delays simultaneously, the selected wavebands were detected by a single photodetector to generate a microwave transversal filter response whose FSR and shape factor (measured by the network analyzer) can be used to calculate the amplitudes and delay times of the wavebands.

	Scenario 1	Scenario 2	Scenario 3	Scenario 4	Scenario 5	Scenario 6	Scenario 7
Measured FSR (GHz)	1.51	0.98	0.72	0.60	0.50	0.43	0.39
Time delay (ns) calculated using Eq.6	0.66	1.02	1.39	1.67	2.00	2.32	2.58
Measured waveband separation (nm)	1.74	2.64	3.66	4.32	5.16	5.88	6.84
Time delay (ns) calculated using Eq.7	0.67	1.02	1.41	1.66	1.99	2.26	2.63

Table. 1. Measured free spectral ranges, waveband separations, and their corresponding time delays calculated by Eq. 6 and Eq. 7.

This experiment demonstrated that arbitrary single or multiple true-time delays could be synthesized by slicing an RF-modulated broadband optical source and routing arbitrary sliced wavebands, through upload a phase hologram onto an Opto-VLSI processor, to a high-dispersion fiber where they experience RF delays that depend on their centre wavelengths.

5. Opto-VLSI-based tunable beamformer for microwave phased-array antenna

The growing demand for broadband mobile telephone, Internet and multimedia services, combined with the limited radio frequency (RF) spectrum availability, will require a substantial increase in the channel capacity of wireless and mobile communication systems. However, in current wireless network systems, the base station has no information on the position of each mobile user and radiates the RF signal in all directions within a cell in order to provide radio coverage. This results in inefficient utilization of the radiated power during the transmission and causes interference to adjacent cells (co-channel) that use the same frequency. In addition, the antenna receiver detects signals from all directions including noise and interference signals, making the processing of the desired signals complicated, thus limiting the transmission speed and the number of users.

Phased array antennas are rapidly becoming the new frontier of wireless communications because of their ability to overcome user-interference effects effectively and increase channel capacity substantially, together with a more efficient use of transmitted energy. Every phased array antenna requires a key element, "the beamformer", which locates and tracks the various users associated to a base station and adapts the relative amplitude and phase shift at each antenna element so that the main lobe (highest gain) of the antenna is directed toward an active user, while nulls (i.e., where the directivity is zero) are adapted towards co-channel interferers. Phased array antennas refer to antenna arrays need adaptive signal processing capability to maximize the wanted signals and also minimize the interfering signals. Therefore, spatially-separated users can be served in the same base-station sector using the same frequency/time slot, which is called space division multiple access (SDMA) system.

The requirements for future phased array antenna beamformers include (i) the ability to perform wideband signal processing, (ii) short reaction time, (iii) small footprint (small size and light weight), (iv) increased coverage range and increased resolution, (v) reliability and maintainability, and (vi) low cost.

Currently, two beamforming approaches are mainly deployed in all-electronic phased array antennas. The first approach is based on the use of analogue RF phase shifters to adapt the antenna's directivity pattern (Sorensen et al., 2004), while the second approach is based on digitising the antenna signals using analogue-to-digital converters and using digital signal processing (DSP) to control the mainlobe as well as the nulls of the antenna (Jian et al., 2008). Both analogue and digital beamforming approaches have the advantages of flexibility; however, both are inherently narrowband because of the limited instantaneous bandwidths of RF phase shifters and analogue-to-digital converters.

Photonics-based broadband phased-array antenna beamformers have been extensively investigated over the last decade for applications ranging from modern microwave radar to wireless communication systems. A broadband phased-array antenna requires the generation of variable true-time delays at each antenna element to realize beam or null steering. Several approaches have been adopted to realise tunable true-time delay units, including the use of in-fibre chirped Bragg gratings (FBGs) (Italia *et al.*, 2005a), free-space in conjunction with white cells (Mital et al., 2006a), integrated optical waveguides (Flamand et al., 2000), optically-switched fibre delay structures (Tong&Wu, 1998). However, none of these reported photonics-based true-time delay units has the flexibility to either tune the true-time delay continuously or generate multiple tunable true-time delays for each antenna element simultaneously. Furthermore, the limited flexibility, reconfigurability, and tunability of current photonic beamformers make them impractical for realising broadband null steering.

Broadband null-steering beamformers are much more difficult to realise than beam-steering beamformers. Theoretical analysis of broadband null steering of phased-array antennas shows multiple variable true-time delays are needed for each antenna element, while only one variable true-time delay for an antenna element is required for broadband beam steering. An N-element smart antenna can synthesise (N–1) nulls only, and this requires the beamformer to simultaneously generate (2^{N-1}–1) delayed versions of the RF signal received by the antenna.

Fig. 10(a) shows a typical N-element phased-array antenna architecture, whose array factor (or directional response) is given by (Zmuda *et al.*, 1998):

$$AF_N(\theta) = \prod_{n=1}^{N-1}(x - x_n) = \sum_{m=0}^{N-1} W_m x^m \tag{9}$$

where $x = \exp\left[jkd\sin(\theta)\right]$, d is the antenna element spacing, k = wave number = ω/c, and x_n = x (θ_n) is a zero of the polynomial AF_N corresponding to an antenna null at the angular coordinate θ_n. Note that a change of even one zero affects all the weights, Wm. Note also that with N antenna elements, the phased-array antenna can synthesize only (N–1) nulls, as evident from Equation (9).

Without loss of generality, considering a 4-element phased array antenna, with its main lobe at an angle θ and nulls located along angular coordinates, θ_1, θ_2, and θ_3, the array factor takes the form (Zmuda *et al.*, 2000):

$$AF_4(\theta) = \sum_{m=0}^{3} W_m e^{jmkd\sin(\theta)} = \left(e^{jkd\sin(\theta)} - e^{jkd\sin(\theta_1)}\right)\left(e^{jkd\sin(\theta)} - e^{jkd\sin(\theta_2)}\right)\left(e^{jkd\sin(\theta)} - e^{jkd\sin(\theta_3)}\right) \tag{10}$$

By expanding Eq. (10), we obtain

$$AF_4(\theta) = x^3 - x^2\left(e^{j\omega\tau_{21}} + e^{j\omega\tau_{22}} + e^{j\omega\tau_{23}}\right) + x\left(e^{j\omega\tau_{11}} + e^{j\omega\tau_{12}} + e^{j\omega\tau_{13}}\right) - e^{j\omega\tau_{01}} \tag{11}$$

From Eq. (11), it can be observed that for a 4-element phased array antennas, 2^{4-1}–1 = 7 delay taps need to be generated by the true time delay unit in order to synthesis three nulls, and that the time delays required to be synthesized are:

$$\begin{aligned} &\tau_{21} = \frac{d}{c}\sin(\theta_1), \tau_{22} = \frac{d}{c}\sin(\theta_2), \tau_{23} = \frac{d}{c}\sin(\theta_3) \\ &\tau_{11} = \frac{d}{c}\left[\sin(\theta_1) + \sin(\theta_2)\right], \tau_{12} = \frac{d}{c}\left[\sin(\theta_1) + \sin(\theta_3)\right], \tau_{13} = \frac{d}{c}\left[\sin(\theta_2) + \sin(\theta_3)\right] \\ &\tau_{01} = \frac{d}{c}\left[\sin(\theta_1) + \sin(\theta_2) + \sin(\theta_3)\right] \end{aligned} \tag{12}$$

Generally, for an N-element broadband phased array, the synthesis of (N–1) broadband nulls can be achieved if the beamformer of the antenna can adaptively generate and combine (2^{N-1} –1) delayed versions of the RF signals received by the antenna elements, as illustrated in Fig. 10(b).

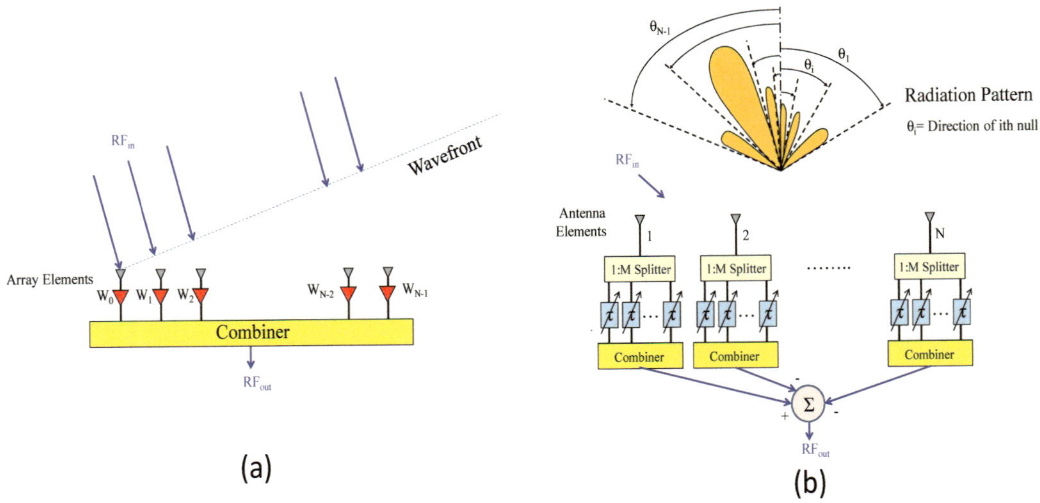

Fig. 10. (a) Typical phased-array antenna architecture. (b) Phased array antenna architecture for broadband null steering.

Recently a novel holographic-based broadband beamformer was proposed and demonstrated, employing an Opto-VLSI processor, a broadband light source, and high dispersion fibres to simultaneously generate arbitrary multiple true-time delays for each antenna element. This beamformer enables the realisation of adaptive multi-element antennas that significantly increase the capacity of next-generation wireless systems. The proposed beamformer has a number of novel features. First, it can adaptively achieve broadband beam- and null-steering through software; second, it incorporates microelectronics and photonics (Opto-VLSI) for RF signal processing, thus adding the flexibility, tunability, accuracy, and reconfigurability of microelectronics to the broadband capability of photonics; and third, it provides a cost-effective and compressed-hardware solution to multi-element antenna beamforming in next-generation wireless systems.

Figure 11 shows the proposed Opto-VLSI-based RF phased-array antennas architecture as well as experimental setup for synthesis of broadband beam steering. The structure was set to tune four fiber lasers, all controlled by a single 2-D Opto-VLSI processor with 8-bit phase level, 512 × 512 pixels of a pixel size of 15μm. Each tunable fiber laser employed an Erbium-doped fiber amplifier (EDFA) operating in C-band, an optical coupler, a polarization controller (PC), and a circulator. The broadband amplified spontaneous emission (ASE) noise resulting from the optical amplifier was split by the optical coupler with a 5/95 power splitting ratio, where 5% of ASE power was used to extract the output of the tunable fiber laser while the remaining 95% was re-circulated in the fiber ring cavity to generate lasing. The polarization controller was used to optimize the diffraction efficiency of the grating plate and to enforce single-polarization lasing. All broadband ASE signals directed to the corresponding collimator array ports via optical circulators were collimated at about 0.5 mm diameter. An optical lens (Lens 1) with 10 cm focal length was used between the collimator array and a diffraction grating plate to focus the collimated ASE beams onto a small spot onto the grating plate. The grating plate, having 1200 lines/mm and a blazed angle of 70° at 1530 nm, spatially de-multiplexed the ASE beams, and spread the ASE spectra into different directions. Another optical lens (Lens 2) with the same focal length, located in the middle position between the grating plate and the Opto-VLSI processor, was used to collimate the dispersed optical beams and map them onto the surface of the 2-D Opto-VLSI processor, which was partitioned into 4 rectangular pixel blocks by software. Each pixel block was assigned to a tunable laser and used to efficiently couple back any part of the ASE spectrum illuminating this pixel block along the incident path into the corresponding collimator port. The Opto-VLSI processor can arbitrarily select any wavebands that are mapped onto its surface using the principle of beam steering described above. The selected wavebands were coupled back into the corresponding fiber collimator port, and then routed back to the gain medium via the corresponding circulator, thus forming an optical loop for single-mode laser generation.

A Labview software was developed to generate and upload the optimized digital phase holograms that simultaneously steer the desired wavebands for each channel and couple back into the corresponding collimators for subsequent recirculation in the fiber loops. Four different wavelengths can independently be selected for lasing within the different fiber loops by uploading appropriate phase holograms (blazed grating) that drive all the pixel blocks of the Opto-VLSI processor. Therefore, this structure enables generation of multiple tunable fiber laser sources, each of which can be independently tuned and output from a specific output port, which offers excellent flexibility to synthesis beam and null steering.

Each RF signal received by the element at the front-end of the phase-array antennas was used to intensity modulate the wavelength channels using an Electro-Optic Modulator (EOM). All the RF-modulated optical signals were coupled into a single fiber and routed into an EDFA for amplification, and then launched into a 10-km Corning LEAF non-zero dispersion shifted optical fiber with dispersion coefficient about 4.2 ps/nm/km and insertion loss of 0.2 dB/km at 1550 nm. Each RF-modulated optical signal experienced a true-time delay that depended on their centre wavelengths, before they were finally detected by a photo-diode that produced the sum of the delayed RF signals. In this way, the TTD between adjacent antenna elements were generated by controlling the wavelength spacing between the various wavelength channels. One of the attractive features of the proposed phased array antenna architecture shown in Fig. 1 is its ability to simultaneously generate all the tunable RF true-time delays for the smart antenna beamformer through optimised phase holograms uploaded onto the Opto-VLSI processor. Various phase holograms, which were synthesized and optimized for specific beam steering scenarios, were stored to enable beamsteering scenarios to be recalled through software.

Experiments were conducted using the setup illustrated in Fig. 11 to evaluate the performance of a 4-element rectangular patch type smart antenna system for flexible beam steering. The antenna elements were separated at a distance of 99 mm, corresponding to half of the RF operating wavelength thus alleviating the effect of side lobes. By refreshing phase holograms generated for the four pixel blocks, the lasing wavelengths can be tuned in terms of their wavelength separations needed to generate the TTD required for RF beam steering.

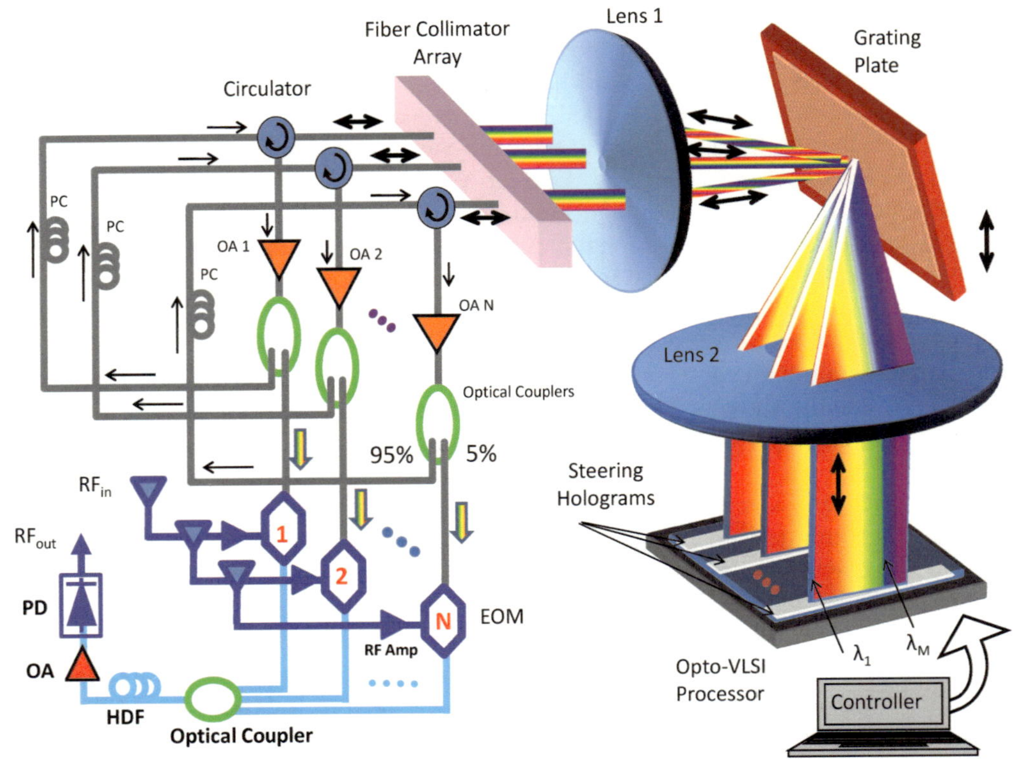

Fig. 11. Opto-VLSI-based phased array antenna architecture for broadband beam steering.

For each beammsteering scenario, the RF signal produced after the photodetection of the delayed RF-modulated optical signals was monitored by an RF power meter. Figures 12 (a-d) show several scenarios of measured antenna radiation patterns (right) corresponding to different equi-spaced wavelength channels (left) generated by the Opto-VLSI processor. For example, in Fig. 12 (a), the Opto-VLSI processor generated a wavelength spacing of 1 nm, which corresponds to 42-ps delay between adjacent antenna elements. The measured beamsteering angle was about 7°, which is in excellent agreement with the theoretical prediction of 7.2°. When the spacing between the laser channels was increased from 1 nm to 4 nm, the measured main lobe was steered from 7° to around 30°, as illustrated in Fig. 12. These experimental results agree very well with the theoretical predictions, demonstrating the excellent capability of the Opto-VLSI processor to realize a phased-array antenna beamformer.

Note that this structure has the capability of null steering if each laser channel is able to generate multiple laser wavelengths. Single or multiple arbitrary wavelengths for each laser channel can be generated and the amplitude of each laser of different wavelength can also be controlled simply by uploading the appropriate steering phase holograms onto the Opto-VLSI processor that can not only select arbitrary multiple wavelengths to lase but also control their cavity losses thus their output powers. Each pixel blocks steers a waveband along its initial path or slightly off-track so that variable optical attenuation (and hence RF attenuation) is achieved for all delayed RF signal simultaneously.

The experiments demonstrated an Opto-VLSI-based tunable beamformer for adaptively steering the radiation patter of RF phased array antennas. By using phase holograms implemented by a single Opto-VLSI processor 4 independent optical wavebands can be generated and their spectral tuned, leading to 4 independent tunable fiber lasers. The RF-modulated lasers are used to generate TTD in 4-element rectangular patch arrayed antennas. Experimental results show the capability of the proposed structure to perform RF beam steering between 0°-30°.

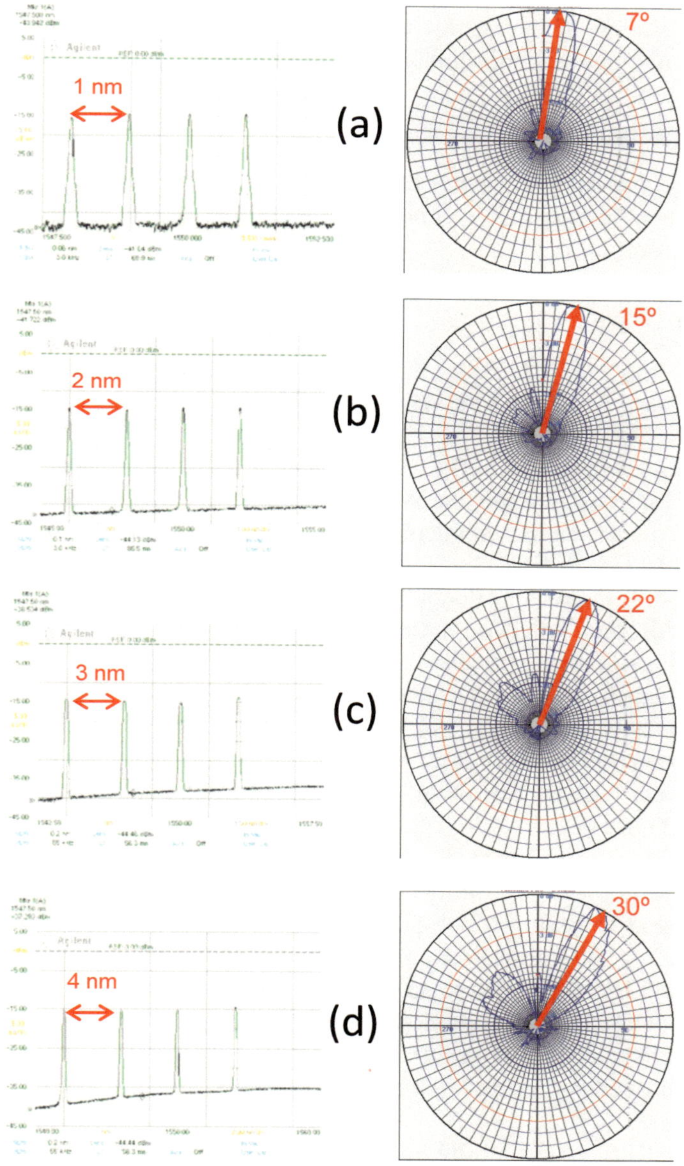

Fig. 12. Opto-VLSI-based tunable laser generations (left), and the corresponding RF beam steering profile measured in the experiments (right).

6. Conclusion

In this chapter, we discussed the reconfigurable phase holograms for the realization of microwave & RF filters, true-time delays, and beamformers, for the first time ever, to overcome the previous disadvantages of poor tunability, inflexibility, and low-resolution of conventional photonic microwave signal processors. By uploading appropriate phase holograms onto an Opto-VLSI processor, (i) full control for tap weights and tap spacings has been achieved to realize a reconfigurable RF filter; (ii) arbitrary single or multiple true-time delays could be synthesized; and (iii) a beamformer for 4-element phased-array antennas has been demonstrated. These work shows that holography is a promising technology to flexibly process microwave & RF signals with high resolution and wide waveband.

7. Acknowledgement

We acknowledge the support of the Department of Nano-bio Materials and Electronics, Gwangju Institute of Science and Technology, Republic of Korea, for the development of the Opto-VLSI processor.

8. References

Anderson, B. L., Rabb, D. J., Warnky, C. M. & Abou-Galala, F. (2006). Binary Optical True-Time Delay Based on the White Cell: Design and Demonstration. J. Lightwave Technol., Vol. 24, pp. 10

Baxter, G. F., S.; Abakoumov, D.; Hao Zhou; Clarke, I.; Bartos, A.; Poole, S.; (2006).Highly programmable Wavelength Selective Switch based on Liquid Crystal on Silicon switching elements. In Proc. OFC/NFOEC, Vol. OTuF2Anaheim, California, USA: OSA.

Blals, S. R. & Yao, J. P. (2008). Tunable photonic microwave filter using a superstructured FBG with two reflection bands having complementary chirps. Ieee Photonics Technology Letters, Vol. 20, pp. 199-201, Issn: 1041-1135.

Capmany, J., Ortega, B. & Pastor, D. (2006). A tutorial on microwave photonic filters. Journal of Lightwave Technology, Vol. 24, No. 1, pp. 201-229, Issn: 0733-8724.

Capmany, J., Ortega, B., Pastor, D. & Sales, S. (2005). Discrete-time optical processing of microwave signals. Journal of Lightwave Technology, Vol. 23, No. 2, pp. 702-723, Issn: 0733-8724.

Capmany, J., Pastor, D. & Ortega, B. (1999). New and flexible fiber-optic delay-line filters using chirped Bragg gratings and laser arrays. Ieee Transactions on Microwave Theory and Techniques, Vol. 47, No. 7, pp. 1321-1326, Issn: 0018-9480.

Chen, C. M., Yi, Y. J., Wang, F., Yan, Y. F., Sun, X. Q. & Zhang, D. M. (2010). Ultra-Long Compact Optical Polymeric Array Waveguide True-Time-Delay Line Devices. Ieee Journal of Quantum Electronics, Vol. 46, No. 5, pp. 754-761, Issn: 0018-9197.

Chen, Y.-W., Yamauchi, S., Wang, N. & Nakao, Z. (2000). A fast kinoform optimization algorithm based on simulated annealing. IEICE Trans. Fundamentals, Vol. E83-A, No. 4, pp. 3

Chen, Y. & Chen, R. T. (2002). A fully packaged true time delay module for a K-band phased array antenna system demonstration. IEEE Photon. Technol. Lett., Vol. 14, pp. 3

Dammann, H. (1979). Spectral characteristics of stepped-phase gratings. Optik, Vol. 53 pp. 9

Feng Xiao, Budi Juswardy, and Kamal Alameh, "Tunable Photonic Microwave Filters Based on Opto-VLSI Processors" IEEE Photonics Technology Letters, vol. 21, pp. 751-753 (2009).
Flamand, G., De Mesel, K., Moerman, I., Dhoedt, B., Hunziker, W., Kalmar, A., Baets, R., Van Daele, P. & Leeb, W. (2000). InP-based PIC for an optical phased-array antenna at 1.06 mu m. Ieee Photonics Technology Letters, Vol. 12, No. 7, pp. 876-878, Issn: 1041-1135.
Frigyes, I. & Seeds, A. J. (1995). Optical generated true-time delay in phased array antennas. IEEE Trans. Microw. Theory Tech., Vol. 43, pp. 9
Hunter, D. B. & Nguyen, L. V. T. (2006). Widely tunable RF photonic filter using WDM and a multichannel chirped fiber grating. Ieee Transactions on Microwave Theory and Techniques, Vol. 54, No. 2, pp. 900-905, Issn: 0018-9480.
Italia, V., Pisco, M., Campopiano, S., Cusano, A. & Cutolo, A. (2005a). Chirped fiber Bragg gratings for electrically tunable time delay lines. Ieee Journal of Selected Topics in Quantum Electronics, Vol. 11, No. 2, pp. 408-416, Issn: 1077-260X.
Italia, V., Pisco, M., Campopiano, S., Cusano, A. & Cutolo, A. (2005b). Chirped Fiber Bragg Gratings for Electrically Tunable Time Delay Lines. IEEE J. Sel. Top. Quantum Electron, Vol. 11, pp. 9
Jian, L., Huali, W. & Guanghui, X. (2008).GA-Based digital multi-beam reconfiguration for satellite phased array antennas. In Microwave and Millimeter Wave Technology, 2008. ICMMT 2008. International Conference on, 1164-1166.
Jiang, Y. Q., Howley, B., Shi, Z., Zhou, Q. J., Chen, R. T., Chen, M. Y., Brost, G. & Lee, C. (2005). Dispersion-enhanced photonic crystal fiber array for a true time-delay structured X-band phased array antenna. Ieee Photonics Technology Letters, Vol. 17, No. 1, pp. 187-189, Issn: 1041-1135.
Juswardy, B., Xiao, F. & Alameh, K. (2009). Opto-VLSI-based photonic true-time delay architecture for broadband adaptive nulling in phased array antennas. Optics Express, Vol. 17, No. 6, pp. 4773-4781, Issn: 1094-4087.
Manolis, I. G., Wilkinson, T. D., Redmond, M. M. & Crossland, W. A. (2002). Reconfigurable multilevel phase holograms for Optical switches. IEEE Photon. Technol. Lett. , Vol. 14, No. 6, pp. 801-803
Meijerink, A., Roeloffzen, C., Meijerink, R., Zhuang, L. M., Marpaung, D., Bentum, M. J., Burla, M., Verpoorte, J., Jorna, P., Hulzinga, A. & van Etten, W. (2010). Novel Ring Resonator-Based Integrated Photonic Beamformer for Broadband Phased Array Receive Antennas-Part I: Design and Performance Analysis. Journal of Lightwave Technology, Vol. 28, No. 1, pp. 3-18, Issn: 0733-8724.
Minasian, R. A. (2006). Photonic signal processing of microwave signals. Ieee Transactions on Microwave Theory and Techniques, Vol. 54, No. 2, pp. 832-846, Issn: 0018-9480.
Mital, R., Wamky, C. M. & Anderson, B. L. (2006a). Design and demonstration of an optical true-time-delay device based on an octic-style white cell. Journal of Lightwave Technology, Vol. 24, No. 2, pp. 982-990, Issn: 0733-8724.
Mital, R., Warnky, C. M. & Anderson, B. L. (2006b). Design and Demonstration of an Optical True-Time-Delay Device Based on an Octic-Style White Cell. J. Lightw. Technol., Vol. 24, pp. 9
Ning, G., Shum, P. & Zhou, J. Q. (2007). Chromatic dispersion effect on microwave photonic filter with a tunable linearly chirped fiber bragg grating. Microwave and Optical Technology Letters, Vol. 49, No. 9, pp. 2131-2133, Issn: 0895-2477.

Ortigosa-Blanch, A., Mora, J., Capmany, J., Ortega, B. & Pastor, D. (2006). Tunable radio-frequency photonic filter based on an actively mode-locked fiber laser. Optics Letters, Vol. 31, No. 6, pp. 709-711, Issn: 0146-9592.

Pastor, D., Ortega, B., Capmany, J., Sales, S., Martinez, A. & Munoz, P. (2003a). Optical microwave filter based on spectral slicing by use of arrayed waveguide gratings. Optics Letters, Vol. 28, No. 19, pp. 1802-1804, Issn: 0146-9592.

Pastor, D., Ortega, B., Capmany, J., Sales, S., Martinez, A. & Munoz, P. (2003b). Optical microwave filter based on spectral slicing by use of arrayed waveguide gratings. Opt. Lett., Vol. 28, pp. 1802-1804

Polo, V., Vidal, B., Corral, J. L. & Marti, J. (2003). Novel tunable photonic microwave filter based on laser arrays and N x N AWG-based delay lines. Ieee Photonics Technology Letters, Vol. 15, No. 4, pp. 584-586, Issn: 1041-1135.

Raz, O., Rotman, R., Danziger, Y. & Tur, M. (2004). Implementation of photonic true time delay using high-order-mode dispersion compensating fibers. Ieee Photonics Technology Letters, Vol. 16, No. 5, pp. 1367-1369, Issn: 1041-1135.

Rideout, H. R., Seregelyi, J. S. & Yao, J. P. (2007). A true time delay beamforming system incorporating a wavelength tunable optical phase-lock loop. Journal of Lightwave Technology, Vol. 25, No. 7, pp. 1761-1770, Issn: 0733-8724.

Seeds, A. J. & Williams, K. J. (2006). Microwave photonics. Journal of Lightwave Technology, Vol. 24, No. 12, pp. 4628-4641, Issn: 0733-8724.

Sorensen, R. K., Iskander, M. F. & Lee, J. J. (2004). Low-cost nonplanar microstrip-line ferrite phase shifter utilizing circular polarization. Ieee Microwave and Wireless Components Letters, Vol. 14, No. 1, pp. 25-27, Issn: 1531-1309.

Taylor, J. D., Chen, L. R. & Gu, X. J. (2007). Simple reconfigurable photonic microwave filter using an arrayed waveguide grating and fiber Bragg gratings. Ieee Photonics Technology Letters, Vol. 19, No. 5-8, pp. 510-512, Issn: 1041-1135.

Tong, D. T. K. & Wu, M. C. (1998). Multiwavelength optically controlled phased-array antennas. Ieee Transactions on Microwave Theory and Techniques, Vol. 46, No. 1, pp. 108-115, Issn: 0018-9480.

Vidal, B., Mengual, T., Ibanez-Lopez, C. & Marti, J. (2006). Optical beamforming network based on fiber-optical delay lines and spatial light modulators for large antenna arrays. Ieee Photonics Technology Letters, Vol. 18, No. 21-24, pp. 2590-2592, Issn: 1041-1135.

Xiao, F., Juswardy, B., Alameh, K. & Lee, Y. T. (2008). Novel broadband reconfigurable optical add-drop multiplexer employing custom fiber arrays and Opto-VLSI processors. Optics Express, Vol. 16, No. 16, pp. 11703-11708, Issn: 1094-4087.

Zheng, R., Alameh, K. E. & Wang, Z. L. (2006). A tunable photonic RF notch filter based on opto-VLSI technology. Microwave and Optical Technology Letters, Vol. 48, No. 6, pp. 1011-1015, Issn: 0895-2477.

Zmuda, H., Toughlian, E. N., Jones, M. A. & Payson, P. M. (2000). Photonic architectures for broadband adaptive nulling with linear and conformal phased array antennas. Fiber and Integrated Optics, Vol. 19, No. 2, pp. 137-154, Issn: 0146-8030.

Zmuda, H., Toughlian, E. N., Payson, P. & Klumpe, H. W. (1998). A photonic implementation of a wide-band nulling system for phased arrays. Ieee Photonics Technology Letters, Vol. 10, No. 5, pp. 725-727, Issn: 1041-1135.